Learning in Non-Stationary Environments

Moamar Sayed-Mouchaweh • Edwin Lughofer
Editors

Learning in Non-Stationary Environments

Methods and Applications

Springer

Editors
Moamar Sayed-Mouchaweh
Ecole des Mines de Douai
Département Informatique et
Automatique
941, Rue Charles Bourseul
59508 Douai cedex
BP 10838
France

Edwin Lughofer
Department of Knowledge-Based
Mathematical Systems
Science Park II
Johannes Kepler University Linz
Altenbergerstrasse 69
A-4040 Linz
Austria

ISBN 978-1-4899-9340-3 ISBN 978-1-4419-8020-5 (eBook)
DOI 10.1007/978-1-4419-8020-5
Springer New York Dordrecht Heidelberg London

© Springer Science+Business Media New York 2012
Softcover reprint of the hardcover 1st edition 2012
All rights reserved. This work may not be translated or copied in whole or in part without the written permission of the publisher (Springer Science+Business Media, LLC, 233 Spring Street, New York, NY 10013, USA), except for brief excerpts in connection with reviews or scholarly analysis. Use in connection with any form of information storage and retrieval, electronic adaptation, computer software, or by similar or dissimilar methodology now known or hereafter developed is forbidden.
The use in this publication of trade names, trademarks, service marks, and similar terms, even if they are not identified as such, is not to be taken as an expression of opinion as to whether or not they are subject to proprietary rights.

Printed on acid-free paper

Springer is part of Springer Science+Business Media (www.springer.com)

To our families and friends

Preface

During the last two decades, rapid technological developments and breakthroughs in automatization processes with the support of modern machines and computers lead to a significant increase in system complexity and state changes, in information sources, in requirements regarding the speed of data handling as well as in the integration of environmental influences. Intelligent systems, equipped with a taxonomy of data-driven system identification and machine learning algorithms, are able to handle these problems partially. Typically, different sorts of models are identified based on historic data samples and employed to handle control problems, to support human beings in decision making processes (e.g., medicine, patient supervision or forecasts in metrology and financial domains), to monitor the quality of manufactured parts in on-line production lines or to perform predictions on important objectives and process variables, saving expenses of high-cost sensors.

Conventional learning algorithms in a batch off-line setting fail whenever dynamic changes of the process appear due to non-stationary environments and external influences, usually leading to extensions of system states and its definition space, to drifts and shifts in underlying data generating processes, or to arising new operating conditions and modes. Re-designing cycles are impracticable as they are usually to slow to cope with online requirements or are demanding significant development costs for re-calibration of the models in additional off-line stages decoupled from the online process. Thus, in such circumstances, techniques which are equipped with the concepts of incremental and evolving learning engines are the only reasonable choice to update the models within a reasonable time frame and to react on dynamic changes quickly, thus to guarantee high process safety. Usually, they contain algorithms for permanent adaptation of model parameters as well as of structural parts of the model, including components responsible for dynamic model expansion and contraction on demand and on-the-fly.

The necessity of such methodologies in theory and practice is on the one hand reflected in several developments during the last 10 to 15 years, where many designs for incremental learning engines containing evolving methodologies emerged in parallel within the fields of soft computing (termed as *evolving intelligent systems*, *evolving connectionist systems*, and *evolving fuzzy systems*), machine learning (termed as *knowledge discovery or incremental learning from data streams*) and pattern recognition (termed as *dynamic data mining* or *incremental, evolving clustering*); and, on the other hand in a densification of organizational events, such as special sessions at the international conferences IEEE International Conference on Fuzzy Systems (FUZZ-IEEE) 2007, IEEE International Conference on Systems, Man and Cybernetics (IEEE-SMC) 2007, European Society for Fuzzy Logic and Technology (EUSFLAT) 2009 and 2011, International Conference on Information Processing and Management of Uncertainty in Knowledge-Based Systems (IPMU) 2010, IEEE World Congress on Computational Intelligence (WCCI) 2010, IEEE International Conference on Machine Learning and Applications (ICMLA) 2010 and 2011, IPMU 2012 (planned), special issues in international journals (IEEE Transactions on Fuzzy Systems, Information Sciences, Neurocomputing, etc.) as well as the workshops and symposia Evolving Fuzzy Systems (EFS) 2006, Genetic and Evolving Fuzzy Systems (GEFS) 2008, Evolving and Self-Developing Intelligent Systems (ESDIS) 2009, International Conference on Adaptive and Intelligent Systems (ICAIS) 2009 and 2011, Evolving Intelligent Systems (EIS) 2010 and 2011, IEEE conference on Evolving and Adaptive Intelligent Systems (EAIS) 2012 (planned); and finally establishing an own journal at Springer termed 'Evolving Systems'.

The aim of this book is to provide a broad picture of recent developments and important methodologies within the field of learning in nonstationary environments, covering aspects from the three major lines of research and therefore intending to address the attention of audiences in the field of machine learning, soft computing, pattern recognition, and data mining. The subdivision of the book follows the spirit of the natural main problems in the context of any form of data-driven methodologies: *dynamic learning in unsupervised problems*, *dynamic learning in supervised classification*, and *dynamic learning in supervised regression problems*. This is completed by a fourth part dedicated to applications where dynamic learning methods serve as key stones for achieving models with high accuracy and assuring process safe and high qualitative industrial systems. The book does not have been particularly written in a mathematical theorem/proof style, but more in a way where ideas, concepts, and algorithms are highlighted by numerous figures, tables, examples, and applications together with their explanations. The intended audience ranges from mathematicians via machine learning and soft computing gurus to technicians from engineering and industrial practice, and finally to students.

Finally, the editors are very grateful to all authors and reviewers for contributing with substantial and very valuable material to make this volume become alive and to set another corner stone in the research and publications history of dynamic and evolving learning concepts. We also acknowledge Ms. Brett Kurzman for establishing the contract with Springer and supporting us in any organizational aspects. We hope that the volume will be a useful basis for further fruitful investigations and fresh ideas as well as a motivation and inspiration for newcomers to join this promising and still emerging field of research.

Douai, France Moamar Sayed-Mouchaweh
Linz, Austria Edwin Lughofer

Contents

Contributors

Omar Ayad Centre de Recherche en STIC (CReSTIC), UFR Sciences Exactes et Naturelles, Universit de Reims Champagne-Ardenne (URCA), Reims, France

John Berezowski Alberta Agriculture and Rural Development, Edmonton, AB, Canada

Abdelhamid Bouchachia Institute of Informatics-Systems, University of Klagenfurt, Klagenfurt, Austria

Hendrik Van Brussel Department of Mechanical Engineering, Katholieke Universiteit Leuven, Leuven, Belgium

Pyramo Costa Graduate Program in Electrical Engineering, Pontifical Catholic University of Minas Gerais, Belo Horizonte, Brazil

Christian Eitzinger Profactor GmbH, Steyr-Gleink, Austria

Fernando Gomide University of Campinas, School of Electrical and Computer Engineering, Sao Paulo, Brazil

Carlos Guardiola CMT-Motores Térmicos/Universidad Politécnica de Valencia, Valencia, Spain

Laurent Hartert CReSTIC, University of Reims Champagne-Ardenne (URCA), Reims, France

Eyke Hüllermeier Department of Mathematics and Computer Science, Philipps-Universität Marburg, Marburg, Germany

Iqbal Jamal AQL Management Consulting Inc., Edmonton, AB, Canada

Nikola Kasabov The Knowledge Engineering and Discovery Research Institute, Auckland, New Zealand

Frank Klawonn Department of Computer Science, Ostfalia University of Applied Sciences, Wolfenbuettel, Germany

Daniel Leite University of Campinas, School of Electrical and Computer Engineering, Sao Paulo, Brazil

Edwin Lughofer Department of Knowledge-Based Mathematical Systems, Johannes Kepler University Linz, Linz, Austria

Sofiane Mazeghrane Centre de Recherche en STIC (CReSTIC), UFR Sciences Exactes et Naturelles, Universit de Reims Champagne-Ardenne (URCA), Reims, France

Nadhir Messai Centre de Recherche en STIC (CReSTIC), UFR Sciences Exactes et Naturelles, Universit de Reims Champagne-Ardenne (URCA), Reims, France

Jean-Michel Papy Flanders' Mechatronics Technology Centre, Leuven, Belgium

Russel Pears The Knowledge Engineering and Discovery Research Institute, Auckland, New Zealand

Witold Pedrycz Department of Electrical & Computer Engineering, University of Alberta, Edmonton, AB, Canada

System Research Institute, Polish Academy of Sciences, Warsaw, Poland

Markus Prossegger Carinthia University of Applied Sciences, Spittal an der Drau, Austria

Hai-Jun Rong MOE Key Laboratory of Strength and Vibration, School of Aerospace, Xi'an Jiaotong University, Xi'an, ShaanXi, China

Davy Sannen Department of Mechanical Engineering, Katholieke Universiteit Leuven, Leuven, Belgium

Moamar Sayed-Mouchaweh Ecole des Mines de Douai, Computer Science and Automatic Control Lab EMDouai-IA, Douai, France

Ammar Shaker Department of Mathematics and Computer Science, Philipps-Universität Marburg, Marburg, Germany

James Edward Smith Department of Computer Science and Creative Technologies, University of the West of England, Bristol, UK

Muhammad Atif Tahir School of Computing, Engineering and Information Sciences, University of Northumbria, Newcastle, UK

Stefan Thumfart Profactor GmbH, Steyr-Gleink, Austria

Katharina Tschumitschew Department of Computer Science, Ostfalia University of Applied Sciences, Wolfenbuettel, Germany

Steve Vandenplas Flanders' Mechatronics Technology Centre, Leuven, Belgium

Harya Widiputra The Knowledge Engineering and Discovery Research Institute, Auckland, New Zealand

Chapter 1
Prologue

Moamar Sayed-Mouchaweh and Edwin Lughofer

Abstract This introductory chapter intends to provide a general overview about the most essential requirements, demands and challenges with respect to dynamic learning of data-driven models in non-stationary environments and applications. It outlines the main lines of research investigated during the last decade in order to cope with the requirements, inter alia to handle high system dynamics, online data streams recorded with a high frequency, drifting system states and very large data bases within fast sample-wise and single-pass model updates conducted on-the-fly and in incremental manner. The last part of this chapter outlines a compact summary of the contents of the book by providing a paragraph about each of the single contributions.

1.1 Modeling in Dynamic Environments: Requirements, Demands, and Challenges

The computerization of many life activities and the advances in data collection and storage technology lead to obtain mountains of data. They are collected to capture information about a phenomena or a process behavior. These data are rarely of direct benefit. Thus, a set of techniques and tools are used to extract useful information for decision support, prediction, exploration, and understanding of phenomena governing the data sources.

M. Sayed-Mouchaweh (✉)
Ecole des Mines de Douai, Computer Science and Automatic Control Lab EMDouai-IA, Douai, France
e-mail: moamar.sayed-mouchaweh@mines-douai.fr

E. Lughofer (✉)
Department of Knowledge-Based Mathematical Systems, Johannes Kepler University Linz, Altenbergerstr. 69, A-4040 Linz, Austria
e-mail: edwin.lughofer@jku.at

M. Sayed-Mouchaweh and E. Lughofer (eds.), *Learning in Non-Stationary Environments: Methods and Applications*, DOI 10.1007/978-1-4419-8020-5_1,
© Springer Science+Business Media New York 2012

Learning methods use historical data points about a process past behavior to build a model, which can be in form of a classifier, an approximation surface, a cluster partition, etc. The model is used as an old experience to classify new samples, to provide decisions on new query points or to predict the process future behavior. For instance, consider a stock market exchange: based on repeated patterns observed in the time series data of a specific stock index during a past time period, some future trends may be predicted; thereby, the interpretation of the patterns is conducted implicitly in the model. Learning methods and techniques become efficient solutions when the relationship between system inputs and output(s) is difficult to understand and cannot be easily described using mathematical models in closed analytical form. In very complex systems, it is either impossible to deduce models from physical knowledge (for instance, consider a water power plant which is affected by many environmental influences) or the development time of the models is that huge such that the costs regarding manpower, etc. would exceed the company's budget. Expert knowledge based on long-term working experience may help to reduce the development effort; however, it usually lacks of sufficient accuracy due to vague and contradictory statements/opinions among the experts and/or operators working with the system.

Therefore, in such cases, the models are constructed by using a set of real data (pairs of input–output vectors), which can be fully automatized in large parts. Figure 1.1 demonstrates a convenient data-driven modeling framework, including various data-driven learning concepts and algorithms based on which the models are finally obtained. Depending on the chosen model architecture, the models may be considered as *black-box*, *dark-grey*, up to *light gray models* [27]. The latter may be achieved by linguistically interpretable fuzzy systems [17, 29] or by hybrid modeling approaches [1], using data-driven mechanisms for estimating parameters in analytical models.

However, it is usually very hard to obtain an exhaustive or completed learning data set that can cover all the characteristics of the real system in all the possible contexts, in particular in dynamic and continuously evolving environments. Indeed, everything that exists changes over time. A typical example of changing environments is the spam detection and filtering. The descriptions of the two classes "spam" and "nonspam" evolve over time due to the changes of user preferences and "spammers" techniques to trick spam classifiers. Thus, classifiers establishing a decision boundary for discriminating between spam and nonspam emails need to adjust their parameters and structure to take into account the changes in their operating environments. This self-adaptation is necessary to preserve the classification accuracy. Another application example which requires dynamic process models is the control of a water power plant. A significant influence of its behavior is due to changing environmental influences such as different weather situations or changing water levels at feeder rivers. Models (e.g., used for control purposes) have to be capable of self-adaptation in order to incorporate such different environmental conditions. Environments' changes can be represented by two concepts. The first concept is "concept-drift." In this concept, the underlying data distribution changes gradually over time. The second concept is "concept-evolution." In this concept, a

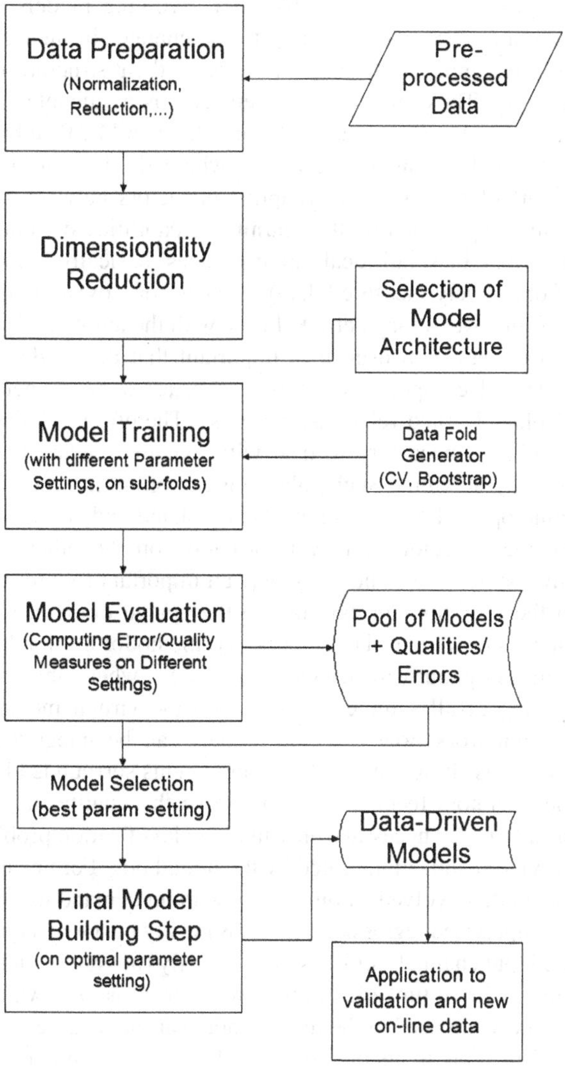

Fig. 1.1 Data-driven modeling framework including dimensionality reduction, model training and evaluation, model selection and a final model training steps based on optimal parameter setting

sudden change in the underlying data distribution can manifest. Thus, the model needs to adjust itself (self-correction or adaptation) as new events happen or new conditions occur. The goal is to ensure an accurate prediction of process behavior according to the changes in new incoming data characteristics.

Models in dynamic and evolving environments are called *dynamic models*. They may change their parameters and adapt them to obtain better performance or change their structure as well in order to add new outputs (classes). Therefore, one can distinguish two main types of dynamic models: *adaptive* and *evolving models*. Adaptive models conserve their structure and adapt their parameters continuously according to newly available data. Evolving models adapt their parameters and structure over time according to the changes in their environments. In this sense, evolving models permanently expand their structure on-the-fly and enrich their memory. Thus, they can be even seen as a valuable contribution within the field of artificial and computational intelligence [4]. Combined with modeling aspects in the field of neural networks, where the human brain is roughly modeled by means of neurons and synaptic connectors between these [14], they are in fact mimicking somehow the learning capabilities of human beings (in fact, within the scope of small ideal circumstances, basically reduced to information in form of objectively measured data). Furthermore, dynamic models require a continuous learning over a long period of time with the ability to forget data becoming obsolete and useless. However, it is important that the model updates its parameters and structural components without a "catastrophic forgetting" (undesired forgetting of older learned relations, patterns). Therefore, a balance between plasticity and stability is necessary to deal with dynamic environments, referred as plasticity–stability dilemma. Such balance is important in order to guarantee convergence to some optimality criterion on the one hand and still to achieve sufficient flexibility for the inclusion of new information on the other hand. Learning in dynamic environments does not only play an important role for changing data characteristic in the time domain, but also in the spatial domain. In fact, learning in dynamic environments may be a necessary methodology for handling large databases as well as spatially distributed data sites, where the joint data is that huge that loading of all samples at once into the virtual memory is simply not possible. A framework how dynamic models can be integrated into an online modeling process is shown in Fig. 1.2. Components surrounded by dotted lines are optional, the operators feedback may improve the quality of the dynamic models, but is usually really necessary only in case classification problems (real class labels to be provided), thus highlighted by the dotted box. For unsupervised learning problems, the models evolved automatically without demanding any manual supervised input, for supervised regression problems mostly the target concepts are by-measured with the input channels, either synchronously (in system identification problems) or at a latter point of time (e.g. in time series forecasting), which, when de facto available, can be again used in the incremental learning engine.

This book treats the problem of learning from data issued of time/spatial-based complex nonstationary (dynamic and evolving) processes, following the framework shown in Fig. 1.2. It draws a round picture of efficient techniques, methods, and tools able to manage, to exploit, and to interpret correctly the increasing amount of data in environments that are continuously changing. In each of these, the goal is to build a model for quantifying, predicting, and classifying the future system behavior, able to tackle and to govern the high variability of complex non-stationary and large-scale systems.

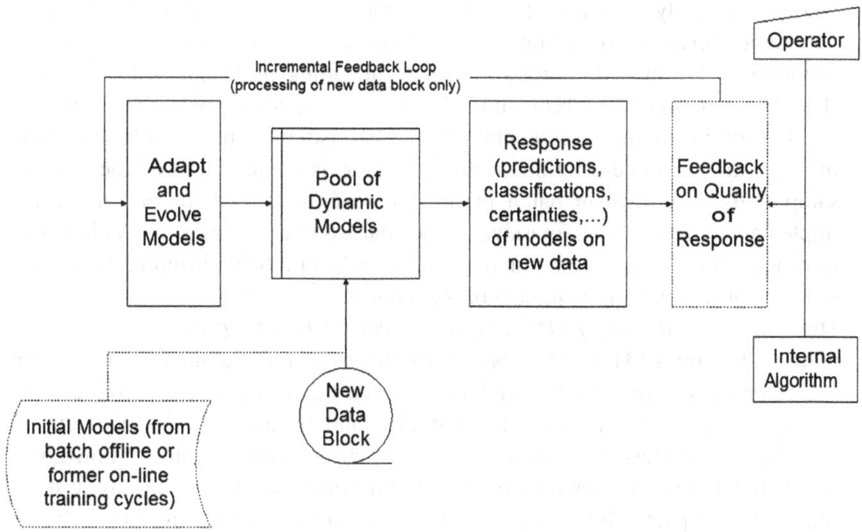

Fig. 1.2 Modeling framework including incremental training procedure for adapting and evolving dynamic models

1.2 General Principles of Learning in Dynamic Environments

In order to cope with all the requirements and demands as mentioned above, three major lines of research emerged during the last 10 to 15 years:

- *Evolving intelligent systems (EIS)* [3]: these types of systems enjoy a great attraction in the soft computing community since several years and are mostly based on evolving neural networks (ENN) [15] and evolving fuzzy systems (EFS) [21] techniques as well as a hybridation of both, providing the basic aspects for the concept of evolving connectionist systems [16]. They allow the update of neuron weights, antecedent, and consequent parameters as well as a dynamic expansion and shrinkage of the structural components (neurons in case of ENN, fuzzy rules in case of EFS), usually in single-pass and incremental manner. While ENN are exclusively focussing on precise modeling, i.e., aiming for models with highest possible accuracies on new unseen data, latest developments of EFS are also concentrated on guiding the evolved models to more interpretable power [22]. This should make EFS very attractive in the near future, as opening the possibility of an enriched human–machine interaction [23]. A comprehensive survey of EIS approaches can be found in [3].
- *Incremental Machine learning (IML)*: a line of research which emerged during the 1990s within the machine learning (ML) community in order to cope with huge or infinite data streams (see below) for building up data-driven

models, basically in form of classifiers and dynamically changing decision boundaries between two or more classes. The recently published monograph [12] summarizes the most important concepts, including incremental tree learners (Hoeffding trees [11]), incremental Naïve Bayesian methods, online Oza bagging and boosting [28] and incremental SVMs [10] using adiabatic (exact) updates, meaning that the same solution of support vectors is found as with all the training samples used at once in batch mode. Most of these methods and more are implemented in the MOA (massive online analysis) framework [7], which is able to process huge data streams incrementally and to perform different evaluation schemes during the online update of the models.

- *Dynamic Data Mining (DDM) and Incremental Clustering*: the concept of dynamic data mining [31,33,35] is based on extracting groups of data (clusters) over a horizon of spatially distributed data sites. Each data site can be incrementally added to the pool of already existing data sites and its data samples are integrated into the actual cluster structure by solving successively a joint optimization problem, where data sites may be assigned different weights according to their importance/impact. The concept applies a collaborative optimization scheme [30], taking explicit advantage of evolving the models blockwise. The native concept of *dynamic clustering* conducts grouping of data over time in the temporal domain, may update centers and ranges of influence, evolving and split clusters as well as merge clusters in order to increase the flexibility of the models for dynamically changing environments—see [9] for a survey of important dynamic, evolving clustering methods.

The general principle in all these lines of research is to observe a change in some statistical properties of data characteristics to decide in which state (stable, warning or action) the system is. These states correspond respectively to no change, gradual change and abrupt change. Thus, the model parameters and structure will be respectively unchanged, slightly adapted or strongly updated. Re-learning is usually omitted as requiring very high computation time and often causing update cycles which are not terminating in real-time within a reasonable online timeframe. Therefore, incremental and sequential learning from data streams are the essential used concepts (1) in order to avoid time-intensive re-training phases and account for the systems dynamics/changing data characteristics with low computational effort (enhancing online performance); and (2) to allow step-wise model building phase with low virtual memory usage, widening the applicability of data-driven modeling techniques to large-scale databases/sites. This is because data is processed in sample-wise and single-pass manner, in particular, a data stream is characterized by the following aspects [12]:

- The data samples or data blocks are continuously arriving online over time. The frequency depends on the frequency of the measurement recording process.
- The data samples are arriving in a specific order, over which the system has no control.

- Data streams are usually not bounded in a size; i.e., a data stream is alive as long as some interfaces, devices, or components at the system are switched on and are collecting data.
- Once a data sample/block is processed, it is usually discarded immediately, afterward.

Change detectors may be an important add-on in case of tracking drifts in the data stream, which are usually changing the underlying distribution of the target concept significantly. Misclassification and errors rates are one of the most used statistical properties to observe a change. In this case, data are divided into batches and their true classes are known in advance; so, the misclassification error is easy to calculate. If this rate decreases significantly after receiving a batch of patterns, then the system is in drift state and the model parameters and structure must be adapted more strongly by gradually outdating older learning relationships. Other change detectors rely on statistical criteria such as the Page–Hinkley test [25] or ADWIN [6]. Another important topic is the flexibility of the models not only according to structural components (by evolving, shrinking, merging operations) but also according to the input space: incremental feature weighting issues (see e.g. Chap. 9, second part) may outweigh less important features and overweigh more important ones, finally achieving a sort of soft (and smooth) dimensionality reduction. This is necessary as an abrupt change in the input structure (exchange a concrete feature with another one) would cause a discontinuous learning behavior.

Old school approaches for learning in dynamic environments are based on sliding time windows or a template containing a fixed number of selected patterns according to their age and usefulness, thus perform the adaptation and evolution of models based on blocks, batches of data; see e.g. [5,13,26,32]. New school approaches are based on samplewise single-pass operations, where the most important information about relations between time series/variables within a data stream are stored in global models and/or statistical criteria, thus models are always up-to-date as early as possible (after each single sample). This increases the robustness and process-safety of the evolved models as long as the methods converge quickly, ideally within each incremental learning cycle. The recursive least squares [20] and incremental support vector update [10] are methods ensuring such convergence, hence often used as learning engines in various dynamic, evolving learning methods. For instance in [2, 8, 19], the former two are using a multiprototype Gaussian modeling of nonconvex classes, where the activation function of each hidden neuron determines the membership degree of an observation to one prototype of a class, and the latter employing a potential function for tracking the most dense samples in order to associate them as potential clusters = rule centers. Other approaches in the context of classification are based on the use of an ensemble of classifiers to track changes in the environment [18,24,34]. The classification of a pattern is achieved by either selecting the best base classifier or by combining all classifiers' predictions. The concept drift tracking is achieved by either taking a weighted majority vote among the classifiers in the ensemble. Each classifier has a weight according to its classification accuracy. With weighted voting, lower weights are assigned to classifiers that fail to track drift. These weights are updated by evaluating the misclassification error of classifiers to keep track on whose classification accuracy is currently the most trustworthy. They can be determined either over a block of data or on the neighborhood of the present instance.

1.3 Contents of the Book

This book addresses the problems of modeling, prediction, classification, data understanding, and processing in nonstationary and unpredictable environments. It presents a comprehensive survey of important recent methods and approaches for the design of systems being able to learn and to fully adapt its structure and to adjust its parameters according to the changes in their environments. This book includes various applications of dynamic models such as business, industrial control, fault detection and diagnosis, quality control, surface inspection, system identification, decision support systems, and security, etc. Naturally, the book may be divided according to the main lines of research as outlined in the previous section (EIS, IML, dynamic data mining (DDM) and clustering). However, the boundaries between these approaches were washing more and more up during the last years (e.g., there exist hybrids of support vector machines (SVM) training and fuzzy model architectures, neural networks with online bagging and boosting or incremental decision trees with statistical-oriented naive Bayes leaves), such that we decided to follow the classical aims and purposes of data-driven models in real-world problems, so to divide the book into

- Dynamic methods for unsupervised problems,
- Dynamic methods for supervised classification problems,
- Dynamic methods for supervised regression problems,
- Applications of learning in nonstationary environments.

In the following, we provide a short summary of each chapter, Chaps. 2 to 4 denote contributions to the unsupervised learning problems (Part 1), Chaps. 5 to 8 to the supervised classification part (Part 2), Chaps. 10 to 12 to the supervised regression part (Part 3) and Chaps. 13 to 15 to applications of dynamic models (Part 4).

Chapter 2

Chapter 2 presents a set of statistical measures (mean, variance, skewness, kurtosis, Pearson correlation coefficient, quantile (median and interquartile range)) to be used as estimators for model parameters generating the underlying data. These measures describe the characteristics (properties) of data distributions. The data arrives continuously in time and its volume increases each day (data stream). To deal with data stream, this chapter proposes an incremental form of statistical measures.

The goal is to carry out computations online with a reduced and limited memory size. To take into account the nonstationary aspect of data, the chapter proposes to use statistical hypothesis tests (χ^2-test and t-test) to detect a change, althrough incremental computation of the tests or window techniques, and to adapt online the statistical models.

Chapter 3

Chapter 3 handles the analysis of spatiotemporal data at various levels of abstraction according to the user perspective (needs) to view the data. The goal is to make the data analysis more humancentric. The spatiotemporal data is represented as a fuzzy relationship between the spatial coordinates and the temporal data. The fuzzy relationship is described in terms of vocabularies (codebooks). Each vocabulary is represented by a fuzzy set. The convolution of the fuzzy sets representing the data vocabularies (coordinates and temporal data) using the t-max composition provides the possibility measure. The latter quantifies the degree of "activation" of the Cartesian product of the codebooks by the fuzzy relationship. The authors propose to reconstruct the fuzzy relation based on the use of t-norm composition of the fuzzy sets as well as their degree of activation of the fuzzy relationship. Then, the chapter proposes the use of the well-known clustering method fuzzy c-means (FCM) to form the elements of the codebooks (considered as the clusters prototypes). The evolvable aspect of systems is reflected by the dependencies between the codebooks used to describe the components of two successive data slices. The level of variability between codebooks used in two successive data slices reflects the changes in the system. This level of variability is considered as the level of uncertainty quantified as the entropy measure.

Chapter 4

Chapter 4 handles the problem of grouping online data samples into clusters using incremental spectral clustering algorithm. This algorithm is based on the use of fuzzy c varieties to cluster data with nonconvex shapes. It considers a cluster as a graph of connected nodes. It assigns a high weight for the nodes forming the clusters and very low weights for the nodes between these clusters. In addition, the algorithm adjusts continuously its structure over time. This incremental property is useful when the available cluster partition at the present time is not enough to discover the complete structure of the data or when this structure evolves due to a change in the environment conditions. The performance of the proposed algorithm is evaluated using a set of simulated data sets.

Chapter 5

Chapter 5 presents a dynamic version of the well-known classification method fuzzy K nearest neighbors (FKNN). This variant of FKNN detects a change of the characteristics of a class by monitoring the change of its gravity center and variance. After the classification of each new pattern in a class, the class gravity center and variance are updated in incremental manner. Then, the difference between the old gravity center and variance and the updated ones is accumulated after each new pattern classification in this class. When this difference becomes greater than a predefined threshold, a drift of this class is confirmed. The drift confirmation allows an adaptation step aiming to update the characteristics of the evolved class using only the recent patterns representing this drift. In addition, when the drift of a class leads it to be too close of another class, this variant of FKNN allows the fusion of these two classes into one class. Then the characteristics of the new merged class are calculated using only the patterns of the evolved and the other merged classes. The determination of the closeness of these two classes is achieved using a similarity measure based on the number of patterns in the ambiguity area between two classes as well as their membership values to these two classes. When this similarity becomes greater than a threshold, the two classes are merged. This variant of FKNN is applied to monitor the quality ("bad", "good") of welled pieces based on the analysis of acoustic signals characterizing the noises of the welding operation.

Chapter 6

Chapter 6 studies the techniques used to predict the model (classifier) estimation accuracy for unseen databases on the use of the existing training set. Indeed, the N-fold cross-validation techniques cannot predict what accuracy might be achievable for unseen data. The goal of this prediction is to show whether further learning is useful to refine the model estimation (and its accuracy) and more investments of uses worth to be achieved to create more samples, which costs money due to annotation efforts (i.e., providing class labels for the additional samples). The proposed techniques divide the observed error into bias and variance terms. Then, an algorithm is used to build the classification model (classifier) using only few samples of a data set and its components (bias and variance) are estimated after all the samples of the data set are used. The predicted error (using all the samples of the data set) will be written as a linear function (regression) of the estimated error using only a few samples. This regression aims to find the parameters that fit the data (estimated errors using increasing size of samples) in order to construct a linear model that can be used to predict the future error rates. In the last section of the chapter it is outlined how these concepts can be applied to detecting drifts in dynamic environments, so to give indicators whether a model needs to be updated or even re-trained (in case of significantly increasing bias and variance error rates).

Chapter 7

Chapter 7 presents a true strictly incremental classifier fusion approach employing an ensemble of updateable classifiers. The proposed approach produces its decision based on the decisions of multiple classifiers. The interest is to produce a decision more robust and more accurate than the ensemble's member classifiers thanks to the exploitation of their diversity. The individual classifiers are incremental Naïve Bayes (NB), k nearest neighbors k-NN (with updating the reference data base) and evolving vector quantization for classification (eVQ-Class). However, any further classifiers can be easily integrated into the ensemble due to the cascadability of the fusion methods. The proposed ensemble is nongenerative trainable ensemble (the number and kind of individual classifiers remain unchanged). It is trainable because its final decision is based on a trainable fusion of individual classifiers decisions. The trainable fusion is a class-indifferent classifier fusion; it decides the support for a class by considering all classes supports according to all the individual classifiers. Both the individual classifiers and the ensemble members are updated in an incremental manner, the latter even with adiabatic (exact) updates, leading to the same solutions as when using the whole batch of data. The incremental aspect of the proposed approach makes it efficient and robust for applications with huge data sets and it can achieve high quality predictions when the systems are fully operational and even when one of the single classifiers fails. The incremental classifier fusion methods are applied to two different dynamic application tasks: inspection of CD Imprints (classifying images into "good" and "bad") and the prediction of maintenance actions for copiers (in particular, the Toner Transfer Fusing TTF belt cleaner) using a large database containing information about the replacing of different components by technicians.

Chapter 8

Chapter 8 presents an instance-based (case-based) learning approach to induce a model from data streams (IBL-DS) basically for the classification in dynamically evolving environments. Some concepts for regression tasks (lazy learning) are also outlined. It determines or predicts the class (the value) of a sample based on the class (values) of its k nearest neighbors. It determines these neighbors using a distance function based on an incremental version of the simplified value difference metric (SVDM). Using SVDM, two values should be considered similar if they make similar class predictions and dissimilar if their class predictions diverge. The proposed approach (IBL-DS) uses three indicators (temporal relevance, spatial relevance, and consistency) to update the model in response to a concept drift. All samples, which are not recent, are considered redundant if they do not change the nearest neighbors regions of any query, i.e., the samples are very close in the instance space and thus they have a similar region of influence. In this case, they are removed. While a sample in a sparsely region is considered very relevant and stored in the case base (instance memory). This approach is efficient in the case where the number of queries is smaller than the number of data streams. This is because this algorithm achieves predictions based on the most nearest neighbors of a query. This algorithm is implemented using the open source software for mining and analyzing large data sets in a stream-like manner massive online analysis (MOA). The proposed algorithm was evaluated using two synthetic data sets (hyperplane generator and random tree generator). The examples were generated by using the concept driftstream procedure to mix streams produced by two different hyperplanes (simulating a rotation) or by two random tree generators. In addition, two real data sets (shuttle and wine) were used to evaluate the proposed algorithm with two different settings and to compare its performance in classification and in regression with respectively Hoeffding and flexible evolving fuzzy inference systems (FLEXFIS).

Chapter 9

Chapter 9 presents a review of the samplewise streaming data driven family of approaches called flexible evolving fuzzy inference systems (FLEXFIS) family. These systems are flexible because they are able to react to changes in system environment conditions by updating (1) the model parameters (updating the already existing rules), and (2) the model structure (extending the model to unexplored regions in the feature space by generating new rules). The native FLEXFIS algorithm is designed for regression problems using Takagi–Sugeno fuzzy model architecture, extensions to the classification case are presented using three different model architectures (first part of the chapter): single model (SM), multimodel (MM) and all-pairs (AP). In the second part, two extensions of the FLEXFIS family are detailed. These extensions include:

- Detection of and reaction on drifts in data streams,
- Reducing the model complexity by merging of redundant rules,
- Reducing the dimensionality of feature space by smoothly and softly outweighting the least important features with respect to changes in the target concept,
- Some concepts how to integrate uncertainty in model predictions using conflict and ignorance models for classification and adaptive local error bars for regression problems,
- Some consideration toward interpretability of the evolved models.

Chapter 10

Chapter 10 presents a sequential adaptive fuzzy inference system (SAFIS) to learn a model from one-by-one observations coming from nonstationary environments. SAFIS is structured in a neural network of five layers to realize a fuzzy rule model. The number of rules and their premise and consequent parameters are adjusted based on the concept of fuzzy rules influence. The latter is determined as the contribution of the rule to the overall output. SAFIS starts with no fuzzy rule. It translates the first input into a first rule. The latter starts with the largest influence in the input space and then this influence decreases exponentially to allow more fuzzy rules to learn the system dynamics. The rule parameters are adjusted based on the use of the winner rule strategy, i.e., only the parameters of the nearest rule to the input data in Euclidean sense are updated using an extended Kalman filter (EKF): this is applied for linear and nonlinear parameters using extended gradient information for the latter. A rule is removed if its influence is less than a certain pruning threshold. The influence of a rule is measured by its statistical contribution to the final model output over time.

Chapter 11

Chapter 11 presents an interval-based evolving modeling (IBeM) approach to adapt the structure and parameters of rule-based models in incremental manner. The use of intervals or granules is efficient in the case of lacking precise information. The proposed approach starts learning from scratch with no prior knowledge about data distribution and properties. IBeM learns online from the sequence of input–output pairs to approximate the target function. It creates granules and rules whenever stream pairs arrive. Then these granules and rules are adapted in recursive manner to consider new unseen pairs of input–output when they are located inside the granule expansion region. If the new pair is outside the expansion region of all available granules, then a new granule is created to extend the current collection of granules. IBeM is applied to mining the level of rain precipitation in different European regions and to build up an evolving predictor for daily fluctuation of the price of an economic index. The results clearly underline how the method works, i.e., how are the granules and intervals evolved, and the improved performance of the method compared over other state-of-the art approaches such as multilayer perceptron (MLP), evolving Takagi–Sugeno model (eTS) and eXtended evolving Takagi–Sugeno model (xTS).

Chapter 12

Chapter 12 presents a review of dynamic methods to learn models from single and multiple time-series data in nonstationary environments. These models are used to capture the dynamics of individual variables generating the single time-series data as well as the dynamic interactions and relationships between them. The key advantage of these methods is that they can adapt (evolve) their structure online to the environment new conditions as new information becomes available. This adaptation would keep the performance of the model estimation over time. The chapter classifies these methods into two categories: inductive (global modeling) and transductive reasoning methods. Inductive methods create a model from all available data representing the entire problem space. While transductive methods estimate the value of a potential model at a single point of the problem space using some additional information related to this point. These methods are then compared in order to show their advantages and drawbacks. The chapter details a well-known family of inductive methods called dynamic evolving neuro-fuzzy inference systems (DENFIS). Then, it presents the instance-based learning method neuro-fuzzy inference (NFI) as an example of transductive methods. Thereafter, it presents an algorithm for extracting profiles (shapes of trends extracted during snapshots) of relationship of multiple time series data. This algorithm is based on variable clustering than sample clustering. Finally a case study of dynamic learning of 10 stock market indexes in the Asia Pacific region is used to show and to compare the performances of the presented approaches in the chapter.

Chapter 13

Chapter 13 describes a step for feature parameter optimization to adapt the parameters inside the feature calculation process when extracting features from images. Thereby, the optimized calculations of two specific types of parameterized image features, namely Gabor features and blob features, are studied in detail. The goal of this optimization is to put emphasis on those parameters which maximize the separation between the different classes and thus to reduce the number of misclassifications. This optimization can be achieved offline and online. In the offline optimization, feature parameters are optimized by looking for the ones that minimize within-class scatter in relation to the between-class scatter. This feature adaptation maximizes the distance between the classes perpendicular to the decision boundary. In the on-line feature optimization, i.e., the target function, within-class scatter in relation to the between-class scatter, and its gradient according to both, the feature and classifier parameters are adapted, i.e., updated, in the direction of the negative gradient of the target function. Two examples of application are used to show the interest of the feature optimization. The first example is an artificial database of images related to the quality control task where different types of defects need to be distinguished. The second example is a texture classification problem. The classification results show improved performance when optimizing the feature calculation part over conventional extraction of features with default fixed parameters.

Chapter 14

Chapter 14 is the application of the FLEXFIS family as demonstrated in Chap. 9 of this book to the quality inspection of production items. This application is divided into post-supervision visual inspection and online quality control by processing directly the process measurement data. The first is based on classifiers built upon features extracted from images showing the surface of production items. The classifiers are applied to decide (1) whether an image denotes a faulty or nonfaulty item (binary classification problem) and (2) to which types of faults potential regions of interests (objects) belong (multi-class classification problem). The second one identifies high-dimensional system models based on implicit dependencies and relations between system variables (measurement channels) and uses these models to decide whether new measurements are denoting faults or non-faults by calculating the degree of deviation to the identified models. Therefore, no annotation effort for operators for labeling samples is required. In the Experimental section, the authors have used ten different classification methods for comparison and nine data sets (five real world visual surface inspection problems from the and European project). The comparison between static classifiers (trained in offline mode) and their evolved version (updated online) shows clearly that the prediction accuracy has been improved in the evolved version of classifiers. The results for online quality control based on measurement data also underline the necessity of evolving models to increase the performance (detection rate and the area under the ROC curve), significantly.

Chapter 15

Chapter 15 presents an online clustering approach for the identification of a temporally switched hybrid dynamic systems (HDS). HDS are characterized by the interaction between continuous time dynamics and discrete events or logic rules. Temporally switched HDS represent a particular class of HDS in which the transition from one operating or function mode to another one is achieved at particular time instants. This class of HDS can be modeled using either switched autoregressive (AR) or AR with eXogenous inputs (ARX) dynamic models. The identification of HDS aims to estimate the parameters of each operating mode. This requires the determination of the number of operating modes of HDS. In Pattern Recognition (PR) approaches, an operating mode can be represented as a class in the feature space. Thus, the proposed approach in this chapter finds the number of classes based on the use of past input-output observations. A least square method with a sliding time window is applied on these observations in order to estimate the coefficients of the AR and ARX dynamic models. These coefficients represent the features of the feature space. The patterns belonging to each class will be then used to estimate the parameters of its associated operating mode. The chapter evaluates the proposed approach using a simulation and real examples. The simulation example is a system switching in different time instants among three different operating modes represented by three different discrete time transfer functions. The second example aims to characterize the noises generated by the steam generator of nuclear power generators prototype fast reactor (PFR). These noises represent the normal operating mode as well as the faulty one in response to a leakage in the steam generator. Two acoustic sensors are used to record these noises.

References

1. Abraham, A., Dote, Y.: Engineering Hybrid Soft Computing Systems. Springer, New York (2010)
2. Angelov, P., Filev, D.: An approach to online identification of Takagi–Sugeno fuzzy models. IEEE Transactions on Systems, Man and Cybernetics, part B: Cybernetics **34**(1), 484–498 (2004)
3. Angelov, P., Filev, D., Kasabov, N.: Evolving Intelligent Systems—Methodology and Applications. John Wiley & Sons, New York (2010)
4. Angelov, P., Kasabov, N.: Evolving computational intelligence systems. In: Proceedings of the 1st International Workshop on Genetic Fuzzy Systems, pp. 76–82. Granada, Spain (2005)
5. Angstenberger, L.: Dynamic fuzzy pattern recognition. Ph.D. thesis, Fakultät für Wirtschaftswissenschaften der Rheinisch-Westfälischen Technischen Hochschule (2000). Aachen, Germany
6. Bifet, A., Gavalda, R.: Learning from time-changing data with adaptive windowing. In: Proceedings of the SIAM International Conference on Data Mining, pp. 443–448. Minneapolis, MN (2007)
7. Bifet, A., Holmes, G., Kirkby, R., Pfahringer, B.: MOA: Massive online analysis. Journal of Machine Learning Research **11**, 1601–1604 (2010)
8. Boubacar, A., Lecoeuche, H., Maouche, S.: Audyc neural network using a new gaussian densities merge mechanism. In: 7th International Conference on Adaptive and Natural Computing Algorithms, pp. 155–158. Coimbra, Portugal (2005)
9. Bouchachia, A.: Evolving clustering: an asset for evolving systems. IEEE SMC Newsletter **36** (2011)
10. Diehl, C., Cauwenberghs, G.: SVM incremental learning, adaptation and optimization. In: Proceedings of the International Joint Conference on Neural Networks Vol. 4, pp. 2685–2690. Boston (2003)
11. Domingos, P., Hulten, G.: Mining high-speed data streams. In: Proceedings of the Sixth ACM SIGKDD International Conference on Knowledge Discovery and Data Mining, pp. 71–80. Boston, MA (2000)
12. Gama, J.: Knowledge Discovery from Data Streams. Chapman & Hall/CRC, Boca Raton, Florida (2010)
13. Hartert, L., Sayed-Mouchaweh, M., Billaudel, P.: A semi-supervised dynamic version of Fuzzy K-Nearest Neighbours to monitor evolving systems. Evolving Systems, **1** (1), 3–15 (2010)
14. Haykin, S.: Neural Networks: A Comprehensive Foundation (2nd Edition). Prentice Hall Inc., Upper Saddle River, New Jersey (1999)
15. Huang, G., Saratchandran, P., Sundararajan, N.: A generalized growing and pruning rbf (ggap-rbf) neural network for function approximation. IEEE Transactions on Neural Networks **16**(1), 57–67 (2005)
16. Kasabov, N.: Evolving Connectionist Systems: The Knowledge Engineering Approach—Second Edition. Springer Verlag, London (2007)
17. Kruse, R., Gebhardt, J., Palm, R.: Fuzzy Systems in Computer Science. Verlag Vieweg, Wiesbaden (1994)
18. Kuncheva, L.: Classifier ensembles for changing environments. In: Proceedings of the 5th International Workshop in Multiple Classifier Systems, pp. 1–15. Cagliari, Italy (2004)
19. Lecoeuche, S., Lurette, C.: Auto-adaptive and dynamical clustering neural network. In: Proceedings of the International Conference on Artificial Neural Networks, pp. 350–358. Istanbul, Turkey (2003)
20. Ljung, L.: System Identification: Theory for the User. Prentice Hall PTR, Prentic Hall Inc., Upper Saddle River, New Jersey (1999)
21. Lughofer, E.: Evolving Fuzzy Systems—Methodologies, Advanced Concepts and Applications. Springer, Berlin Heidelberg (2011)

22. Lughofer, E., Bouchot, J.L., Shaker, A.: On-line elimination of local redundancies in evolving fuzzy systems. Evolving Systems **2**(3), 165–187 (2011)
23. Lughofer, E.: Human-inspired evolving machines—the next generation of evolving intelligent systems? IEEE SMC newsletter **36** (2011)
24. Minku, F., White, A., Yao, X.: The impact of diversity on on-line ensemble learning in the presence of concept drift. IEEE Transactions on Knowledge and Data Engineering **22**, 730–742 (2010)
25. Mouss, H., Mouss, D., Mouss, N., Sefouhi, L.: Test of Page-Hinkley, an approach for fault detection in an agro-alimentary production system. In: Proceedings of the Asian Control Conference, Volume 2, pp. 815–818 (2004)
26. Nakhaeizadeh, G., Taylor, C., Kunisch, G.: Dynamic supervised learning. Some basic issues and application aspects. Classification and knowledge organization. pp. 123–135. Springer Verlag, Berlin Heidelberg (1997)
27. Nelles, O.: Nonlinear System Identification. Springer, Berlin (2001)
28. Oza, N.C., Russell, S.: Online bagging and boosting. Artificial Intelligence and Statistics, pp. 105–112 (2001)
29. Pedrycz, W., Gomide, F.: Fuzzy Systems Engineering: Toward Human-Centric Computing. John Wiley & Sons, Hoboken, New Jersey (2007)
30. Pedrycz, W., Rai, P.: Collaborative clustering with the use of fuzzy c-means and its quantification. Fuzzy Sets and Systems **159**(18), 2399–2427 (2008)
31. Pedrycz, W., Weber, R.: Special issue on soft computing for dynamic data mining. Applied Soft Computing **8**(4), 1281–1282 (2008)
32. Sayed-Mouchaweh, M.: Semi Supervised Classification Method for Dynamic Applications. Fuzzy Sets and Systems, **161**(4), 544–563 (2010)
33. Sayed-Mouchaweh, M., Messai, N.: A clustering-based approach for the identification of a class of temporally switched linear systems. Elsevier, Pattern Recognition Letters (PRL), **33**(2), 144–151 (2012)
34. Tsymbal, A., Pechenizkiy, M., Cunningham, P., Puuronen, S.: Handling local concept drift with dynamic integration of classifiers: domain of antibiotic resistance in nosocomial infections. In: Proc. 19th IEEE Int. Symposium on Computer-Based Medical Systems CBMS 2006, pp. 679–684. Maribor, Slovenia (2006)
35. Weber, R.: Dynamic data mining. Encyclopedia of Data Warehousing and Mining pp. 722–728 (2009)

Part I
Dynamic Methods for Unsupervised Learning Problems

Chapter 2
Incremental Statistical Measures

Katharina Tschumitschew and Frank Klawonn

Abstract Statistical measures provide essential and valuable information about data and are needed for any kind of data analysis. Statistical measures can be used in a purely exploratory context to describe properties of the data, but also as estimators for model parameters or in the context of hypothesis testing. For example, the mean value is a measure for location, but also an estimator for the expected value of a probability distribution from which the data are sampled. Statistical moments of higher order than the mean provide information about the variance, the skewness, and the kurtosis of a probability distribution. The Pearson correlation coefficient is a measure for linear dependency between two variables. In robust statistics, quantiles play an important role, since they are less sensitive to outliers. The median is an alternative measure of location, the interquartile range an alternative measure of dispersion. The application of statistical measures to data streams requires online calculation. Since data come in step by step, incremental calculations are needed to avoid to start the computation process each time new data arrive and to save memory so that not the whole data set needs to be kept in the memory. Statistical measures like the mean, the variance, moments in general, and the Pearson correlation coefficient render themselves easily to incremental computations, whereas recursive or incremental algorithms for quantiles are not as simple or obvious. Nonstationarity is another important aspect of data streams that needs to be taken into account.

K. Tschumitschew (✉)
Department of Computer Science, Ostfalia University of Applied Sciences,
Salzdahlumer Str. 46/48, D-38302 Wolfenbuettel, Germany
e-mail: katharina.tschumitschew@ostfalia.de

F. Klawonn
Department of Computer Science, Ostfalia University of Applied Sciences,
Salzdahlumer Str. 46/48, D-38302 Wolfenbuettel, Germany

Bioinformatics and Statistics, Helmholtz Centre for Infection Research,
Inhoffenstr. 7, D-38124 Braunschweig, Germany
e-mail: f.klawonn@ostfalia.de; frank.klawonn@helmholtz-hzi.de

M. Sayed-Mouchaweh and E. Lughofer (eds.), *Learning in Non-Stationary Environments: Methods and Applications*, DOI 10.1007/978-1-4419-8020-5_2,
© Springer Science+Business Media New York 2012

This means that the parameters of the underlying sampling distribution might change over time. Change detection and online adaptation of statistical estimators is required for nonstationary data streams. Hypothesis tests like the χ^2- or the t-test can be a basis for change detection, since they can also be calculated in an incremental fashion. Based on change detection strategies, one can derive information on the sampling strategy, for instance the optimal size of a time window for parameter estimations of nonstationary data streams.

2.1 Introduction

Statistics and statistical methods are used in almost every aspect of modern life, like medicine, social surveys, economy, and marketing, only to name few of application areas. A vast number of sophisticated statistical software tools can be used to search and test for structures and patterns in data. Important information about the data generating process is provided by the simple summary statistics. Characteristics of the data distribution can be described by summary statistics like the following one.

- Measures of location: The mean and quantiles provide information about location of the distribution. Mean and median are representatives for the center of the distribution.
- Measures of spread: Common measures for the variation in the data are standard deviation, variance, and interquartile range.
- Shape: The third and fourth moments provide information about the skewness and the kurtosis of a probability distribution.
- Dependence: For instance, the Pearson correlation coefficient is a measure for the linear dependency between two variables. Other common measures for statistical dependency between two variables rank correlation coefficients like Spearman's rho or Kendall's tau.

Apart from providing information about location and spread of the data distribution, quantiles also play an important role in robust data analysis, since they are less sensitive to outliers.

Summary statistics can be used in a purely exploratory context to describe properties of the data, but also as estimators for model parameters of an assumed underlying data distribution.

More complex and powerful methods for statistical data analysis are for instance hypothesis tests. Statistical hypothesis testing allows us to discover the current state of affairs and therefore help us to make decisions based on the gained knowledge. Hypothesis test can be applied to a great variety of problems. We may need to test just a simple parameter or the whole distribution of the data.

However, classical statistics operates with a finite, fixed data set. On the other hand, nowadays it is very important to continuously collect and analyze data sets increasing with time, since the (new) data may contain useful information.

Sensor data as well as the seasonal behavior of markets, weather, or animals are in the focus of diverse research studies. The amount of recorded data increases each day. Apart from the huge amount of data to be dealt with, another problem is that the data arrive continuously in time. Such kind of data is called data stream. A data stream can be characterized as an unlimited sequence of values arriving step by step over time. One of the main problems for the analysis of data streams is limited computing and memory capabilities. It is impossible to hold the whole data set in the main memory of a computer or computing device like an ECU (electronic control unit) that might also be responsible for other tasks than just analyzing the data. Moreover, the results of the analysis should be presented in acceptable time, sometimes even under very strict time constraints, so that the user or system can react in real time. Therefore, the analysis of data streams requires efficient online computations. Algorithms based on incremental or recursive computation schemes satisfy the above requirements. Such methods do not store all historical data and do not need to browse through old data to update an estimator or an analysis, in the ideal case, each data value is touched only once.

Consequently the application of statistical methods to data streams requires modifications to the standard calculation schemes in order to be able carry out the computations online. Since data come in step by step, incremental calculations are needed to avoid to start the computation process from scratch each time new data arrive and to save memory, so that not the whole data set must be kept in the memory. Statistical measures like the sample mean, variance and moments in general and the Pearson correlation coefficient render themselves easily incremental computation schemes, whereas, for instance, for standard quantiles computations the whole data is needed. In such cases, new incremental methods must be developed that avoid sorting the whole data set, since sorting requires in principal to check the whole data set. Several approaches for the online estimation of quantiles are presented for instance in [1, 9, 19, 25].

Another important aspect in data stream analysis is that the data generating process does not remain static, i.e., the underlying probabilistic model cannot be assumed to be stationary. The changes in the data structure may occur over time. Dealing with nonstationary data requires change detection and on-line adaptation. Different kinds of nonstationarity have been classified in [2]:

- Changes in the data distribution: the change occurs in the data distribution. For instance, mean or variance of the data distribution may change over time.
- Changes in concept: here concept drift refers to changes of a target variable. A target variable is a variable, whose values we try to predict based on the model estimated from the data, for instance for linear regression it is the change of the parameters of the linear relationship between the data.
  - Concept drift: concept drift describes gradual changes of the concept. In statistics, this usually called structural drift.
  - Concept shift: concept shift refers to an abrupt change which is also referred to as structural break.

Hence change detection and online adaptation of statistical estimators are required for nonstationary data streams. Various strategies to handle nonstationarity are proposed, see for instance [11] for a detailed survey of change detection methods. Statistical hypothesis tests may also be used for change detection. Since we are working with data streams, it is required that the calculations for the hypothesis tests can be carried out in an incremental way. For instance, the χ^2-test and the t-test[1] render themselves easily to incremental computations. Based on change detection strategies, one can derive information on the sampling strategy, for instance the optimal size of a time window for parameter estimations of nonstationary data streams [3, 26].

This chapter is organized as follows. Incremental computations of the mean, variance, third and fourth moments and the Pearson correlation coefficient are explained in Sect. 2.2. Furthermore two algorithms for the on-line estimation of quantiles are described in Sect. 2.3. In Sect. 2.4 we provide on-line adaptations of statistical hypothesis test and discuss different change detection strategies.

2.2 Incremental Calculation of Moments and the Pearson Correlation Coefficient

Statistical measures like sample central moments provide valuable information about the data distribution. So the sample mean or empirical mean (first sample central moment) is the measure of the center of location of the data distribution, the measure of variability is sample variance (second sample central moment). The third and fourth central moments are used to compute skewness and kurtosis of the data sample. Skewness provides us the information about the asymmetry of the data distribution and kurtosis give us an idea about the degree of peakedness of the distribution.

Another important statistic is the correlation coefficient. The correlation coefficient is a measure for linear dependency between two variables.

In this section, we introduce incremental calculations for these statistical measures.

In the following, we consider a real-valued sample $x_1, \ldots, x_t, \ldots$ ($x_i \in \mathbb{R}$ for all $i \in \{1, \ldots, t, \ldots\}$).

Definition 2.1. Let $x_1, \ldots, x_t$ be a random sample from the distribution of the random variable X.

The sample or empirical mean of the sample of size t, denoted by $\bar{x}_t$, is given by the formula

$$\bar{x}_t = \frac{1}{t} \sum_{i=1}^{t} x_i. \tag{2.1}$$

[1]For precise definitions, see Sect. 2.4.

Equation (2.1) cannot be applied directly in the context of data streams, since it would require to consider all sample values at each time step. Fortunately, (2.1) can be easily transformed into an incremental scheme.

$$\begin{aligned}
\bar{x}_t &= \frac{1}{t}\sum_{i=1}^{t} x_i \\
&= \frac{1}{t}\left(x_t + \sum_{i=1}^{t-1} x_i\right) \\
&= \frac{1}{t}\left(x_t + (t-1)\bar{x}_{t-1}\right) \\
&= \bar{x}_{t-1} + \frac{1}{t}\left(x_t - \bar{x}_{t-1}\right). \qquad (2.2)
\end{aligned}$$

The incremental update (2.2) requires only three values to calculate the sample mean at time point t:

- The mean at time point $t-1$.
- The sample value at time point t.
- The number of sample values so far.

The empirical or sample variance can be calculated in an incremental fashion in a similar way.

Definition 2.2. Let $x_1,\ldots,x_t$ be a random sample from the distribution of the random variable X. The empirical or sample variance of a sample of size t is given by

$$s_t^2 = \frac{1}{t-1}\sum_{i=1}^{t}(x_i - \bar{x}_t)^2 \qquad (2.3)$$

Furthermore, $s_t = \sqrt{s_t^2}$ is called the sample standard deviation.

In order to simplify the calculation, we use following notation:

$$\tilde{m}_{2,t} = \sum_{i=1}^{t}(x_i - \bar{x}_t)^2 \qquad (2.4)$$

In the following, the formula for incremental calculation is derived from (2.4) using (2.2).

$$\begin{aligned}
\tilde{m}_{2,t} - \tilde{m}_{2,t-1} &= \sum_{i=1}^{t} x_i^2 - t\bar{x}_t^2 - \sum_{i=1}^{t-1} x_i^2 + (t-1)\bar{x}_{t-1}^2 \\
&= x_t^2 - t\bar{x}_t^2 + (t-1)\bar{x}_{t-1}^2 \\
&= x_t^2 - \bar{x}_{t-1}^2 + t\left(\bar{x}_{t-1}^2 - \bar{x}_t^2\right)
\end{aligned}$$

$$
\begin{aligned}
&= x_t^2 - \bar{x}_{t-1}^2 + t\,(\bar{x}_{t-1} - \bar{x}_t)\,(\bar{x}_{t-1} + \bar{x}_t) \\
&= x_t^2 - \bar{x}_{t-1}^2 + t\left(\bar{x}_{t-1} - \bar{x}_{t-1} - \frac{1}{t}\,(x_t - \bar{x}_{t-1})\right)(\bar{x}_{t-1} + \bar{x}_t) \\
&= x_t^2 - \bar{x}_{t-1}^2 + (\bar{x}_{t-1} - x_t)\,(\bar{x}_{t-1} + \bar{x}_t) \\
&= (x_t - \bar{x}_{t-1})\,(x_t + \bar{x}_{t-1} - \bar{x}_{t-1} - \bar{x}_t) \\
&= (x_t - \bar{x}_{t-1})\,(x_t - \bar{x}_t)\,.
\end{aligned}
$$

Consequently, we obtain the following recurrence formula for the second central moment:

$$\tilde{m}_{2,t} = \tilde{m}_{2,t-1} + (x_t - \bar{x}_{t-1})\,(x_t - \bar{x}_t) \tag{2.5}$$

The unbiased estimator for the variance of the sample according to (2.5) is given by

$$s_t^2 = \frac{1}{t-1}M_{2,t} = \frac{(t-2)\,s_{t-1}^2 + (x_t - \bar{x}_{t-1})\,(x_t - \bar{x}_t)}{t-1}. \tag{2.6}$$

Definition 2.3. Let $x_1,\ldots,x_t$ be a random sample from the distribution of the random variable X. Then the k-th central moment of a sample of size t is defined by

$$m_{k,t} = \frac{1}{t}\sum_{i=1}^{t}(x_i - \bar{x}_t)^k. \tag{2.7}$$

In order to simplify the computations and to facilitate the readability of the text, we use the following expression for the derivation.

$$\tilde{m}_{k,t} = \sum_{i=1}^{t}(x_i - \bar{x}_t)^k, \tag{2.8}$$

therefore $\tilde{m}_{k,t} = t \cdot m_{k,t}$.

For the third- and fourth-order moments, which are needed to calculate skewness and kurtosis of the data distribution, incremental formulae can be derived in a similar way, in the form of pairwise update equations for $\tilde{m}_{3,t}$ and $\tilde{m}_{4,t}$.

$$
\begin{aligned}
\tilde{m}_{3,t} &= \sum_{i=1}^{t-1}(x_i - \bar{x}_t)^3 + (x_t - \bar{x}_t)^3 \\
&= \sum_{i=1}^{t-1}\left(x_i - \bar{x}_{t-1} - \frac{1}{t}\,(x_t - \bar{x}_{t-1})\right)^3 + \left(x_t - \bar{x}_{t-1} - \frac{1}{t}\,(x_t - \bar{x}_{t-1})\right)^3 \\
&= \sum_{i=1}^{t-1}((x_i - \bar{x}_{t-1}) - b)^3 + (tb - b)^3
\end{aligned}
$$

$$
\begin{aligned}
&= \sum_{i=1}^{t-1} \left((x_i - \bar{x}_{t-1})^3 - 3b\,(x_i - \bar{x}_{t-1})^2 + 3b^2\,(x_i - \bar{x}_{t-1}) - b^3 \right) + (t-1)^3\, b^3 \\
&= \tilde{m}_{3,t-1} - 3b\tilde{m}_{2,t-1} - ((t-1)\, b^3 + (t-1)^3\, b^3 \\
&= \tilde{m}_{3,t-1} - 3b\tilde{m}_{2,t-1} + t\,(t-1)\,(t-2)\, b^3 \qquad (2.9)
\end{aligned}
$$

where $b = \frac{x_t - \bar{x}_{t-1}}{t}$.

From (2.9), we obtain a one-pass formula for the third-order centered statistical moment of a sample of size t:

$$\tilde{m}_{3,t} = \tilde{m}_{3,t-1} - 3\frac{(x_t - \bar{x}_{t-1})}{t}\tilde{m}_{2,t-1} + \frac{(t-1)\,(t-2)}{t^2}\,(x_t - \bar{x}_{t-1})^3 . \qquad (2.10)$$

The derivation for the fourth-order moment is very similar to (2.9) and thus is not detailed here.

$$
\begin{aligned}
\tilde{m}_{4,t} = \tilde{m}_{4,t-1} &- 4\frac{(x_t - \bar{x}_{t-1})}{t}\tilde{m}_{3,t-1} + 6\left(\frac{x_t - \bar{x}_{t-1}}{t}\right)^2 \tilde{m}_{2,t-1} \\
&+ \frac{(t-1)\,(t^2 - 3t + 3)}{t^3}\,(x_t - \bar{x}_{t-1})^4 . \qquad (2.11)
\end{aligned}
$$

The results presented above offer the essential formulae for efficient, one-pass calculations of statistical moments up to the fourth order. Those are important when the data stream mean, variance, skewness, and kurtosis should be calculated. Although these measures cover the needs of the vast majority of applications for data analysis, sometimes higher-order statistics should be used. For the computation of higher-order statistical moments, see for instance [6].

Now we derive a formula for the incremental calculation of the sample correlation coefficient.

Definition 2.4. Let $x_1,\ldots,x_t$ be a random sample from the distribution of the random variable X and $y_1,\ldots,y_t$ be a random sample from the distribution of the random variable Y. Then the sample Pearson correlation coefficient of the sample of size t, denoted by $r_{xy,t}$, is given by the formula

$$r_{xy,t} = \frac{\sum_{i=1}^{t} (x_i - \bar{x}_t)\,(y_i - \bar{y}_t)}{(t-1)\, s_{x,t} s_{y,t}} \qquad (2.12)$$

where $\bar{x}_t$ and $\bar{y}_t$ are the sample means of X and Y and $s_{x,t}$ and $s_{y,t}$ are the sample standard deviations of X and Y, respectively.

The incremental formula for the sample standard deviation can be easily derived from the incremental formula for sample variance (2.6). Hence, only the numerator of (2.12) needs to be considered further. Furthermore, the numerator of (2.12) represents the sample covariance $s_{xy,t}$.

Definition 2.5. Let $x_1,\ldots,x_t$ be a random sample from the distribution of the random variable X and $y_1,\ldots,y_t$ be a random sample from the distribution of the random variable Y. Then the sample covariance $s_{xy,t}$ of the sample of size t is given by t

$$s_{xy,t} = \frac{\sum_{i=1}^{t}(x_i-\bar{x}_t)(y_i-\bar{y}_t)}{t-1} \tag{2.13}$$

where $\bar{x}_t$ and $\bar{y}_t$ are the sample means of X and Y and $s_{x,t}$ and $s_{y,t}$ are the sample standard deviations of X and Y, respectively.

The formula for the incremental calculation of the covariance is given by

$$\begin{aligned}(t-1)\,s_{xy,t} &= \sum_{i=1}^{t-1}(x_i-\bar{x}_t)(y_i-\bar{y}_t)+(x_t-\bar{x}_t)(y_t-\bar{y}_t)\\ &= \sum_{i=1}^{t-1}((x_i-\bar{x}_{t-1})-b_x)((y_i-\bar{y}_{t-1})-b_y)+(t-1)^2\,b_xb_y\\ &= (t-2)\,s_{xy,t-1}+t\,(t-1)\,b_xb_y \end{aligned} \tag{2.14}$$

where $b_x = \frac{(x_t-\bar{x}_{t-1})}{t}$ and $b_y = \frac{(y_t-\bar{y}_{t-1})}{t}$. Hence, the incremental formula for the sample covariance is

$$s_{xy,t} = \frac{(t-2)}{(t-1)}s_{xy,t-1}+\frac{1}{t}(x_t-\bar{x}_{t-1})(y_t-\bar{y}_{t-1}) \tag{2.15}$$

Therefore, to update the Pearson correlation coefficient, we have to compute the sample standard deviation and covariance first and subsequently use (2.12).

Above in this section, we presented incremental calculations for the empirical mean, empirical variance, third and fourth sample central moments and sample correlation coefficient. These statistical measures can also be considered as estimators of the corresponding parameters of the data distribution. Therefore, we are interested in the question how many values x_i do we need to get a "good" estimation of the parameters. Of course, as we deal with a data stream, in general we will have a large amount of data. However, some application are based on time window techniques. For instance, for change detection methods presented in the section (Sect. 2.4). Here we need to compare at least two samples of data; on that account, the data have to be split into smaller parts. To answer the question about the optimal amount of data for statistical estimators, we have to analyze the variances of the parameter estimators. The variance of an estimator shows how efficient this estimator is.

Here we restrict our considerations to a random sample from a normal distribution with expected value 0. Let $X_1,\ldots,X_t$ be independent and identically distributed (i.i.d.) random variables following a normal distribution, $X_i \sim N\left(0,\, \sigma^2\right)$ and $x_1,\ldots,x_t$ are observed values of these random variables.

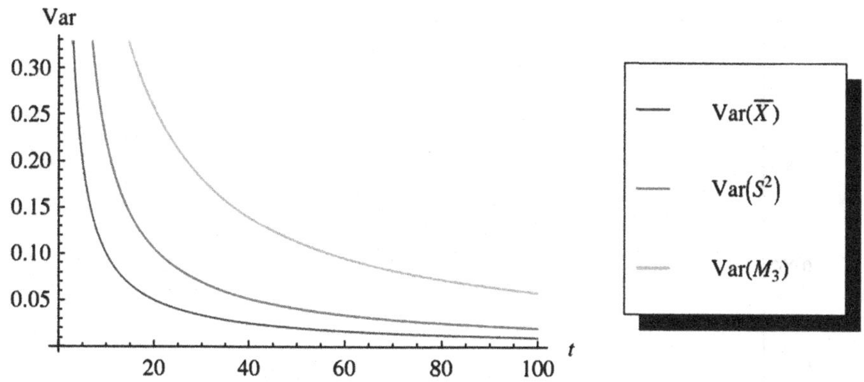

Fig. 2.1 Variances from bottom to top of parameter estimators for the expected value, the variance and the third moment of a standard normal distribution

The variance of the estimator of the expected value[2] $\bar{X}_t = \frac{1}{t}\sum_{i=1}^{t} X_i$ is given by

$$Var(\bar{X}_t) = \frac{\sigma^2}{t}. \tag{2.16}$$

The variance of the unbiased estimator of the variance $S^2 = \frac{1}{t-1}\sum_{i=1}^{t}(X_i - \bar{X}_t)^2$ is given by

$$Var\left(S_t^2\right) = \frac{2}{(t-1)}\sigma^4. \tag{2.17}$$

The variance of the distribution of the third moment is shown in (2.18) (see [6] for more detailed information)

$$Var(M_{3,t}) = \frac{6(t-1)(t-2)}{t^3}\sigma^6. \tag{2.18}$$

Figure 2.1 shows (2.16), (2.17), and (2.18) as functions in t for $\sigma^2 = 1$ (standard normal population). It is obvious that for small amounts of data, the variance of the estimators is quite large, consequently more values are needed to obtain a reliable estimation of distribution parameters. Furthermore, the optimal sample size depends on the statistic to be computed. For instance, for the sample mean and a sample of size 50, the variance is already small enough, whereas for the third moment estimator to have the same variance, many more observations are needed.

We apply the same considerations to the sample correlation coefficient. Let X and Y be two random variables following normal distributions and let $X_1, \ldots, X_t$ and $Y_1, \ldots, Y_t$ be i.i.d. samples of X and Y, respectively: $X_i \sim N\left(0, \sigma_x^2\right)$ and

[2]We use capital letters here to distinguish between random variables and real numbers that are denoted by small letters.

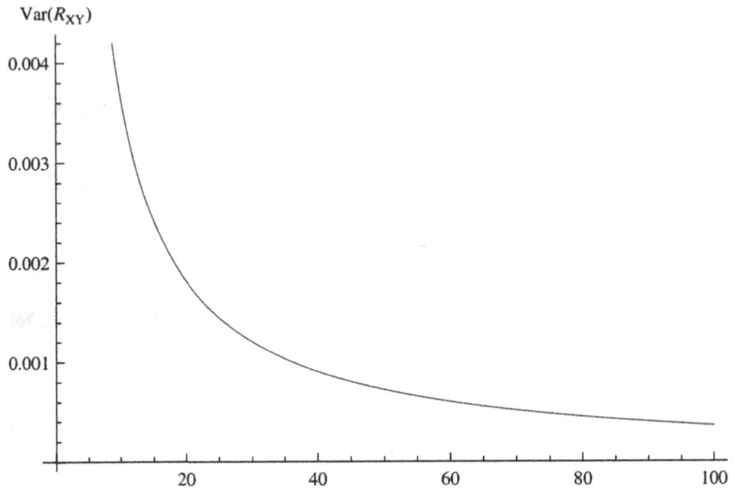

Fig. 2.2 Asymptotic variance of the sample correlation coefficient

$Y_i \sim N\left(0,\, \sigma_y^2\right)$. Assume the correlation between X and Y is equal to ρ_{XY}. Then the asymptotic variance of the sample correlation coefficient is given by (see [7])

$$Var\left(R_{XY,t}\right) \approx \frac{\left(1-\rho_{XY}^2\right)^2}{t}. \tag{2.19}$$

Attention should be paid to the asymptotic nature of (2.19). This formula can be used only for sufficiently large t (see [7]). Equation (2.19) is illustrated in Fig. 2.2 as a function in t for $\rho_{XY} = 0.9$. Since for different values of ρ_{XY}, the plots are very similar, they are not shown here.

In this section, we have provided equations for incremental calculation of the sample mean, sample variance, third and fourth moments and the Pearson correlation coefficient. These statistics allow us to summarize a set of observations analytically. Since we assume that the observations reflect the population as a whole, these statistics give us an idea about the underlying data distribution. Other important summary statistics are sample quantiles. Incremental approaches for quantiles estimation are described in the next section.

2.3 Incremental Quantile Estimation

Quantiles play an important role in statistics, especially in robust statistics, since they are not or less sensitive to outliers. For $q \in (0,1)$, the q-quantile has the property that $q \cdot 100\%$ of the data are smaller and $(1-q) \cdot 100\%$ of the data are larger than this value. The median, i.e., the 50% quantile, is a robust measure of location and the interquartile range[3] is a robust measure of spread. Incremental or recursive techniques for quantile estimation are not as obvious as for statistical moments, since for the sample quantile computation the entire sorted data are needed. Nevertheless, there are techniques for incremental quantile estimation. In this section, we describe two different approaches. First approach is restricted to continuous symmetric unimodal distributions. Therefore, this method is not very useful for all real world data. The second approach is not restricted to any kind of distribution and is not limited to continuous random variables. We also provide experimental results for both algorithms for different kinds of distributions.

2.3.1 Incremental Quantile Estimation for Continuous Random Variables

Definition 2.6. For a random variable X with cumulative distribution function F_X, the q-quantile ($q \in (0,1)$) is defined as $\inf\{x \in \mathbb{R} \mid F_X(x) \geq q\}$. If x_q is the q-quantile of a continuous random variable, this implies $P(X \leq x_q) = q$ and $P(X \geq x_q) = 1 - q$.

For continuous random variables, an incremental scheme for quantile estimation is proposed in [10]. This approach is based on the following theorem.

Theorem 2.1. *Let $\{\xi_t\}_{t=0,1,\ldots}$ be a sequence of identically distributed independent (i.i.d.) random variables with cumulative distribution function F_ξ. Assume that the density function $f_\xi(x)$ exists and is continuous in the α-quantile x_α for an arbitrarily chosen α $(0 < \alpha < 1)$. Further let the inequality*

$$f_\xi(x_\alpha) > 0 \tag{2.20}$$

be fulfilled. Let $\{c_t\}_{t=0,1,\ldots}$ be a (control) sequence of real numbers satisfying the conditions

$$\sum_{t=0}^{\infty} c_t = \infty, \quad \sum_{t=0}^{\infty} c_t^2 < \infty. \tag{2.21}$$

Then the stochastic process X_t defined by

$$X_0 = \xi_0, \tag{2.22}$$

$$X_{t+1} = X_t + c_t Y_{t+1}(X_t, \xi_{t+1}), \tag{2.23}$$

[3]The interquartile range is the midrange containing 50% of the data and it is computed as the difference between the 75%- and the 25%-quantiles: $IQR = x_{0.75} - x_{0.25}$.

with

$$Y_{t+1} = \begin{cases} \alpha - 1 & \text{if } \xi_{t+1} < X_t, \\ \alpha & \text{if } \xi_{t+1} \geq X_t, \end{cases} \tag{2.24}$$

almost surely converges to the quantile x_α.

The proof of the theorem is based on stochastic approximation and can be found in [18]. A standard choice of the sequence $\{c_t\}_{t=0,1,\ldots}$ is $c_t = 1/t$. However, convergence might be extremely slow for certain distributions. Therefore, techniques to choose a suitable sequence $\{c_t\}_{t=0,1,\ldots}$, for instance, based on an estimation of the probability density function of the sampled random variable, are proposed in [10, 17].

Although this technique of incremental quantile estimation has only minimum memory requirement, it has certain disadvantages.

- It is only suitable for continuous random variables.
- Unless the sequence $\{c_t\}_{t=0,1,\ldots}$ is well chosen, convergence can be extremely slow.
- When the sampled random variable changes over time, especially when the c_t are already close to zero, the incremental estimation of the quantile will remain almost constant and the change will be unnoticed.

In the following, we present an algorithm to overcome these problems.

2.3.2 Incremental Quantile Estimation

Here we provide a more general approach which is not limited to continuous random variables. First we describe an algorithm for incremental median estimation, which can be generalized to arbitrary quantiles. Since this algorithm is not very suitable for noncentral quantiles, we modify this approach in such a way that it yields good results for all quantiles.

2.3.2.1 Incremental Median Estimation

Before we discuss the general problem of incremental quantile estimation, we first focus on the special case of the median, since we will need the results for the median to develop suitable methods for arbitrary quantiles.

For the incremental computation of the median we store a fixed number, a buffer of m sorted data values $a_1, \ldots, a_m$ in the ideal case the $\frac{m}{2}$ closest values left and the $\frac{m}{2}$ closest values right of the median, so that the interval $[a_1, a_m]$ contains the median. We also need two counters L and R to store the number of values outside the interval $[a_1, a_m]$, counting the values left and right of the interval separately. Initially, L and R are set to zero.

Table 2.1 A small example data set

t	1	2	3	4	5	6	7	8	9
Data	3.8	5.2	6.1	4.2	7.5	6.3	5.4	5.9	3.9

The algorithm works as follows. The first m data points $x_1,\ldots,x_m$ are used to fill the buffer. They are entered into the buffer in increasing order, i.e., $a_i = x_{[i]}$ where $x_{[1]} \leq \ldots \leq x_{[m]}$ are the sorted values $x_1,\ldots,x_m$. After the buffer is filled, the algorithm handles the incoming values x_t in the following way:

1. If $x_t < a_1$, i.e., the new value lies left of the interval supposed to contain the median, then $L^{\text{new}} := L^{\text{old}} + 1$.
2. If $x_t > a_m$, i.e., the new value lies right of the interval supposed to contain the median, then $R^{\text{new}} := R^{\text{old}} + 1$.
3. If $a_i \leq x_t \leq a_{i+1}$ $(1 \leq i < m)$, x_t is entered into the buffer at position a_i or a_{i+1}. Of course, the other values have to be shifted accordingly and the old left bound a_1 or the old right bound a_m will be dropped. Since in the ideal case, the median is the value in the middle of the buffer, the algorithm tries to achieve this by balancing the number of values left and right of the interval $[a_1, a_m]$. Therefore, the following rule is applied:

   a. If $L < R$, then remove a_1, increase L, i.e. $L^{\text{new}} := L^{\text{old}} + 1$, shift the values $a_2,\ldots,a_i$ one position to the left and enter x_t in a_i.
   b. Otherwise remove a_m, increase R, i.e. $R^{\text{new}} := R^{\text{old}} + 1$, shift the values $a_{i+1},\ldots,a_{m-1}$ one position to the right and enter x_t in a_{i+1}.

In each step, the median $\hat{q}_{0.5}$ can be easily calculated from the given values in the buffer and the counters L and R by

$$\hat{q}_{0.5} = \begin{cases} a_{\frac{L+m+R}{2}-L} & \text{if } t \text{ is odd,} \\ \dfrac{a_{\frac{L+m+R-1}{2}-L} + a_{\frac{L+m+R+1}{2}-L}}{2} & \text{if } t \text{ is even.} \end{cases} \tag{2.25}$$

It should be noted that it can happen that at least one of the indices $\frac{L+m+R}{2} - L$, $\frac{L+m+R-1}{2} - L$ and $\frac{L+m+R+1}{2} - L$ are not within the bounds $1,\ldots,m$ of the buffer indices and the computation of the median fails. The interval length $a_m - a_1$ can only decrease and at least for continuous distributions X with probability density function $f_X(q_{0.5}) > 0$, where $q_{0.5}$ is the true median of X, it will tend to zero with increasing sample size. In an ideal situation, the buffer of m stored values contains exactly the values in the middle of the sample. Here, we assume that at this point in time the sample consists of $m+t$ values (Table 2.1).

Table 2.2 illustrates how this algorithm works with an extremely small buffer of size $m = 4$ based on the data set given in Table 2.1.

In the following, we generalize and modify the incremental median algorithm proposed in the previous section and analyze the algorithm in more detail.

Table 2.2 The development of the buffer and the two counters for the small example data set in Table 2.1

t	L	a_1	a_2	a_3	a_4	R
4	0	3.8	4.2	5.2	6.1	0
5	0	3.8	4.2	5.2	6.1	1
6	0	3.8	4.2	5.2	6.1	2
7	1	4.2	5.2	5.4	6.1	2
8	2	5.2	5.4	5.9	6.1	2
9	3	5.2	5.4	5.9	6.1	2

2.3.2.2 An Ad hoc Algorithm

This algorithm for incremental median estimation can be generalized to arbitrary quantiles in a straightforward manner. For the incremental q-quantile estimation $(0 < q < 1)$, only case 3 requires a modification. Instead of trying to get the same values for the counters L and R, we now try to balance the counters in such a way that $qR \approx (1-q)L$ holds. This means, step 3a is applied if $L < (1-q)t$ holds, otherwise step 3b is carried out. t is the number of data sampled after the buffer of length m has been filled.

Therefore, in the ideal case, when we achieve this balance, a proportion of q of the data points lies left and a proportion of $(1-q)$ lies right of the interval defined by the buffer of length m.

Now we are interested in the properties of the incremental quantile estimator presented above. Since we are simply selecting the k-th order statistic of the sample, at least for continuous random variables and larger pre-sampling sizes, we can provide an asymptotic distribution of the order statistic and therefore for the estimator.

Assume, the sample comes from a continuous random variable X and we are interested in an estimation of the q-quantile x_q. Assume furthermore that the probability density function f_X is continuous and positive at x_q. Let ξ_k^t $(k = \lfloor tq \rfloor + 1)$ denote the k-th order statistic from an i.i.d. sample. Then ξ_k^t has an asymptotic normal distribution [7]

$$N\left(x_q;\ \frac{q(1-q)}{tf^2(x_q)}\right) \tag{2.26}$$

From (2.26), we can obtain valuable information about the quantile estimator.

In order to have a more efficient and reliable estimator, we want the variance of (2.26) to be as small as possible. Under the assumption that we know the data distribution, we can compute the variance of ξ_k^t.

Let X be a random variable following a standard normal distribution and assume we have a sample $x_1, \ldots, x_t$ of X, i.e., these values are realizations of the i.i.d. random variables $X_i \sim N(0,\ 1)$. We are interested in the median of X. According to (2.26), the sample median $\xi_{\lfloor 0.5t \rfloor+1}^t$ follows asymptotically a normal distribution:

$$\xi_{\lfloor 0.5t \rfloor+1}^t \sim N\left(0;\ \frac{\pi}{2t}\right). \tag{2.27}$$

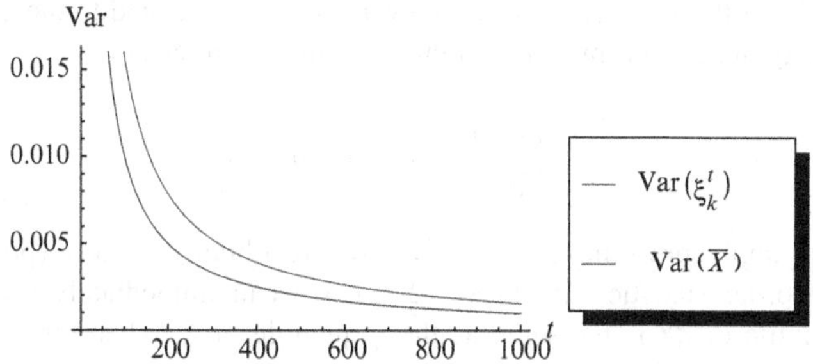

Fig. 2.3 Variance from bottom to top of $\bar{X}$ and ξ_k^t under the assumption of a standard normal distribution of X

Figure 2.3 shows the variance of the order statistic $\xi^t_{\lfloor 0.5t \rfloor+1}$ as a function in t when the chosen quantile is $q = 0.5$, i.e., the median, and the original distribution from which the sample comes is a standard normal distribution $N(0;1)$. The second curve in the figure corresponds to the variance of the sample mean.

The variance of the sample mean $\bar{X}$ is only slightly better than that of the order statistic $\xi^t_{\lfloor 0.5t \rfloor+1}$, nevertheless we should keep in mind the asymptotic character of the distribution (2.26).

Furthermore, from (2.26) we obtain the other nice property of the incremental quantile estimator: It is an asymptotically unbiased estimator of sample quantiles. It is even a consistent estimator.

Unfortunately, as it was shown in [25], the probability for the algorithm to fail is much smaller for the estimation of the median than for arbitrary quantiles. Therefore, despite the nice properties of this estimator this simple generalization of the incremental median estimation algorithm to arbitrary quantiles is not very useful in practice. In order to amend this problem, we provide a modified algorithm based on pre-sampling.

2.3.2.3 Incremental Quantile Estimation With Presampling iQPres

Here we introduce the algorithm iQPres (incremental quantile estimation with pre-sampling) [25]. As already mentioned above, the failure probability for the incremental quantile estimation algorithm in Sect. 2.3.2.2 is lower for the median than for extreme quantiles. Therefore, to minimise the failure probability we introduce an incremental quantile estimation algorithm with pre-sampling.

Assume we want to estimate the q-quantile. We pre-sample n values and we simply take the l-th smallest value $x_{(l)}$ from the pre-sample for some fixed

$l \in \{1,\ldots,n\}$. At the moment, l does not even have to be related to the q-quantile. The probability that $x_{(l)}$ is smaller than the q-quantile of interest is

$$p_l = \sum_{i=0}^{l} \binom{n}{i} \cdot q^i \cdot (1-q)^{n-i}. \tag{2.28}$$

So when we apply pre-sampling in this way, we obtain the new (presampled) distribution (order statistic) ξ_l^n. From (2.28), we can immediately see that the $(1-p_l)$-quantile of ξ_l^n is the same as the q-quantile of X. Therefore, instead of estimating the q-quantile of X, we estimate the $(1-p_l)$-quantile of ξ_l^n. Of course, this is only helpful, when l is chosen in such a way that the failure probabilities for the $(1-p_l)$-quantile are significantly lower than the failure probabilities for the q-quantile. In order to achieve this, l should be chosen in such a way that $(1-p_l)$ is as close to 0.5 as possible.

We want to estimate the q-quantile $(0 < q < 1)$. Fix the parameters m, l, n. (For an optimal choice see [25].)

1. Presampling: n succeeding values are stored in increasing order in a buffer b_n of length n. Then we select the l-th element in the buffer. The buffer is emptied afterwards for the next presample of n values.
2. Estimation of the $(1-p_l)$-quantile based on the l-th element in the buffer for presampling: this is carried out according to the algorithm described in Sect. 2.3.2.2.

The quantile is then estimated in the usual way, i.e.,

$$\begin{aligned} k &= \lceil (m+L+R) * (1-p_l) - l + 0.5 \rceil, \\ r &= (m+L+R) * (1-p_l) - l + 0.5 - k, \\ \hat{q} &= (1-r) \cdot a_{k-R} + r \cdot a_{k-R+1} \qquad \text{(quantile estimator)}. \end{aligned}$$

Of course, this does only work when the algorithm has not failed, i.e., the corresponding index k is within the buffer of m values.

2.3.3 Experimental Results

In this section, we present an experimental evaluation of the presented algorithms iQPres and the algorithm described in Sect. 2.3.1. The evaluation is based on artificial data sets.

First, we consider estimations of the lower and upper quartile as well as the median for different distributions:

- Exponential distribution with parameter $\lambda = 4$ (Exp(4))
- Standard normal distribution (N(0;1))
- Uniform distribution on the unit interval (U(0,1))

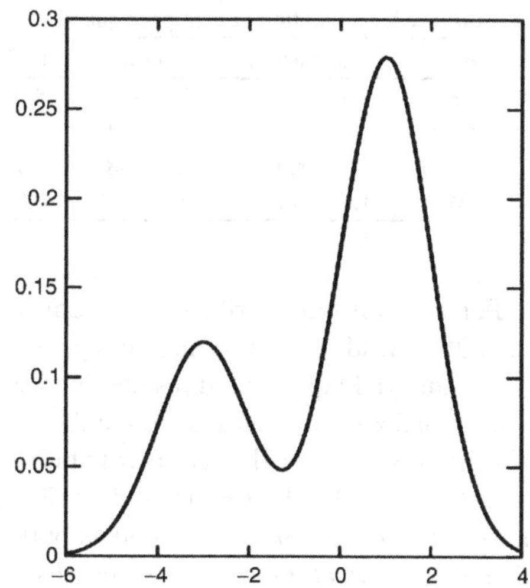

Fig. 2.4 An example for an asymmetric, bimodal probability density function

Table 2.3 Estimation of the lower quartile $q = 0.25$

Distr.	True quantile	iQPres	(2.23)	MSE (iQPres)	MSE (2.23)
Exp(4)	1.150728	1.152182	1.718059	2.130621E-5	2.675568
N(0;1)	−0.674490	−0.672235	−0.678989	5.611009E-6	0.008013
U(0,1)	0.250000	0.250885	0.250845	1.541123E-6	4.191695E-5
GM	−2.043442	−2.042703	0.185340	1.087618E-5	5.331730

Table 2.4 Estimation of the median $q = 0.5$

Distr.	True quantile	iQPres	(2.23)	MSE (iQPres)	MSE (2.23)
Exp(4)	2.772589	2.7462635	5.775925	7.485865E-4	10.906919
N(0;1)	0.000000	6.8324E-4	−0.047590	1.786715E-5	0.009726
U(0,1)	0.500000	0.495781	0.499955	1.779917E-5	2.529276E-6
GM	0.434425	0.434396	0.117499	2.365156E-6	0.451943

- An asymmetric bimodal distribution given by a Gaussian mixture model (GM) of two normal distributions. The cumulative distribution function of this distribution is given by

$$F(x) = 0.3 \cdot F_{\mathrm{N(-3;1)}} + 0.7 \cdot F_{\mathrm{N(1;1)}}$$

where $F_{\mathrm{N}(\mu;\sigma^2)}$ denotes the cumulative distribution function of the normal distribution with expected value μ and variance σ^2. Its probability density function is shown in Fig. 2.4.

The quantile estimations were carried out for samples of size of 10,000 that were generated from these distributions. We have repeated each estimation 1,000 times. Tables 2.3–2.5 show the average over all estimations for our algorithm (iQPres with a memory size of $M = 150$) and for the technique based on Theorem 2.1 where we used the control sequence $c_t = \frac{1}{t}$. The mean squared error over the 1,000 repeated runs is also shown in the tables.

Table 2.5 Estimation of the upper quartile $q = 0.75$

Distr.	True quantile	iQPres	(2.23)	MSE (iQPres)	MSE (2.23)
Exp(4)	5.545177	5.554385	5.062660	1.054132E-4	0.919735
N(0;1)	0.674490	0.674840	0.656452	3.600748E-7	0.003732
U(0,1)	0.750000	0.750883	0.749919	8.443136E-7	2.068730E-5
GM	1.366114	1.366838	0.027163	1.193377E-6	2.207112

For the uniform distribution, incremental quantile estimation based on (2.23) and iQPres leads to very similar and good results. For the normal distribution, both algorithms yield quite good results, but iQPres seems to be slightly more efficient with a smaller mean square error. For the bimodal distribution based on the Gaussian mixture model and a skewed distribution such as the exponential distribution, the estimations for the algorithm based on (2.23) are more or less useless, at least when no specific effort is invested to find an optimal control sequence $\{c_t\}_{t=0,1,...}$. iQPres does not have any problems with these distributions. As already mentioned before, it is also not required for iQPres that the sampling distribution is continuous whereas it is a necessary assumption for the technique based on (2.23).

2.4 Hypothesis Tests and Change Detection

In this section we demonstrate how hypothesis testing can be adapted to an incremental computation scheme for the cases of the χ^2-test and the t-test. Moreover, we discuss the problem of nonstationary data and explain various change detection strategies with the main focus on the use of statistical tests.

2.4.1 Incremental Hypothesis Tests

Statistical test are methods to check the validity of hypotheses about distributions or properties of distributions of random variables. Since statistical tests rely on samples, they cannot definitely verify or falsify a hypothesis. They can only provide probabilistic information supporting or rejecting the hypothesis under consideration.

Statistical tests usually consider a null hypothesis H_0 and an alternative hypothesis H_1. The hypotheses may concern parameters of a given class of distributions, for instance unknown expected value and variance of a normal distribution. Such tests are called parameter tests. In such cases, the a priori assumption is that the data definitely originate from a normal distribution. Only the parameters are unknown. In contrast to parameter tests, nonparametric tests concern more general hypothesis, for example, whether it is reasonable at all to assume that the data come from a normal distribution.

The error probability that the test will erroneously reject the null hypothesis, given the null hypothesis is true, is used as an indicator of the reliability of the test.

Sometimes a so-called p-value is used. The p-value is smallest error probability that can be admitted, so that the test will still reject the null hypothesis for a given sample. Therefore, a low p-value is a good indicator for rejecting the null hypothesis. Usually, the acceptable error probability α (α-error) should be specified in advance, before the test is carried out. The smaller α is chosen, the more reliable is the test when the outcome is to reject the null hypothesis. However, when α is chosen too small, then the test will not tend to reject the null hypothesis, although the sample might not speak in favor of it.

Some of the hypothesis tests can be applied to data streams, since they can be calculated in an incremental fashion. We discuss in this section the incremental adaptation of two statistical tests, the χ^2-test and the t-test. Note, that the application of hypothesis tests to data streams, using incremental computation or window techniques, requires the repeated execution of the test. This can cause the problem of multiple testing. The multiple testing problem is described later in this section.

2.4.1.1 χ^2-test

The χ^2-test has various applications. The principal idea of the χ^2-test is the comparison of two distributions. One can check whether two samples come from the same distribution, a single sample follows a given distribution or also whether two samples are independent.

Example 2.1. A die is thrown 120 times and the observed frequencies are as follows: 1 is obtained 30 times, 2–25, 3–18, 4–10, 5–22, and 6–15. We are interested in the question whether the die is fair or not.

The null hypothesis H_0 for the χ^2-test claims that the data follow a certain (cumulative) probability distribution $F(x)$. The distribution of the null hypothesis is than compared to the distribution of the data. The null hypothesis can for instance be a given distribution, e.g., a uniform or a normal distribution, and the χ^2-test can give an indication, whether the data strongly deviate from this expected distribution. For an independence test for two variables, the joint distribution of the sample is compared to the product of the marginal distributions. If these distributions differ significantly, this is an indication that the variables might not be independent.

The main idea of the χ^2-test is to determine how well the observed frequencies fit the theoretical/expected frequencies specified by the null hypothesis. Therefore, the χ^2-test is appropriate for data from categorical or nominally scaled random variables. In order to apply the test to continuous numeric data, the data domain should be partitioned into r categories first.

First we discus the χ^2 goodness of fit test. Here we assume to know from which distribution the data come. Then the H_0 and H_1 hypotheses can be stated as follows:

H_0: The sample comes from the distribution F_X
H_1: The sample does not come from the distribution F_X

Therefore the problem from Example 2.1 can be solved with the help of the χ^2 goodness of fit test. Consequently, the H_0 and H_1 hypotheses are chosen as follows:

H_0: $P(X=1)=p_1=\frac{1}{6},\dots,P(X=6)=p_6=\frac{1}{6}$
H_1: $P(X=i)\neq\frac{1}{6}$ for at least one value $i\in\{1,\dots,6\}$

Let $X_1,\dots,X_n$ be i.i.d. continuous random variables and $x_1,\dots,x_n$ the observations from these random variables. Then the test statistic is computed as follows

$$\chi^2=\sum_{i=1}^{r}\frac{(O_i-E_i)^2}{E_i} \tag{2.29}$$

where O_i are the observed frequencies and E_i are the expected frequencies.

Since we are dealing with continuous random variables, to compute the observed and expected frequencies we should carry out a discretisation of the data domain.

Let $F_X(x)$ be the assumed cumulative distribution function. The x-axis have to be split into r pairwise disjoint sets or bin S_i. Then the expected frequency in bin S_i is given by

$$E_i=n\left(F_X\left(a_{i+1}\right)-F_X\left(a_i\right)\right), \tag{2.30}$$

where $\left[a_i,a_{i+1}\right)$ is interval corresponding to bin S_i.

Furthermore, for the observed frequencies we obtain

$$O_i=\sum_{x_{k_i}\in S_i}1. \tag{2.31}$$

O_i is therefore the amount of observations in the i-th interval.

The statistic (2.29) has an approximate χ^2-distribution with $(r-1)$ degrees of freedom under the following assumptions: First, the observations are independent from each other. Second, the categories—the bins S_i—are mutually exclusive and exhaustive. This means that no categories may have an expected frequency of zero, i.e. $\forall i\in 1,\dots,r: E_i>0$. Furthermore, no more than 20% of the categories should have an expected frequency less than five. If this is not the case, categories should be merged or redefined. Note that this might also lead to a different number of degrees of freedom.

Therefore, the hypothesis H_0 that the sample comes from the particular distribution F_X is rejected if

$$\sum_{i=1}^{r}\frac{(O_i-E_i)^2}{E_i}>\chi^2_{1-\alpha}, \tag{2.32}$$

where $\chi^2_{1-\alpha}$ is the $(1-\alpha)$-quantile of the χ^2-distribution with $(r-1)$ degrees of freedom.

Table 2.6 summarizes the observed and expected frequencies and computations for Example 2.1. All E_i are greater than zero, even greater than 4. Therefore, there is no need to combine categories. The test statistic is computed as follows:

$$\sum_{i=1}^{r}\frac{(O_i-E_i)^2}{E_i}=5+1.25+0.2+5+0.2+1.25=12.9 \tag{2.33}$$

Table 2.6 Example 2.1

Number i on the die	E_i	O_i	$\frac{(O_i-E_i)^2}{E_i}$
1	20	30	5
2	20	25	1.25
3	20	18	0.2
4	20	10	5
5	20	22	0.2
6	20	15	1.25

The obtained result $\chi^2 = 12.9$ should be evaluated with $(1-\alpha)$-quantile of the χ^2-distribution. For that purposes table of the χ^2-distribution ([7]). The corresponding degrees of freedom are computed as explained above $(r-1) = (6-1) = 5$. For $\alpha = 0.05$ the tabled critical value for 5 degrees of freedom is $\chi^2_{0.95} = 11.07$, which is smaller than computed test statistic. Therefore, the null hypothesis is rejected at the 0.05 significance level. For significance level 0.02, the critical value is $\chi^2_{0.98} = 13.388$ and therefore the null hypothesis cannot be rejected at this level. This result can be summarized as follows: $\chi^2 = 12.9$ with 5 degrees of freedom can be rejected for all significance levels bigger than 0.024. This indicates that the die is unfair.

In order to adapt the χ^2 goodness of fit test to incremental calculation, the observed frequencies should be computed in an incremental fashion.

$$O_i^{(t)} = \begin{cases} O_i^{(t-1)} + 1 & \text{if } x_t \in S_i, \\ O_i^{(t-1)} & \text{otherwise.} \end{cases} \tag{2.34}$$

The expected frequency should also be recalculated corresponding to the increasing amount of observations.

$$E_i^{(t)} = \frac{E_i^{(t-1)}}{(t-1)} t. \tag{2.35}$$

Another very common test is the χ^2 independence test. This test evaluates the general hypothesis that two variables are statistically independent from each other.

Let X and Y be two random variables and $(x_1, y_1), \ldots, (x_n, y_n)$ are the observed values of these variables. For continuous random variables, the data domains should be partitioned into r and q categories, respectively. Therefore, the observed values of X can be assigned to one of the categories $S_1^X, \ldots, S_r^X$ and the observed values of Y to one of the categories $S_1^Y, \ldots, S_q^Y$. Then O_{ij} is the frequency of occurrence of the observation (x_{k_i}, y_{k_j}), where $x_{k_i} \in S_i^X$ and $y_{k_j} \in S_j^Y$. Furthermore,

$$O_{i\bullet} = \sum_{j=1}^{q} O_{ij} \tag{2.36}$$

and

$$O_{\bullet j} = \sum_{i=1}^{r} O_{ij} \tag{2.37}$$

denote the marginal observed frequencies.

Table 2.7 Contingency table

$X \setminus Y$	S_1^Y	...	S_j^Y	...	S_q^Y	Marginal of X
S_1^X	O_{11}	...	O_{1j}	...	O_{1q}	$O_{1\bullet}$
$\vdots$	$\vdots$	$\vdots$	$\vdots$	$\vdots$	$\vdots$	$\vdots$
S_i^X	O_{i1}	...	O_{ij}	...	O_{iq}	$O_{i\bullet}$
$\vdots$	$\vdots$	$\vdots$	$\vdots$	$\vdots$	$\vdots$	$\vdots$
S_r^X	O_{r1}	...	O_{rj}	...	O_{rq}	$O_{r\bullet}$
Marginal of Y	$O_{\bullet 1}$	...	$O_{\bullet j}$	...	$O_{\bullet q}$	n

Table 2.7 illustrates the observed absolute frequencies. The total number of observations in the table is n. The notation O_{ij} represents the number of observations in the cell with index ij (i-th row and j-th column), $O_{i\bullet}$ the number of observations in the i-th row and $O_{\bullet j}$ the number of observations in the j-th column. This table is called contingency table.

It is assumed that the random variables X and Y are statistically independent. Let p_{ij} be the probability of being in the i-th category of the domain of X and the j-th category of the domain of Y. $p_{i\bullet}$ and $p_{\bullet j}$ are the corresponding marginal probabilities. Then, corresponding to the assumption of independence for each pair

$$p_{ij} = p_{i\bullet} \cdot p_{\bullet j} \tag{2.38}$$

holds. Equation (2.38) defines statistical independence. Therefore, the null and the alternative hypotheses are as follows:

H_0: $p_{ij} = p_{i\bullet} \cdot p_{\bullet j}$
H_1: $p_{ij} \neq p_{i\bullet} \cdot p_{\bullet j}$

Thus, if X and Y are independent, then the expected absolute frequencies are given by

$$E_{ij} = \frac{O_{i\bullet} \cdot O_{\bullet j}}{n}. \tag{2.39}$$

The test statistic, again checking the observed frequencies against the expected frequencies under the null hypothesis, is as follows.

$$\chi^2 = \sum_{i=1}^{r} \sum_{j=1}^{q} \frac{(O_{ij} - E_{ij})^2}{E_{ij}} \tag{2.40}$$

The test statistic has an approximate χ^2-distribution with $(r-1)(s-1)$ degrees of freedom. Consequently, the hypothesis H_0 that X and Y are independent can be rejected if

$$\sum_{i=1}^{r} \sum_{j=1}^{q} \frac{(O_{ij} - E_{ij})^2}{E_{ij}} \geq \chi^2_{1-\alpha} \tag{2.41}$$

Table 2.8 Contingency table

Values\variables	X_1	...	X_j	...	X_m	Σ
S_1	O_{11}	...	O_{1j}	...	O_{1m}	$O_{1\bullet}$
$\vdots$	$\vdots$	$\vdots$	$\vdots$	$\vdots$	$\vdots$	$\vdots$
S_i	O_{i1}	...	O_{ij}	...	O_{im}	$O_{i\bullet}$
$\vdots$	$\vdots$	$\vdots$	$\vdots$	$\vdots$	$\vdots$	$\vdots$
S_r	O_{r1}	...	O_{rj}	...	O_{rm}	$O_{r\bullet}$
Σ	$O_{\bullet 1}$	...	$O_{\bullet j}$	...	$O_{\bullet m}$	n

where $\chi^2_{1-\alpha}$ is the $(1-\alpha)$-quantile of the χ^2-distribution with $(r-1)(s-1)$ degrees of freedom.

For the incremental computation of $O_{i\bullet}$, $O_{\bullet j}$, and O_{ij} corresponding formulae must be developed. For the time point t and the new observed values (x_t, y_t), the incremental formulae are given by

$$O_{i\bullet}^{(t)} = \begin{cases} O_{i\bullet}^{(t-1)} + 1 & \text{if } x_t \in S_i^X, \\ O_{i\bullet}^{(t-1)} & \text{otherwise.} \end{cases} \tag{2.42}$$

$$O_{\bullet j}^{(t)} = \begin{cases} O_{\bullet j}^{(t-1)} + 1 & \text{if } y_t \in S_j^Y, \\ O_{\bullet j}^{(t-1)} & \text{otherwise.} \end{cases} \tag{2.43}$$

$$O_{ij}^{(t)} = \begin{cases} O_{ij}^{(t-1)} + 1 & \text{if } x_t \in S_i^X \wedge y_t \in S_j^Y, \\ O_{ij}^{(t-1)} & \text{otherwise.} \end{cases} \tag{2.44}$$

The χ^2 goodness of fit test can be extended to a χ^2 homogeneity test ([22]). Whereas the χ^2 goodness of fit test can be used only for a single sample, the χ^2 homogeneity test is used to compare whether two or more samples come from the same population.

Let $X_1, \ldots, X_m$ $(m \geq 2)$ be discrete random variables, or continuous random variables discretised into r categories $S_1, \ldots, S_r$. The data for each of the m samples from random variables $X_1, \ldots, X_m$ (overall n values) are entered in a contingency table. This table is similar to the one for the χ^2 independence test.

The samples are represented by the columns and the categories by the rows of Table 2.8. We assume that each of the samples is randomly drawn from the same distribution. The χ^2 homogeneity test checks whether m samples are homogeneous with respect to the observed frequencies. If the hypothesis H_0 is true, the expected frequency in the i-th category will be the same for all of the m random variables. Therefore, the null and the alternative hypotheses can be stated as follows:

H_0: $p_{ij} = p_{i\bullet} \cdot p_{\bullet j}$
H_1: $p_{ij} \neq p_{i\bullet} \cdot p_{\bullet j}$.

From H_0 follows that the rows are independent of the column.

Therefore, the computation of an expected frequency can be summarized by

$$E_{ij} = \frac{O_{i\bullet} \cdot O_{\bullet j}}{n}. \tag{2.45}$$

Although the χ^2 independence test and χ^2 homogeneity test evaluate different hypothesis, they are computed identically. Therefore, the incremental adaptation of the χ^2 independence test can also be applied to the χ^2 homogeneity test.

Commonly in case of two samples the Kolmogorov–Smirnov test is used, since it is an exact test and in contrast to the χ^2-test can be applied directly without previous discretisation of continuous distributions. However, the Kolmogorov–Smirnov test does not have any obvious incremental calculation scheme. The Kolmogorov–Smirnov test is described in Sect. 2.4.2.2.

2.4.1.2 The *t*-Test

The next hypothesis test for which we want to provide incremental computation is the t-test. Different kinds of the t-test are used. We restrict our considerations to the one sample t-test and the t-test for two independent samples with equal variance.

The one sample t-test evaluates whether a sample with particular mean could be drawn from the population with known expected value μ_0. Let $X_1, \ldots X_n$ be i.i.d. and $X_i \sim N\left(\mu; \sigma^2\right)$ with unknown variance σ^2. The null and the alternative hypotheses for two-sided test are:

H_0: $\mu = \mu_0$, the sample comes from the normal distribution with expected value μ_0.

H_1: $\mu \neq \mu_0$, the sample comes from a normal distribution with an expected value differing from μ_0.

The test statistic is given by

$$T = \sqrt{n}\frac{\bar{X} - \mu}{S}, \tag{2.46}$$

where $\bar{X}$ is the sample mean and S the sample standard deviation. The statistic (2.46) is t-distributed with $(n-1)$ degrees of freedom. H_0 is rejected if

$$t < -t_{1-\alpha/2} \text{ or } t > t_{1-\alpha/2} \tag{2.47}$$

where $t_{1-\alpha/2}$ is the $(1-\alpha/2)$-quantile of the t-distribution with $(n-1)$ degrees of freedom and t is the computed value of the test statistic (2.46), i.e. $t = \sqrt{n}\frac{\bar{x}-\mu_0}{s}$.

One-sided tests are given by the following null and alternative hypotheses:

H_0: $\mu \leq \mu_0$ and $H_1: \mu > \mu_0$. H_0 is rejected if $t > t_{1-\alpha}$.

H_0: $\mu \geq \mu_0$ and $H_1: \mu < \mu_0$. H_0 is rejected if $t < -t_{1-\alpha}$.

This test can be very easily adapted to incremental computation. For this purpose, the sample mean and the sample variance have to be updated as in (2.2) and (2.6), respectively, as described in Sect. 2.2. Note that the degrees of freedom of the t-distribution should be updated in each step as well.

$$t_{n+1} = \sqrt{n+1}\frac{\bar{x}_{n+1} - \mu_0}{s_{n+1}} \tag{2.48}$$

Unlike previous notations we use here $n+1$ for the time point, since the letter t is already used for the computed test statistic. Furthermore, as mentioned above the $(1-\alpha/2)$-quantile of the t-distribution with n degrees of freedom should be used to evaluate the null hypothesis. However for $n \geq 30$, the quantiles of the standard normal distribution could be used as approximation of the quantiles of the t-distribution.

The t-test for two independent samples is used to evaluate whether two independent sample come from two normal distributions with the same expected value. The two sample means $\bar{x}$ and $\bar{y}$ are used to estimate the expected values μ_X and μ_Y of the underlying distributions. If the result of the test is significant, we assume that the samples come from two normal distributions with different expected values. Furthermore, we assume that the variances of the underlying distributions are unknown.

The t-test is based on the following assumptions:

- The samples are drawn randomly.
- The underlying distribution is a normal distribution.
- The variances of the underlying distributions are equal, i.e. $\sigma_X^2 = \sigma_Y^2$.

Let $X_1, \ldots X_{n_1}$ i.i.d. and $X_i \sim N\left(\mu_X; \sigma_X^2\right)$ and $Y_1, \ldots Y_{n_2}$ i.i.d. and $Y_i \sim N\left(\mu_Y; \sigma_Y^2\right)$ with unknown expected values and unknown variances and $\sigma_X^2 = \sigma_Y^2$.

The null and the alternative hypothesis can be defined as follows:

H_0: $\mu_X = \mu_Y$, the samples come from the same normal distribution.
H_1: $\mu_X \neq \mu_Y$, the samples come from normal distributions with different expected values.

In this case, a two-sided test is carried out; however, similar to the one sample t-test also a one-sided test can be defined.

The test statistic is computed as follows.

$$T = \frac{\bar{X} - \bar{Y}}{\sqrt{\frac{(n_1-1)S_X^2 + (n_2-1)S_Y^2}{n_1+n_2-2}}} \sqrt{\frac{n_1 n_2}{n_1 + n_2}} \tag{2.49}$$

where S_X^2 and S_Y^2 are the unbiased estimators for the variances of X and Y, respectively.

Equation (2.49) is a general equation for the t-test for two independent samples and can be used in both cases of equal and unequal sample sizes.

The statistic (2.49) has a t-distribution with (n_1+n_2-2) degrees of freedom. Let

$$t = \frac{\bar{x}-\bar{y}}{\sqrt{\frac{(n_1-1)s_X^2+(n_2-1)s_Y^2}{n_1+n_2-2}}}\sqrt{\frac{n_1 n_2}{n_1+n_2}} \tag{2.50}$$

be the computed value of the statistic (2.49). Then the hypothesis H_0 that the samples come from the same normal distribution is rejected if

$$t < -t_{1-\alpha/2} \text{ or } t > t_{1-\alpha/2}, \tag{2.51}$$

where $t_{1-\alpha/2}$ is the $(1-\alpha/2)$-quantile of the t-distribution with (n_1+n_2-2) degrees of freedom.

Similar to the one sample t-test, the t-test for two independent samples can be easily computed in an incremental fashion, since the sample means and the variance can be calculated in an incremental way. Here the degrees of freedom should also be updated with the new observed values.

2.4.1.3 Multiple Testing

Multiple testing refers to the application of number of tests simultaneously. Instead of a single null hypothesis, a tests for a set of null hypotheses H_0, $H_1,\ldots,H_n$ are considered. These null hypotheses do not have to exclude each other.

An example for multiple testing is a test whether m random variables $X_1,\ldots X_m$ are pairwise independent. This means, the null hypotheses are $H_{1,2},\ldots,H_{1,m},\ldots,$ $H_{m-1,m}$ where $H_{i,j}$ states that X_i and X_j are independent.

Multiple testing leads to the undesired effect of cumulating the α-error. The α-error α is the probability to reject the null hypothesis erroneously, given it is true. Choosing $\alpha = 0.05$ means that in 5% of the cases the null hypothesis would be rejected, although it is true. When k tests are applied to the same sample, then the error probability for each test is α. Under the assumption that the null hypotheses are all true and the tests are independent, the probability that at least one test will reject its null hypothesis erroneously is

$$P(\ell \geq 1) = 1 - P(\ell = 0) \tag{2.52}$$

$$= 1-(1-\alpha)\cdot(1-\alpha)\ldots\cdot(1-\alpha) \tag{2.53}$$

$$= 1-(1-\alpha)^k. \tag{2.54}$$

ℓ is the number of tests rejection the null hypothesis.

A variety of approaches have been proposed to handle the problem of cumulating the α-error. In the following, two common methods will be introduced shortly.

The simplest and most conservative method is Bonferroni correction [21]. When k null hypotheses are tested simultaneously and α is the desired overall α-error for all tests together, then the corrected α-error for each single test should be chosen as $\tilde{\alpha} = \frac{\alpha}{k}$. The justification for this correction is the inequality

$$P\left(\bigcup_i A_i\right) \leq \sum_i P(A_i). \tag{2.55}$$

For Bonferroni correction, A_i is the event that the null hypothesis H_i is rejected, although it is true. In this way, the probability that one or more of the tests rejects its corresponding null hypothesis is at most α. In order to guarantee the significance level α, each single test must be carried out with the corrected level $\tilde{\alpha}$.

Bonferroni correction is a very rough and conservative approximation for the true α-error. One of its disadvantages is that the corrected significance level $\tilde{\alpha}$ becomes very low, so that it becomes almost impossible to reject any of the null hypotheses.

The simple single step Bonferroni correction has been improved by Holm [12]. The Bonferroni–Holm method is a multistep procedure in which the necessary corrections are carried out stepwise. This method usually yields larger corrected α-values than the simple Bonferroni correction.

When k hypotheses are tested simultaneously and the overall α-error for all tests is α, for each of the tests the corresponding p-value is computed based on the sample x and the p-values are sorted in ascending order.

$$p_{[1]}(x) \leq p_{[2]}(x) \leq \ldots \leq p_{[k]}(x) \tag{2.56}$$

The null hypotheses H_i are ordered in the same way.

$$H_{[1]}, H_{[2]}, \ldots, H_{[k]} \tag{2.57}$$

In the first step, $H_{[1]}$ is tested by comparing $p_{[1]}$ with $\frac{\alpha}{k}$. If $p_{[1]} > \frac{\alpha}{k}$ holds, then $H_{[1]}$ and the other null hypotheses $H_{[2]}, \ldots, H_{[k]}$ are not rejected. The method terminates in this case. However, if $p_{[1]} \leq \frac{\alpha}{k}$ holds, $H_{[1]}$ is rejected and the next null hypothesis $H_{[2]}$ is tested by comparing the p-value $p_{[2]}$ and the corrected α-value $\frac{\alpha}{k-1}$. If $p_{[2]} > \frac{\alpha}{k-1}$ holds, $H_{[2]}$ and the remaining null hypotheses $H_{[3]}, \ldots, H_{[k]}$ are not rejected. If $p_{[2]} \leq \frac{\alpha}{k-1}$ holds, $H_{[2]}$ is rejected and the procedure continues with $H_{[3]}$ in the same way.

The Bonferroni–Holm method tests the hypotheses in the order of their p-values, starting with $H_{[1]}$. The corrected α_i-values $\frac{\alpha}{k}, \frac{\alpha}{k-1}, \ldots \alpha$ are increasing. Therefore, the Bonferroni–Holm method rejects at least those hypotheses that are also rejected by simple Bonferroni correction, but in general more hypotheses can be rejected.

2.4.2 Change Detection Strategies

Detecting changes in data streams has become a very important area of research in many application fields, such as stock market, web activities, or sensors

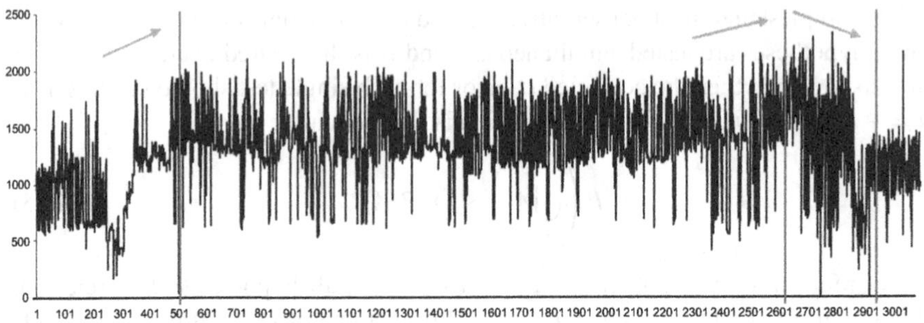

Fig. 2.5 An example of change detection for time series data from a waste water treatment plant

measurements, just to name a few. The main problem for change detection in data streams is limited memory capacity. It is unrealistic to store the full history of the data stream. Therefore, efficient change detection strategies tailored to the data stream should be used. The main requirements for such approaches are: low computational costs, fast change detection, and high accuracy. Moreover it is important to distinguish between true changes and false alarms. Abrupt changes as well as slow drift in the data generating process can occur. Therefore, a "good" algorithm should be able to detect both kinds of changes.

Various strategies are proposed to handle this problem, see for instance [11] for a detailed survey of change detection methods. Most of these approaches are based on time window techniques [2, 15]. Furthermore, several approaches are presented for evolving data streams as they are discussed in [8, 13, 14].

In this section, we introduce two types of change detection strategies: incremental computation and window technique-based change detection. Furthermore, we put the main focus on statistical tests. We assume to deal with numeric data streams. As already mentioned in the introduction, two types of change are identified: concept change and change of data distribution. We do not differentiate in this work between both of them, since the distribution of the target variable will be changed in both cases.

2.4.2.1 iQPres for Change Detection

The incremental quantile estimator iQPres from Sect. 2.3.2.3 can be used for change detection [25]. In case, the sampling distribution changes, having a drift of the quantile to be estimated as a consequence, such changes will be noticed, since the simple version of iQPres without shifted parallel estimations will fail in the sense that it is not able to balance the counters L and R any more.

In order to illustrate how iQPres can be applied to change detection, we consider daily measurements for gas production in a waste water treatment plant over a period of more than eight years. The measurements are shown in Fig. 2.5.

iQPress has been applied to this data set to estimate the median with a memory size of $M = 30$. The optimal choice for the sizes of the buffers for pre-sampling and median estimation is then $n = 3$ and $m = 27$, respectively. At the three time points 508, 2,604, and 2,964, the buffer cannot be balanced anymore, indicating that the median has changed. These three time points are indicated by vertical lines in Fig. 2.5. The arrows indicate whether the median is increased or decreased. An increase corresponds to an unbalanced buffer with the right counter R becoming too large, whereas a decrease leads to an unbalanced buffer with the left counter L becoming too large. The median increases at the first point at 508 from 998 before and 1,361 after this point. At time point 2,604, the median increases to 1,406 and drops again to 1,193 at time point 2,964.

Note that algorithms based on Theorem 2.1 mentioned in Sect. 2.3.1 are not suitable for change detection.

By using iQPres for change detection in the data distribution, we assume that the median of the distribution changes with the time, however, if this is not the case and only another parameter like the variance of the underlying distribution changes, other strategies for change detection should be used.

2.4.2.2 Statistical Tests for Change Detection

The theory of hypothesis testing is the main background for change detection. Several algorithms for change detection are based on hypothesis tests.

Hypothesis tests could be applied to change detection in two different ways:

- Change detection through incremental computation of the tests: by this approach the test is computed in an incremental fashion, for instance, as it is explained in Sect. 2.4.1. Consequently the change can be detected if the test starts to yield different results as before.
- Window techniques: by this approach the data stream divided into time windows. A sliding window could be used as well as nonoverlapping windows. In order to detect potential changes, we need either to compare data from an earlier window with data from newer one or to test only the new data (for instance, whether the data follow a known or assumed distribution). When the window size is not too large, it is not necessary to be able to compute the tests in an incremental fashion. Therefore, we are not restricted to tests that render themselves to incremental computations, but many other tests could be used. Hybrid approaches combining both techniques are also possible. Of course, window techniques with incremental computations within the window will lead to less memory consumptions and faster computations.

We will not give a detailed description for change detection based on incremental computation here, since the principles of these methods are explained in Sect. 2.4.1. However, the problem of multiple testing as discussed in Sect. 2.4.1 should be taken into account when a test is applied again and again over time. Even if the underlying distribution does not change over time, any test will erroneously reject the null

```
Initialise window W, i = 0
for each new x_t do
    if i < step then
        W ← W ∪ {x_t} (i.e., add x_t to the W)
        W ← W \ w_0 (i.e., remove oldest element in W)
        i = i + 1
        if i = step then
            i = 0
            split W into W_0 and W_1
            test W_0 and W_1 for change
            if change detected then
                report change at time t
            end if
        end if
    end if
end for
```

Fig. 2.6 General scheme of a change detection algorithm based on time windows and statistical tests

hypothesis of no change in the long run if we only carry out the test often enough. Different approaches to solve this problem are presented in Sect. 2.4.1.3. Another problem of this approach is the "burden of old data". If a large amount of data has been analyzed already and the change is not very drastic, it may happen that the change will be detected with large delay or not detected at all when a very large window is used. On that account it may be useful to re-initialize the test from time to time.

To detect changes with by window technique, we need to compare two samples of data and have to decide whether the hypothesis H_0 that they come from the same distribution is true.

First we will present a general meta-algorithm for change detection based on a window technique, without any specific fixed test. This algorithm is presented in Fig. 2.6. The constant `step` specifies, after how many new values the change detection should checked again.

This approach follows an simple idea: when the data from two subwindows of W are judged as "distinct enough", the change is detected. Here "distinct enough" is specified by the selected statistical test for distribution change. In general, we assume the splitting of W into two subwindows of equal size. Nevertheless, any "valid" splitting can be used. Valid is meant in terms of the amount of data that is needed for the test to be reliable.

However, by a badly selected cut point the change can be detected with large delay as Fig. 2.7 shows. The rightmost part indicates a change in the data stream. As the change occurs almost at the end of the subwindow W_1, it is most likely that the change remains at first undetected. Of course, since the window will be moved

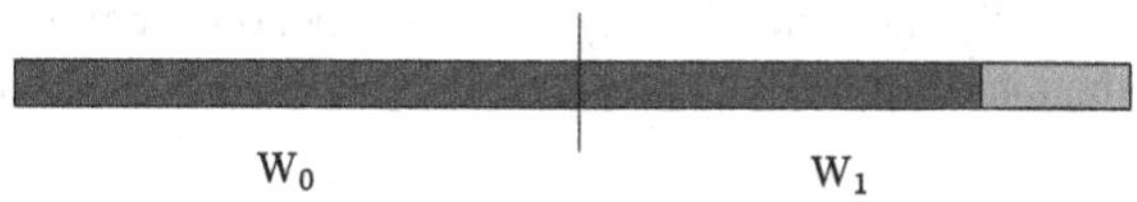

Fig. 2.7 Subwindows problem

Fig. 2.8 Modification of the algorithm for change detection to avoid the sub-windows problem

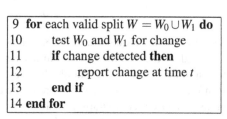

```
9  for each valid split W = W0 ∪ W1 do
10     test W0 and W1 for change
11     if change detected then
12         report change at time t
13     end if
14 end for
```

forward with new data points arriving, at some point the change will be detected, but it may be from essential interest, to detect the change as early as possible.

To solve this problem, we modify the algorithm in Fig. 2.6 in the following way: instead of splitting window W only once, the splitting is carried out several times. Figure 2.8 shows the modified part of the algorithm in Fig. 2.6 starting at step 9.

How many times the window should be split, should be decided based on the required performance and precision of the algorithm. We can run the test for each sufficiently large subwindow of W, although the performance of the algorithm will decrease, or we can carry out fixed number of splits. Note that also for the windows technique-based approach, attention should be paid to the problem of multiple testing (see Sect. 2.4.1.3). Furthermore, we do not specify here the effect of the detected change. The question whether the window should be re-initialized depends on the application. A change in the variance of the data stream might have a strong effect on the task to be fulfilled with the online analysis of the data stream or it might have no effect as long as the mean value remains more or less stable.

For the hypothesis test in step 10, of the algorithm, any appropriate test for the distribution change can be chosen. Since we do not necessarily have to apply an incremental scheme for the hypothesis test, the Kolmogorov–Smirnov test can also be considered for change detection. The Kolmogorov–Smirnov test is designed to compare two distribution, whether they are equal or not. Therefore, two kinds of questions could be answered with the help of the Kolmogorov–Smirnov test:

- Does the sample arise from a particular known distribution?
- Do two samples coming from different time windows have the same distribution?

We are particularly interested in the second question. For this purpose, the two sample Kolmogorov–Smirnov goodness-of-fit test should be used.

Let $X_1,\dots,X_n$ and $Y_1,\dots,Y_m$ be two independent random samples from distributions with cumulative distribution functions F_X and F_Y, respectively. We want to test the hypothesis $H_0: F_X = F_Y$ against the hypothesis $H_1: F_X \neq F_Y$. The Kolmogorov–Smirnov statistic is given by

$$D_{n,m} = \sup_t |S_{X,n}(x) - S_{Y,m}(x)|, \tag{2.58}$$

where $S_{X,n}(x)$ and $S_{Y,m}(x)$ are corresponding empirical cumulative distribution function[4] of the first and second sample. H_0 is rejected at level α if

$$\sqrt{\frac{nm}{m+n}} D_{n,m} > K_\alpha \tag{2.60}$$

where K_α is the α-quantile of the Kolmogorov distribution.

To adapt the Kolmogorov–Smirnov test as a change detection algorithm, first the significance level α should be chosen (we can also use for instance the Bonferroni correction to avoid the multiple testing problem). The value of K_α needs either numerical computation or should be stored in a table.[5] Furthermore, values from the subwindows W_0 and W_1 represent two samples $x_1, \ldots, x_n$ and $y_1, \ldots, y_m$. Then the empirical cumulative distribution functions $S_{X,n}(x)$ and $S_{Y,m}(x)$ and the Kolmogorov–Smirnov statistic should be computed. Note that for the computation of $S_{X,n}(x)$ and $S_{Y,m}(x)$ in case of unique splitting the samples have to be sorted only initially, afterward the new values have to be inserted and the old values must be deleted from the sorted lists. In case of multiple splitting we have to decide either to sort each time from scratch or to save sorted lists for each kind of splitting.

An implementation of the Kolmogorov–Smirnov test is for instance available in the R statistics library (see [4] for more information).

Algorithm 2.8 based on the Kolmogorov–Smirnov test as the hypothesis test in step 10 has been implemented in Java using R-libraries and has been tested with artificial data. For the data generation process, the following model was used:

$$Y_t = \sum_{i=1}^{t} X_i. \tag{2.61}$$

We assume the random variables X_i to be normally distributed with expected value $\mu = 0$ and variance σ^2, i.e. $X_i \sim N(0, \sigma^2)$. Here Y_t is a one dimensional random walk [24]. To make the situation more realistic, we consider the following model:

$$Z_t \sim N(y_t, 1). \tag{2.62}$$

The process (2.62) can be understood as a constant model with drift and noise, the noise follows a normal distribution whose expected value equals the actual value of the random walk and whose variance is 1.

---

[4]Let $x_{r_1}, x_{r_2}, \ldots x_{r_n}$ be a sample in ascending order from the random variables $X_1, \ldots, X_n$. Then the empirical distribution function of the sample is given by

$$S_{X,n}(x) = \begin{cases} 0 & \text{if } x \le x_{r_1}, \\ \frac{k}{n} & \text{if } x_{r_k} < x \le x_{r_{k+1}}, \\ 1 & \text{if } x > x_{r_k}. \end{cases} \tag{2.59}$$

[5]This applies also to the t-test and the χ^2-test.

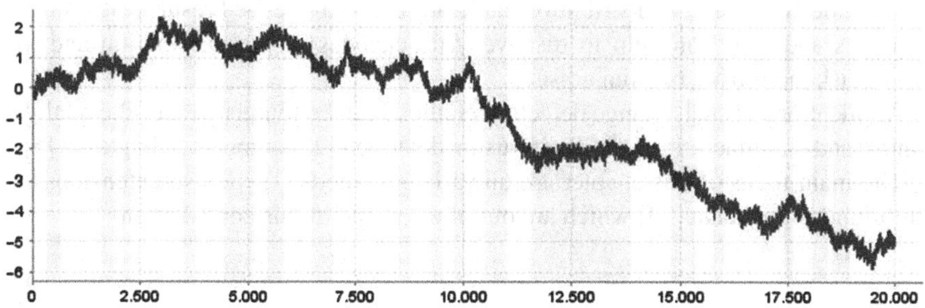

Fig. 2.9 An example of change detection for the data generated by the process (2.62)

The data were generated with the following parameters: $\sigma_1 = 0.02$, $\sigma_2 = 0.1$. Therefore, the data have a slow drift and are furthermore corrupted with noise.

Algorithm 2.8 has been applied to this data set. The size of the window W was chosen to be 500. The window is always split into two subwindows of equal size, i.e., 250. The data are identified by the algorithm as nonstationary. Only very short sequences are considered to be stationary by the Kolmogorov–Smirnov test. These sequences are marked by the darker areas in Fig. 2.9. In the interval, $[11,14,414,445]$ stationary parts are mixed with occasionally occurring small nonstationary parts. For easier interpretation, we joined these parts to one larger area. Of course, since we are dealing with the window, the real stationary areas are not exactly the same as shown in the figure. The quality of change detection depends on the window. For slow gradual changes in the form of concept drift a larger window is a better choice, whereas for abrupt changes in terms of a concept shift a smaller window is of advantage.

2.5 Conclusions

We have introduced incremental computation schemes for statistical measures or indices like the mean, the median, the variance, the interquartile range, or the Pearson correlation coefficient. Such indices provide information about the characteristics of the probability distribution that generates the data stream. Although incremental computations are designed to handle large amounts of data, it is not extremely useful to calculate the above mentioned statistical measures for extremely large data sets, since they quickly converge to the parameter of the probability distribution they are designed to estimate as can be seen in Figs. 2.1–2.3. Of course, convergence will only occur when the underlying data stream is stationary.

It is therefore very important to use such statistical measure or hypothesis tests for change detection. Change detection is a crucial aspect for nonstationary data streams or "evolving systems." It has been demonstrated in [26] that naïve adaption without taking any effort to distinguish between noise and true changes of the underlying sample distribution can lead to very undesired results. Statistical measures and tests can help to discover true changes in the distribution and to distinguish them from random noise.

Applications of such change detection methods can be found in areas like quality control and manufacturing [16,20], intrusion detection [27] or medical diagnosis [5].

The main focus of this chapter are univariate methods. There also extensions to multidimensional data [23] which are out of the scope of this contribution.

References

1. Aho, A.V., Ullman, J.D., Hopcroft, J.E.: Data Structures and Algorithms. Addison Wesley, Boston (1987)
2. Basseville, M., Nikiforov, I.: Detection of Abrupt Changes: Theory and Application (Prentice Hall information and system sciences series). Prentice Hall, Upper Saddle River, New Jersey (1993)
3. Beringer, J., Hüllermeier, E.: Effcient instance-based learning on data streams. Intelligent Data Analysis **11**, 627–650 (2007)
4. Crawley, M.: Statistics: An Introduction using R. Wiley, New York (2005)
5. Dutta, S., Chattopadhyay, M.: A change detection algorithm for medical cell images. In: Proc. Intern. Conf. on Scientific Paradigm Shift in Information Technology and Management, pp. 524–527. IEEE, Kolkata (2011)
6. Fischer, R.: Moments and product moments of sampling distributions. In: Proceedings of the London Mathematical Society, Series 2, 30, pp. 199–238 (1929)
7. Fisz, M.: Probability Theory and Mathematical Statistics. Wiley, New York (1963)
8. Ganti, V., Gehrke, J., Ramakrishnan, R.: Mining data streams under block evolution. SIGKDD Explorations **3**, 1–10 (2002)
9. Gelper, S., Schettlinger, K., Croux, C., Gather, U.: Robust online scale estimation in time series: A model-free approach. Journal of Statistical Planning & Inference **139**(2), 335–349 (2008)
10. Grieszbach, G., Schack, B.: Adaptive quantile estimation and its application in analysis of biological signals. Biometrical journal **35**, 166–179 (1993)
11. Gustafsson, F.: Adaptive Filtering and Change Detection. Wiley, New York (2000)
12. Holm, S.: A simple sequentially rejective multiple test procedure. Scandinavian Journal of Statistics **6**, 65–70 (1979)
13. Hulten, G., Spencer, L., Domingos, P.: Mining time changing data streams. In: Proceedings of the seventh ACM SIGKDD international conference on Knowledge discovery and data mining (2001)
14. Ikonomovska, E., Gama, J., Sebastião, R., Gjorgjevik, D.: Regression trees from data streams with drift detection. In: 11th int conf on discovery science, LNAI, vol 5808, pp. 121–135. Springer, Berlin (2009)
15. Kifer, D., Ben-David, S., Gehrke, J.: Detecting change in data streams. In: Proc. 30th VLDB Conf., pp. 199–238. Toronto, Canada (2004)
16. Lai, T.: Sequential changepoint detection in quality control and dynamic systems. Journal of the Royal Statistical Society, Series B **57**, 613–658 (1995)
17. Möller, E., Grieszbach, G., Schack, B., Witte, H.: Statistical properties and control algorithms of recursive quantile estimators. Biometrical Journal **42**, 729–746 (2000)
18. Nevelson, M., Chasminsky, R.: Stochastic approximation and recurrent estimation. Verlag Nauka, Moskau (1972)
19. Qiu, G.: An improved recursive median filtering scheme for image processing. IEEE Transactions on Image Processing **5**, 646–648 (1996)

20. Ruusunen, M., Paavola, M., Pirttimaa, M., Leiviska, K.: Comparison of three change detection algorithms for an electronics manufacturing process. In: Proc. 2005 IEEE International Symposium on Computational Intelligence in Robotics and Automation, pp. 679–683 (2005)
21. Shaffer, J.P.: Multiple hypothesis testing. Ann. Rev. Psych **46**, 561–584 (1995)
22. Sheskin, D.: Handbook of Parametric and Nonparametric Statistical Procedures. CRC-Press, Boca Raton, Florida (1997)
23. Song, X., Wu, M., Jermaine, C., Ranka, S.: Statistical change detection for multi-dimensional data. In: Proceedings of the 13th ACM SIGKDD international conference on Knowledge discovery and data mining, pp. 667–676. ACM, New York (2007)
24. Spitzer, F.: Principles of Random Walk (2nd edition). Springer, Berlin (2001)
25. Tschumitschew, K., Klawonn, F.: Incremental quantile estimation. Evolving Systems **1**, 253–264 (2010)
26. Tschumitschew, K., Klawonn, F.: The need for benchmarks with data from stochastic processes and meta-models in evolving systems. In: N.K.P. Angelov D. Filev (ed.) International Symposium on Evolving Intelligent Systems. SSAISB, Leicester, pp. 30–33 (2010)
27. Wang, K., Stolfo, S.: Anomalous payload-based network intrusion detection. In: E. Jonsson, A. Valdes, M. Almgren (eds.) Recent Advances in Intrusion Detection, pp. 203–222. Springer, Berlin (2004)

Chapter 3
A Granular Description of Data: A Study in Evolvable Systems

Witold Pedrycz, John Berezowski, and Iqbal Jamal

Abstract A human-centric way of data analysis, especially when dealing with data distributed in space and time, is concerned with data representation in an interpretable way where a perspective from which the data are analyzed is actively established by the user. Being motivated by this essential feature of data analysis, in the study we present a granular way of data analysis where the data and relationships therein are described through a collection of information granules defined in the spatial and temporal domain. We show that the data, expressed in a relational fashion, can be effectively described through a collection of Cartesian products of information granules forming a collection of semantically meaningful data descriptors. The design of the codebooks (vocabularies) of such information granules used to describe the data is guided through a process of information granulation and degranulation. This scheme comes with a certain performance index whose minimization becomes instrumental in the optimization of the codebooks. A description of logical relationships between elements of the codebooks used in the granular description of spatiotemporal data present in consecutive time frames is elaborated on as well.

W. Pedrycz (✉)
Department of Electrical & Computer Engineering University of Alberta, Edmonton AB T6R 2G7 Canada, and System Research Institute, Polish Academy of Sciences, Warsaw, Poland
e-mail: wpedrycz@ualberta.ca

J. Berezowski
Ross University School of Veterinary Medicine, St Kitts, West Indies
e-mail: john.berezowski@gov.ab.ca

I. Jamal
AQL Management Consulting Inc., Edmonton, AB, Canada
e-mail: iqbaljamal@aqlmc.com

M. Sayed-Mouchaweh and E. Lughofer (eds.), *Learning in Non-Stationary Environments: Methods and Applications*, DOI 10.1007/978-1-4419-8020-5_3,
© Springer Science+Business Media New York 2012

3.1 Introduction

Spatiotemporal data become visible in numerous applications. We envision data distributed in time and space: readings associated with oil wells distributed over space and reported in time, recordings of sensors distributed over some battlefield area or disaster region, counts of animal disease recorded in several counties over time, etc.—all of those are compelling examples of spatiotemporal data.

Spatial and temporal phenomena (systems) can be effectively analyzed and described at various levels of abstraction. The selected level of abstraction is of paramount importance as it helps realize a user-friendly approach to data analysis. In essence, the user can establish the most suitable perspective to view the data (identifying a suitable level of detail to be included in the analysis); making an overall analysis more humancentric. This view inherently brings a concept of information granules into the picture. Information granules, no matter how they are formalized (either as sets, fuzzy sets, rough sets, or others) are regarded as a collection of conceptual landmarks using which one looks at the data and describes them. We show that the relational way of capturing data naturally invokes the mechanisms of relation calculus where the concepts of granulation and degranulation are crucial to the optimization of the vocabulary (codebook) of information granules by means of which the main dependencies in data become revealed and quantified. The key objective of this study is to introduce a relational way of data analysis through their granulation and degranulation and show a constructive way of forming information granules leading to the best granular data representation. Furthermore, we introduce some way of characterizing relationships between granular codebooks pertaining to the description of data collected in successive time frames.

The presentation of the material is organized in a top–down fashion. We start with an illustrative practical example (Sect. 3.2) and then move on to the problem statement (Sect. 3.3) by stressing the relational character of data to be processed. A mechanism of granulation of information and a way of representing data in the language of information granules is covered in Sect. 3.4, which leads to a concept of a so-called granular signature of data (Sect. 3.5). The fundamental concept of information granulation–degranulation is introduced in Sect. 3.6 where we also show how a vocabulary (codebook) of information granules can be optimized so that the reconstruction of data implies a minimal value of the reconstruction error. With regard to this optimization, it is discussed how fuzzy clustering leads to the realization of information granules (Sect. 3.7). The issue of evolvability of the relational description of data emerging in presence of a series of data frames reported in successive time slices is covered in Sect. 3.8.

3.2 From Conceptual Developments of Information Granules to Applications

As an example of a real-world problem whose formulation and solutions can be naturally cast within the conceptual framework to be discussed here, relates to a comprehensive analysis of spatiotemporal data and associated event detection schemes formed directly on a basis of information granules. An interesting system where we encounter a wealth of spatiotemporal data is found at http://www1.agric.gov.ab.ca/department/$deptdocs.nsf/all/cl12944. It concerns a set of data collection stations distributed across the province of Alberta, see Fig. 3.1. This system helps graph, compare, and download data in almost real-time mode for more than 270 stations where data themselves go back to April 2005. There are some other data sets like precipitation and temperature (normal), temperature extremes, solar radiation, soil temperature, and other important weather characteristics. In a nutshell, we encounter a collection of spatial data—stations identified by some x–y coordinates. For each station, there is an associated suite of time series (say, dealing with the temperature, precipitation, radiation), see Fig. 3.2. An analysis

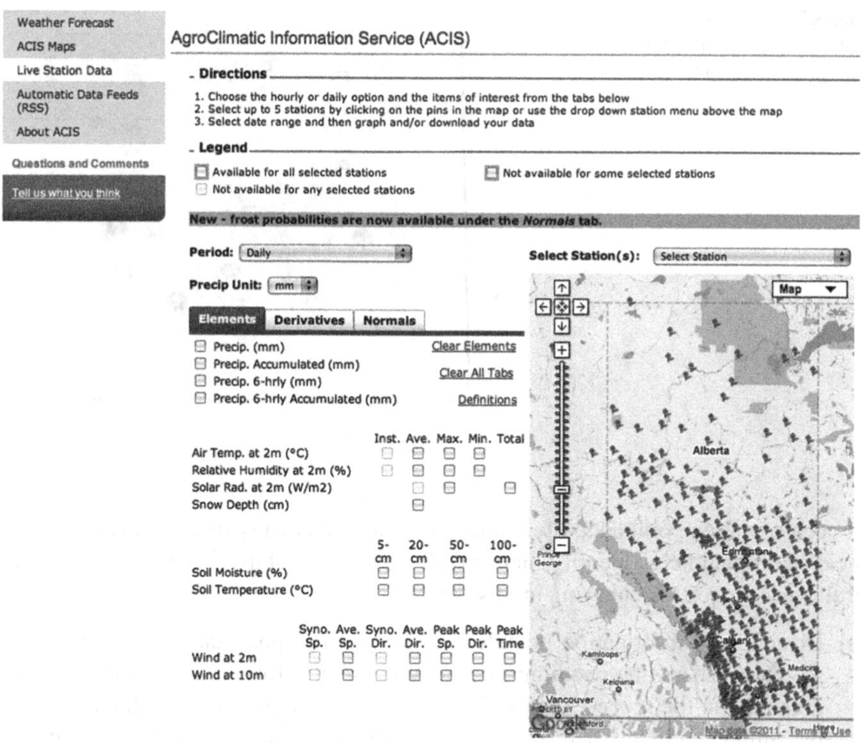

Fig. 3.1 A screen snapshot of the AgroClimatic Information Service (ACIS)

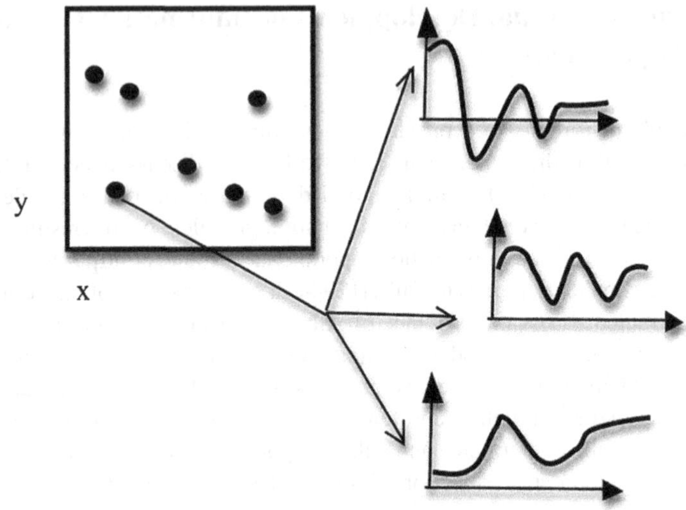

Fig. 3.2 Collection stations along with associated time series

Fig. 3.3 From collection stations to spatiotemporal information granules (clusters)

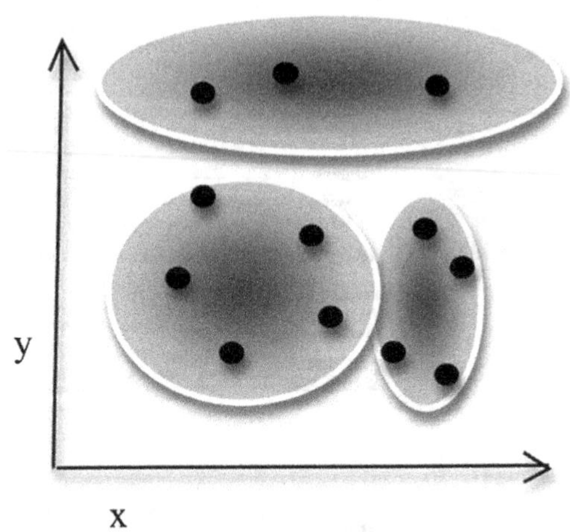

of such spatiotemporal data offers interesting and practically relevant insights into their nature and helps identify main trends and relationships. Several challenging problems are worth identifying in this context:

(a) Detection of information granules in the spatiotemporal data space, Fig. 3.3. As shown there, the resulting information granules form regions (information granules) in the space x–y by bringing together points (stations) characterized by similar temporal data (time series) and being in close vicinity to each other. From the algorithmic perspective, one can envision using a clustering algorithm (which has to be carefully designed to capture the spatial as well as

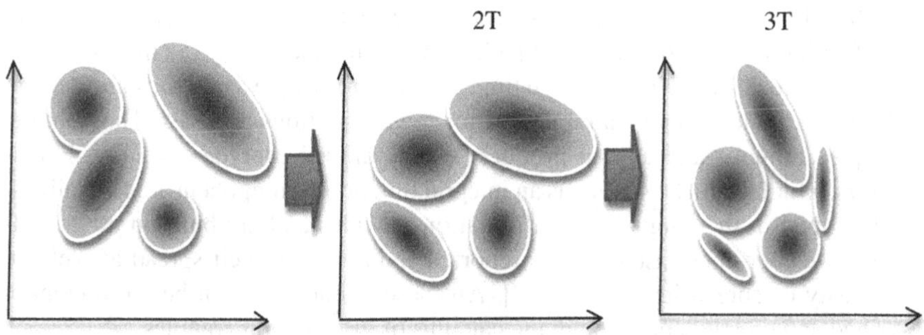

Fig. 3.4 Evolution of information granules through revealing dependencies among information granules observed in successive discrete time moments (determined by the predetermined length of the time horizon T)

temporal component). This implies that we have to carefully specify a distance function used to quantify the closeness of elements in the spatiotemporal data space. From the perspective of practical application, by forming the information granules the user can get a better insight into the regions of the province, their location, and size which exhibit some level of inner homogeneity while being distinct from the others. Such findings could be of immediate value to decision-makers in agriculture sector.

(b) Alluding to the constructs noted above, it is apparent that the temporal component of the data depends directly on the length of the corresponding time series. In particular, this impacts a way in which time series are represented (as the time series could be either stationary or nonstationary, which is usually the case when dealing with longer time horizons for which the information granules are to be formed). This brings a concept of the temporal scale (which again triggers some temporal granularity). The information granules are constructed over some time horizon. They evolve so in successive time moments, say T, 2T, … we obtain a collection of information granules, see Fig. 3.4.

These information granules are related. By describing relationships (dependencies) between the information granules arising in time horizons T, 2T, etc., an interesting dynamic pattern of changes (dynamics) at a more abstract level can be revealed. It could be regarded as a generalization of the dynamic process where now the dynamics is captured at a certain nonnumeric level and with the varying number of granules, we witness changes in the level of specificity of the granular descriptors (which becomes reflective of the varying level of complexity of the associated time series).

The data collected in the system described above can be used in conjunction with other sources of data (e.g., related with livestock and humans in the realization of event detection systems). Those are essential to prevention of disasters and effective risk management for livestock production, public health, and rural communities. The need for animal health surveillance systems and ensuing decision-making architectures capable of rapid disease detection is of paramount relevance. This is exacerbated by a number of factors. The speed of transportation has never been faster, facilitating the rapid spread of pathogens around the world, and makes timely detection more difficult. The average time to transport people and goods around the world is less than the incubation period for most infectious diseases. The volume of people, animals, and animal products that is moved everyday is large. Event detectors, which could be sought as an integral functional component are formalized and described in the language of information granules and are built on a basis of reported weather data, disease cases reported and a way of their spread as well as an intensity of spread itself [3–5, 7, 8]. All these phenomena can be conveniently described and operated on through the calculus of information granules.

3.3 Problem Formulation

The data we are concerned with are of spatiotemporal character. The temporal data (time series) are associated with some geographical x–y coordinates. Their description is provided in the form of a certain vector z positioned in the p-dimensional unit hypercube. We can encounter here a genuine diversity of available ways used to describe time series such as parameters of the autoregressive time series model (AR model), coefficients of the Fourier expansion, cepstral coefficients, components of discrete wavelet transforms (DWT), parameters of fuzzy rule-based system (in case of fuzzy modeling of the series), and others. For instance, in case of the AR model we have $z = [a_0\ a_1\ a_2\ \dots\ a_s]^T$ where the model comes in the form $z(K+1) = \sum_{j=0}^{s} a_j z(K-j)$. One could also encounter models that are more specialized to fit a certain category of signals such as this is the case in ECG signals modeled in terms of Hermite expansion. In all these cases, we assume that the values of the descriptors have been normalized so that the results are located in the $[0,1]^p$ hypercube.

The available data come in the form of the triples $(x_k, y_k, \mathbf{z}_k)$ with $\mathbf{x}_k = [x_k\ y_k]^T$ being the spatial coordinates of the temporal data $\mathbf{z}_k, k = 1, 2, \dots, N$. The equivalent representation that is convenient in our considerations, comes in a form of a fuzzy relation $R(x, y, \mathbf{z})$ assuming vector values in $[0,1]^p$ for specific values of $\mathbf{x}_k$ and y_k. In other words, the above notation underlines that the pair (x_k, y_k) is associated with a vector of temporal data $\mathbf{z}_k$. An illustration of an example fuzzy relation is included in Fig. 3.5.

Given the relational data R, we are interested in the following two important characterization and design issues:

(a) A description of R realized in a terms of a certain vocabulary of granular descriptors, see Fig. 3.6. We form a certain focused view at the data provided by the vocabulary. It has to be noted, as shown in Fig. 3.6, that different vocabularies (codebooks) facilitate different, sometimes complementary views at the available experimental evidence.
(b) Analysis of granulation and degranulation capabilities of the vocabularies. The intriguing question is about capabilities of the given vocabulary to capture the essence of the data (granular descriptors). Subsequently based upon the granular

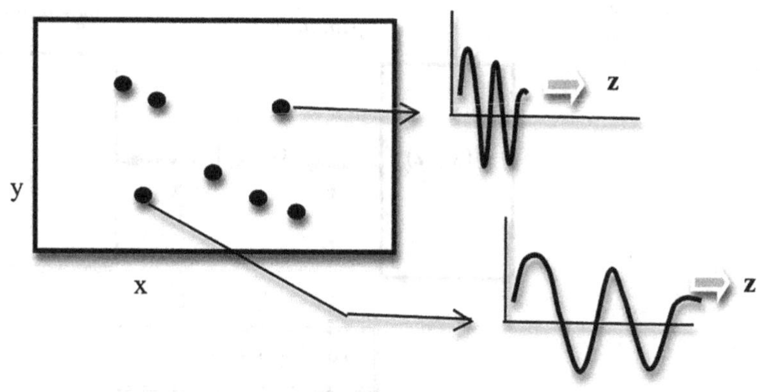

Fig. 3.5 Visualization of the fuzzy relation R; associated with some combination of the (x,y)-coordinates are sine—like time series characterized, e.g., by a collection of their Fourier expansion coefficients

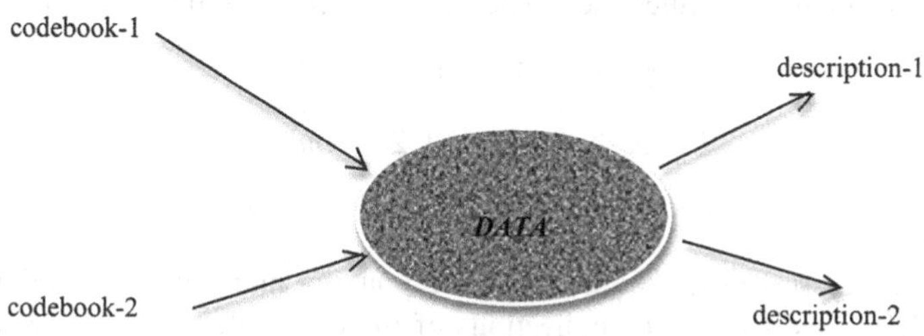

Fig. 3.6 Formation of various views at relational data supported by the use of different vocabularies

representation of the data, we discuss how to reconstruct (degranulate) the original data, see Fig. 3.7. The quality of this reconstruction (as we anticipate that the process of reconstruction could impart some losses) can be quantified and the figure of merit associated with this process can be used in the optimization of the codebook.

3.4 Granulation of Information and Granular Data Representation

We characterize the available data R in terms of a collection of information granules-semantically meaningful entities [1, 2, 6, 17] that are formed in the spatial and temporal coordinates. Formally, information granules can be captured in different ways, say in the form of sets, rough sets, fuzzy sets, to name a few alternatives. Here we are concerned with their realization by means of fuzzy sets. We define a

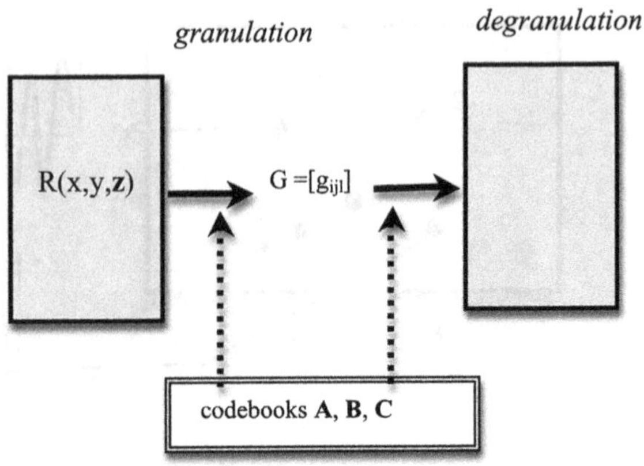

Fig. 3.7 The concept of granulation and degranulation realized with the aid of a vocabulary of information granules

collection of spatial information granules (located in the x- and y-coordinates) and temporal information granules used to describe the spatiotemporal signal

$$A_i : \mathbb{R} \to [0,1], i = 1,2,\ldots,n$$
$$B_j : \mathbb{R} \to [0,1], j = 1,2,\ldots,m$$
$$C_l : [0,1]^p \to [0,1]^p, l = 1,2,\ldots,r$$

Those are fuzzy sets described by the corresponding membership functions, that is $A_i(x), B_j(y)$, and $C_l(\mathbf{z})$. The collections of these granules are organized in the form of families of fuzzy sets A, B, C, respectively. For instance, we have $A = \{A_1, A_2, \ldots, A_n\}$, $B = \{B_1, B_2, \ldots, B_m\}$ and $C = \{C_1, C_2, \ldots, C_r\}$. These fuzzy sets come with a well-defined meaning. For instance, we may have linguistic descriptors such as

- A_i—east region of x-coordinate, B_j—north region of the y-coordinate.
- C_l—high frequency and low amplitude of the first component of the Fourier expansion of the signal.

See Fig. 3.8. We form Cartesian products of the granular descriptors. For instance, the Cartesian product of A_i and B_j defined above, the Cartesian product $A_i \times B_j$, can be regarded as a descriptor of the region of $x - y$ space located in the north-east region of the plane.

The Cartesian product of the above information granules can be treated as a granular descriptor—a linguistic "probe" used to describe the available data. A collection of all probes is used to describe/perceive the essence of data. Taken altogether they can be sought as a granular codebook (vocabulary) supporting a concise description of the collected experimental evidence.

The number of information granules for the spatial aspect of data could be reduced if we consider treating both coordinates en block meaning that a family of fuzzy sets is with respect to x- and y-coordinate. This gives rise to the membership functions of both arguments, say $A_1(x,y), A_2(x,y), \ldots$.

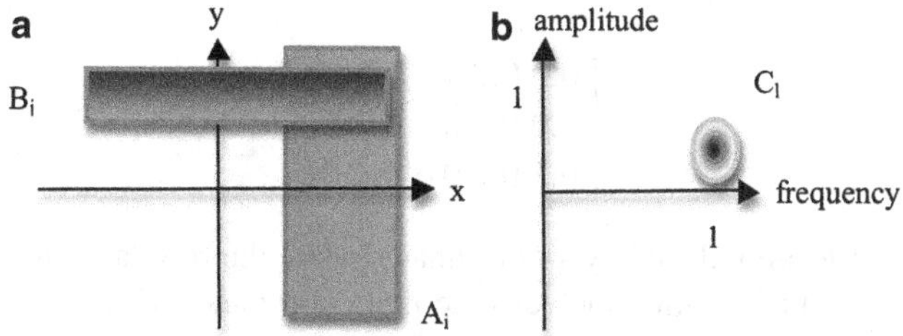

Fig. 3.8 Examples of linguistic descriptors used in the granular representation of relational data: spatial granular descriptors A_i and B_j (**a**) and spatial granular descriptor C_l (**b**)

3.5 A Granular Signature (Descriptor) of Relational Data

Given the collection of information granules—vocabularies $\{A_i, B_j, C_l\} i = 1,2,\ldots,n; j = 1,2,\ldots,m; l = 1,2,..,r$ they are used to describe (represent) the relational data R. This description is realized by performing a convolution of the elements of the vocabularies A, B, and C with the fuzzy relation of data, cf. [14]. This convolution is realized in a way it is quite commonly present in signal processing as well as the technology of fuzzy sets [16] that is

$$g_{ijl} = (A_i \times B_j \times C_l) \circ R \tag{3.1}$$

$i = 1,2\ldots,n; j = 1,2,\ldots,m; l = 1,2,\ldots,r$. Expressing the fuzzy sets of the codebooks in terms of the corresponding membership functions we obtain the values of g_{ijl} as a result of the max-t composition operator [14] of the fuzzy sets and the fuzzy relation

$$g_{ijl} = \max_{x_k, y_k, z_k} [A_i(x_k) t B_j(y_k) t C_l(\mathbf{z}) t R(x_k, y_k, \mathbf{z}_k)] \tag{3.2}$$

where t stands for a certain t-norm [12] (the minimum operator can be viewed as a particular case).

This max-t composition comes with a useful interpretation: g_{ijl} is then a possibility measure of the relational data with respect to the selected information granules indexed (i, j, l), $Poss(A_i, B_j, C_l; R)$. The possibility measures arise here in an explicit manner: they describe a degree of activation of a certain element of the Cartesian product of the codebooks by the fuzzy relation R. As an example, let us consider the fuzzy relation R with the following entries

$$\begin{bmatrix} 1.0\ 0.7\ 0.2\ 0.3 \\ 0.5\ 0.9\ 1.0\ 0.0 \\ 0.0\ 0.4\ 0.8\ 1.0 \end{bmatrix}$$

and

$$\begin{bmatrix} 0.0\ 0.2\ 0.7\ 1.0 \\ 1.0\ 0.3\ 0.6\ 0.3 \\ 0.3\ 0.5\ 0.5\ 0.2 \end{bmatrix}$$

The columns of R are indexed by x-coordinates while the rows are linked with the y-coordinate. The first relation includes the first coordinates of $\mathbf{z}_k, z_{k1}$. The second coordinates of $\mathbf{z}_k$ form the second matrix. With each entry of the matrix associated are the successive descriptors of the time series; the length of the descriptor vectors of the time series is equal to 2. For instance, considering the location at $(3,3)$—entry of the relation R, the descriptor of the time series positioned there is $z_k = [0.8\ \ 0.5]^T$ while $z_k = [0.0\ \ 0.3]^T$ for the entry $(4,2)$.

The vocabularies formed by information granules are as follows:

- For A with $n = 2$, $A_1 = [1.0\ \ 0.5\ \ 0.2\ \ 0.0]$, $A_2 = [0.3\ \ 1.0\ \ 0.7\ \ 0.6]$
- For B with $m = 2$, $B_1 = [1.0\ \ 0.6\ \ 0.3]$, $B_2 = [0.0\ \ 0.5\ \ 1.0]$
- For C with $r = 2$, $C_1 = [1.0\ \ 0.2]$, $C_2 = [0.1\ \ 1.0]$

Let us now compute g_{111}. Following the formula (3.2), we proceed with the successive calculations, $g_{111} = \max_{x,y}[\min(\min(A_1(x), B_1(y)), \max_z(\min(C_1(z), R(x,y,z)))]$. The inner expressions result in the following:

$$(A_1 \times B_1)(x,y) = \begin{bmatrix} 1.0\ 0.5\ 0.2\ 0.0 \end{bmatrix} \begin{bmatrix} 1.0\ 0.6\ 0.3 \end{bmatrix}^T = \begin{bmatrix} 1.0\ 0.5\ 0.2\ 0.0 \\ 0.6\ 0.5\ 0.2\ 0.0 \\ 0.0\ 0.0\ 0.0\ 0.0 \end{bmatrix}$$

$$\max_z(\min(C_1(z), R(x,y,z))) = \max\left(\begin{bmatrix} 1.0\ 0.7\ 0.2\ 0.3 \\ 0.5\ 0.9\ 1.0\ 0.0 \\ 0.0\ 0.4\ 0.8\ 1.0 \end{bmatrix}, \begin{bmatrix} 0.0\ 0.2\ 0.2\ 0.2 \\ 0.2\ 0.2\ 0.2\ 0.2 \\ 0.2\ 0.2\ 0.2\ 0.2 \end{bmatrix}\right)$$

$$= \begin{bmatrix} 1.0\ 0.7\ 0.2\ 0.3 \\ 0.5\ 0.9\ 1.0\ 0.2 \\ 0.2\ 0.4\ 0.8\ 0.0 \end{bmatrix}$$

Finally, we find the minimum of the corresponding entries of $\begin{bmatrix} 1.0\ 0.5\ 0.2\ 0.0 \\ 0.6\ 0.5\ 0.2\ 0.0 \\ 0.0\ 0.0\ 0.0\ 0.0 \end{bmatrix}$ and $\begin{bmatrix} 1.0\ 0.7\ 0.2\ 0.3 \\ 0.5\ 0.9\ 1.0\ 0.2 \\ 0.2\ 0.4\ 0.8\ 0.0 \end{bmatrix}$, which yields the relation $\begin{bmatrix} 1.0\ 0.5\ 0.2\ 0.0 \\ 0.5\ 0.5\ 0.2\ 0.0 \\ 0.0\ 0.0\ 0.0\ 0.0 \end{bmatrix}$. The maximum taken over all entries of this relation returns $g_{111} = 1.0$.

The result conveyed by (3.2) can be interpreted as a granular characterization of R. Obviously, when we consider some other granular codebook, we arrive at a different description or view of the same data. The flexibility available here is essential; we customize the analysis of the data and the mechanism of customization

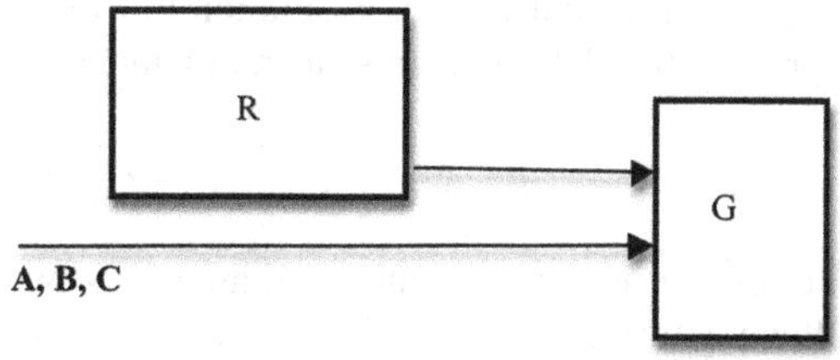

Fig. 3.9 Realization of the granular signature G of relational data R

is offered in the form of the vocabulary. The elements of g_{ijl} being arranged into a single fuzzy relation $G = [g_{ijl}]$ can be treated as a granular signature of R realized for a certain collection of generic information granules (codebook) A, B, C, refer to Fig. 3.9. It is worth noting that G supplies an important compression effect: the original data are reduced to granular data of dimensionality $n * m * r$. Obviously, the reduction depends on the size of the codebook. The Cartesian products of elements of $A \times B \times C$ of information granules can be ordered linearly depending upon the values of $G = [g_{ijl}]$; higher values of its entries indicate that the corresponding entries are more consistent, supported by experimental evidence conveyed by R. In this way, we can identify the most essential relational descriptors of R. More specifically, the relational data R translate into a family of triples of descriptors coming from the Cartesian product of the vocabularies $A \times B \times C$ indexed by the associated values of possibilities:

$$R = \{(A_{i1}, B_{j1}, C_{l1}), g_{i1j1l1}\} \tag{3.3}$$

We organize the entries of the granular representations g_{i1j1l1} in a nonincreasing order; the first entries of the sequence provides the essential granular descriptors of R, that is $g_I \leq g_J \leq g_K$ where $I = (i_1, j_1, l_1)$, $J = (i_2, j_2, l_2)$, etc.

3.6 The Concept of Granulation–Degranulation: A Way of Describing Quality of Vocabulary

Given the granular signature of the relational data captured by G as well as the elements of the vocabulary itself, we can reconstruct R. The result is an immediate consequence that (3.2) is a fuzzy relational equation and for each combination of the elements of A and B, we arrive at an individual fuzzy relational equation indexed by the elements of the codebook of A and B [11, 14]. The degranulation problem associates with a solution to the system of relational equations (3.1) in which the fuzzy sets A_i, B_j, C_l as well as the entries of the fuzzy relation g_{ijl} are given while the fuzzy relation R is to be determined. The solution to the problem comes as the maximal fuzzy relation expressed as follows [10]:

$$\hat{R} = |_{i,j,l}((A_i \times B_j \times C_l)\Phi g_{ijl}) \tag{3.4}$$

One can show that there is a solution to the system of fuzzy relational equations expressed by (3.1). In terms of the membership values of the fuzzy relation, we obtain

$$\hat{R}(x,y,z) = \min_{i,j,1}[(A_i(x)tB_j(y)tC_1(z))\Phi g_{ijl}] \tag{3.5}$$

The pseudo-residual operator [10] associated with a given t-norm is defined in the following form:

$$a\Phi b = \sup\{c \in [0,1] | atc \leq b\} \tag{3.6}$$

The expression (3.8) requires some attention as we encounter here the vector-valued membership function $Cl(z)$. The calculations realized there are considered coordinate-wise that is we compute consecutive entries of $\hat{R}$ for the arguments $z_1, z_2, \ldots, z_p$ by computing the corresponding expressions. For z_1

$$\hat{R}(x,y,z) = \min_{i,j}[(A_i(x)tB_j(y)tC_1(z_1))\Phi g_{ij1}] \tag{3.7}$$

for z_2

$$\hat{R}(x,y,z) = \min_{i,j}[(A_i(x)tB_j(y)tC_1(z_2))\Phi g_{ij2}] \tag{3.8}$$

In case we use the minimum operator, the corresponding pseudo-inverse (3.6) reads as follows:

$$a\Phi b = \sup\{c \in [0,1] | \min(a,c) \leq b\} = \begin{cases} b & a > b \\ 1 & a \leq b \end{cases} \tag{3.9}$$

Observe that in all cases the inclusion relationship holds, namely $R \subseteq \hat{R}$. The quality of the granulation–degranulation is expressed by comparing how much the reconstruction of R given by (3.8), namely $\hat{R}$, coincides with the relational data themselves (R). The associated performance index expressing the quality of reconstruction is defined as follows:

$$Q = \sum_{K=1}^{N} \|R(x_k,y_k,\mathbf{z}_k) - \hat{R}(x_k,y_k,\mathbf{z}_k)\|^2 \tag{3.10}$$

Again in light of the vector values assumed by the fuzzy relation, the above distance is computed by taking distances over the successive coordinates of $\mathbf{z}$. For the Euclidean distance one explicitly specifies the components of the fuzzy relation, which yields $Q = \sum_{j=1}^{p} \sum_{K=1}^{N} \|R(x_k,y_k,z_{kj}) - \hat{R}(x_k,y_k,z_{kj})\|^2$. The associated optimization process may then link with a formation of information granules minimizing the distance expressed by (3.10). This can be taken into consideration when designing fuzzy sets (membership functions) as well as selecting the number of fuzzy sets. Referring to the original expression for the max-t composition, the calculations are carried out coordinate-wise, that is we compute $\min(C_l(z_1), R(x,y,z_1))$, $\min(C_l(z_2), R(x,y,z_2))$, etc.

In summary, the presented approach offers a general methodology of describing data in the language of information granules, which can be schematically captured as $D \rightarrow G(D)$, where G denotes a granulation mechanism exploiting a collection of information granules in the corresponding vocabularies. In view of flexibility offered by different views at the data, we can highlight this effect in Fig. 3.7: different granulation schemes cast an analysis of D in different perspectives and give rise to different sequences of dominant granular descriptors.

3.7 Construction of Information Granules

Fuzzy clustering is a vehicle of information granulation. As such it can be used here to form the elements of the codebook. As an example, we use here the well-known method of the fuzzy c-means (FCM) [6]. The results of clustering come in the form of the prototypes of the clusters and the partition matrix, which describes an allocation of elements (membership degrees) to the clusters.

Let us discuss in detail on how the corresponding codebooks A, B, and C are formed. For the spatial component, the data for the construction of elements of A and B are one dimensional. The membership functions $A_1, A_2, \ldots, A_n$ are expressed in the form

$$A_i(x) = \frac{1}{\sum_{i=1}^{n} \left(\frac{x-v_i}{x-v_j}\right)^{2/(\zeta-1)}} \tag{3.11}$$

$i = 1, 2, \ldots, n$, where $v_1, v_2, \ldots, v_n \in R$ are the prototypes constructed by running the FCM using the data $\{x_1, x_2, \ldots, x_N\}$. The fuzzification coefficient is denoted by ζ, $\zeta > 1$. In the same way formed are the elements of the second codebook, B_1, B_2, $\ldots$, B_m by clustering one-dimensional data $\{y_1, y_2, \ldots, y_N\}$. The codebook forming the elements of C is constructed by clustering multidimensional data $z_1, z_2, \ldots, z_N$. Here, the membership functions of the elements of C are expressed in the form

$$C_1(\mathbf{z}) = \frac{1}{\sum_{i=1}^{r} \left(\frac{\mathbf{z}-\mathbf{v}_i}{\mathbf{z}-\mathbf{v}_j}\right)^{2/(\zeta-1)}}, \tag{3.12}$$

where $\mathbf{v}_1, \mathbf{v}_2, \ldots, \mathbf{v}_r$ are the prototypes located in $[0,1]^p$. The successive coordinates of the membership function C_l, $C_l(\mathbf{z})$ can be expressed explicitly by projecting the prototypes onto the corresponding axes of the unit hypercube and computing the membership functions based upon these projected prototypes, see Fig. 3.10.

Take the i_0-th coordinate of z, that is z_{i0}. Project the original prototypes $v_1, v_2, \ldots, v_c$ on this coordinate obtaining the real numbers $v_{1i0}, v_{2i0}, \ldots, v_{ci0}$. By making use of these scalar prototypes, we obtain the membership functions of the i_0-th coordinate of z.

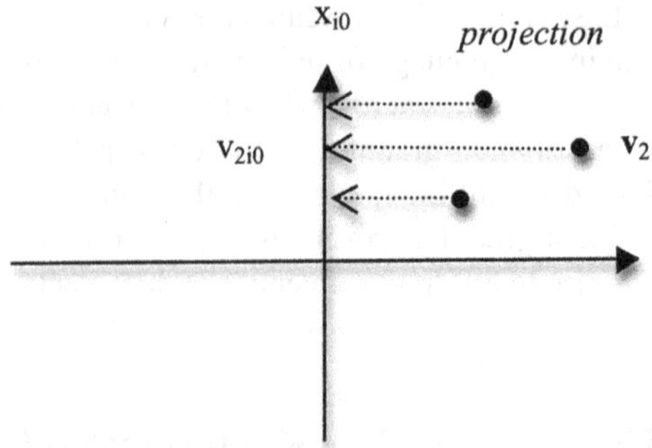

Fig. 3.10 Computing membership values based on i_0-th coordinate of the prototypes

There are several strategies of organizing a clustering process; the three main categories are worth highlighting here:

(a) Clustering realized separately for the spatial data and temporal data. In this approach, we exercise two separate clustering algorithms, which do not interact. The results are used together. The advantage of this design strategy is that one can consider a number of well-established clustering techniques. A disadvantage comes from a lack of interaction when forming the clusters.
(b) Clustering temporal and spatial data realized by running a single clustering method. The interplay between the temporal and spatial data is realized by forming an aggregate distance function composed of two parts quantifying closeness of elements in the spatial domain and the space in which temporal data are described. The approach by [15] is an example of the clustering strategy falling within this realm. Further more advanced approaches in which the distance functions are quite different are worth pursuing here (say, an Euclidean distance used for spatial data and Itakura distance with regard to the temporal data).
(c) Clustering temporal and spatial data realized by two clustering methods with a granular interaction realized at the structural level of data. While the clustering methods run individually, they receive feedback about the structure discovered for the other data set (spatial or temporal). The interaction is typically realized at the level of proximity matrices induced at the locally formed partition matrices [13].

There are several parameters of the codebook formed by the information granules that could be effectively used in the optimization of the description of the relational data. This list includes:

(a) The number of information granules defined for the corresponding spatial and temporal coordinates.
(b) The values of the fuzzification coefficients (ζ) used in the construction of information granules. As discussed in the literature, by changing the value of the fuzzification coefficient, we directly affect the shape of the membership function. The values close to 1 yield membership functions whose shape is

close to the characteristic functions of sets. The commonly used value of the fuzzification coefficient is equal to 2. Higher values of this parameter result in spike-like membership functions.

The first group of the parameters is of combinatorial nature. In general, there is a general monotonicity relationship: higher number of clusters can help reduce the value of the optimization criterion (3.10). The optimization is of a combinatorial nature, though so as to assure an effective design, one has to engage some techniques of global optimization such as, e.g., evolutionary optimization. In light of the monotonicity relationship, it could be more legitimate to look at the allocation of the number of information granules to the vocabularies of the information granules formed for the spatial and temporal dimensions of the variables. Let c be a number of available information granules used in the granulation process and specified in advance. We choose the number of granules for the x-, y-, and temporal coordinates of the variables, say c_1, c_2, and c_3 so that they sum is equal to c. How to do this allocation, comes as a result of combinatorial optimization. The parametric aspects of the optimization are implemented by selecting suitable values of the fuzzification coefficients; those could be assume different values for the corresponding codebooks, say ζ_1 for A, ζ_2 for B and ζ_3 for C.

3.8 Describing Relationships Between Granular Descriptors of Data

The spatiotemporal data discussed so far along with their granular description (signature) may be viewed as a certain snapshot of a certain phenomenon reported over some period of time, say a week or a month. The data could be acquired over consecutive time periods (time slices) thus forming a sequence of spatiotemporal data $D(T)$, $D(2T)$, $\ldots D(MT)$. As each data is subject to its granular description (and behind each of these description comes a certain codebook), it is of interest to look at the relationships between $D(T)$, $D(2T)$, etc. or better to say between $G(D(T))$ and $G(D(2T))$. The latter task translates into a formation of relationships between the components of the codebooks being used for the successive data "slices" say, $D(T)$, $D(2T)$, etc. Another way of expressing such dependencies is to look at the characterization of information granules forming a certain vocabulary at time slice T by the elements of the vocabulary pertaining to the successive $2T$th time slice. We discuss these two approaches in more detail.

3.8.1 Determination of Relationships Elements of Vocabularies

We construct the relationships between $\{A_i(2T), B_j(2T), C_k(2T)\}$ and $A_i(T), B_j(T), C_k(T)$, so the evolution of the underlying system is expressed in terms of evolution

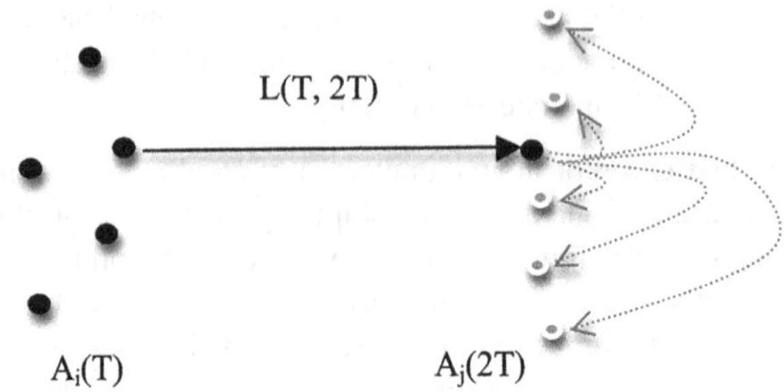

Fig. 3.11 Computing the degrees of activation of the elements of the codebook $A(2T)$ by the elements of $A(T)$; $v_i(T)$ and $v_j(2T)$ (*black* and *grey dots*) are the prototypes of the corresponding fuzzy sets in the T and $2T$ time periods

of information granules describing the data structure. The number of elements in the vocabularies $A_i(T)$ and $A_i(2T)$ need not to be same as the granularity of the vocabularies can evolve to reflect the complexity of the data and structures being encountered. The relationships between $A_i(T)$ and $A_l(2T)$ can be expressed in terms of logic dependencies. In general, we formulate a logic relationship in the form of the fuzzy relational equation:

$$a(2T) = a(T) \circ L(T, 2T) \tag{3.13}$$

where $\circ$ stands for a certain composition operator, say a max-t or a max–min one (being already used in the previous constructs) and $L(T, 2T)$ captures the relationships between coordinates of the vectors $a(T)$ and $a(2T)$. The vectors $a(T)$ and $a(2T)$ being the elements of the unit hypercubes, contain the degrees of activation of the elements of the vocabulary A. The dimensionality of $a(T)$ and $a(2T)$ is equal to $n(T)$ and $n(2T)$, respectively. As noted, the dimensionality of these two vectors might not be the same. The estimation of the fuzzy relation can be realized in a number of different ways depending upon a formation of the input–output data to be used in the estimation procedure of $L(T, 2T)$. We note that the above relational description is a nonstationary model as the fuzzy relation L depends on the time indexes (T and $2T$). One could form a stationary model where L is time independent that is we have $a(nT) = a(nT - T) \circ L, n = 2, 3, \ldots$.

A certain way, using which this could be realized in an efficient way is to consider the data composed of the pairs (here we are considering the elements of the first codebook, A)

$$a_1(T) = [1\ 0\ \ldots\ 0] \quad a_1(2T) = [\lambda_{11}\ \lambda_{12}\ \ldots\ \lambda_{1n(2T)}] = \boldsymbol{\lambda_1} \tag{3.14}$$

Here $\lambda_1(T)$ is a degree of activation of the vocabulary at $2T$ by the 1st element of the vocabulary used at T. The degrees of activation λ_{1i} are computed as follows:

$$\lambda_{1i} = \frac{1}{\sum_{j=1}^{n(2T)} \left(\frac{\|v_1(T) - v_i(2T)\|}{\|v_1(T) - v_j(2T)\|} \right)^{2/(\zeta - 1)}} \tag{3.15}$$

Refer to Fig. 3.11 for a more detailed explanation.

Likewise, we have the remaining $n(T)$ input–output pairs

$$a_2(T) = [0\ 1\ \ldots\ 0] \quad a_2(2T) = [\lambda_{21}\ \lambda_{22}\ \ldots\ \lambda_{2n(2T)}] = \lambda_2 \tag{3.16}$$

$$a_{n(T)}(T) = [0\ 0\ \ldots\ 1] \quad a_{n(T)}(2T) = [\lambda_{n(T)1}\ \lambda_{n(T)2}\ \cdots\ \lambda_{n(T)n(2T)}] = \lambda_2 \tag{3.17}$$

We consider all pairs of data $\{(a_1(T), a_1(2T)), \ldots, (a_{n(T)}(T), a_{n(T)}(2T)\}$ and solve a system of relational equations (3.13) with regard to $L(T,2T)$. In case of the max-t composition, the result is straightforward; the solution does exist and the largest one is expressed in the form

$$\hat{L}(T,2T) = [1\ 0\ \ldots\ 0]^T \Phi[\lambda_{11}\ \lambda_{12}\ \ldots\ \lambda_{1n(2T)}] \bigcap [0\ 1\ \ldots\ 0] \Phi[\lambda_{21}\ \lambda_{22}\ \ldots\ \lambda_{2n(2T)}]$$
$$\times \bigcap \ldots \bigcap [0\ 0\ \ldots\ 1] \Phi[\lambda_{n(T)1}\ \lambda_{n(T)2}\ \cdots\ \lambda_{n(T)n(2T)}] \tag{3.18}$$

In other words, the fuzzy relation computed in this way comes in the form

$$L(T,2T) = |\lambda_1\ \lambda_2\ \ldots\ \lambda_{n(T)}|^T \tag{3.19}$$

We can quantify the changes when moving from one vocabulary to the next one by expressing uncertainty of expressing the elements of the codebook at $2T$, $A(2T)$ in terms of the elements of the codebook formed for the data at T.

3.8.2 *Uncertainty Quantification of Variable Codebooks*

We use the entropy measure [9] to quantify the level of uncertainty. More specifically, we look at the successive rows of $L(T,2T)$ and for each of them compute the entropy. For instance, for the i-th row we have

$$H\lambda_i = -\sum_{j=1}^{n(2T)} \lambda_{ij} \log_2(\lambda_{ij}) \tag{3.20}$$

The highest uncertainty (viz. difficulty of expressing the elements of one codebook by another one) occurs when all elements of i are the same and equal to $1/n(2T)$. At a more aggregate level, we consider an average entropy that is $\frac{1}{n(T)} \sum_{i=1}^{n(T)} H(\lambda_i)$. Its value is reflective of the level of variability between the codebooks used in successive time epochs and uncertainty when expressing the elements of $A(T)$ in terms of the components of $A(2T)$. In the same way, we compute the entropy associated with the evolution of the codebooks B and C. This type of evaluation is useful in the assessment of the evolution of the granularity of the descriptors of data. The granular description of the system presented in this way highlights and quantifies two important facets of evolvability, that is evolvability being reflective of the changes of the system itself and evolvability being an immediate result of the adjustable perception of the data caused directly by changes made to the vocabularies used in the description of the data. These two are interlinked.

In case we fix our perception of the system that is the codebooks are left unchanged $A(T) = A(2T) = \ldots B(T) = B(2T) = \ldots$, etc., the matrices of granular descriptors of the data, $G(T)$, $G(2T), \ldots$, etc. are essentially reflective of the dynamics of the system. The dynamics being captured in this way depends upon a level of specificity of the codebooks. What is intuitive, the less detailed the codebooks, the closer the granular descriptors become and some minor evolvability trends present in the system are not taken into consideration. When the codebook (vocabulary) itself changes (because of the structural complexity of the data themselves or perception at the system, which has been changed), we envision an aggregate effect of evolvability stemming from the dynamics of the system as well as its varying level of perception.

3.9 Conclusions

There is an ongoing quest in data analysis to make the results more user friendly (interpretable) as well as make the role of the user more proactive in the processes of data analysis. The study offered in this paper arises as one among interesting possibilities with this regard. It should be noted that the closeness among data is expressed by taking into account a certain aggregate of closeness in the spatial domain and temporal domain. While the x–y coordinates could be left unchanged in successive time slices (which reflects a situation of fixed measuring stations/sensors), one can easily handle a situation of mobile sensors whose distribution varies over time slices.

It has been demonstrated that the concept of information granularity not only helps conceptualize a view at the data but also construct effective algorithms supporting the development of relational models.

The nonstationarity phenomenon might be inherent to the data themselves (viz. the system or phenomenon producing the data). The evolution effect could be also related to the varying position one assumes when studying the system. These two facets of dynamical manifestation of the system/perception are captured and quantified in terms of the relational descriptors and dependencies between them.

Acknowledgments Support from Natural Sciences and Engineering Research Council (NSERC) is greatly appreciated.

References

1. Bargiela, A., Pedrycz, W.: Granular Computing: An Introduction. Kluwer Academic Publishers, Dordrecht (2003)
2. Bargiela, A., Pedrycz, W.: Toward a theory of granular computing for human-centered information processing. IEEE Transactions on Fuzzy Systems **16**(2), 320–330 (2008)

3. Berezowski, J.: Animal health surveillance. In: The Encyclopedia of Life Support Systems (EOLSS). EOLSS Publishers Co. Ltd (2009)
4. Berezowski, J.: Alberta veterinary surveillance network, advances in pork production. In: Proceedings of 2010 Banff Pork Seminar vol. 21, pp. 87–93. Quebec, Canada (2010)
5. Berezowski, J., Renter, D., Clarke, R., Perry, A.: Electronic veterinary practice surveillance network for alberta's cattle population. In: Proceedings of the Twenty Third World Buiatrics Congress. Quebec, Canada (2004)
6. Bezdek, J.: Pattern Recognition with Fuzzy Objective Function Algorithms. Kluwer Academic/Plenum Publishers, U.S.A. (1981)
7. Burkom, H., Murphy, S., Shmueli, G.: Automated time series forecasting for biosurveillance. Statistics in Medicine **26**, 4202–4218 (2007)
8. Checkley, S., Berezowski, J., Patel, J., Clarke, R.: The alberta veterinary surveillance network. In: Proceedings of the 11th International Symposium for Veterinary Epidemiology and Economics. Cairns, Australia (2009)
9. Cover, T.M., Thomas, J.A.: Elements of Information Theory. John Wiley & Sons, New York (1991)
10. Di-Nola, A., Sessa, S., Pedrycz, W., Sanchez, E.: Fuzzy Relational Equations and Their Applications in Knowledge Engineering. Kluwer Academic Publishers, Dordrecht (1989)
11. Hirota, K., Pedrycz, W.: Fuzzy relational compression. IEEE Transactions on Systems, Man, and Cybernetics, Part B: Cybernetics **29**(3), 407–415 (1999)
12. Klement, E., Mesiar, R., Pap, E.: Triangular Norms. Kluwer Academic Publishers, Dordrecht Norwell New York London (2000)
13. Loia, V., Pedrycz, W., Senatore, S.: Semantic web content analysis: A study in proximity based collaborative clustering. IEEE Transactions on Fuzzy Systems **15**(6), 1294–1312 (2007)
14. Nobuhara, H., Pedrycz, W., Sessa, S., Hirota, K.: A motion compression/reconstruction method based on max t-norm composite fuzzy relational equations. Information Sciences **176**(17), 2526–2552 (2006)
15. Pedrycz, W., Bargiela, A.: Fuzzy clustering with semantically distinct families of variables: descriptive and predictive aspects. Pattern Recognition Letter **31**(13), 1952–1958 (2010)
16. Pedrycz, W., Gomide, F.: Fuzzy Systems Engineering: Toward Human-Centric Computing. John Wiley & Sons, Hoboken, NJ (2007)
17. Zadeh, L.: Towards a theory of fuzzy information granulation and its centrality in human reasoning and fuzzy logic. Fuzzy Sets and Systems **90**, 111–117 (1997)

Chapter 4
Incremental Spectral Clustering

Abdelhamid Bouchachia and Markus Prossegger

Abstract In the present contribution, a novel algorithm for off-line spectral clustering algorithm is introduced and an online extension is derived in order to deal with sequential data. The proposed algorithm aims at dealing with nonconvex clusters having different forms. It relies on the notion of communicability that allows to handle the contiguity of data distribution. In the second part of the paper, an incremental extension of the fuzzy c-varieties is proposed to serve as a building block of the incremental spectral clustering algorithm (ISC). Initial simulations are presented towards the end of the contribution to show the performance of the ISC algorithm.

4.1 Introduction

Spectral clustering is often considered as an efficient clustering capable of dealing with different cluster shapes compared to other clustering algorithms such as (fuzzy) K-means, Gaussian mixture model, K-median, deterministic annealing, etc. Indeed spectral clustering has been proven to be efficient in many applications related to artificial vision, image processing and web mining that exhibit complex cluster shapes.

The goal is to find partitions of the graph formed from the data such that the edges between these partitions have very low weight, while the edges within a

A. Bouchachia (✉)
Institute of Informatics-Systems, University of Klagenfurt, Klagenfurt, Austria
e-mail: hamid@isys.uni-klu.ac.at

M. Prossegger
Carinthia University of Applied Sciences, Spittal an der Drau, Austria
e-mail: m.prossegger@cuas.at

M. Sayed-Mouchaweh and E. Lughofer (eds.), *Learning in Non-Stationary Environments: Methods and Applications*, DOI 10.1007/978-1-4419-8020-5_4,
© Springer Science+Business Media New York 2012

partition have high weight. This refers to the mincut problem which consists of solving the following minimization problem. For a given number k of subsets, the mincut method seeks to find a partition $P_1, \cdots, P_k$ which minimizes:

$$\text{mincut}(P_1 \cdots P_k) = \frac{1}{2}\sum_{i=1}^{k} W(P_i, \overline{P_i}) \text{ such that: } W(P,Q) = \sum_{i \in P,\ j \in Q} w_{ij}, \tag{4.1}$$

where P_i and $\overline{P_i}$ are the ith partition (cluster = a set of nodes) and its complement (the rest of nodes in the graph), w_{ij} is the weight on the edge (i, j). The weights correspond to pairwise affinities. We seek to find the minimum cut of the graph which specifies the optimal partitioning of the data.

Widely used criteria are the RatioCut [18] and the normalized cut, Ncut [31] which should be minimized.

$$\text{RatioCut}(P_1 \cdots P_k) = \frac{1}{2}\sum_{i=1}^{k} \frac{W(P_i, \overline{P_i})}{|P_i|} \tag{4.2}$$

$$\text{Ncut}(P_1 \cdots P_k) = \frac{1}{2}\sum_{i=1}^{k} \frac{W(P_i, \overline{P_i})}{\text{vol}(P_i)} \text{ such that: } \text{vol}(P) = \sum_{i \in P}\sum_{j=1}^{n} w_{ij}, \tag{4.3}$$

where $\text{vol}(P)$ indicates the volume of the set of nodes belonging to the Pth cluster and $|P_i|$ is the number of vertices in P_i.

Both criteria RatioCut and Ncut are however NP-hard. Spectral clustering is considered as a less harder problem in the sense that eigenvectors corresponding to the smallest eigenvalues of the normalized Laplacian (4.7) approximate the solution of the normalized cut criterion, while the eigenvectors corresponding to the smallest eigenvalues of the unnormalized Laplacian (4.5) approximate the solution of the ratiocut criterion. Usually the eigenvector corresponding to the second smallest eigenvalue indicates the structure of the data can be inferred by thresholding.

There exist several spectral clustering algorithms which rely on the same generic clustering steps involving the spectral properties of the graph representing the data and differ mainly in tuning procedures to deal with data sets having particular geometric forms (shells, lines, etc.). The common generic clustering steps are shown in Algorithm 1.

In the first step, there is no principled or formalized way to generate affinity matrix, but usually the Gaussian kernel is applied:

$$W(x_i, x_j) = \exp(-||x_i - x_j||^2 / 2\sigma^2) \tag{4.4}$$

Algorithm 1 Generic steps of spectral clustering

1: Let X^t be the set of points (i.e., graph vertices) to be clustered at time t: $X = \{x_1, \cdots x_n\}$, $x_i \in \mathbb{R}^d$ and k the number of clusters
2: Compute the affinity (or weight) matrix W of X^t
3: Transform W into a Laplacian matrix (L)
4: Compute the first k smallest eigenvalues $(e_1, \cdots, e_k)$ and the corresponding eigenvectors of L
5: Form the matrix $V = [v_1, \cdots, v_k]$ containing the corresponding eigenvectors (arranged column wise) whose number of rows is n (the size of the data)
6: Apply a clustering algorithm to cluster V (i.e., the k eigenvectors)
7: Assign the vertices x_i ($i = 1 \cdots n$ data points) to cluster j if and only if the rows of the vectors v_l were assigned to cluster j in step 6.

In the second step, many forms of Laplacian matrices can be derived and used to generate the eigenvectors. Here also there is no consistent definition of Graph Laplacian. We may use:

- Unnormalized Laplacian:
$$L = D - W \tag{4.5}$$
where D is the diagonal degree matrix whose diagonal elements $d_i = \sum_j w_{ij}$ are the vertex degrees of W.
- Symmetric normalized Laplacian:
$$L_{\text{sym}} = D^{-1/2} L D^{-1/2} = I - D^{-1/2} W D^{-1/2} \tag{4.6}$$
- Normalized Laplacian (the random walk normalization or discrete Laplace operator):
$$L_{rw} = D^{-1} L = I - D^{-1} W \tag{4.7}$$

In the sixth step, any clustering algorithm can be used. Very often k-means (of fuzzy c-means) is applied [9, 24–26], but some authors proposed other algorithms like k-lines [14].

Often the number of clusters is determined using the notion of eigengap, which is the difference between two consecutive eigenvalues λ_k and λ_{k-1}. The most stable clustering is generally given by the value k that maximizes the expression

$$\Delta_k = |\lambda_k - \lambda_{k-1}| \tag{4.8}$$

Recently there has been an increasing interest in incremental clustering [5, 6]. However, little work has been done in regards to spectral clustering. A concise overview of incremental spectral clustering is presented in Sect. 4.2. In a recent contribution by Ning et al. [27], an interesting idea in incremental spectral clustering has been presented. It aims at estimating the eigenspace incrementally as new data are added. This idea can easily be applied to step 4 of Algorithm 1. However, doing so the spirit of incrementality is not fully captured in the sense that the eigenspace will be kept which may be obviously large.

In the present contribution, we will try to embed incrementality at a later level, namely step 6 of Algorithm 1. The basic motivations for this approach are:

- The spectrum of a block diagonal matrix of the form:

$$S = \begin{bmatrix} S_1 & 0 & \dots & 0 \\ 0 & S_2 & \dots & 0 \\ \vdots & \vdots & \ddots & \vdots \\ 0 & 0 & \dots & S_n \end{bmatrix}$$

  S is given by the union of the spectra of S_i. Here, S_i corresponds to the spectrum of the i-th batch of data.
- Embedding incrementality at step 6 of Algorithm 1 allows handing the data that continuously arrive over time in an efficient manner.

To realize this solution, we introduce in this paper an incremental version of the famous fuzzy c-varieties clustering algorithm.

Before delving into the details of the proposed solution, the structure of this contribution is presented as follows. Section 4.2 introduces a short overview of incremental spectral clustering. Section 4.3 provides the description of incremental fuzzy c-varieties which serves to explain the overall incremental spectral clustering in Sect. 4.4. In Sect. 4.5, a set of experiments are discussed to show the effectiveness of the proposed approach. Section 4.6 highlights the contributions and future work.

4.2 Related Work

Because of the good performance of spectral clustering algorithms, they gain more and more attention from the scientific community. Most of the standard spectral clustering algorithms are dealing with offline data and have been proven to be successful in various applications like speech recognition [1], image analysis and pattern recognition [10, 32, 35], and interdisciplinary data analysis frameworks [28, 29, 36].

In the emerging field of high dynamic online/web-based resources like social networks, there is the need to enhance the algorithms to enable incrementally clustering of data without the need to run the clustering from scratch each time new data arrive. The incremental approach does not require to keep a huge and even growing data set and the costly computation of the affinity matrix can be limited to consider new entries only. So far there have been very few attempts to accommodate spectral clustering to an online setting.

For instance in [33] a general purpose incremental algorithm relying on spectral clustering for dealing with online topological mapping has been presented. Similar approach for dealing with the problem of appearance-based localization in outdoor environments has been also introduced in [34]. Such an approach is based on the clustering algorithm proposed in [26] and aims at estimating iteratively cluster representatives. Using a similarity threshold, the number of clusters is increased if the smallest allowed distance is less than the threshold and the cluster representative replaces all entries of the origin data points in the affinity matrix. During the online map updating of the robot, the incremental spectral clustering eliminates the need of computing the similarities between the current and all previous input (images).

Most of the incremental spectral clustering algorithms are designed to support the insertion and deletion of new data points only, while the critical issue is to react on similarity changes between the data points. In [27], the incremental clustering is based on the standard spectral clustering enhanced by the incidence vector/matrix (described in [4]) representing both kinds of dynamics. Decomposing the Laplacian matrix into two incidence matrices, each similarity change between data points is regarded as an incidence vector appended to the original incidence matrix. Considering the influence of the data points outside the spatial neighborhood of the new data points, the Laplacian and incidence matrices may change which induces an equivalent change of the eigenvalues and eigenvectors.

All mentioned algorithms above have reached similar accuracy but with much lower computation costs compared to standard spectral clustering.

4.3 Incremental Fuzzy C-Varieties

Fuzzy c-varieties proposed in [2] is a modified version of fuzzy c-means (FCM) with the purpose of dealing with linear varieties. It has served to inspire other extensions of FCM such as c-elliptotypes [2, 23] and c-shells [11, 12]. A general overview of extensions of FCM proposed in the literature is summarized in [8] as shown in Table 4.1.

The motivation behind applying the fuzzy c-varieties algorithms stems from the fact that the eigenvectors generated from the data have the shape of lines. Testing the fuzzy c-varieties on an artificial data set consisting of four lines illustrates

Table 4.1 A family of shell clustering algorithms

Algorithm	Cluster shape
Fuzzy c-varieties [2, 3]	Line segments and lines
Fuzzy c-shells [3, 12, 15]	Circles and ellipses
Fuzzy c-spherical shells [22]	Circles and spheres
Fuzzy c-rings [16]	Circles
Fuzzy c-means [2]	Spheres
Gustafson–Kessel [17]	Ellipsoids and possibly lines
Gath–Geva [16]	Ellipsoids and possibly lines
Fuzzy c-elliptotype [30]	Ellipsoids and possibly lines
Fuzzy c-quadric shells [20, 21]	Linear and quadric shell
Fuzzy c-rectangular shells [19]	Rectangles/polygons (approximation of circle, lines, ellipses)

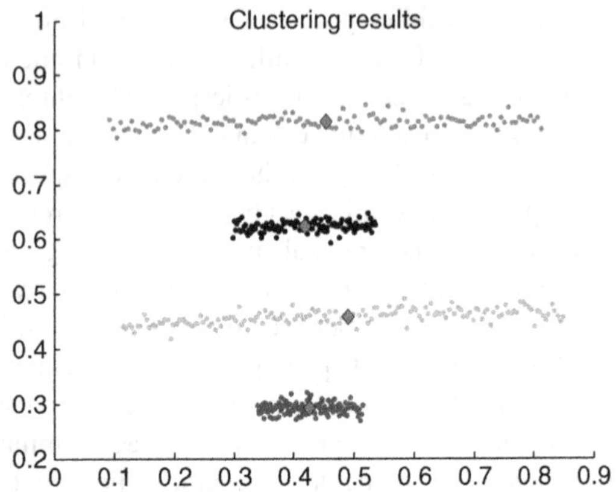

Fig. 4.1 Clustering results using FCV

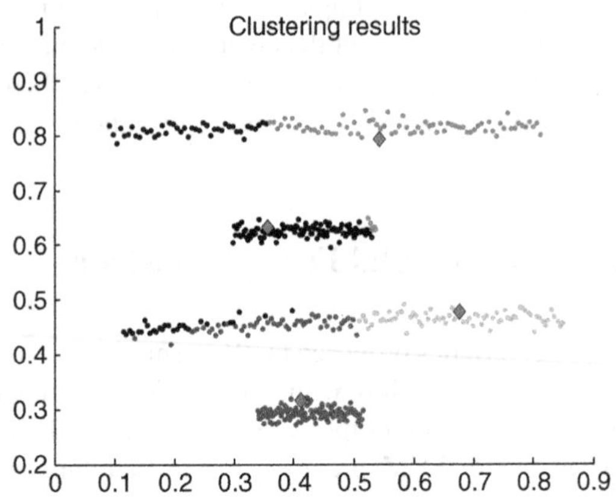

Fig. 4.2 Clustering results using FCM

that the algorithm is capable of discovering each line as a cluster (Fig. 4.1). When applying the standard FCM (Fig. 4.2), the results are understandably worse. Clearly the choice of FCV is well founded.

FCV is based on linear varieties (affine spaces) which are characterized by their dimensions. A linear variety of dimension r ($0 \leq r < s$) in $\Re^s$ through the point v and spanned by the linearly independent vectors $\{d_1, d_2, \cdots, d_r\} \subset \Re^s$ by:

$$V_r(v, d_1, d_2, \cdots, d_r) = \left\{ y \in \Re^s | y = v + \sum_{j=1}^{r} t_j d_j;\ t \in R \right\} \tag{4.9}$$

Note that if $r = 0$, then $V_0 = \{v \in \Re^s\}$ which is a point. If $r = 1$, $V_1 = \{v + td\}$ which represents a line and if $r = 2$, V_2 is plane in $\Re^2$ defined by $\{d_1, d_2\}$. Generally speaking, for $r = s - 1$, V_{s-1} is a hyperplane in $\Re^s$.

The objective function of FCV has the same form as that of FCM. It measures the weighted sum of squared errors from the data points $x_k \in \Re^s$ and the linear varieties as shown in (4.10):

$$J_{V_{rm}}(U,V) = \sum_{i=1}^{c}\sum_{k=1}^{N} u_{ik}^m D_{ik}^2 \tag{4.10}$$

where D_{ik} is the orthogonal distance between, the prototype V_i which is an r-dimensional linear variety and the data point x_k. The distance is given as follows:

$$D_{ik} = \left(||x_k - v_i||^2 - \sum_{j=1}^{r} \left[(x_k - v_i)^T d_j^{(i)} \right]^2 \right)^{1/2} \tag{4.11}$$

The orthogonal distance is obtained by projecting $(x_k - v_i)$ onto the linear subspace of $\Re^s$ spanned by the set $\{d_1, d_2, \cdots, d_r\} \subset \Re^s$ and computing the length of $(x_k - v_i)$ minus its best least squares approximation.

The objective function (4.10) is subject to the following constraints:

$$u_{ik} \in [0,1]; \ \ \sum_{i=1}^{c} u_{ik} = 1, \forall k; \ \ 0 < \sum_{k=1}^{N} u_{ik} < N, \forall i \tag{4.12}$$

Minimization of the objective function with respect to U and V subject those constraints in (4.12) yields the following expressions:

$$u_{ik} = \begin{cases} D(x_k,V_i)^{(2/(1-m))} \left[\sum_{j=1}^{c} D(x_k,V_i)^{(2/(1-m))} \right]^{-1} & I_k = \emptyset \\ \begin{cases} 0 & i \in \overline{I_k} \\ \sum_{i \in I_k} u_{ik} = 1 & i \in I_k \end{cases} & I_k \neq \emptyset \end{cases} \tag{4.13}$$

where $I_k = \{i | 1 \leq i \leq c; D_{ik} = 0\}$ and $\overline{I} = \{1, 2, \cdots, c\} \setminus I_k$. The prototypes are computed using the following:

$$v_i = \frac{\sum_{k=1}^{N} u_{ik}^m x_k}{\sum_{k=1}^{N} u_{ik}^m} \tag{4.14}$$

The $d_j^{(i)}$'s in (4.9) are the eigenvectors corresponding to the largest eigenvalues of the fuzzy scatter matrix of the i-th cluster given by:

$$S_i = \sum_{k=1}^{N} u_{ik}^m (x_k - v_i)(x_k - v_i)^T \tag{4.15}$$

Algorithm 2 : Steps of FCV for static clustering

Given c (number of clusters), m (fuzziness factor, $m \geq 1$), r (the dimension of the linear varieties, $r \leq s$) and ε (stopping threshold)
Let $X = \{x_1, \cdots x_n\}$ be the set of points to be clustered
Initialize $U^{(0)}$ taking the constraints (4.12) into account and set $t = 1$;
repeat
    Compute the centers $v_i^{(t)}$ using (4.14) and (4.13)
    Compute the fuzzy scatter matrices $S_i^{(t)}$ of the c clusters using (4.15), (4.14) and (4.13)
    Compute the eigenvectors corresponding to the r largest eigenvalues
    Compute the new partition matrix $U^{(t)}$ using (4.13) and (4.11)
until $||U^{(t)} - U^{(t-1)}|| \leq \varepsilon$

For a static clustering problem, the FCV algorithm consists of the steps shown in Algorithm 2.

Given a dynamic setting where data arrive over time, the FCV algorithm must be adapted. In the present contribution, we consider an aggregative approach that consists of two steps:

1. Cluster the new batch of data B_p yielding $C_1^{(p)}, \cdots C_c^{(p)}$ for a given number of clusters c.
2. Merge and cluster the obtained $C_i^{(p)}$ clusters and the existing clusters obtained from previous merge operation (after the arrival of B_{p-1}).

To realize step 2, a weighted version FCV is applied to generate an approximation of partitioning of the data so far seen by clustering the existing clusters and the clusters newly generated after B_p. The weighted version of the FCV is given by the following objective function:

$$J'_{V_{rm}}(U,V) = \sum_{i=1}^{c} \sum_{k=1}^{N} w_k u_{ik}^m D_{ik}^2 \tag{4.16}$$

Here a weight is assigned to each data point x_k (which actually represents either a prototype or a true data point). In the case of clusters' prototypes, the weight reflects the proportion of the data of a batch covered the cluster of those prototypes. Formally the weight of a cluster (generated from a batch B_p) is given in a straightforward manner from the partition matrix associated with B_p clustering:

$$w_i = \frac{\sum_{k=1}^{N_p} u_{ik}}{N_p} \tag{4.17}$$

where N_p is the size of the batch B_p. The denominator serves as a normalization factor over the size of the batch. Thus by removing the denominator from (4.17) the length will impact the formulation of the partition matrix associated with the objective function $J'_{V_{rm}}$ (4.16). In the present study, we adopt (4.17) so that the current prototypes obtained from the previous aggregation stage and which summarize the previous batches $(1\cdots p-1)$ have a weight 1. However, this would mean that the aggregated prototypes will have higher weight than the ones generated from the last batch B_p, which leads to a conservative approach in the update of the clusters mean. In order to keep the approach flexible with respect to data drift and shift [7], a persistence factor $\zeta_k \in [0,1]$ may be appended to the current prototypes. Hence, in the objective function (4.16) can be rewritten as follows:

$$J''_{V_{rm}}(U,V) = \sum_{i=1}^{c}\sum_{k=1}^{N} w_k \zeta_k u_{ik}^m D_{ik}^2 \tag{4.18}$$

For the new clusters obtained after B_p, $\zeta_k = 1$ and for those obtained so far by aggregation, $\zeta_k < 1$ will stand as parameters. In the context of this study, we will set all ζ_k for the aggregated prototypes to 0.95 to keep the algorithm stable but evolving.

Minimization of the objective function in (4.18) with respect to U and V subject to the constraints in (4.12) yields the following expressions:

$$u_{ik} = \begin{cases} D(x_k,V_i)^{(2/(1-m))}\left[\sum_{j=1}^{c} D(x_k,V_i)^{(2/(1-m))}\right]^{-1} & I_k = \emptyset \\ \begin{cases} 0 & i \in \overline{I_k} \\ \sum_{i\in I_k} u_{ik} = 1 & i \in I_k \end{cases} & I_k \neq \emptyset \end{cases} \tag{4.19}$$

$$v_i = \frac{\sum_{k=1}^{N} u_{ik}^m w_k \zeta_k x_k}{\sum_{k=1}^{N} u_{ik}^m w_k \zeta_k} \tag{4.20}$$

The eigenvectors $d_j^{(i)}$ needed to compute the distance $D(x_k,V_i)$ are computed using the fuzzy scatter matrices resulting from the new and old prototypes (and eventually true data points).

For efficiency reasons, one may consider the following setup. Since the number of clusters is always small, we could do the clustering of prototypes via the WFCV batchwise (after the arrival of each batch), but we could reduce the frequency, say after b batches. This gives more stability and accuracy to the clustering process.

To assign data of the recently arriving batch B_p to clusters, we compute the membership values $u_{..}$ using the formulae in (4.19) and (4.11) with respect to the final aggregated prototypes and varieties.

4.4 Incremental Spectral Clustering

After explaining step 6 and 7 of Algorithm 1, we turn now to briefing the details of the spectral clustering algorithm—much of it has already been mentioned.

The spectral clustering algorithm implemented in this paper relies on the notion of communicability introduced in [13] for dealing with complex networks such as

Fig. 4.3 Critical data for spectral clustering

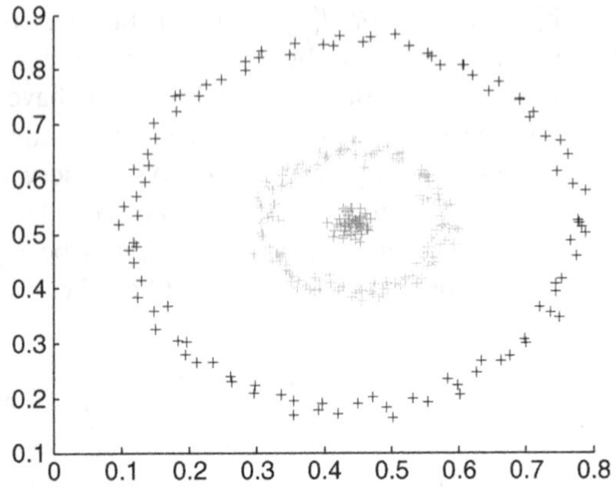

Fig. 4.4 Affinity matrix of the ring data

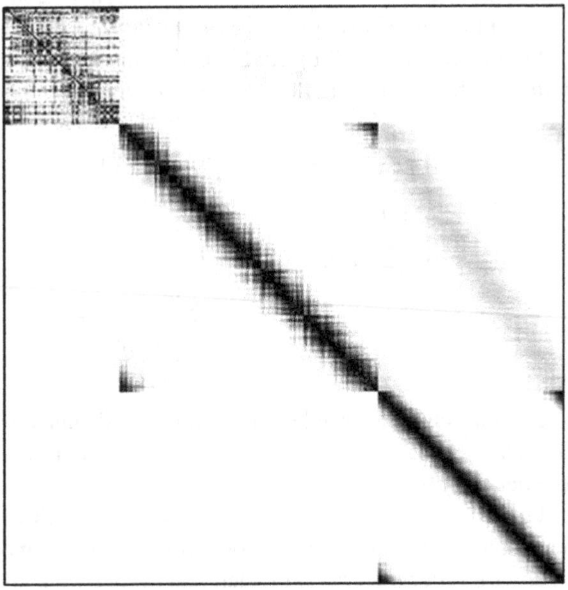

social networks. This idea is motivated by the fact that in general spectral clustering is efficient if the affinity matrix (4.4) is block diagonal reflecting clear structure of the clusters. However, this happens if the data presents clear structure with dense clusters separated by low density boundaries. As noted in [14], a data with nonconvex clusters as that shown in Fig. 4.3 in the form of two rings around a central cluster where the points at the opposite sides of the external ring are far from each other, further than from points from the central cluster, it is difficult for a spectral clustering to detect the right structure of the data. The affinity matrix in Fig. 4.4 shows that no clear structure is visible except probably in top left corner.

The solution lies in devising mechanisms to achieving a clear structure through some transformation of the affinity matrix. The idea consists of enhancing the similarity between the data points that are directly or transitively connected.

In the previous example points of a ring should be considered close to each other whatever their coordinates are. Similar to the idea presented in [14] relying on the notion of conductivity, in this contribution we use the communicability concept introduced in the context of complex networks [13].

Communicability between two points x and y is defined as the number of shortest paths, P^s_{xy}, of length s between x and y plus the number of walks, W_{xy}, of length $k > s$. Hence, communicability is expressed as follows:

$$G_{xy} = \alpha_s P^s_{xy} + \sum_{k>s}^{\infty} \beta_k W^k_{xy} \tag{4.21}$$

α_s and β_k must be selected such as the communicability converges. By setting $\alpha_s = 1/s!$ and $\beta_k = 1/k!$ we obtain:

$$G_{xy} = \sum_{k=0}^{\infty} \frac{(A^k)_{xy}}{k!} = (e^A)_{xy} \tag{4.22}$$

This can be further be rewritten as:

$$G_{xy} = \sum_{j=0}^{n} \phi_j(x)\phi_j(y)e^{\lambda_j} \tag{4.23}$$

where $\phi_j(x)$ is the xth element of the jth orthonormal eigenvector of the adjacency matrix associated with the eigenvalue λ_j.

It is worthwhile to notice that the communicability is computed based on the adjacency matrix. In other words, in order to apply it to weighted graphs it is necessary to introduce some adaptation. For this purpose, we apply the symmetric normalization of the Laplacian formulated in (4.6), that is:

$$C_{ij} = \frac{G_{ij}}{\sqrt{D_i D_j}},\ D_i = \sum_k G_{ik}. \tag{4.24}$$

The resulting affinity matrix G is not symmetric and to render it so, we apply the following:

$$C_{ij} = \min(C_{ij}, C_{ji}) \tag{4.25}$$

Having formulated a new affinity matrix that takes the continuity of data into account, the remaining stages of the overall algorithm are identical to those outlined in Algorithm 1, step 5 to step 7. In step 3, instead of the transformation into L, we implement C. The difference here is that in step 4 we compute the k largest eigenvalues instead of the smallest since because we are not using the unnormalized Laplacian matrix (4.5) in the normalization stage (4.24), but directly an affinity matrix. This is similar to the method of Ng [26].

Algorithm 3 : Steps of the proposed offline spectral algorithm

1: Let X^t be the set of points (i.e., graph vertices) to be clustered at time t: $X = \{x_1, \cdots x_n\}$, $x_i \in \mathbb{R}^d$ and k the number of clusters
2: Compute the affinity (or weight) matrix W of X^t
3: Compute the communicability matrix C from W
4: Compute the first k largest eigenvalues $(e_1, \cdots, e_k)$ and the corresponding eigenvectors of G
5: Form the matrix $V = [v_1, \cdots, v_k]$ containing the corresponding eigenvectors (arranged column wise) whose number of rows is n (the size of the data)
6: Apply a clustering algorithm to cluster V (i.e., the k eigenvectors)
7: Assign the vertices x_i ($i = 1 \cdots n$ data points) to cluster j if and only if the rows of the vectors v_l were assigned to cluster j in step 6.

Algorithm 4 : Steps of the proposed online spectral algorithm

```
Let {B}_t be the set of batches to be clustered at time t = 1...T and k the number of clusters.
  Prot is variable that contains the set of prototypes and initially is set to ∅.
for t = 1 to T do
    Apply the offline spectral clustering algorithm (Algorithm 3) and retain the prototypes V =
    {v_1 ··· v_c} resulting from the weighted fuzzy c-varieties (WFCV) clustering of the selected
    eigenvectors of the communicability matrix
    if t=1 then
        prot=V;
    else
        prot = prot ⋃ V
        apply WFCV on prot as input data to obtain c clusters
    end if
end for
```

Algorithms 3 and 4 show the steps of the offline and online spectral clusterings algorithms.

4.5 Simulations

To evaluate the algorithm proposed in this contribution, we will look mainly at two aspects following the main contributions:

- Clustering performance on data with different shapes in an offline setting
- Efficiency of clustering when data arrive over time. For such a purpose, a number of benchmark data sets are used.

4.5.1 Offline Clustering of Data

Because the quality of the incremental spectral algorithm highly depends on the offline spectral clustering component, it is necessary to check the quality of the clustering algorithm in the offline setting. Figure 4.5 shows the capacity of the algorithm to accurately cluster data having different shapes. In particular, the conti-

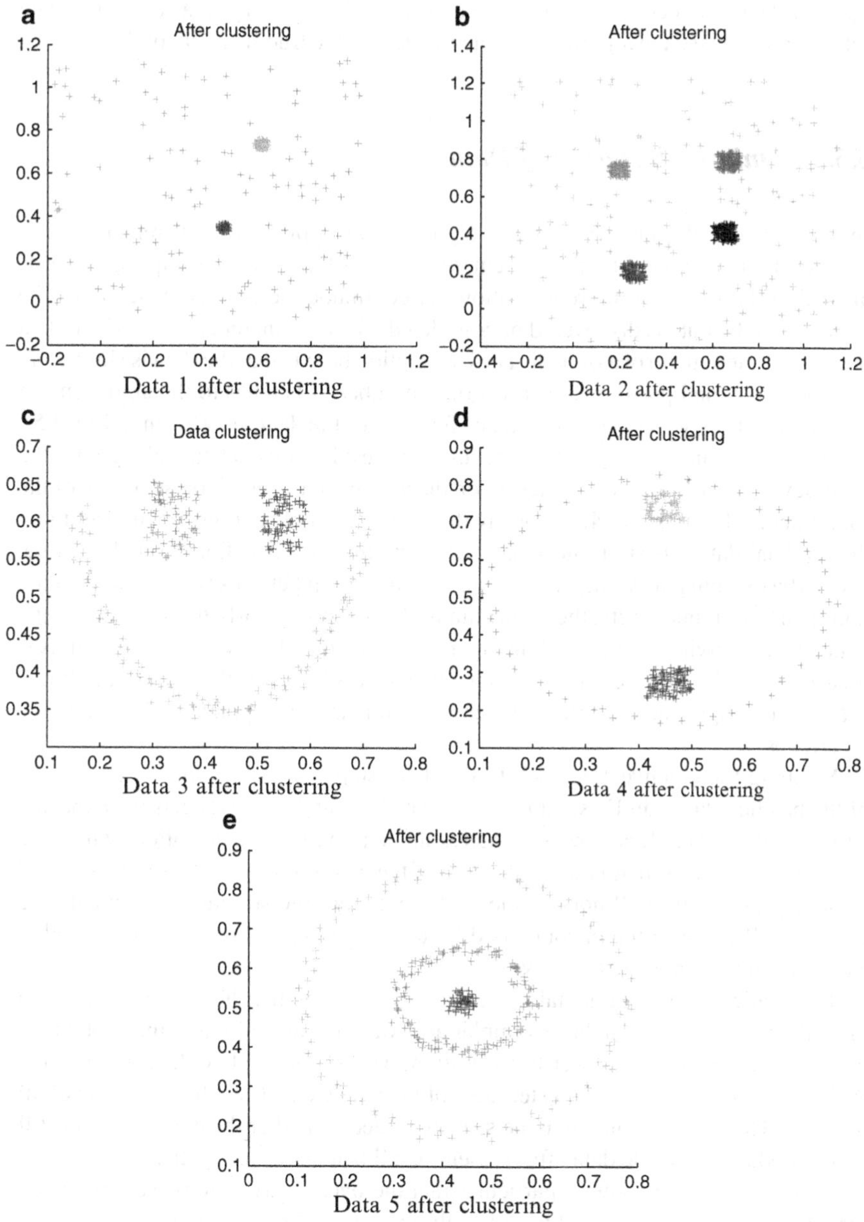

Fig. 4.5 Efficiency of the algorithm on different shapes

guity of data has been observed by the algorithm. Even though many of these data sets are nonconvex, the algorithm is able to uncover the true structure of the clusters.

4.5.2 Online Clustering of Data

In this section, the behavior of the algorithm in coping with incrementality is discussed. In response to the advent of arrival of new data samples, the ISC algorithm should adapt in order to allow for continuous self-adjustment so that the structure of the data is discovered or completed. To study incrementality, we look at the case of random arrival of data. The assumption here is that the data is shuffled in a random way and presented over a certain number of batches to the algorithm. In this contribution, we consider 4 batches for each of the data sets shown in Fig. 4.5.

In the following, we highlight the clustering results of each batch along with the accuracy of the algorithm when tested on the whole data after the presentation of the four batches. Figures 4.6–4.10 portray the final results of running the algorithm on the artificial data sets split into 4 batches. In particular, Figs. 4.6a, 4.7a, 4.8a, 4.9a, and 4.10a explain partly why the overall ISC algorithm behaves well. Thanks to the notion of communicability, the algorithm is able to show clearly the structure of the data clusters relying on the leading eigenvectors of the affinity matrix. This makes it easy to use the fuzzy c-varieties to detect clusters. It is worth noticing the linear shape of the eigenvectors entries which has motivated the application of the fuzzy c-varieties.

Mapping the eigenvectors back to the data (step 7 of Algorithm 1), the results of the batches shown in Figs. 4.6b, 4.7b, 4.8b, 4.9b, and 4.10b reflects the expected structure of the data. It is, however, interesting to remark that in situations where the structure is not clear, that is when the data of the batch is loosely dense, the spectral clustering algorithm will not be able to discover the true structure. A typical case where the ISC algorithm encounters difficulties is Data 4 (Fig. 4.9b). In the other cases, the algorithm works very well.

The challenge in incremental learning is usually whether the results obtained in the offline mode (when all data examples are available) are similar to those obtained in the online mode. To answer this question, we use the final configuration of the ISC algorithm to predict the cluster label of the whole data (i.e., the collection of all batches). This way of doing may be seen as a check whether the algorithm can still accommodate the whole data after presenting all batches sequentially.

Figures 4.11–4.15 show in particular that the overall structure is well predicted for all data sets. However, a kind of "salt and pepper" can be noticed, especially in the case of Data 3 (Fig. 4.13) and Data 2 (Fig. 4.12). Such noisy examples are probably due to the accumulation and aggregation of clusters through the FCV algorithm. The points in the eigenspace that are at the boundary of clusters may be tricky to handle because centers in the new batches may slightly generate a shift.

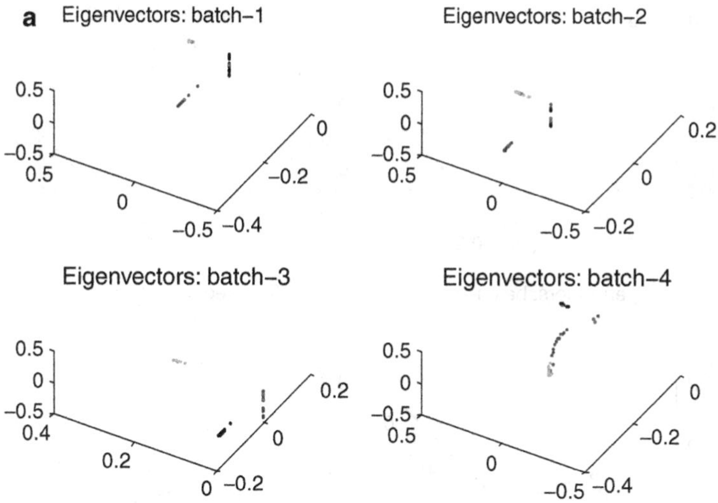

Clustering the eigenvectors of Data 1

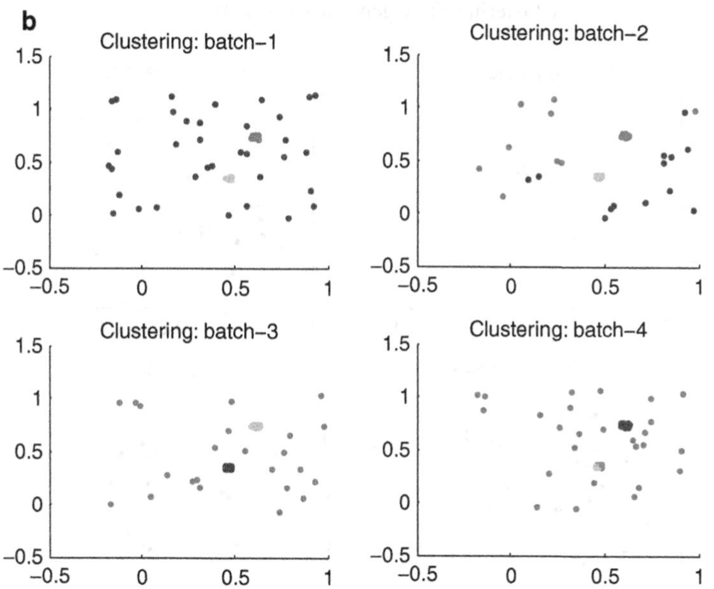

Batches of Data 1 clustered

Fig. 4.6 Efficiency of the algorithm on Data 1 in the online setting

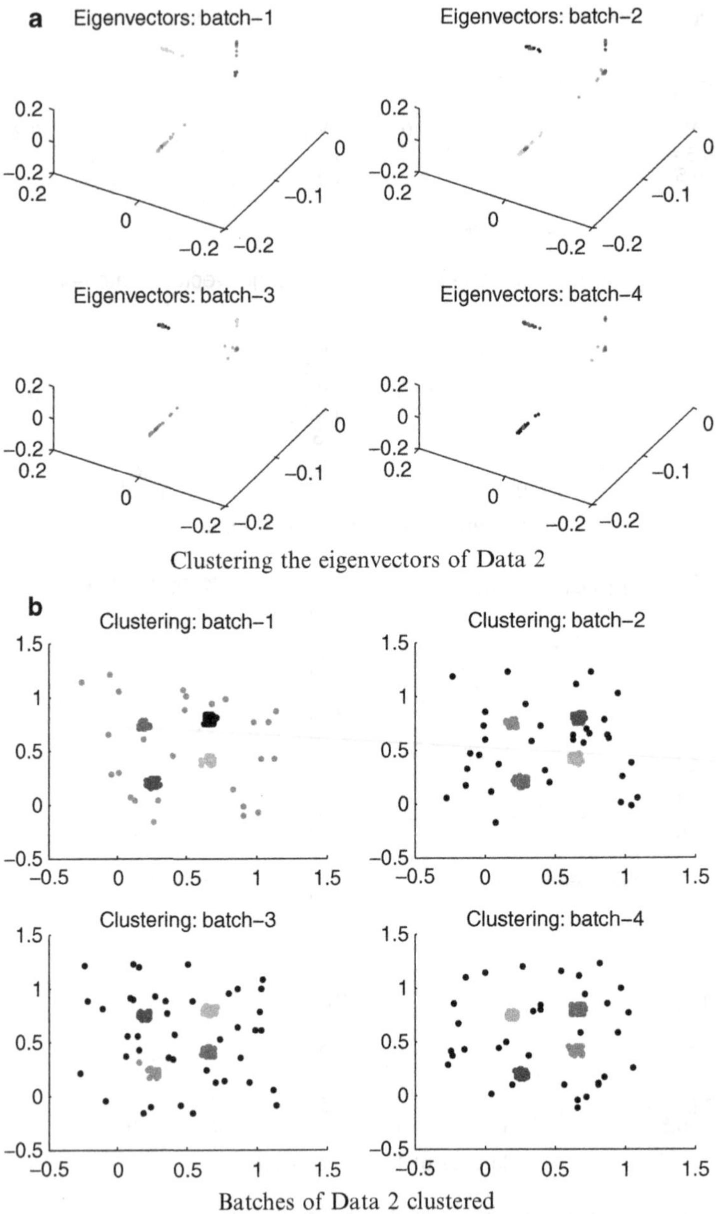

Fig. 4.7 Efficiency of the algorithm on Data 2 in the online setting

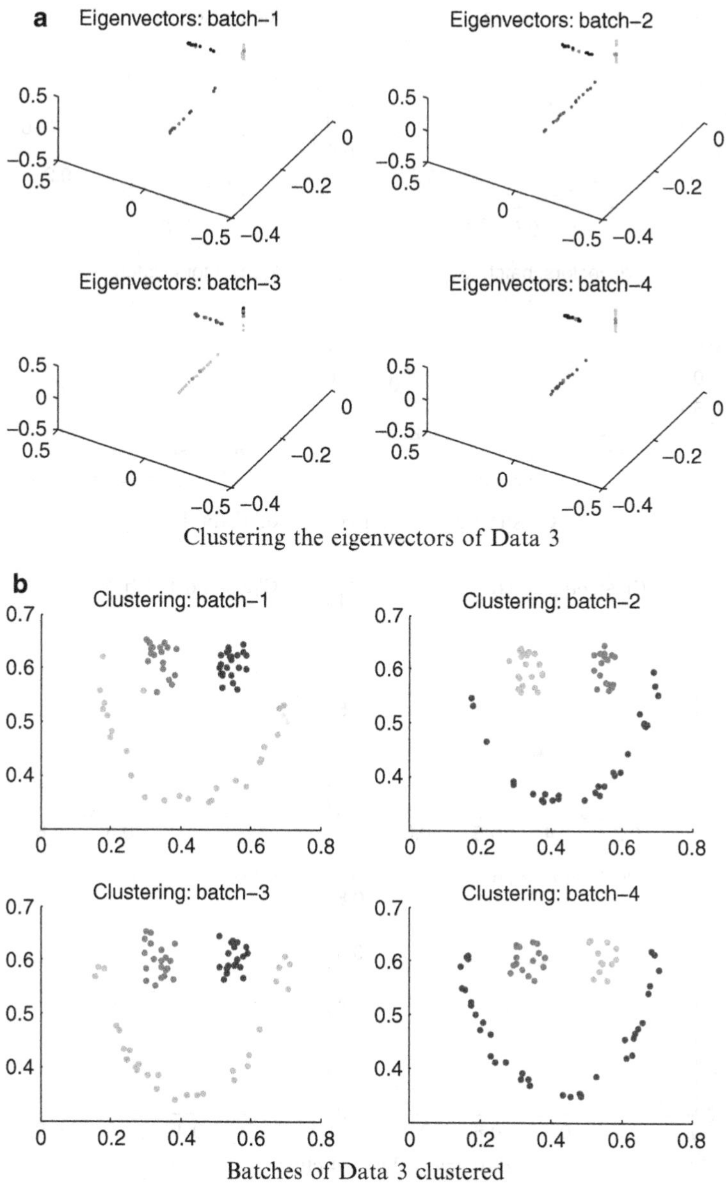

Fig. 4.8 Efficiency of the algorithm on Data 3 in the online setting

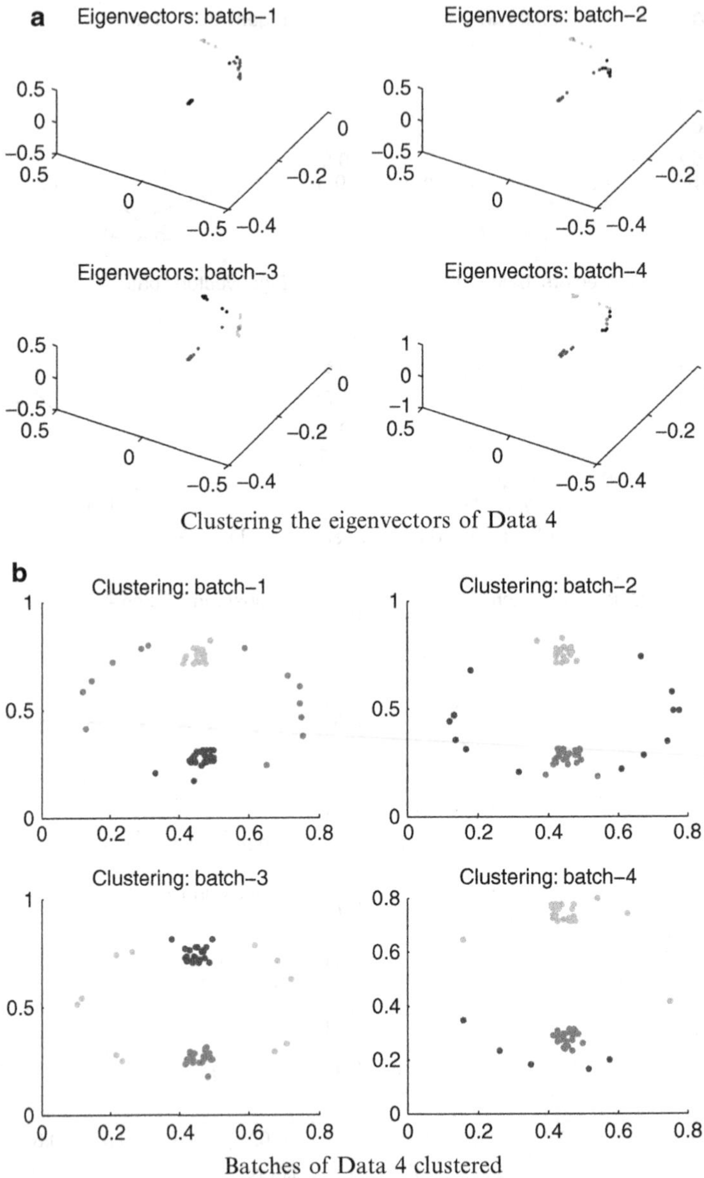

Fig. 4.9 Efficiency of the algorithm on Data 4 in the online setting

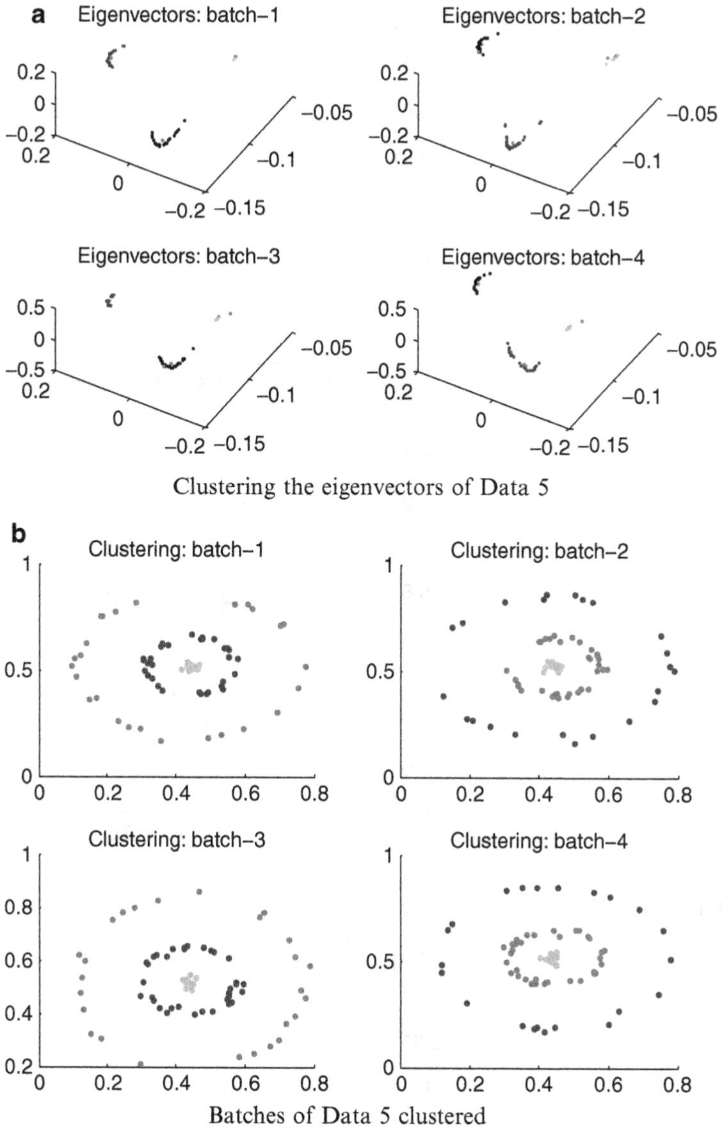

Fig. 4.10 Efficiency of the algorithm on Data 5 in the online setting

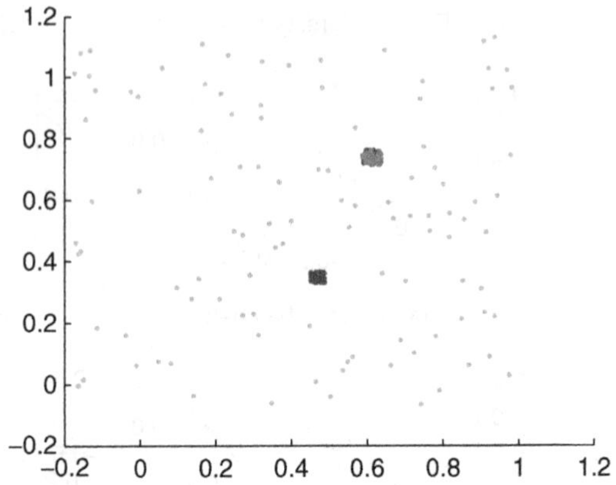

Fig. 4.11 Prediction of Data 1

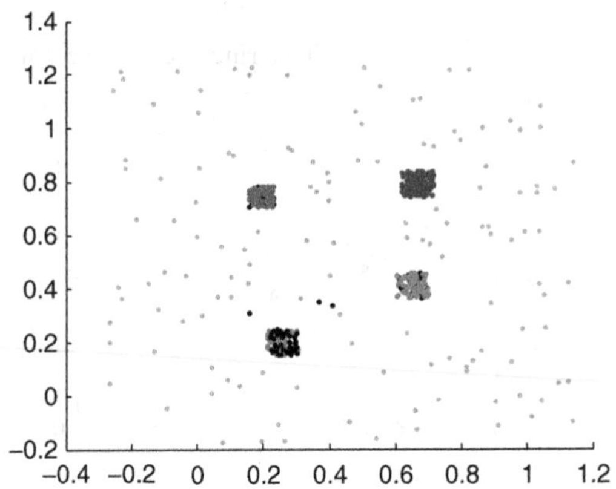

Fig. 4.12 Prediction of Data 2

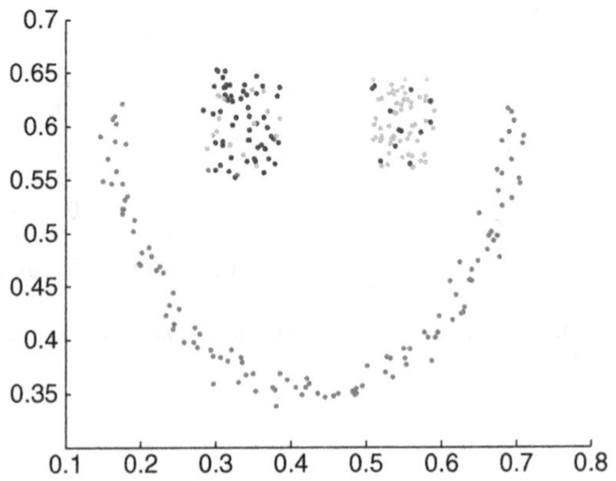

Fig. 4.13 Prediction of Data 3

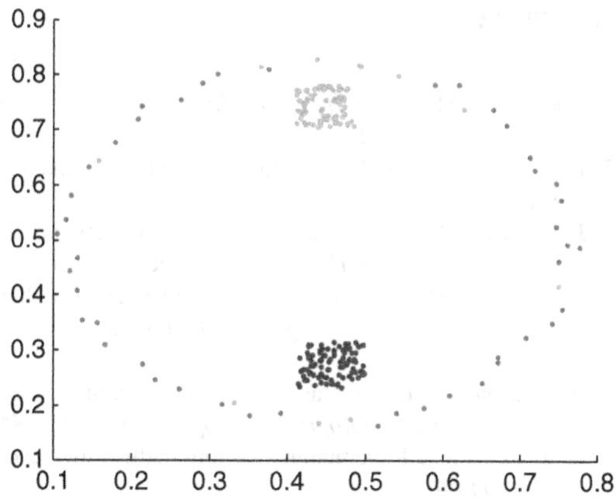

Fig. 4.14 Prediction of Data 4

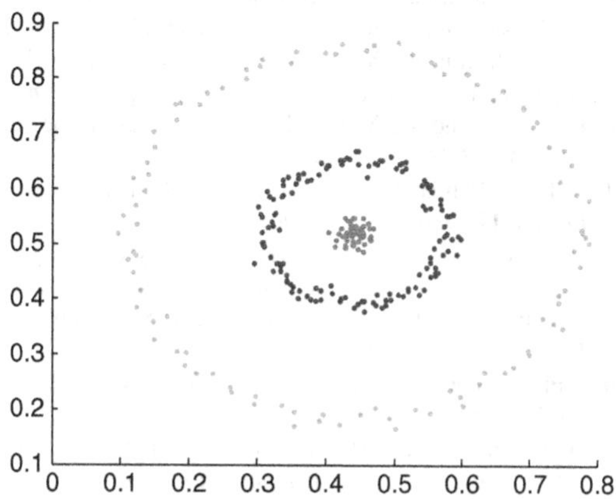

Fig. 4.15 Prediction of Data 5

4.6 Conclusion

The present contribution presents a novel incremental spectral clustering algorithm that is based on an incremental version of fuzzy c-varieties. The aim is to handle nonconvex data sets which are usually difficult to cluster. The evaluation of the proposed algorithm is however preliminary and many issues need to be further pursued.

While the algorithm works nicely in most of the cases presented, in some situations where data samples are loosely connected, it behaves incorrectly. This might, however, be the case of most clustering algorithms. Our future investigation will be concerned with the study of such difficult cases. Moreover, we intend to look into the salt and pepper problem more precisely at both levels: modeling the affinity matrix and the fuzzy c-varieties which are the central components of the ISC algorithm. It is also worth noting that the current version of the algorithm does not open new clusters online, but can handle single samples. These are further aspects that we will look into in the future.

References

1. F. Bach and M. Jordan. *Spectral Clustering for Speech Separation*, pages 221–250. John Wiley and Sons, Ltd. (2009)
2. J. Bezdek. Pattern Recognition with Fuzzy Objective Function Algorithms. *Plenum Press* (1981)
3. J. Bezdek and R. Hathaway. Numerical Convergence and Interpretation of the Fuzzy C-shells Clustering Algorithm. *IEEE Trans. on Neural Networks*, **3**, 787–793 (1992)
4. B. Bollobas. *Modern Graph Theroy*. Springer, New York, USA (1998)
5. A. Bouchachia. An evolving classification cascade with self-learning. *Evolving Systems*, **1**(3), 143–160 (2010)
6. A. Bouchachia. Fuzzy classification in dynamic environments. *Soft Computing—A Fusion of Foundations, Methodologies and Applications*, **15**, 1009–1022 (2011)
7. A. Bouchachia. Incremental learning with multi-level adaptation. *Neurocomputing*, **74**(11), 1785–1799 (2011)
8. A. Bouchachia and W. Pedrycz. Enhancement of fuzzy clustering by mechanisms of partial supervision. *Fuzzy Sets and Systems*, **157**(13), 1733–1759 (2006)
9. A. Bouchachia and M. Prossegger. A hybrid ensemble approach for the steiner tree problem in large graphs: a geographical application. *Applied Soft Computing* **11**(8), 5745–5754 (2011)
10. H. Chang and D. Yeung. Robust path-based spectral clustering. *Pattern Recognition*, **41**, 191–203 (2008)
11. R. Dave. Validating fuzzy partitions obtained through c-shells clustering. *Pattern Recognition Letters*, **17**(6), 613–623 (1996)
12. R. Dave and K. Bhaswan. Adaptive fuzzy c-shells clustering and detection of ellipses. *IEEE Transactions on Neural Networks*, **3**(5), 643–662 (1992)
13. E. Estrada and J. Rodríguez-Velázquez. Subgraph centrality in complex networks. *Phys. Rev. E*, **71**(5), 056103 (2005)
14. I. Fischer and I. Poland. New methods for spectral clustering. Technical Report IDSIA-12-04, IDSIA (2004)
15. H. Frigui and R. Krishnapuram. A comparison of fuzzy shell-clustering methods for the detection of ellipses. *IEEE Transactions on Fuzzy Systems*, **4**(2), 193–199 (1996)
16. I. Gath and D. Hoory. Fuzzy Clustering of Elliptic Ring-shaped Clusters. *Pattern Recognition Letters*, **16**(7), 727–741 (1995)
17. D. Gustafson and W. Kessel. Fuzzy Clustering with a Fuzzy Covariance Matrix. In *Proc. of the IEEE Conf. on Decision and Control*, pp. 761–766 (1979)
18. L. Hagen and A. Kahng. New spectral methods for ratio cut partitioning and clustering. *IEEE Trans. Computer-Aided Design*, **11**(9), 1074–1085 (1992)
19. F. Hoeppner. Fuzzy shell clustering algorithms in image processing: fuzzy c-rectangular and 2-rectangular shells. *IEEE Transactions on Fuzzy Systems*, **5**(4), 599–613 (1997)
20. R. Krishnapuram, H. Frigui, and O. Nasraoui. Fuzzy and possibilistic shell clustering algorithms and their application to boundary detection and surface approximation. i. *IEEE Transactions on Fuzzy Systems*, **3**(1), 29–43 (1995)
21. R. Krishnapuram, H. Frigui, and O. Nasraoui. Fuzzy and possibilistic shell clustering algorithms and their application to boundary detection and surface approximation. ii. *IEEE Transactions on Fuzzy Systems*, **3**(1), 44–60 (1995)
22. R. Krishnapuram, O. Nasraoui, and H. Frigui. The fuzzy c spherical shells algorithm: a new approach. *IEEE Transactions on Neural Networks*, **3**(5), 663–671 (1992)
23. J. Leski. Fuzzy c-varieties/elliptotypes clustering in reproducing kernel hilbert space. *Fuzzy Sets and Systems*, **141**(2), 259–280 (2004)
24. C. Lim, S. Bohacek, J. Hespanha, and K. Obraczka. Hierarchical max-flow routing. In *IEEE Conference on Global Telecommunications Conference*, pp. 550–556 (2005)
25. M. Meila and J. Shi. Learning segmentation by random walks. In *Neural Information Processing Systems, NIPS*, 873–879 (2001)

26. A. Ng, M. Jordan, and Y. Weiss. On spectral clustering: Analysis and an algorithm. In *Advances in Neural Information Processing Systems*, 849–856. MIT Press (2001)
27. H. Ning, W. Xu, Y. Chi, Y. Gong, and T. Huang. Incremental spectral clustering by efficiently updating the eigen-system. *Pattern Recognition*, **43**(1), 113–127 (2010)
28. D. Niu, J. Dy, and M. Jordan. Multiple non-redundant spectral clustering views. In *Proceedings of the 27th International Conference on Machine Learning*, pp. 831–838 (2010)
29. M. Prossegger and A. Bouchachia. Ant colony optimization for steiner tree problems. In *Proceedings of the 5th international conference on Soft computing as transdisciplinary science and technology*, pp. 331–336. ACM (2008)
30. T. Runkler and R. Palm. Identification of nonlinear systems using regular fuzzy c-elliptotype clustering. In *Proceedings of the Fifth IEEE International Conference on Fuzzy Systems*, volume 2, pp. 1026–1030 (1996)
31. J. Shi and J. Malik. Normalized cuts and image segmentation. *IEEE Trans. Pattern Anal. Mach. Intell.*, **22**(8), 888–905 (2000)
32. F. Tung, A. Wong, and D. Clausi. Enabling scalable spectral clustering for image segmentation. *Pattern Recogn.*, **43**, 4069–4076 (2010)
33. C. Valgren, T. Duckett, and L. Lilienthal. Incremental spectral clustering and its application to topological mapping. In *Proc. IEEE International Conference on Robotics and Automation*, pp. 4283–4288 (2007)
34. C. Valgren and A. Lilienthal. Incremental spectral clustering and seasons: Appearance-based localization in outdoor environments. In *Robotics and Automation, 2008. ICRA 2008. IEEE International Conference on*, pp. 1856 –1861 (2008)
35. L. Wang, C. Leckie, R. Kotagiri, and J. Bezdek. Approximate pairwise clustering for large data sets via sampling plus extension. *Pattern Recognition*, **44**, 222–235 (2011)
36. D. Yan, L. Huang, and M. Jordan. Fast approximate spectral clustering. In *Proceedings of the 15th ACM SIGKDD international conference on Knowledge discovery and data mining*, pp. 907–916. ACM (2009)

Part II
Dynamic Methods for Supervised Classification Problems

Chapter 5
Semisupervised Dynamic Fuzzy *K*-Nearest Neighbors

Laurent Hartert and Moamar Sayed-Mouchaweh

Abstract This chapter presents a semi-supervised dynamic classification method to deal with the problem of diagnosis of industrial evolving systems. Indeed, when a functioning mode evolves, the system characteristics change and the observations, i.e. the patterns representing observations in the feature space, obtained on the system change too. Thus, each class membership function must be adapted to take into account these temporal changes and to keep representative patterns only. This requires an adaptive method with a mechanism for adjusting its parameters over time. The developed approach is named Semi-Supervised Dynamic Fuzzy *K*-Nearest Neighbors (SS-DFKNN) and comprises three phases: a detection phase to detect and confirm classes evolutions, an adaptation phase realized incrementally to update the evolved classes parameters and to create new classes if necessary and a validation phase to keep useful classes only. To illustrate this approach, the diagnosis of a welding system is realized to detect the weldings quality (good or bad), based on acoustic noises issued of weldings operations.

5.1 Introduction

Evolving systems are functioning in a dynamic environment. With the occurrence of new events, evolving systems change, and their corresponding classes and patterns characteristics evolved in the feature space. Indeed, the functioning mode of an

L. Hartert (✉)
University of Reims Champagne-Ardenne, CReSTIC, Moulin de la Housse, BP 1039, 51687 Reims Cedex, France
e-mail: laurent.hartert@univ-reims.fr

M. Sayed-Mouchaweh
Ecole des Mines de Douai, Computer Science and Automatic Control Lab, EMDouai-IA, F-59500 Douai, France
e-mail: moamar.sayed-mouchaweh@mines-douai.fr

M. Sayed-Mouchaweh and E. Lughofer (eds.), *Learning in Non-Stationary Environments: Methods and Applications*, DOI 10.1007/978-1-4419-8020-5_5,

© Springer Science+Business Media New York 2012

evolving system can evolve from normal to faulty in response to the occurrence of a fault, such as a leak, to the wear of a tool or to a bad setting. To realize the diagnosis of these systems, Pattern Recognition (PR) methods need to adjust their parameters by doing automatic corrections or by warning an operator that will adjust himself the classifier parameters. When a self-adaptation of the classifier parameters is wanted, the method has to monitor the evolution of a system over time. In this case, dynamic learning is necessary to update the feature space characteristics. Then, the PR method has to use the informative patterns only to adjust the class structure. In the literature, several PR methods [9, 10, 17] are used to monitor the functioning modes evolutions of dynamic systems, to realize the fault diagnosis of complex systems or to accomplish the fault prognosis. Indeed, these methods are particularly adapted when the prior knowledge about the system behavior is not sufficient to construct an analytical model of the process.

5.1.1 Pattern Recognition

PR methods use exclusively a set of measurements, i.e., quantitative observations, about process operating modes to build a mapping from the observation space into a decision space, called the feature space. In PR, historical patterns or observations about system functioning modes are divided into groups of similar patterns, called classes. Each class is associated to a functioning mode (normal or faulty). Classes and patterns are represented by a set of d attributes, so they can be viewed as d-dimensional vectors, or points, in the feature space. The PR principle consists in classifying the new patterns by using a classifier. According to the a priori information available on the system, three types of PR methods can be used: supervised PR methods, unsupervised PR methods and semi-supervised PR methods. When labeled patterns, i.e., patterns with their class assignment, can be obtained the PR is supervised [28]. These methods use the known labeled patterns, i.e., the learning set, to build a classifier that best separates the different known classes in order to minimize the misclassification error. The model of each class can be represented by a membership function which determines the membership value of a pattern to a class. On the contrary when no information is available on the classes of a system, PR is unsupervised [6, 11, 12, 29]. The unsupervised PR methods, or clustering methods, are based on similarity functions, so that when patterns with the same characteristics occur they are classified in the same class, and when patterns with different characteristics occur a new class is created to classify them. Once the classifier has learned the classes membership functions, new incoming patterns are assigned to the class for which they have the maximum membership value. The third type of PR methods, the semisupervised one [8, 13] uses the supervised information, the known labeled patterns and classes, to estimate the classes characteristics and the unsupervised learning is used to detect new classes and to learn their membership functions.

5.1.2 Evolving Systems

In the case of evolving systems [2, 3, 18, 22, 23], classes are dynamic and their characteristics change in the course of time. Classes can evolve slowly or abruptly to a new position in the feature space, according to the system parameters which evolve over time. Thus, each class membership function must be adapted to take into account these temporal changes. This requires an adaptive classifier with a mechanism for adjusting its parameters over time. Hence, some of the new incoming patterns reinforce and confirm the information contained in the previous ones, but the other ones can bring new information (creation, drift, fusion, splitting of classes, etc.). This new information can concern a change in operating conditions, the development of a fault or simply more significant changes in the system's dynamic. Angstenberger [4] and Nakhaeizadeh et al. [25] act on the classifier parameters, by substituting or adding some recent and representatives patterns to the learning set according to the state (stable, warning, action) in which the system is. This adaptation is based only on the most recent batch of patterns selected by a time window [25] or by an estimation of the patterns usefulness [15]. Other approaches providing a global model rather than a local model on demand are based on the use of evolving neural networks [1, 4, 7]. In [2], a potential function based on the distance between data points is defined for the new points. According to the potential obtained for new data points, the point can reinforce or confirm the information contained in the previous ones, or a new rule can be added. In [1], the neural network is based on a multi-prototype Gaussian modeling of nonconvex classes. The activation function of each hidden neuron determines the membership degree of an observation to a prototype of a class. According to the membership degree of new acquisitions, the prototype, i.e., the hidden neuron can be adapted, deleted or a new prototype can be created. Data analysis can be realized on data coming from evolving systems in order to obtain the most informative parameters of a system that will be necessary to discriminate classes using a PR method. In this chapter, we use the statistical characteristics to supply spatial information like the number of peaks present in a signal, the standard deviation value, the root mean square value, the maximum value, the kurtosis value, etc. Some information can be computed on different parts of each signal or on entire signals. The set of characteristics, i.e. parameters, found by these methods represents the attributes which permit to characterize each signal obtained on a system. Using these informative parameters, signals are transformed into patterns in the feature space. If the parameters are well determined, classes are well discriminated and they are represented in different regions of the feature space.

5.1.3 Dynamic Learning and Classification

In this chapter, a semi-supervised dynamic method based on Fuzzy K-Nearest Neighbors (FKNN) [19] is developed. It was interesting to develop this method for the case of evolving systems since FKNN is a simple but efficient well known classification method. However, FKNN becomes inefficient when the size of the learning set is too important or when k is not well chosen. k is generally determined by experimentation, but it is still a parameter difficult to determine. A criterion often used is [9], where N is the number of patterns in the learning set. Several other versions of KNN exist in the literature (KNN with prototype, Adaptive KNN [26], etc.). In [21], a version of KNN pre-assigns a class to several subregions of the feature space in order to classify more rapidly the new patterns. In [30], a hierarchical research algorithm is developed to find the k-nearest neighbors using a nonmetric measure in a binary feature space. This measure is a similarity measure computed between the binary values representing the patterns. In [14, 20], respectively high dimensional and k-dimensional trees are used to find the most interesting parts of the feature space where to find the k-nearest neighbors. Only some branches of the trees have to be browsed to find the k neighbors, but the trees branches can be fast unbalanced. Another version of FKNN, called Instance-Based Learning on Data Streams (IBL-DS) [5], detects changes in the data streams by using a prediction error and the standard deviation of the 100 first patterns. If a change is detected, the 20 latest classified patterns are used to estimate the evolution realized. Based on the used indicators, a percentage of patterns initially defined is deleted from the reference base according to their spatial location and to their temporal behavior. Song et al. [27] uses two informative measures to find patterns susceptible to be the k most informative neighbors. These measures are based on probability measures calculated locally and globally. Another version of KNN [24] uses kernel-based dimensionality reduction methods to improve the classification results. These methods are used to solve some challenging problems like the application of [4] which concerns the credit scoring. The authors aims to decide whether a new customer is a good or a bad risk according to changes in his consumption. Guedalia et al. [16] deals with the problem of classification of the quality of fruits according to the damage resulting from bad weather or other external events. In [1], the authors aim to detect and to follow up the progressive evolution of the functioning modes of a thermal regulator due to the age of its components or to other temporal factors in its environment. Cohen et al. [7] treats dynamic traffic data streams in order to reduce the waiting time of drivers at the road intersections.

In this chapter, we have chosen to develop the FKNN approach since it is well known and often used in machine learning (ML) applications. The developed Semisupervised Dynamic Fuzzy K-Nearest Neighbors method is semi-supervised in order to consider the known information of a system, even when only a few observations are available, and in order to detect unknown classes and to estimate their characteristics. Semisupervised methods are particularly well adapted to evolving systems for which all classes can not be known in advance. The developed method is presented to realize the monitoring of evolving systems. The method is applied on a real industrial system in order to detect the weldings quality and to monitor their progressive evolutions. The chapter is organized as follows. In Sect. 5.2, the functioning of the proposed approach is detailed and illustrated. Then, in Sect. 5.3, the approach is applied and evaluated using the application. Finally, conclusions and perspectives end this chapter.

5.2 Semisupervised Dynamic Fuzzy *K*-Nearest Neighbors (SS-DFKNN)

The selection of a PR method has to be realized according to the system on which the method is applied. Indeed, according to the application several parameters change as the number of patterns available for the learning set, the number of classes, the system dynamic, the number of dimensions, i.e., attributes, of the feature space, etc. In this section, we develop the PR method Fuzzy *K*-Nearest Neighbors (FKNN) in order to detect classes evolutions and to adapt these latter according to the dynamic of their evolutions. The proposed version is semisupervised, in order to:

- take into account an initial learning set X representative of the known information of a system.
- improve the classes characteristics estimation by using the new patterns
- detect new classes or subclasses according to the evolutions of the system characteristics

In this chapter, the SemiSupervised Dynamic Fuzzy KNN (SS-DFKNN) developed method permits to consider patterns evolutions even in the area of the feature space where no pattern was learned. The objectives of this approach are to follow classes evolutions by taking into account the patterns usefulness, and well estimate the new functioning modes of a system according to the estimated adapted classes characteristics. SS-DFKNN is composed of several phases which are presented in the following parts, and the method is illustrated with an example.

5.2.1 *Learning and Classification Phases*

In the learning phase of SS-DFKNN, all labeled patterns and classes are learned. The learning set X must contain a minimum of two patterns in order to calculate the initial center of gravity and standard deviation of each class, which are used in the indicators of evolution computed by SS-DFKNN. These values are calculated as follows:

- the current center of gravity $CG_{A_{\text{curr}}}$ of each class C according to each attribute A.
- the initial standard deviation $\sigma_{A_{\text{init}}}$ of each class C according to each attribute A.

These values permit to consider the dispersion of a class and its drift in the feature space. The center of gravity and the standard deviation values can be calculated for all types of classes. However, in the case of complex classes, we consider that these latter can be estimated using Gaussian subclasses. In the classification phase of SS-DFKNN, each new pattern is classified sequentially according to the class of its k-nearest neighbors. So, as for FKNN, the parameter k has to be defined initially. Once a new pattern is classified in one of the known classes, the detection of classes evolutions can be realized based on two indicators.

5.2.2 *Detection of a Class Evolution*

The classification of a new pattern x in one of the known classes determines the class, which can be evolving. Indeed, after the classification of x in the class C, only the class C has to be updated. In this phase of SS-DFKNN, the detection phase, the new characteristics of the class C are calculated to detect a class evolution. The current value of the standard deviation $\sigma_{A_{\text{curr}}}$ and of the current center of gravity $CG_{A_{\text{curr}}}$ of C are incrementally updated by:

$$\sigma_{A_{\text{curr}}} = \sqrt{\frac{N_C - 1}{N_C} \times \sigma^2_{A_{\text{curr}-1}} + \frac{\left(x - CG_{A_{\text{curr}-1}}\right)^2}{N_C + 1}}, \tag{5.1}$$

$$CG_{A_{\text{curr}}} = \frac{CG_{A_{\text{curr}-1}} \times N_C}{N_C + 1} + \frac{x}{N_C + 1}, \tag{5.2}$$

where N_C is the number of patterns in C before the classification of x. $\sigma_{A_{\text{curr}-1}}$ and $CG_{A_{\text{curr}-1}}$ are respectively the variance and the center of gravity of the class, according to the attribute A, before the classification of x. Based on the computed values of $CG_{A_{\text{curr}}}$, $\sigma_{A_{\text{curr}}}$, and $\sigma_{A_{\text{init}}}$ two drift indicators are used to monitor the temporal changes of a system.

- the first indicator i_{1A} represents the change of compactness of the class for each attribute A of the feature space:

$$i_{1A} = \frac{\sigma_{A_{\text{curr}}} \times 100}{\sigma_{A_{\text{init}}}} - 100. \tag{5.3}$$

  i_{1A} is given in percentage. If at least one attribute A has a value of i_{1A} greater than a threshold th1 then the class C has begun to change its characteristics according to this attribute. th_1 can be fixed to a small value when it is interesting to follow small evolutions of the class. For example, fixed to 5, it represents an evolution of 5% of the class characteristics. On the contrary when only important evolutions have to be detected, a greater value of th_1 can be necessary.

- the second indicator i_{2A} represents the distance between x_A and $CG_{A_{\text{curr}}}$ according to the current standard deviation $\sigma_{A_{\text{curr}}}$ for each attribute A of the feature space:

$$i_{2A} = \frac{|(x_A - CG_{A_{\text{curr}}})| \times 100}{\sigma_{A_{\text{curr}}}} - 100. \tag{5.4}$$

  i_{2A} is given in percentage. If at least one attribute A has a value of i_{2A} greater than th_1, then the point is not situated in the same area of the feature space than the other patterns of the class C.

However, one single pattern can involve changes for the center of gravity and for the standard deviation of a class. In some cases, this pattern can be a noise instead of a class evolution so a minimum number of successive evolved patterns $NbMin$ has to be detected in order to confirm the evolution. If $NbMin$ is fixed to a high number, then the delay detection of the class evolution can be important. This number has to be defined according to a ratio between the noise present in the patterns and the delay detection of a class evolution. The class evolution is confirmed when $NbMin$ successive values of the two indicators i_{1A} and i_{2A} are greater than th_1. The adaptation phase which permits to adapt classes based on their evolution, is explained in the next section.

5.2.3 *Adaptation of an Evolving Class after Validation of its Evolution*

SS-DFKNN integrates a mechanism to adjust the evolved class parameters in the adaptation phase, when serious changes in a class' characteristics are detected during the detection phase. When a class evolution is confirmed, a new class or subclass is created based on useful patterns only. This adaptation is realized in several parts:

- a new class or subclass C' is created and the most representative patterns of the evolution are selected. Since the last classified pattern x of the class represents one of the evolved patterns of the class, x is selected. x also represents the most recent change in the class evolution. The other informative selected patterns are the $k-1$ nearest neighbors of x. No distance has to be calculated to find these patterns since they were already determined during the classification of x. Indeed, to classify x, the classifier has computed several distances between the known patterns to find its nearest neighbors. These k selected patterns represent the only new patterns of a new class C'. Indeed, only k patterns are selected to represent an evolved class since the parameter k corresponds to the number of patterns judged as sufficiently representative to classify a new pattern.
- the k selected patterns are deleted from the class C.

- the new center of gravity $CG_{A_{\mathrm{curr}}}$ of the class C is calculated and the current standard deviation $\mathrm{Std}_{A_{\mathrm{curr}}}$ of the class is computed.
- $CG_{A_{\mathrm{curr}}}$ and $\sigma_{A_{\mathrm{init}}}$ are computed for the class C'. These values are computed rapidly since the number of patterns in the evolved class is equal to k.
- the number of classes is updated.

This adaptation permits to online follow the evolution of classes with a constant and low adaptation time. Then, new patterns are classified in their corresponding class. Using this approach, all patterns and classes are kept in the feature space and an evolving class C generates at least one new class or subclass C'. If C is considered as useless, i.e., not anymore representative of a class, it can be interesting to delete this class C in order to avoid the problem of growing size of the data set. This approach permits to update and reinforce the known classes using new patterns. This is for classes for which no evolution has occurred but they are still informative and useful classes. The approach also permits to create new classes when an evolution of the system characteristics occurs. The solution presented in this chapter to deal with useless classes is presented in the next section.

5.2.4 Validation of the Existing Classes

The noise is taken into account by SS-DFKNN since a sufficient number $NbMin$ of evolving patterns is needed to consider a class evolution. However, in some cases, the noise or other events can lead to delete useless classes:

- when a short time living class is created based on few patterns, it represents only a transitory functioning mode. This temporary functioning mode can appear during the evolution of a system characteristics, changing this latter from a normal functioning mode to an abnormal functioning mode. Transitory classes are not representatives of any system functioning mode, so they can be deleted.
- when a class considered as noisy is created.
- when a class containing very few information is kept.

SS-DFKNN deletes classes corresponding to these cases, when:

- an insufficient number n_1 of patterns is contained in the class ($n_1 > k$),
- and when no pattern has been classified in the class while a sufficient number n_2 of patterns has been classified in the others classes.

However, it is not an obligation to define these two parameters if the suppression of classes is not necessary. For example, for the application considering critical data, it can be better to keep all characteristics patterns of all classes. Sometimes, classes do not need to be deleted but to be merged. Indeed, according to classes which can be created and to the evolutions of classes, it is necessary to measure over time the overlapping of classes. If several classes are created and if they drift or grow toward a common direction, then these classes have to be merged. To decide if two classes have become sufficiently close to be merged, a similarity measure has been used [11]. This measure considers the overlapping or the closeness between classes based on the membership values of the classified patterns.

$$\delta_{iz} = 1 - \frac{\sum_{x \in C_i \vee x \in C_z} |\pi_i(x) - \pi_z(x)|}{\sum_{x \in C_i} \pi_i(x) + \sum_{x \in C_z} \pi_z(x)}, \tag{5.5}$$

where $\pi_i(x)$ and $\pi_z(x)$ are respectively the membership values of x according to C_i and C_z. δ_{iz} is the similarity measure between two classes. More the similarity value is close to 1, more the two classes are similar and have to be merged. The maximal value represents two classes completely overlapped so it is not needed to wait until the similarity value is equal to 1 to merge two classes. After each new classifier pattern, this measure is calculated, if it is greater than a threshold $rmth_{\text{Fusion}}$ between two classes then they must be merged.

5.2.5 SS-DFKNN Algorithm

In Fig. 5.1, the algorithm describing all parts of SS-DFKNN is presented.

5.2.6 Hints for the Definition of SS-DFKNN Parameters

SS-DFKNN needs several parameters which can be defined according to each application characteristics. The defined parameters influence the classifier performances, however we can propose some default values which are generally adapted to dynamic systems:

- k corresponds to the number of neighbors considered by the k-NN methods to realize the classification of a pattern. It is the most common parameter of the k-NN methods. It should be defined according to the size of the data set, to the noise of a system and to the closeness between classes.
- th_1 is one of the most important parameter of the method. It permits to detect the evolution of a class. A class which does not evolve will have almost always the same characteristics, even if noise occurs. So, if an evolution is realized, abruptly or gradually, its characteristics will change. To allow small changes of class without waiting for an important evolution, a value equal of th_1 equal to 5 is a good compromise.
- $NbMin$ permits to validate an evolution. It should be defined at least equal to $k, (k \geq 1)$ in order to wait for a sufficient number of representative patterns permitting to well estimate the characteristics of a new class. Moreover it must not be too high to delay the evolution detection. $NbMin$ should be defined

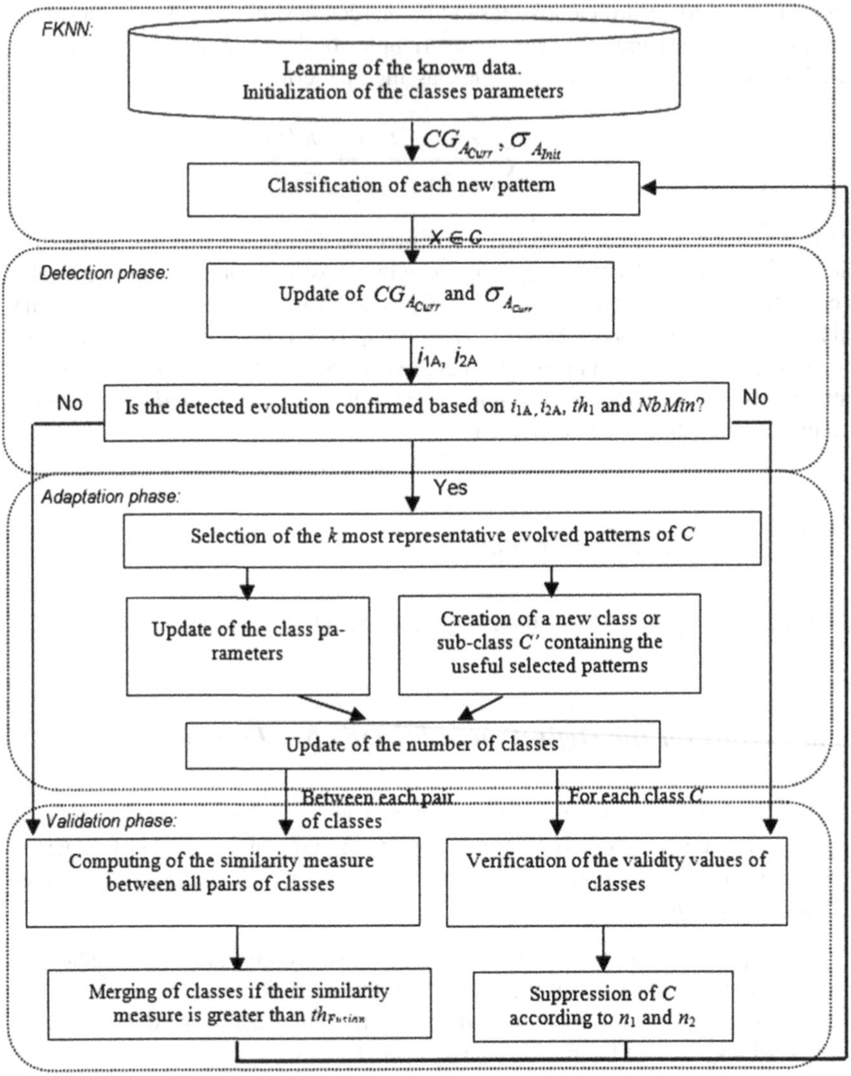

Fig. 5.1 Description of the semisupervised dynamic fuzzy k-NN method

between k and $k+5$, respectively if k is high or small. If k and $NbMin$ are small, the risk to obtain false alarms becomes bigger.

- th_{Fusion} is an optimization parameter. Indeed, even if no fusion occurs, the simple occurrence of a class means an evolution of the system has been realized. In that case, an alarm should be raised on the system to call a human operator which will verify the system state. A th_{Fusion} value between 0.05 and 0.2 permits to merge classes which begin to have the same characteristics.

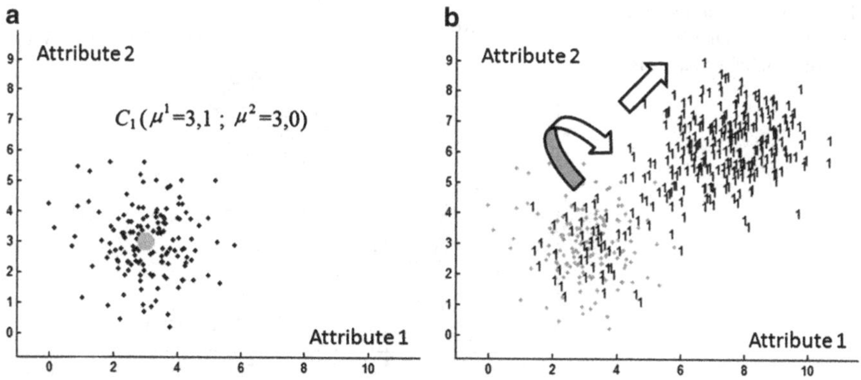

Fig. 5.2 (**a**) Learning set; (**b**) class evolution

- n_1 should be defined greater than $k, (n_1 > k)$ since a class will contain at least k patterns (at its creation). A default value of n_1 should be $k * 2$.
- n_2 should not be defined too small since after the creation of a class, it can be necessary to wait in order to classify more patterns in the created class. On the contrary, if no new pattern is classified in a new class after a large number of classified patterns, then the class is not useful. It is probably a noisy class or an ephemeral problem has occurred on the system. Then, the value of n_2 should be defined around 20. It means than 1 pattern on 20 should be classified in a new class, in order to confirm progressively its usefulness. For the others classes, even if they received no more patterns for a long time they will not be deleted since they have already confirmed their usefulness by having a sufficient number of patterns.

5.2.7 *Illustrative Example*

This example presents the dynamic evolution of a class. A progressive drift is generated according to the following equations:

- $t = 0$: One hundred and fifty patterns are used as a learning set. Only the initial class is known. The values of mean and standard deviation of the class are for the attribute 1, $\mu^1 = 3$ and $\sigma^1 = 1$, and for the attribute 2, $\mu^2 = 3$ and $\sigma^2 = 1$ (Fig. 5.2a).
- $t = 1$–50: Fifty new patterns appear with the same characteristics than the initial ones; so, there is still no evolution or drift.
- $t = 51$–200: A sudden change appears in the mean values of the class according to each attribute j, $j \in \{1,2\}$. This change is followed by a progressive drift of the class mean according to each attribute (Fig. 5.2b):

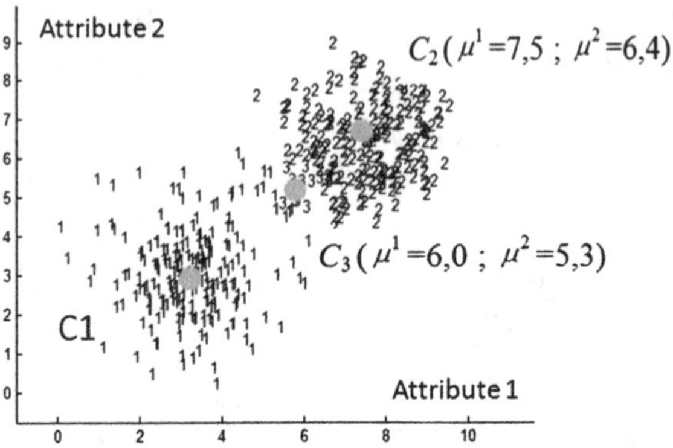

Fig. 5.3 Classification result obtained by SS-DFKNN after classification of all patterns

$$\mu^{1'}(t) = \mu^1 + 2 + \frac{4\times(t-50)}{150}, \tag{5.6}$$

$$\mu^{2'}(t) = \mu^2 + 2 + \frac{2\times(t-50)}{150}, \tag{5.7}$$

whenever $51 \le t \le 200$.

- $t = 201$–300: One hundred new patterns appear. They have the same characteristics than the ones of the final class.

During the classification of the evolving patterns, several classes have been created. Then, some of them have been merged and others have been deleted. The final classification result obtained by SS-DFKNN is presented in Fig. 5.3. The method has finally obtained 3 classes: one corresponds to the initial class, one corresponds to the final location of the class, and one corresponds to a transition class which could have been deleted. Then, the method has succeeded in detecting the class evolution. The initial class $C1$ has kept its characteristics and the class $C2$ well corresponds to the expected class. The classification results of SS-DFKNN were obtained using the following parameters ($k = 5$; $\text{th}_1 = 5$; $NbMin = 5$; $\text{th}_{\text{Fusion}} = 0.2$; $n_1 = 10$; $n_2 = 20$) and a delay detection of 4 patterns has occurred in order to detect the class evolution. The maximum classification time obtained was equal to 5×10^{-2} s and the mean classification time was equal to 5×10^{-3} s. In the next part, SS-DFKNN is applied to a welding system in order to realize the diagnosis of the system and to follow the classes evolutions.

5.3 Application Results

In this section, we use SS-DFKNN to deal with the problem of weldings quality monitoring on an industrial welding system (Fig. 5.4) used by the company Turquais (Raucourt-et-Flaba, France).

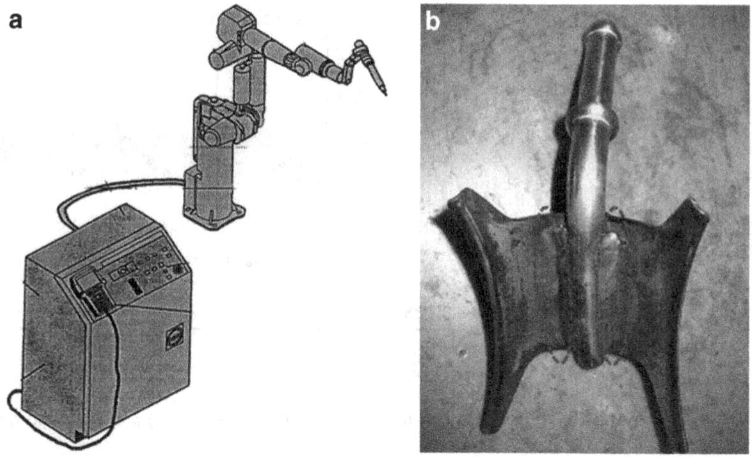

Fig. 5.4 The monitored welding system with its control system (**a**), and the two weldings realized between two metal pieces (**b**)

5.3.1 Application and Acquisition of Acoustic Noises

The welding system is able to realize the weldings of different types of metals in few seconds in order to obtain several welded pieces in a row. In this chapter, we monitor the weldings quality obtained between two metal pieces (Fig. 5.4b). The interest to monitor this system is to online detect all bad welded pieces in order to correct as soon as possible the system parameters or to change one of its welding tools. The proposed SS-DFKNN method has to detect every change of welding quality and it has to warn the human operator if a welded piece is considered as bad quality. The approach is based on the analysis, on the interpretation and on the classification of the acoustic signals issued of the weldings between two metal pieces. Currently, the human expert operator in charge of the welding machine detects weldings qualities according to the welding noise he hears. Based on this observation, we have installed an acquisition system using a microphone which is sensitive for the audible sound range that a human ear can hear. This sound represents the noises issued of the welding operation. The microphone is placed near the welding system, there is approximately 50 cm between the microphone and the metal pieces being welded. This permit to obtain more accurate sounds and to reduce significantly the welding system environment noises. The sampling frequency was fixed initially to 15 KHz for the set of measures. This frequency has been fixed in order to contain all sounds that a human can hear and to respect the Shannon's law which imposes a sampling frequency at least twice higher than the frequency of the event to study. A signal is obtained for each welding realized by the system. Two examples of noisy weldings obtained on the system are presented in Fig. 5.5. In Fig. 5.5a, a good quality welding is presented, its shape is almost constant and even if a lot of noises is present, no discontinuity is observed. On the contrary, in

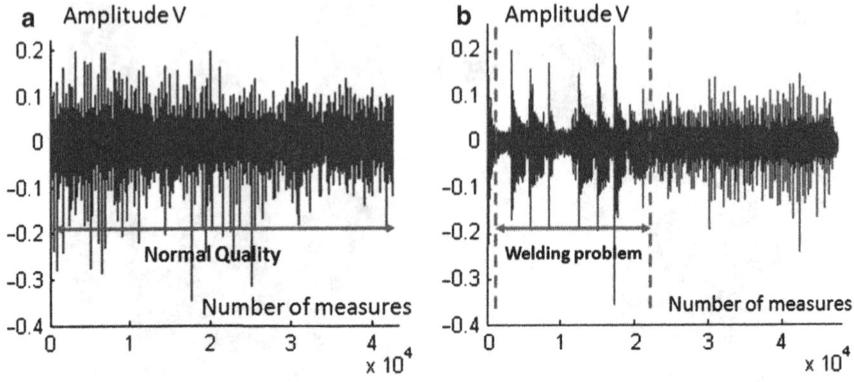

Fig. 5.5 Examples of noisy signals obtained for a good quality welding (**a**) and for a bad quality one (**b**)

Fig. 5.5b a bad quality welding is presented. The quality of this welding is initially bad, and then the welding becomes good. So, the evolution of welding quality can be distinguished by observing changes in some characteristics of the emitted acoustic signals. The quality of a welding can evolve so quickly that even when only a part of a welding is a bad quality, the global welding quality is considered as a bad quality. The acquisition of multiple acoustic signals was realized during the functioning of the welding system in order to construct a learning data set and a test set of the good and bad welded pieces. In the next part, the data analysis of these signals is realized to find informative parameters which can be used to discriminate the weldings qualities.

5.3.2 Signal Analysis and Feature Space

The ratio signal to noise is poor on signals issue of this industrial system in the Turquais company. To be able to select the interesting frequencies of signals, we have begun to search the main informative frequencies used during the realization of a welding. To do this, we have calculated the Energy Spectral Density (ESD) of each signal Fig. 5.6.

In Fig. 5.6, we can see some frequencies which are particularly present, for example the ones from 2,000 Hz to 4,000 Hz and from 6,000 Hz to 7,000 Hz. From a global point of view, the set of informative frequencies seems to be situated below 7,000 Hz. In order to follow the evolution of each welding over time, we have used a sliding window. It is important that this window contains enough observations in order to obtain representative patterns. We have studied experimentally different sizes, containing between 20 and 500 patterns. A window too small did not permit to characterize the functioning modes since it did not contain enough observations,

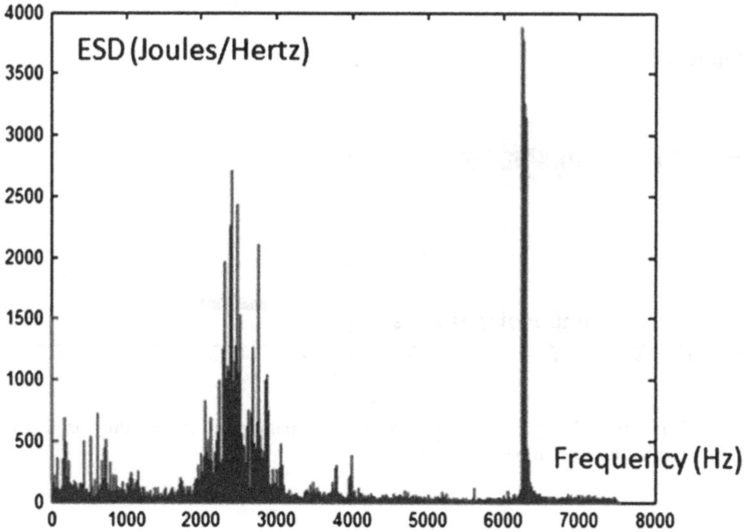

Fig. 5.6 Energy spectral density of an acoustic welding signal

while a too large window creates a delay to detect evolutions and the classification result was lower. This is a time window including 200 observations which has been selected. A window with this size was sufficiently informative and did not generate a delay in the computing of the parameters of the feature space. The window shifts with 200 new patterns. For each one of these windows, we have calculated the energy spectral density and several statistical parameters (mean, maximum, RMS, Kurtosis, dissymmetry coefficient, standard deviation, etc.). We have selected the statistical parameters which permitted to discriminate classes of good quality weldings from the ones of bad quality weldings. Two parameters were kept to establish the feature space:

- the value of dissymmetry coefficient (skewness), noted p_1, calculated for the first derivative of each time window.
- the RMS value of the spectral density for the frequencies between 6,000 Hz and 7,000 Hz, noted p_2, calculated for each time window. Parameters were only selected for these frequencies they were the most discriminative frequencies to characterize the welding quality.

Each observation window corresponds to a pattern in the feature space. In Fig. 5.7, a good quality welding is represented with its corresponding patterns in the feature space.

In Fig. 5.8a, welding signal of bad quality is represented. On that figure, the beginning of the window of each pattern corresponding to a bad quality welding is represented by "∗." Patterns corresponding to this signal are presented in Fig. 5.8b. In Fig. 5.8, we can see two classes which should be estimated by the classifier.

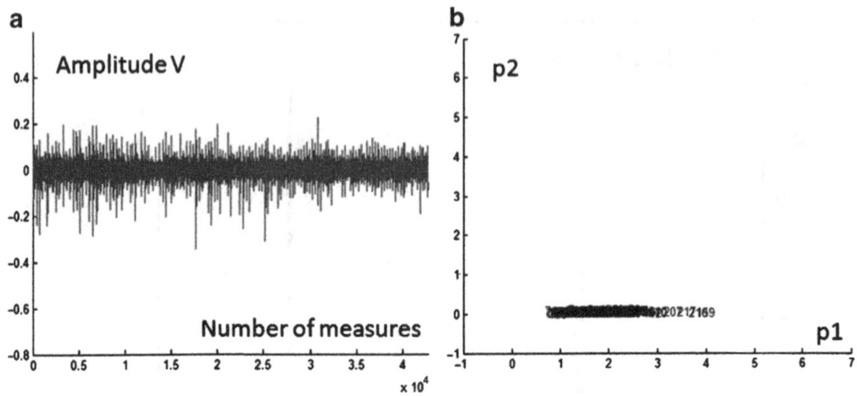

Fig. 5.7 Good quality welding (**a**) and its corresponding patterns in the feature space (**b**). Each pattern is represented by its number of window

We can also see that some patterns realize a transition between the windows corresponding to good and bad weldings qualities. For example, the pattern of window 18 leads the system toward the faulty class, while pattern 26 brings back the system toward the normal functioning mode. Then, when a bad quality appears, several round trips occur between the two classes. The functioning mode evolves according to the temporal and frequential characteristics of the system. In order to show with more precision the round trip of patterns, a zoom is realized on a part of the signal and its corresponding patterns are presented in Fig. 5.9.

5.3.3 Classification Results

In order to be in the same position that a human operator that will use our method, we only consider a single class as known; the class C_1 that contains patterns which correspond to the windows of good quality welding. For the classification of all acquired weldings, we have used a learning set such as the one of Fig. 5.10. From this learning set, we have realized the classification of each welding, i.e., the classification of each signal acquired on the system, one after the other. After the classification of a first welding of bad quality, SS-DFKNN ($k = 5$; $\text{th}_1 = 5$; $NbMin = 5$; $\text{th}_{\text{Fusion}} = 0.1$; $n_1 = 10$; $n_2 = 400$) permits to obtain classes of the Fig. 5.11. After the classification of this first welding, two classes have been estimated by the method. The evolution of the class has been validated at $t = 299$ while the evolution has really started at $t = 295$. A delay detection of $NbMin = 5$ windows has occurred. This delay corresponds to the patterns which can be classified in a transition class. It permits to confirm the evolution of the class C_1 with a small delay while avoiding some false alarms which can occur with the noise present in this system. Then the others weldings of bad quality, coming from the others acquired signals, have

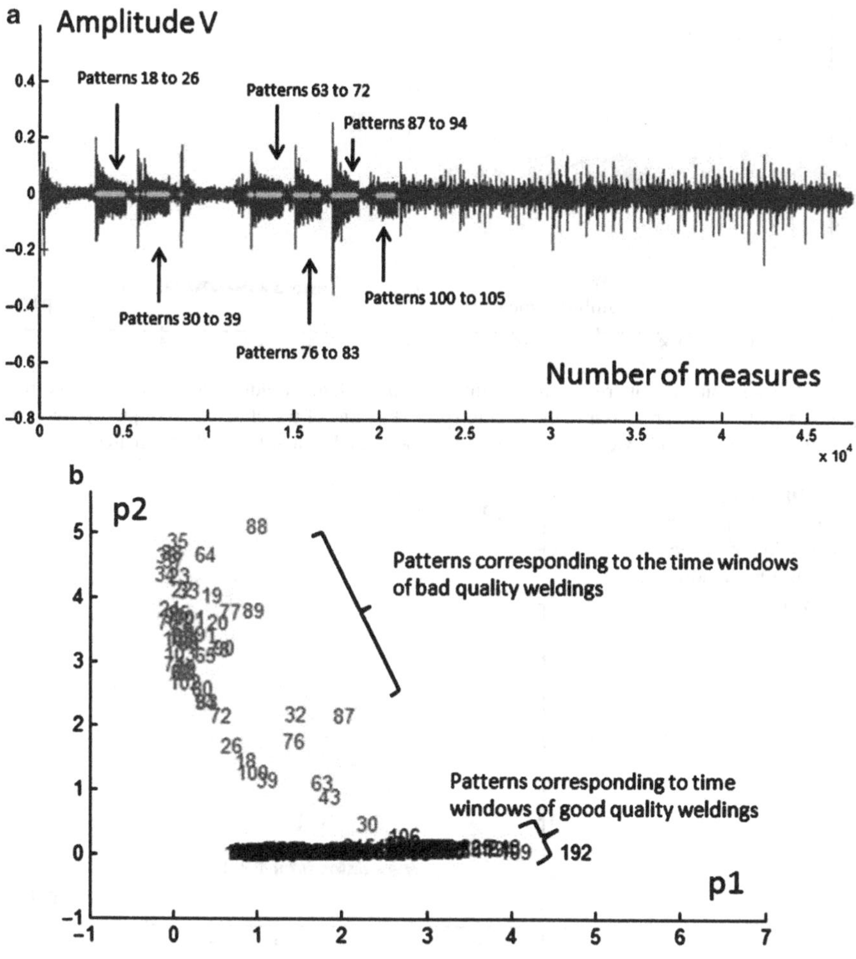

Fig. 5.8 Bad quality welding (**a**) and its corresponding patterns in the feature space (**b**)

also been classified. The classification result of all these weldings is presented in Fig. 5.12. A new class C_3 has been created, it corresponds to a transitive area of average welding quality present between the good welding quality class and the bad quality class. Only patterns which had a sufficient change in their characteristics are classified in class C_2. Then, no false alarm was raised during classification of these patterns. The set of others weldings of good quality has then be classified. All weldings of good quality are classified in C_1 (Fig. 5.13).

After the classification of all acquired patterns, some conclusions can be presented:

- 100% of the good quality weldings are classified in C_1.
- 100% of the bad quality weldings are detected.

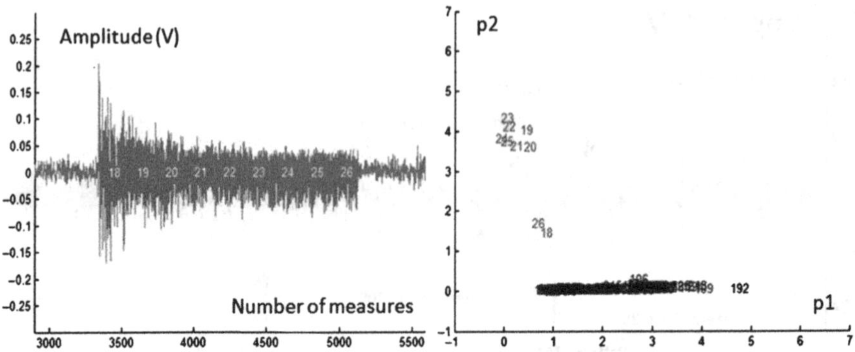

Fig. 5.9 Zoom on the patterns corresponding to a part of the welding problem and its corresponding patterns are presented in the feature space. Patterns of the class of good quality are also presented to have a global view of the two classes and to understand the evolution realized

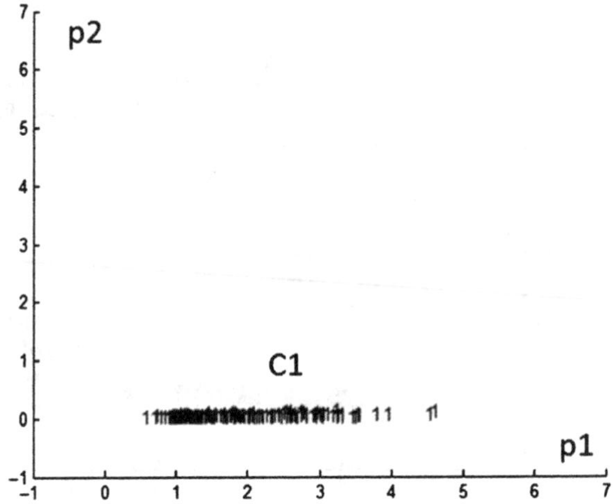

Fig. 5.10 Example of learning set used for the classification of the new weldings. One hundred and seventy five patterns are part of C_1

- misclassified patterns (0.2% of all patterns) correspond to transitive patterns. They only influence the delay detection of some weldings which have a bad welding quality,
- very few information (2 patterns of C_1 at a minimum) are necessary to use SS-DFKNN.
- the delay detection of a bad welding is small (8 ms).

All classification results obtained by SS-DFKNN for this application are presented in Table 5.1.

The set of bad quality weldings has had several patterns classified in C_2 and C_3. Only some bad quality weldings patterns were misclassified in C_1. It concerns only the patterns which have generated delay evolution detection. These patterns were

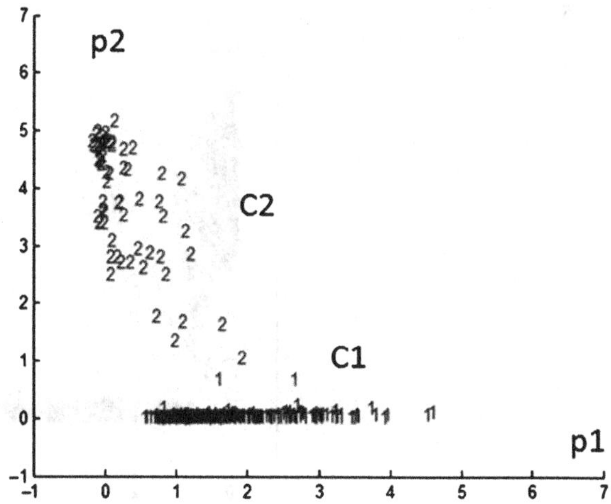

Fig. 5.11 Classification of the first welding of bad quality

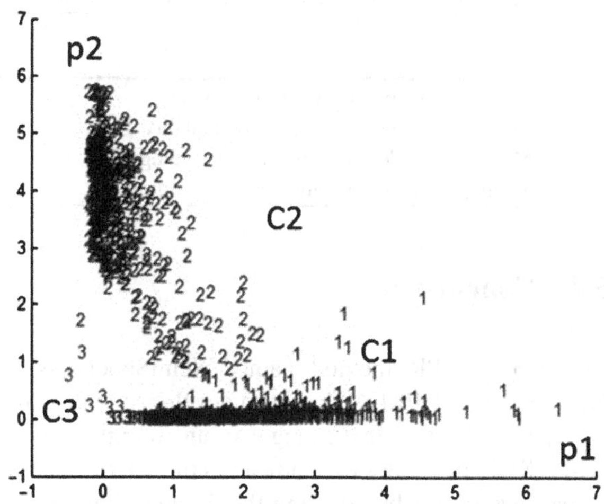

Fig. 5.12 Classification of all weldings of bad quality

then not misclassified consecutively but they correspond to few first patterns of bad quality weldings. No welding pattern of good quality was misclassified. Then, the classification result obtained by SS-DFKNN permits to perfectly distinguish good quality weldings from those of bad quality. Moreover, only small delay detection occurs so that the dynamic PR proposed can be online applied to this system. An acoustic or visible alarm system will be set up in order to warn human operators if a welding problem occurs.

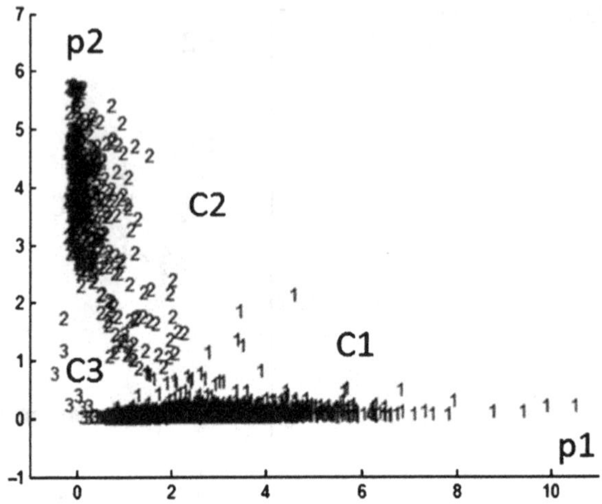

Fig. 5.13 Classification result of all weldings of the database

Table 5.1 Classification results obtained by SS-DFKNN

	Semi-supervised dynamic FKNN
Detection of weldings with a bad quality	100%
Detection of weldings with a good quality	100%
Mean t_{delay} to detect a bad quality welding	5 patterns
Mean t_{delay} to detect a good quality welding	No delay

5.4 Conclusion

The dynamic PR method named SemiSupervised Dynamic Fuzzy *K*-Nearest Neighbors (SS-DFKNN) has been developed in this chapter in order to demonstrate its capacities to realize the diagnosis and monitoring of industrial evolving systems. SS-DFKNN integrates two indicators of patterns usefulness which permit to follow classes evolutions by adapting these latter if an evolution is confirmed. When an evolution is realized, classes or subclasses are created to represent the current functioning mode of a system. These evolved classes can permit to better estimate the current functioning mode of an evolving system according to the time, to well monitor the evolutions of complex classes (defined by several subclasses) and to progressively find which functioning mode a class may reach after its evolution. Indeed without adapting classes, an evolution of the classes characteristics will be detected much later than when classes are adapted. SS-DFKNN can use only a few patterns to initiate the method. However, more the learning set is representative of the classes characteristics, better the detection of evolutions is. The classes characteristics of all classes are refined sequentially with the classification of the new patterns. The update of the evolved classes parameters is realized in a low time so that this method can be applied online. In this chapter, SS-DFKNN has been illustrated by a drift example and applied on an industrial welding system. For each welding operation, an acoustic signal was acquired and used by SS-DFKNN.

SS-DFKNN has well classified these signals which permitted to detect all bad quality weldings and it also detected each one of the welding quality evolution realized by the welding system.

SS-DFKNN uses several parameters to monitor evolving systems. Among these parameters we can particularly estimate that k, th_l, and $NbMin$ have a major importance in the results the method can obtain. According to their values, a delay detection can occur, noisy patterns can be considered as a class evolution and patterns can be misclassified. A new version of this method is being developed in order to progressively adapt the classes parameters and the classifier parameters, in order to obtain better results and to simplify the initial definition of these latter.

References

1. Amadou-Boubacar, H., Lecoeuche, S., Maouche, S.: Self-adaptive kernel machine: Online clustering in RKHS. In: Proceedings of the IEEE IJCNN05. Montreal, Canada (2005)
2. Angelov, P.: A fuzzy controller with evolving structure. Information Sciences **161**(1–2), 21–35 (2004)
3. Angelov, P., Filev, D., Kasabov, N.: Evolving Intelligent Systems—Methodology and Applications. John Wiley & Sons, New York (2010)
4. Angstenberger, L.: Dynamic fuzzy pattern recognition. Ph.D. thesis, Fakultät für Wirtschaftswissenschaften der Rheinisch-Westfälischen Technischen Hochschule (2000). Aachen, Germany
5. Beringer, J., Hüllermeier, E.: Efficient instance-based learning on data streams. Intelligent Data Analysis **11**(6), 627–650 (2007)
6. Bezdek, J.: Pattern Recognition with Fuzzy Objective Function Algorithms. Kluwer Academic/Plenum Publishers, USA (1981)
7. Cohen, L., Avrahami, G., Last, M.: Incremental info-fuzzy algorithm for real time data mining of non-stationary data streams. In: Proceedings of the TDM Workshop. Brighton, UK (2004)
8. Cozman, F., Cohen, I., Cirelo, M.: Semi-supervised learning of mixture models. In: Proceedings of the 20th International Conference on Machine Learning (ICML). Washington DC, USA (2003)
9. Dubuisson, B.: Diagnostic et reconnaissance des formes. Tech. rep., Trait des Nouvelles Technolo-gies, srie Diagnostic et Maintenance, HERMES (1990)
10. Duda, R., Hart, P., Stork, D.: Pattern Classification—Second Edition. Wiley-Interscience (John Wiley & Sons), Southern Gate, Chichester, West Sussex, England (2000)
11. Frigui, H., Krishnapuram, R.: A robust algorithm for automatic extraction of an unknown number of clusters from noisy data. Pattern Recognition Letters **17**, 1223–1232 (1996)
12. Frigui, H., Krishnapuram, R.: Clustering by competitive agglomeration. Pattern Recognition **307**, 1109–1119 (1997)
13. Gabrys, B., Bargiela, A.: General fuzzy min–max neural network for clustering and classification. IEEE Transactions on Neural Networks **11**(3), 769–783 (2000)
14. Garcia, V.: Suivi d'objets d'intrt dans une sequence d'images: des points saillants aux mesures statistiques. Tech. rep., University of Nice (2009)
15. Gibb, W., Auslander, D., Griffin, J.: Adaptive classification of myocardial electrogram waveforms. IEEE Transactions on Biomedical Engineering **41**, 804–808 (1994)
16. Guedalia, I., London, M., Werman, M.: An on-line agglomerative clustering method for non-stationary data. Neural Computation **11**(2), 521–540 (1999)
17. Jain, A., Duin, R., Mao, J.: Statistical pattern recognition: A review. IEEE Transactions on Pattern Analysis and Machine Intelligence **22**(1), 4–37 (2000)

18. Kasabov, N.: Evolving Connectionist Systems: The Knowledge Engineering Approach—Second Edition. Springer Verlag, London (2007)
19. Keller, J., Gray, M., Givens, J.: A fuzzy *k*-nn neighbor algorithm. IEEE Transactions on Systems, Man and Cybernetics **15**(4), 580–585 (1985)
20. Kybic, J.: Incremental updating of nearest neighbor-based high-dimensional entropy estimation. In: Proceedings of the ICASSP 2006, pp. 804–807 (2006)
21. Law, Y., Zaniolo, C.: An adaptive nearest neighbor classification algorithm for data streams. In: Proceedings of the 9th European Conference on Principles and Practice of Knowledge Discovery in Databases (PKDD 2005), pp. 108–120. Porto, Portugal (2005)
22. Lughofer, E.: Evolving Fuzzy Systems—Methodologies, Advanced Concepts and Applications. Springer, Berlin Heidelberg (2011)
23. Lughofer, E., Angelov, P.: Handling drifts and shifts in on-line data streams with evolving fuzzy systems. Applied Soft Computing **11**(2), 2057–2068 (2011)
24. Min, R.: A non-linear dimensionality reduction method for improving nearest neighbour classification. Ph.D. thesis, University of Toronto (2005). Toronto, Canada
25. Nakhaeizadeh, G., Taylor, C., Kunisch, G.: Dynamic supervised learning. Some basic issues and application aspects. classification and knowledge organization, pp. 123–135. Springer Verlag, Berlin Heidelberg (1997)
26. Roncaglia, A., Elmi, I., Dori, L.: Adaptive *K*-NN for the detection of air pollutants with a sensor array. IEEE Sensor Journal **4**(2), 248–256 (2004)
27. Song, Y., Huang, J., Zhou, D.: I*k*-NN: Informative *k*-nearest neighbor pattern classification. In: Proceedings of the PKKD 2007 conference, pp. 248–264 (2007)
28. Therrien, C.: Decision Estimation and Classification: An Introduction to Pattern Recognition and Related Topics. John Wiley & Sons, New York (1989)
29. Vachkov, G.: Online classification of machine operation modes based on information compression and fuzzy similarity analysis. In: Proceedings of the IFSA-EUSFLAT 2009 conference, pp. 1456–1461. Lisbon, Portugal (2009)
30. Zhang, B., Srihari, S.: A fast algorithm for finding *k*-nearest neighbors with non-metric dissimilarity. In: Proceedings of the 8th International Workshop on Frontiers in Handwriting Recognition (IWFHR'02), pp. 13–19 (2002)

Chapter 6
Making Early Predictions of the Accuracy of Machine Learning Classifiers

James Edward Smith, Muhammad Atif Tahir, Davy Sannen, and Hendrik Van Brussel

Abstract The accuracy of machine learning systems is a widely studied research topic. Established techniques such as cross validation predict the accuracy on unseen data of the classifier produced by applying a given learning method to a given training data set. However, they do not predict whether incurring the cost of obtaining more data and undergoing further training will lead to higher accuracy. In this chapter, we investigate techniques for making such early predictions. We note that when a machine learning algorithm is presented with a training set the classifier produced, and hence its error, will depend on the characteristics of the algorithm, on training set's size, and also on its specific composition. In particular we hypothesize that if a number of classifiers are produced, and their observed error is decomposed into bias and variance terms, then although these components may behave differently, their behavior may be predictable. Experimental results confirm this hypothesis, and show that our predictions are very highly correlated with the values observed after undertaking the extra training. This has particular relevance to learning in nonstationary environments, since we can use our characterization of bias and variance to detect whether perceived changes in the data stream arise from sampling variability or because the underlying data distributions have changed, which can be perceived as changes in bias.

J.E. Smith (✉)
Department of Computer Science and Creative Technologies, University of the West of England, Bristol, BS16 1QY, UK
e-mail: james.smith@uwe.ac.uk

M.A. Tahir
School of Computing, Engineering and Information Sciences, University of Northumbria, Newcastle, UK
e-mail: muhammad.tahir@northumbria.ac.uk

D. Sannen • H.V. Brussel
Department of Mechanical Engineering, Katholieke Universiteit Leuven, Leuven, Belgium
e-mail: davysannen@gmail.com; hendrik.vanbrussel@mech.kuleuven.be

M. Sayed-Mouchaweh and E. Lughofer (eds.), *Learning in Non-Stationary Environments: Methods and Applications*, DOI 10.1007/978-1-4419-8020-5_6,

© Springer Science+Business Media New York 2012

6.1 Introduction

Predicting the accuracy of a trained machine learning system when presented with previously unseen test data is a widely studied research topic. Techniques such as cross validation are well established and understood both theoretically and empirically, e.g., [18,34]. However, these techniques predict the accuracy on unseen data *given the existing training set*. For example, N-fold Cross Validation (NCV) averages the fitness estimated from N runs, each using a proportion $1 - 1/N$ of the available data to train a classifier and $1/N$ to evaluate it. Therefore, repeating with different values of N can give the user some indication of how the error rate changed as the training set increased to the current size, since lower values of N effectively equate to smaller training sets. However, NCV does not predict what accuracy might be achievable after further training. Thus if the current accuracy is not acceptable, and obtaining data comes at cost, NCV and similar techniques do not offer any insights into whether it is worth incurring the cost of further training.

This is of more than theoretical interest, because the successful application of machine learning techniques to "real-world" problems places various demands on the collaborators. Not only must the management of the industrial or commercial partner be sufficiently convinced of the potential benefits that they are prepared to invest money in equipment and time but, vitally, there must also be a significant investment in time and commitment from the end-users in order to provide training data from which the system can learn. This poses a problem if the system developed is not sufficiently accurate, as the users and management may view their input as wasted effort, and lose faith with the process.

In some cases this effort may be re-usable if, for example, the user has been labeling training examples that can be stored in their original form, and which come from a fairly stationary distribution. However, this is frequently not the case. For example, in many applications it may not be practical to store the physical training examples rather, it is necessary to characterize them by a number of variables. If the failure of the Machine Learning system in such cases stems from an inappropriate or inadequate choice of descriptors, then the whole process must be repeated. Not only has the user's input been a costly waste of time and effort but there also may be a loss of faith in the process which can manifest in reduced attention and consistency when classifying further samples. To give a concrete example from the field of diagnostic visual inspection (e.g., manufacturing process control or medical images), it frequently turns out that it is not sufficient to store each relevant image—other information is necessary such as process variables, or patients' history. If this data is not captured at the same time, and is not recoverable post-hoc, then the effort of collecting and labeling the database of examples has been wasted.

A significant factor that would help in gaining confidence and trust from end-users would be the ability to quickly and accurately predict whether the learning process was going to be successful. Perhaps more importantly from a commercial viewpoint, it would be extremely valuable to have an early warning that the users can save their effort while the system designer refines the choice of data, algorithms etc.

From the perspective of learning in nonstationary environments, such a tool could provide a number of advantages. Firstly, a deviation from the expected progress can be used as an indicator that there has been a fundamental shift in the nature of the training input provided to the algorithm. This could arise either because the user providing the labels has changed, or the underlying data set is dynamic and requires a classifier that is able to take account of this. In either case, early warning is needed to enable corrective action to be taken.

In this chapter, we investigate a technique for making such early predictions of future error rates. We will consider that we are given n samples, and that the system is still learning and refining its model at this stage. We are interested in predicting what final accuracy might be achievable if the users were to invest the time to create n' more samples. This leads us to focus on two questions. First, what are the most appropriate descriptors of the system's behavior after some limited number n of samples, and then later after an additional n' samples? Second, is it possible to find useful relationships for predicting the second of these quantities from the first?

Theoretical studies, backed up by empirical results, have suggested that the total error rate follows a power–law relationship, diminishing as extra training samples are provided. While these theoretic bounds on error are rather loose, they provide motivation for investigating practical approaches for quickly and reliably estimating the error rate that may be observed after future training. In general the error will be a complicated function, but the hypothesis of this chapter is that we can deal with it more easily if we decompose it into a number of more stable functions. Therefore this chapter concentrates on the use of the well-known bias–variance decomposition [8, 21] as a source of predictors when an algorithm is used to build a classification model from a data set. Specifically, our hypothesis is that if the observed error is decomposed into bias and variance terms, then although these components may behave differently, their behavior may be individually predictable.

To test our hypothesis we first apply a range of algorithms to a variety of data sets, for each combination periodically estimating the error components as more training samples are introduced, until the full data set has been used. All of the data arising from this (rather lengthy) process is merged and regression analysis techniques are applied to produce three sets of predictive models—one each for bias, variance and total error. Each of these models takes as input a measurement obtained from the classifier produced when only a few samples (n) from a data set have been presented to the learning algorithm, and predicts the value after all samples have been applied ($n+n'$). As the data have been merged, the intention is that these models are algorithm-data set independent. We examine the stability and valid range of these models using simple linear regressors. Moving on to consider trainable ensembles of different classifiers, we show how a similar approach can be applied to obtain estimates on the upper bound of the achievable accuracy, which can predict the progression of the ensemble's performance.

The rest of this chapter proceeds as follows. In Sect. 6.2, we review related work in the field, including the bias–variance decomposition of error that we will use. Following that, Sect. 14.2.1 describes the experimental methodology used to collect the initial statistics, and test the resulting models. Section 6.4 describes and discusses the results obtained. In Sect. 6.5, we show how this approach may be extended to predict the future accuracy of trainable ensembles of classifiers. In Sect. 6.6, we discuss how these methods could be applied to detect changes in underlying data distributions that would trigger re-learning in no stationary environments. Finally in Sect. 6.7, we draw some conclusions and suggestions for further work.

6.2 Background

6.2.1 Notation

For the sake of clarity we will use a standard notation throughout this chapter, reinterpreting results from other authors as necessary.

We assume classification tasks, where we are given an instance space X and a predicted categorical variable Y. The "true" underlying function F is a mapping $F: X \to Y$.

Let D be the set of all possible training sets of size n sampled from the instance space X, and $d \in D = \{(x_1,y_1),(x_2,y_2),\dots,(x_n,y_n)\}$.

When a machine learning algorithm C is presented with d it creates a classifier, which we may view as a hypothesis about the underlying mapping: $H_{Cd}: X \to Y$. The subscripts C and d make it explicit that the specific classifier H induced depends on the learning algorithm and the training set. For a specific learning algorithm C, the set of classifiers that it can induce is denoted $\mathscr{H}$.

We consider a $0/1$ misclassification error—in other words the error is zero if H correctly predicts the true class of an item $x \in X$, and 1 otherwise. More formally, the misclassification cost of a single data item x with a specific classifier H is:

$$\mathrm{Cost}\,(H_{Cd},x) = \begin{cases} 0 \; H_{Cd}(x) = F(x) \\ 1 \; H_{Cd}(x) \neq F(x) \end{cases}. \tag{6.1}$$

The expected error of the classifier created from n data points is then given by integrating over X and d, taking into account their conditional likelihood, i.e.:

$$\mathrm{Error}\,(H_{Cn},X) = \int_{x \in X, d \in D} P(x)P(d|n)\mathrm{Cost}(H_{Cd},x), \tag{6.2}$$

where $P(d|n)$ is the probability of generating a specific training set $d \in D$ given the training set size n, and $P(x)$ is the probability of selecting an item $x \in X$ to be classified. In practice of course it is not possible to exactly measure the true error, so approaches such as bootstrapping, hold-out, and cross validation are used to estimate the error, given a finite sized set of examples. In *bootstrapping*, new data sets are repeatedly generated from the original data set using random sampling with replacement. The new data sets, which most likely contain duplicate examples, are then used to train a classifier and the examples that are not part of the data sets are used for testing. *Hold-out* approaches divide the available data into two sets (typically using a 70/30 split), train a classifier using the larger set and then estimate its accuracy using the "unseen" smaller set. As described above, *N-fold cross validation* (NCV) is an improvement on the hold-out approach which aims to avoid the possibility of accidentally selecting an "easy" test set. The available data is split into N (usually equally sized) blocks. Each of the blocks in turn is then used as a test set to estimate the accuracy of a classifier built from the remaining $N-1$ blocks. The average of these is then used as an estimate of the accuracy of the classifier that would be built from all of the available data. We will use the lower case "error" to denote an estimation is being used for the true error.

6.2.2 Relationship to Other Work

Cortes et al. [10] presented an empirical study where they characterized the behavior of classification algorithms using "learning curves". These suggest that the predicted error of the classifier after n samples have been presented will follow a power–law distribution in n:

$$\text{error}(n) = an^{-\alpha} + b, \tag{6.3}$$

where the constants a (the learning rate), α (the decay rate), and b (the asymptotic Bayes error rate) depend on the particular combination of classification algorithm and data set, but α is usually close to, or less than one. This suggests that given a particular classifier–data set combination, it should be possible to commence training, take periodic estimates of the error as n increased, and then use regression to find values for a, b, α that fit the data, and can be used for predicted future error rates. "Progressive sampling" uses training sets ("samples") with progressively larger sizes (i.e., increasing n) until some desired accuracy has been reached. This can be inefficient if a larger number of "samples" is used as each must be evaluated. Using a similar approach to Cortes et al. recent papers have attempted to fit a learning curve to a few samples in order to predict the size needed [23, 29]. Mukerhjee et al. [26] have pointed out a problem with this curve-fitting approach, namely that for low values of n the estimated error rates are subject to high variability, which leads to significant deviations when fitting the power–law curve. They have presented an extension of the method which uses a "significance permutation test" to establish the significance of the observed classifier error prior to curve fitting.

These results fit in with theoretical bounds from "Probably Approximately Correct" (PAC) theory such as those presented by Vapnik in [35]. These begin with the assumption that a training set $d = \{x_k, y_k\}, 1 \leq k \leq n, y_k \in \{-1, 1\}$ is drawn independently and identically distributed (iid) from a data set, and that future training and test data will be drawn from the data set in the same way.

Given the restriction $Y = \{-1, 1\}$, the test error $\mathrm{Error}(H_{Cn})$, (the probability of misclassification) is defined to be:

$$\mathrm{Error}(H_{Cn}) = E\left[\frac{1}{2} \mid F(x) - H_{Cn}(x) \mid\right], \tag{6.4}$$

where in comparison to (6.1), the division by two maps absolute values of differences in Y onto costs (rather than having a fixed cost of 1 for misclassification), and the H_{Cn} denotes that we are taking the expectation for the general case. The current empirically measured training error $\mathrm{error}(H_{Cd})$ is:

$$\mathrm{error}(H_{Cd}) = \frac{1}{n}\sum_{k=1}^{n}\frac{1}{2}\left|y_k - H_{Cd}(x_k)\right|. \tag{6.5}$$

Note that since this estimates the error by classifying the n elements of the training set with a classifier trained on that data, it is calculated as a summation and will underestimate the true error. Vapnik showed that the amount of underestimation can be bounded [35]. If ψ represents the Vapnik–Chervonenkis (VC) dimension, and $0 \leq \eta \leq 1$, then with probability $1 - \eta$:

$$\mathrm{Error}(H_{Cn}) \leq \mathrm{error}(H_{Cd}) + \sqrt{\frac{\psi \log(2n) + \psi(1 - \log \psi) - \log(\eta/4)}{n}}. \tag{6.6}$$

Effectively this equation makes explicit an assumption that machine learning algorithms inherently produce classifiers which overfit the available training data. The VC-dimension ψ is a measure of the capacity of a hypothesis space of classification algorithm C, so may be thought of as the "power" of C. It is the maximum number of points that can be arranged so that C can always "shatter" them—for example, the VC-dimension of a linear classifier such as a perceptron is three, since no straight line can separate the four points of an XOR problem. Equation (6.6) makes it clear that more powerful algorithms (higher ψ) are more likely to over-fit the data, and so it may be used as grounds to select between two algorithms which produce the same training error but have different complexity (related to ψ). It also makes explicit the dependency on n: for a given training set error, the maximum amount by which this will underestimate the true error decreases by approximately $\sqrt{\psi \log n / n}$.

However, in practice these bounds tend to be rather "loose." There have been other more recent developments in Statistical Learning Theory which use a similar approach but exploit Rademacher complexity to provide tighter bounds, such as those in [1–3, 27]. Common to all of these approaches, as with the use of VC-dimension results, is the idea that on the basis of the available training data, an algorithm selects a classifier H_{Cd} from some class $\mathscr{H}$ available to it. To analyze the learning outcomes, the "error" observed when the training data is classified by H_{Cd} is broken down into the Bayes optimal error (which cannot be avoided) plus an amount by which best ($H* \in \mathscr{H}$) in the current class of classifiers would be more than Bayes optimal, plus an amount by which the classifier H_{Cd} currently estimated by the algorithm to be "best" is different to the actual best $H*$. Thus for example, approaches such as *Structural Risk Minimisation* can be thought of as principled methods for increasing the size/complexity of the current class of classifiers $\mathscr{H}$ until it includes the Bayes optimal classifier.

The underlying assumption is that the error is estimated using the current training set, and that this almost certainly overfits the true underlying distribution (i.e. $H_{Cd} \neq H*$) so the current estimates of error for the chosen classifier H_{Cd} will be less than the "true" error that would be seen if it was applied to the whole data distribution. Therefore, bounds are derived which describe the extent to which the error on the training set underestimates the true error. Since this can be described in terms of the search problem of identifying $H* \in \mathscr{H}$, it is understandable that they take into account the amount of information available to the search algorithm—i.e., the size n of the training set.

While this is a valid and worthwhile line of theoretical research, we would argue that it is not currently as useful for the practitioner. Consider the example of a user who is highly skilled in his/her domain, but knows nothing about Machine Learning, and is providing the training examples from which a classifier is constructed. The theory above effectively says: *"Based on what you have told me, I've built a classifier which seems to have an error rate of x%. I can tell you with what probability the "true" error rate is worse than x + y%, for any positive y."* If they have provided enough labeled data items to create what appears to be an accurate classifier, then this is valuable. However, if they are still early on in the process, and the current error rates are high, it gives no clues as to whether they will drop. Instead we attempt to provide heuristics that answer a different question: *"Based on what you have told me, I've built some classifiers and although the current error rate is x% it will probably drop to y%, where* $y \leq x$*"*.

To do this, we note that the analysis above relates the true test error to a specific estimated error from a given training set size, and that the variance in the predicted error depends strongly on n. This has prompted us to examine different formulations that explicitly decompose the error into terms arising from the inherent bias of the algorithm (related to its VC dimension, or to the difference between $H*$ and the Bayes optimal classifier) and the variability arising from the choice of $d \in D$.

6.2.3 Bias–Variance Decomposition

A number of recent studies have shown that the decomposition of a classifier's error into bias and variance terms can provide considerable insight into the prediction of the performance of the classifier [8, 21]. Originally, it was proposed for regression [17] but later, this decomposition has been successfully adapted for classification [8, 21, 31]. While a single definition of bias and variance is adopted for regression, there is considerable debate about how the definition can be extended to classification [5,12,16,19,21,22]. In this chapter, we use Kohavi and Wolpert's [21] definition of bias and variance on the basis that it is the most widely used definition [37,38], and has strictly nonnegative variance terms.

Kohavi and Wolpert define bias, variance and noise as follows [21]:

Squared Bias *"This quantity measures how closely the learning algorithm's average guess (over all possible training sets of the given training set size) matches the target."*

Variance *"This quantity measures how much the learning algorithm's guess bounces around for the different training sets of the given size."*

Intrinsic noise *"This quantity is a lower bound on the expected cost of any learning algorithm. It is the expected cost of the Bayes-optimal classifier."*

Given these definitions, we can restate (6.2) as:

$$\text{Error}(H_{C,n}) = \int_{x \in X} P(x)\left(\sigma_x^2 + \text{Bias}_x^2 + \text{Variance}_x\right). \tag{6.7}$$

Assuming a fixed cardinality for Y (finite set of classes), and noting D has finite cardinality, the summation terms in the integral are:

$$\text{Bias}_x^2 = \frac{1}{2}\sum_{d \in D} P(d|F,n) \sum_{y \in Y} \left[P(F(x)=y) - P(H_{Cd}(x)=y)\right]^2,$$

$$\text{Variance}_x = \frac{1}{2} - \frac{1}{2}\sum_{y \in Y}\sum_{d \in D} P(d|F,n)P(H_{Cd}(x)=y)^2,$$

$$\sigma_x^2 = \frac{1}{2} - \frac{1}{2}\sum_{y \in Y} P(F(x)=y)^2,$$

where the terms $P(F(x)=y), P(H_{Cd}(x)=y), P(d|F,n)$ make explicit that some terms are conditional probability distributions since the Bayes error may be non-zero, the classification output may not be crisp, and the specific choice of training set depends on the underlying function and the number of samples.

In practice, these values are estimated from repeated sampling of training sets to acquire the necessary statistics, which are then manipulated to give the different terms. Thus, the Bias term considers the squared difference between the actual and predicted probabilities that the label is y for a given input x for a given training set d. To calculate its value, the inner term sums over all possible values of y, and then the outer summation averages over all training sets of a given size. By comparison, the Variance term just considers the distribution of predicted values, $P(F(x)=y)$, but reverses the order of the summation to emphasize the effect of different training sets. The intrinsic noise term sums over all possible output values y the squared probabilities that the actual target $F(x)=y$ for a given input x. If the underlying class boundaries are crisp, then $P(F(x)=y)$ will be zero except for one value of y, the summation will be 1 and σ_x^2 will consequently be zero.

6.2.4 *Bias as an Upper Limit on Accuracy*

An alternative perspective on the analysis in Sect. 6.2.3 is that the bias term reflects an inherent limit on a classifier's accuracy resulting from the way in which it forms decision boundaries. For example, an elliptical class boundary can never be exactly replicated by a classifier which divides the space using axis-parallel decisions. A number of studies have been made confirming the intuitive idea that the size of variance term drops as the number of training samples increases, whereas the estimated bias remains more stable, e.g., [8]. Therefore, we can treat the sum of the inherent noise and bias terms as an upper limit on the achievable accuracy for a given classifier. Noting that in many prior works it is assumed that the inherent noise term is zero, and that for a single classifier it is not possible to distinguish between inherent noise and bias, we hereafter adopt the convention of referring to these collectively as bias.

6.3 Experimental Methodology

The hypothesis of the main part of this chapter is that values of the bias and variance components estimated after n training samples can be used to provide accurate predictions for their values after $n+n'$ samples, and hence for the final error rate observed. To do this prediction, we use statistical models built from a range of data set-algorithm combinations. The following sections describe our choice of experimental methodology, algorithms, and data sets.

6.3.1 *Procedure for Building the Models*

Our experimental procedure is as follows:

- For each data set x and classifier i, we estimated the values of error (e_{ixn}), bias (b_{ixn}) and variance (v_{ixn}) components using the first $n \in \{100, 200, \ldots, 1{,}000\}$ samples.
- For each data set, we then estimated the values of error, bias, and variance using all of the samples in the data set. Note that this results in different values of n' for different data sets. Note also that we do not use a separate "test set." We consider that since one is always making estimates of the error on unseen data it is more consistent to relate estimates of the error at different points in training *using the same estimation methodology*.
- After all of the collected data was pooled, we applied linear regression to create models of the form $Q_{(n+n')} = a_{Qn} \cdot Q_n + b_{Qn}$, where Q is one of bias, variance, or total error. In these models Q_n is the independent variable, $Q_{(n+n')}$ the dependent

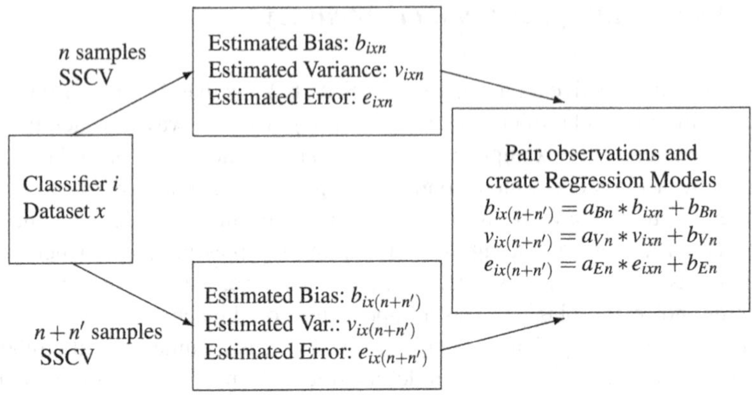

Fig. 6.1 Methodology for creating predictive models. This is repeated for $n \in \{100, 200, \ldots, 1{,}000\}$

variable and the constants a_{Qn} and b_{Qn} are estimated by the linear regression procedure for each variable $Q \in$ {total error, bias, variance} and for each value of n. We compute the coefficient of determination R^2 to measure how well the simple linear model explains the variability of the independent variable, and hence the quality of the predictions—the closer R^2 is to 1, the better is the prediction.

- Note that when used with a new classifier i or data set x, this gives us two ways of predicting the final error $e_{ix(n+n')}$ based on the first n samples: either directly from the observed error or by summing the predictions for bias and variance.
  - In the first case there is one independent variable, so $e_{ix(n+n')} = a_{En} \cdot e_{ixn} + b_{En}$, where a_{En} and b_{En} are taken from our models.
  - In the second case, the two decomposed components (bias b_{ixn} and variance v_{ixn}) are treated as independent variables, i.e. $e_{ix(n+n')} = a_{Bn} \cdot b_{ixn} + a_{Vn} \cdot v_{ixn} + b$, where b $(= b_{Bn} + b_{Vn})$, a_{Bn}, and a_{Vn} are given by our models.

Figure 6.1 shows this process for a single value of n.

We would like to re-iterate for the sake of clarity that we are not building models which relate error, bias, and variance as a function of the number of training samples n. In that case, it would certainly be true that by the two models (bias as a function of n) and (variance as a function of n) could be combined into a single linear model (error as a function of n). As the wealth of theoretical work described above shows, there is ample evidence to suggest that no simple predictive linear model exists. Instead we are building and combining linear models of the form *future bias/variance/error as a function of current bias/variance/error* and seeing how the predictive power of these models changes as a result of the value of n.

These linear models are of course an extremely simple way of modeling the relationship between our various predictors; more sophisticated techniques exist in the fields of statistics and also Machine Learning, and will be examined in a later section. However, as the results will show, linear models are sufficient for our purposes.

6.3.2 Choice of Classifiers

In order to obtain the data for modeling ten different classification algorithms were selected, each with different bias and variance characteristics. These were: Naive Bayes [13], C4.5 [30], Nearest Neighbor [11], Bagging [4], AdaBoost [15], Random Forest [6], Decision Table [20], Bayes Network [13], Support Vector Machine [28], and Ripple-Down Rule learner [39]. Note that this set includes two methods for creating ensembles: AdaBoost (using Decision Stumps as the base classifier) and Bagging (using a decision tree with reduced error pruning). In these cases, since we are solely interested in the outputs, we treat the ensemble as a single entity, rather than attempt a bias–variance–noise–covariance decomposition [9]. For all these classifiers, the implementation in the *WEKA* library [39] is used, and the default parameters in WEKA are used for each classifier. We also used WEKA's Java implementation of Kohavi and Wolpert's definition of Bias and Variance (*weka.classifiers.BVDecompose*).

6.3.3 Data Sets

The data collection required to build the statistical models is carried out on data sets derived from four Artificial and five real-world visual surface inspection problems from the European DynaVis project[1] [14, 25]. Each artificial problem consists of 13,000 contrast images created by a tuneable randomized image generator. Class labels (good/bad) were assigned to the images by using different sets of rules of increasing complexity acting on the generator. The real-world data sets came from CD-imprint and egg inspection problems. There are 1,534 CD images, each labeled by four different operators, and 4,238 labeled images from the egg inspection problem. The same set of image processing routines are applied to segment and measure regions of interest (ROI) in each image. From each set of images are derived two data sets. The first has 17 features describing global characteristics of the image and the ROI it contains. In the second, these are augmented by the maximum value (over all the ROI) for each of 57 ROI descriptors. Adding the labels available provides a total of 18 different data sets with a range of dimensionality and cardinality.

---

[1] www.dynavis.org

To build the models, we used 14 of the data sets: the six derived from the first three artificial image sets, the six from the CD images labeled by the first three operators and the two from the egg data. The remaining four data sets, derived from the fourth artificial image set, and the CD labeled by Operator 4 are reserved for evaluation purposes. In each case we took $n' = \text{total\_set\_size} - 1{,}000$, so n' differs between data sets.

6.3.4 Prediction Methodology and Sampling Considerations

To create the model data, we repeatedly draw training and test sets from the n samples from which we can estimate the total error, together with its bias and variance components. This raises the issue of how we should do this repeated process.

If the variables in X are continuous, or unbounded integers, then the underlying distribution over which the classifier may have to generalize is of course infinite. For bounded integer or categorical variables, the number of potential training sets of size n drawn iid from an underlying distribution of X is of size $|D| = |X|!/n!(|X|-n)!$, so in practice even for nontrivial data sets it is not possible to evaluate all possible training sets d of size n. However the success (or otherwise) of the approach proposed in this chapter depends on the accuracy with which we can predict error components, particularly for when the training set sizes are low. This immediately raises the question of finding the most appropriate methodology for estimating the values of those quantities. To give a simple example of why this is important, a later result in this chapter partially relies on being able to distinguish between those data items that are *always* going to be misclassified by a given classifier, and those which will *sometimes* be misclassified, depending on the choice of training set. Since the well known N-fold cross validation approach only classifies each data item once, it does not permit this type of decomposition and cannot be used. In a preliminary paper [33], we have examined two possible approaches: the "hold-out" method proposed by Kohavi and Wolpert [21] and the "Sub-Samples Cross Validation" (SSCV) method proposed by Webb and Conilione [38]. The latter have argued that the hold-out approach proposed in [21] is fundamentally flawed, partly because it results in small training sets, leading to instability in the estimates it derives. This was confirmed by our results [33] which showed that the stability of the estimates, and hence the accuracy of the resulting prediction was far higher for the SubSampling method. Therefore, we restrict ourselves to this approach.

The SSCV procedure is designed to address weaknesses in to both the hold-out and bootstrap procedures by providing a greater degree of variability between training sets. In essence, this procedure repeats N-fold CV l times, thus ensuring that each sample x from the training set of size n is classified l times by the classifier i. The true b_{ix} and v_{ix} can be estimated as b_{ixn} and v_{ixn} from the resulting set of classifications. The final bias and variance is estimated from the average of all $x \in D$ [37, 38], thus using all n' samples.

6.4 Explanatory Power of the Models

Figures 6.2 and 6.3 show scatter plots of the values for error, bias, and variance as measured after $n \in \{100, 1{,}000\}$ samples and after all samples. Different markers indicate different total numbers of samples. Note that in each case the same range is used for x- and y-axes, so a 1:1 correspondence would from a diagonal from bottom-left to top-right of the plot. In each case we show the results of a linear regression, with 95% confidence intervals. Thus, the values for each classifier–data set pair as estimated after a few, then all, samples constitutes a single point marked on the plot. For each combination, the vertical distance between the actual point and the mean regression line shows the difference between the value as measured from all samples available, and the value predicted on the basis of just n samples .

From Fig. 6.2, we make the following observations:

- The models built from only 100 samples do not fit the data well: the plots are very scattered and the coefficient of determination is low—in other words the linear regression shown would only account for 31–32% of the observed variation in values for the final variables (bias, error, variance).
- The models predict that although the error, bias, and variance will all fall from the values observed after 100 samples: the total error by 65%, the variance by 70%, but the bias only by 50%.
- The models also predict a nonnegative residual component for each—4%, 1% and 3%, respectively, which clearly is incorrect since it suggests that no classifier-data set combination would have zero error.
- From the magnitude of the effects, we can see that the bias terms account for the majority of the observed error.
- Comparing the estimates of variance after $n = 100$ with the final values, the former are much higher. This makes it apparent that the small size of the data sets is leading to considerable noise, which introduces error into the modeling process.
- If we visualize a diagonal line through the plots for variance and total error, in each case the regression line lies below this—so the models show the observed values with $n = 100$ overestimate the final values.
- For the bias plot, the markers for all sized data sets would fall fairly evenly on either side of the diagonal. Thus, the "noise" in the bias plot does not seem to be particularly a function of the data set size.
- By contrast, for the error and variance the markers for $n = n' = 1{,}534$, which fall at the lower end of the scales, would fall around, or often above the 1:1 line, whereas those for the larger data sets would predominantly fall below the line.

This last observation is worthy of further consideration. It shows that the linear regression is a compromise. For the smaller data sets ($n + n' = 1{,}534$) whatever form the variance takes as a function of n, a Taylor expansion would give similar values to those observed after ($n = 100$), whereas for the larger data sets the variance clearly

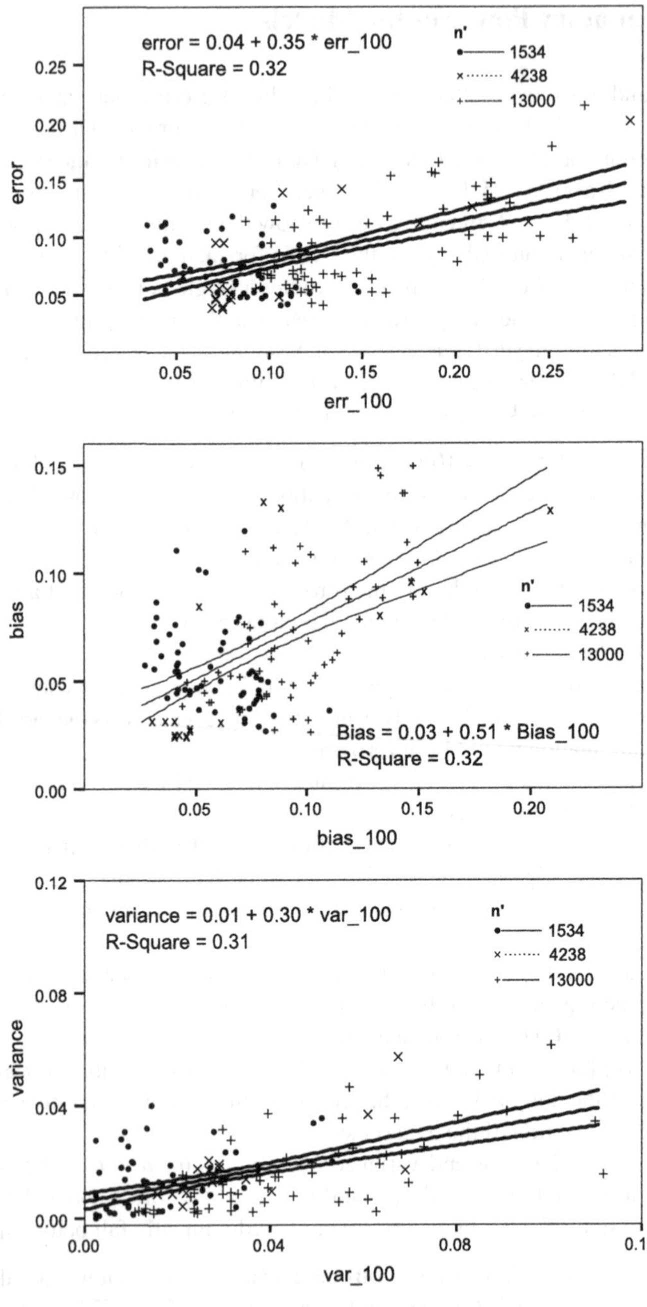

Fig. 6.2 Scatter plots of the Error (*top*), Bias (*middle*) and Variance (*bottom*) estimated after 100 samples (x-axis) and the same descriptors estimated using all samples (y-axis), together with results from linear regression (middle lines) and the 95% confidence intervals (upper and lower lines). Absolute values on individual plots vary, but in each case the x- and y-axes scale over the same range (so a 1:1 correspondence would form a diagonal of the plot)

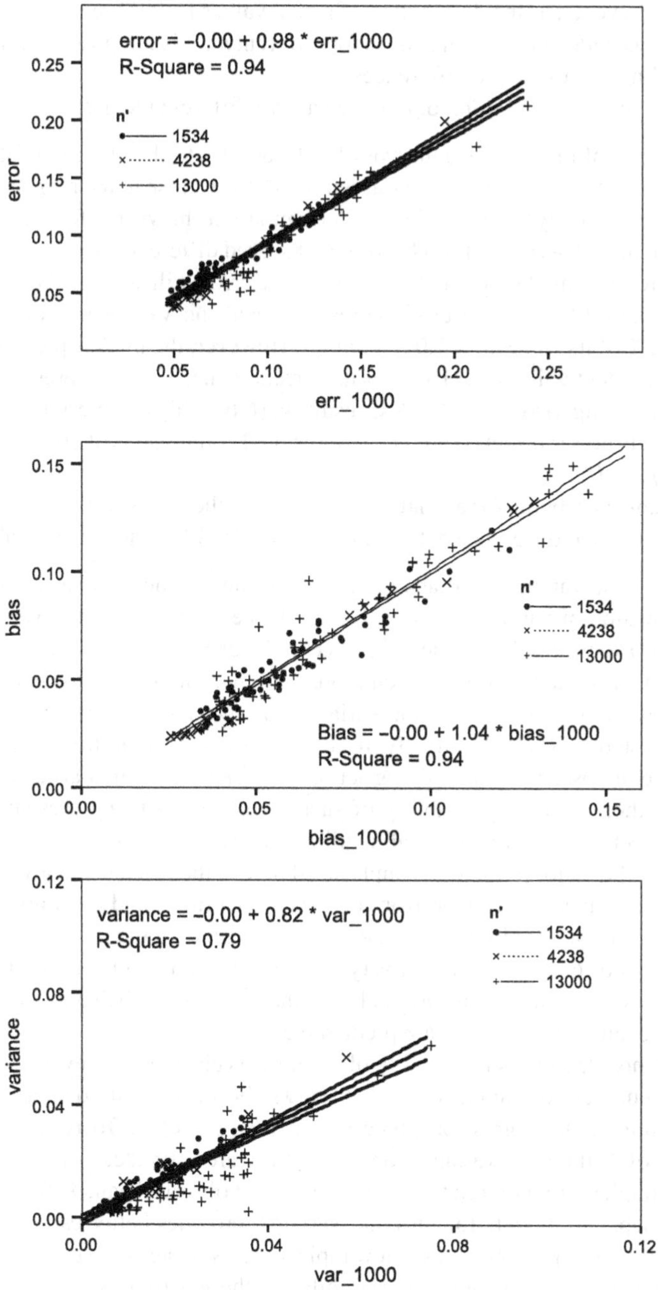

Fig. 6.3 Scatter plots of the Error (*top*), Bias (*middle*) and Variance (*bottom*) estimated after 1,000 (right) samples (x-axis) and the same descriptors estimated using all samples (y-axis), together with results from linear regression (middle lines) and the 95% confidence intervals (upper and lower lines). Absolute values on individual plots vary, but in each case the x- and y-axes scale over the same range (so a 1:1 correspondence would form a diagonal of the plot)

falls away. However, as the distribution of actual values for different data sets of the same size n' is wide, and overlaps those for different n', it is not possible for a single regression line to capture the differences.

Turning our attention to Fig. 6.3, we see a very different picture:

- The close fit of the models built from 1,000 samples to the observed data can be confirmed both visual inspection (all points fall very close to the regression line), and statistical analysis (coefficient of determination shows that for bias and total error, the model accounts for 94% of the observed differences).
- The regression models predicts that the final error will drop to 98% from its current value (direct error–error regression), and the variance component will fall to 82% of its value after 1,000 samples. However, the models predict that the bias component will rise to 104% of its current value. This is interesting since it suggests a comparison with the PAC results in (6.6). This will merit further study.
- The models now (correctly) predict zero residual components of error (i.e., $b_{Bn} = b_{Vn} = b_{En} = 0$).
- The variance accounts for a smaller proportion of the total error.
- There is no clear difference between the results for different values of n'.

This last observation is perhaps the least expected: if our arguments about the Taylor expansion of variance for $n = 100$ hold true, they should do even more so for $n = 1{,}000$ so we might see the difference in the distribution of variance markers for different sized data sets to be even more extreme. The fact that it is not can be explained by the hypothesis that the variance follows some inverse power–law in n—as suggested by (6.3). Intuitively, if elements of this variability are caused by the presence or absence in the training set of samples from particular regions of the data space, then both the probability of such elements not being present, and the averaged effect of their influence, fall nonlinearly as n increases.

However, the major point to be emphasized here is that even using a very simple model that is a linear regression from observed quantities, and does not take into account how far into the future (n') one is trying to predict, the models capture the characteristics of the observed data very closely. The results in Fig. 6.3 thus form strong evidence to confirm our original hypothesis—that the behaviors of the bias and variance, although different are predictable.

To show how the predictive quality of the models changes as they are built from increasing numbers of samples, Fig. 6.4 shows the coefficients of determination computed during the regression process as a function of n. To recapitulate, for each value of n, the bias, variance, and total error are estimated using SSCV, and regression models are built relating these to the final observed values. It is clear that the use of separate models for bias and variance provides better estimates of the predicted error. The plot also shows how rapidly the estimates (and correspondingly the predictive quality of the regression) stabilize in these two cases. What is apparent is that the method will hold well after only a few hundred data samples have been presented. Although the variance does not correlate highly (over 0.9) until closer

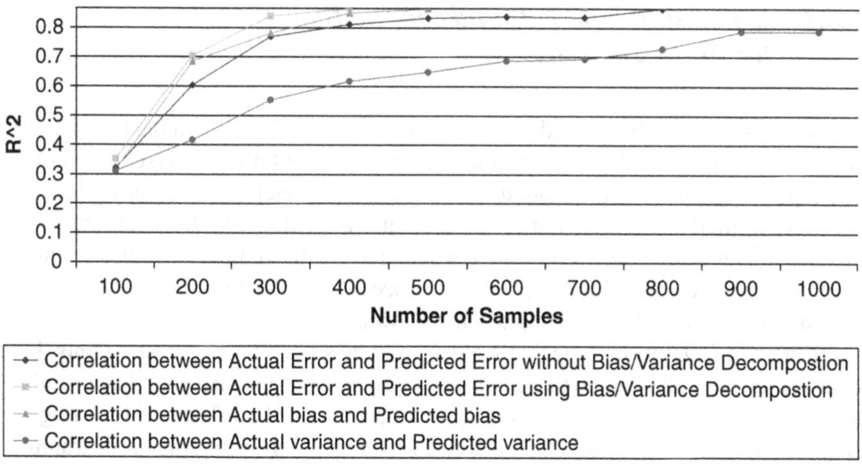

Fig. 6.4 Coefficient of Determination (R^2) between predicted and observed final error, and n, as a function of the number of samples used to build the statistical models

to 1,000 samples have been presented, the total error estimated via decomposition is well correlated because of the relatively greater size, and stability of the bias component.

6.5 Extension to Ensemble Classifiers

The concept of decomposing error into different terms has also been used to help explain the behavior of ensembles of algorithms. When the algorithms concerned are performing regression tasks, decomposing the error of an ensemble into terms representing the mean bias and variance of the individual algorithms, and the covariance between them is fairly straightforward. A good recent survey of both the bias–variance–covariance and ambiguity decompositions may be found in the first few pages of [9]. However, just as defining bias and variance for 0/1 loss functions was nontrivial, and there were several versions before Kohavi and Wolpert [21] created their formulation in which variance is always nonnegative, the extension to handle covariance in a natural way is also problematic. To the best of our knowledge there has not been a successful model decomposing 0/1 loss functions for ensembles of classifiers, so it is not immediately possible to simply extend the approach we took for single classifiers. However, in this section we present some initial findings from an approach in which we treat the entire ensemble as a single classifier. Revisiting the definitions of bias in Sect. 6.2.3, we next develop predictors for upper limits on its attainable accuracy based on simple observations of the behavior of the individual classifiers in the ensemble.

6.5.1 Estimating Lower Bounds on the Bias for Finite Data Sets

The analysis in Sect. 6.2.3 used a very general model predicated on the fact that the data items x could be drawn from a large, potentially infinite universe of samples, corresponding to unlimited future use of the classifiers. Here we are concerned with the more limited case where our future estimates are still drawn from a finite set of size $n+n'$. In particular, we consider whether we can predict the values of those estimates, before completing the training process. In order to achieve this, we can reformulate the models above slightly as follows.

To start with, let us assume that we have a finite set X of sample data points. For consistency with above note that $|X| = n+n'$. Because we are treating the ensemble as a single high-level entity, we need not worry about the effects of Boosting or Bagging approaches to creating ensembles by repeatedly sampling from training sets. Therefore, we assume that at our higher level training sets of size n are created by sampling from X uniformly without replacement. Let D denote the set of training sets created in this way, and d be any member of D, then we note that under these conditions $P(d|X,n) = \frac{1}{|D|} = \frac{(n)!(n')!}{(n+n')!}$.

Now let A^+, A^-, B partition X such that $A^+ \cup A^- \cup B = X$, and $A^+ \cap A^- = A^+ \cap B = A^- \cap B = \emptyset$, where:

- A^+ is the (possibly empty) subset of data items where for all training sets a classifier trained on that set correctly predicts the class of item x.

$$\forall x \in A^+, d, d' \in D, y \in Y \;\; Y_H(y|x,d) = Y_H(y|x,d') = Y_F(y|x).$$

- A^- is the (possibly empty) subset of data items where for all training sets a classifier trained from that set incorrectly predicts the class of item x.

$$\forall x \in A^-, d, d' \in D, y \in Y \;\; Y_H(y|x,d) = Y_H(y|x,d') \neq Y_F(y|x).$$

- B is the (possibly empty) set of data items where $Y_H(y|x,d)$, the hypothesis describing the predicted class of item x depends on the choice of training sets d.

$$\forall x \in B \;\; \exists d, d' \in D \bullet Y_H(y|x,d) \neq Y_H(y|x,d').$$

So now lets look at what this means in terms of our estimates of the bias of the classifier. This will of course depend on the methods used for the estimates. Following well-established previous research, we will assume that each item in the data set is predicted exactly k times. This is true with $k=1$ for N-fold cross validation, and for $k>1$ for the Webb and Conilione approach, in general although interestingly not for the Kohavi approach [21]. This means that when we sum over the data items x in the counterpart of (6.7) each term occurs with equal probability.

Note that bias$_x$ as stated above is composed of terms which themselves depend on the choice of training sets, and that we are assuming a fixed set of data points and a fixed size training sets. We therefore refine the definition of bias to take these into account, and average over all possible training sets.

$$\text{bias}^2 = \frac{1}{2}\sum_{x\in X} P(x) \sum_{d\in D} P(d|X,n) \sum_{y\in Y} [P(F(x)=y) - P(H_{Cd}(x)=y)]^2. \tag{6.8}$$

If we assume we are sampling iid then $P(x) = 1/|X|$ and $P(d|X,n) = 1/|D|$. We now turn our attention to the case where each data item $x \in X$ is unambiguously associated with one of two possible class labels $y \in Y$, and we will further constrain our ensemble to output crisp decisions so that $P(H_{Cd}(x) = y) \in \{0,1\}$. Partitioning the data set X as above, we note that we make use of the following conditions when performing the summation. First, the set A^+ does not contribute to the bias since the predicted class for this subset of items is always correct. Second, $\forall x \in X, C, d \in D, \exists y_1, y_2 \in Y, y_1 \neq y_2 : F(x) = y_1 \wedge H_{CD}(x) = y_2$. This means that within the partition A^- for each combination of x and d, there are exactly two values of y which both contribute $+1$ to the summation. This yields:

$$\begin{aligned}\text{bias}^2 &= \frac{1}{2}\sum_{x\in A^-} P(x) \sum_{d\in D} p(d|X,n)\cdot 2 \\ &+\frac{1}{2}\sum_{x\in B} P(x) \sum_{d\in D} P(d|X,n) \sum_{y\in Y} [P(F(x)=y) - P(H_{Cd}(x)=y)]^2, \qquad (6.9)\\ &= \frac{|A^-|}{|X|} + \frac{1}{2}\frac{|B|}{|X|}\cdot\frac{n!(|X|-n)!}{|X|!}\sum_{x,d}\sum_{y\in Y}[P(F(x)=y) - P(H_{Cd}(x)=y)]^2.\end{aligned} \tag{6.10}$$

The last term will take a value between 0 and $|B|/|X|$ since for each value of y the difference will be 0 for some training sets and 1 for others which the gives bounds:

$$|A^-|/|X| < \text{bias}^2 < (|A^-| + |B|)/|X|. \tag{6.11}$$

This reformulation makes it explicit that considering the proportion of samples which the ensemble always misclassifies will yield a strict underestimate of the bias provided that there exist any items for which the prediction made is dependent on the training set. Furthermore, since according to (6.7) the variance term is always nonnegative, we can say that the quantity $|A^-|/|X|$ constitutes a strict lower bound on the error rate of a classifier—or an ensemble treated as a single entity.

6.5.2 Experimental Approach

Previous sections illustrated the successful use of regression models built from a variety of data set–classifier combinations to predict the error rates that could be attained after future training. However, decomposing the error into different components is not straightforward for ensembles of classifiers [9]. Moreover, this would require running N-fold cross validation a number of times to get accurate estimates of bias and variance components for each combination of data set, algorithm, and n. This becomes computationally expensive when extended to a heterogeneous ensemble, particularly if the ensemble is itself trainable.

For this section, we use a slightly different approach. Previously we pooled the results from many experiments to build regression models relating observations of bias and variance after different values of n training data to the same variables of $n+n'$ items. Here we treat each data set independently, and build regression models to characterize the ensemble's learning curve as a function of n. As noted above, there is theoretical [36] as well as empirical [10, 13, 26] evidence that these learning curves have a power–law dependency on the number of training samples, i.e., they are of the form

$$\text{error}_{\text{ensemble}} = a \cdot n^b + c, \tag{6.12}$$

where a is the learning rate, b the decay rate, and c the Bayes error (the minimum achievable error or, in the error-decomposition framework, the "noise").

In our experiments, the bound on the ensemble's error derived in Sect. 6.5.1, $|A^-|/|X|$, was used as an estimate of the minimum achievable error. When faced with a new data set–ensemble combination, we make observations of $|A^-|$ and the ensemble error at regular intervals, and then feed these into the power–law regression model in order to fine-tune the parameters of the model so that it fits the new data and predicts the future development of the ensemble error, as will be detailed in Sect. 6.5.4. Before elaborating on these results, in Sect. 6.5.3 we analyze the stability of the estimation of the lower bound on the error by using $|A^-|/|X|$.

6.5.3 Analysis of the Stability of Estimators of Lower Bounds on Error

For the experiments performed here, 22 Machine Vision data sets from the DynaVis project were used (2 different feature spaces—17 and 74 features—for each of 5 CD-Print, 5 Artificial, and the Egg image sets). The CART [7] and C4.5 [30] decision trees, the Naive Bayes [13], Nearest Neighbor [11], and eVQ [24] classifiers were used as base classifiers, the decisions of which were combined using the Discounted Dempster–Shafer ensemble training method [32]. For each data set, each classifier, and each value of $n \in \{100, 120, \ldots, 1{,}000\}$ samples, N-fold cross-validation was repeated l times to make l predictions of the class of each item in the training set. From this data, we calculated the values of $|A^-|/|X|$ as a function of n for each data set (i.e., 22 values for each value of n). For clarity, we denote the values $|A^-|/|X|$ hereafter as Or_n.

In order to examine the stability of the predicted bounds as n increased, we plotted Or_n against Or_{final} and used linear regression as before to fit a model of

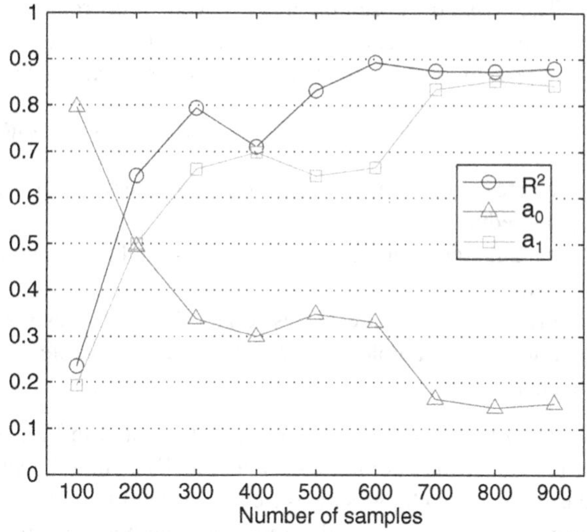

Fig. 6.5 The linear regression components for the correlation of Or_n vs. $\mathrm{Or}_{\mathrm{final}}$ together with the coefficient of determination R^2 as a function of n

the form $\mathrm{Or}_{\mathrm{final}} = a_1 \cdot \mathrm{Or}_n + a_0$, and to estimate the quality of the model via R^2. Figure 6.5 shows the progression of the coefficients a_0 and a_1 and the corresponding values of R^2 as a function of n.

As can be seen in Fig. 6.5, the models generated as n increases do produce predictions which correlate well to the observed values after further training. However, as can be seen by the progression of the coefficients, the nature of the regression models changes. For low values of n the models predict a high constant value for $\mathrm{Or}_{\mathrm{final}}$ with a low component related to the observed value of Or_n—essentially the system has not seen enough "difficult" samples. Since the major component of the predicted value of $\mathrm{Or}_{\mathrm{final}}$ is fixed for $n = 100$, the correlation is fairly low. As n increases and a more representative sample of the data is seen, the situation changes. Thus for training set sizes $n \geq 700$, the predicted value is dominated by the observed value ($a_1 \approx 0.85$) with only a low constant component ($a_0 \approx 0.15$). For these training set sizes, R^2 increases to approximately 0.9.

6.5.4 Empirical Results for Predicting Lower Bounds and Total Errors

The values Or_n for different n can be used for predicting not just a lower bound on, but also an estimate of the error of a trained ensemble. The following procedure can be used:

1. Or_n is measured for different n and a constant regression is performed for these values, i.e., we obtain the constant OR which minimizes the Mean Square Error with the values of Or_n across different values of n. This value forms our estimate of the lower bound on the achievable error.

2. The errors the ensemble makes are also recorded for different n.
3. A power–law regression is performed for the ensemble errors, asymptotically approaching the estimated constant OR as calculated in step 1:

$$\text{error}_{\text{ensemble}} = a \cdot n^b + \text{OR}, \tag{6.13}$$

   where a and b are the regression parameters which are optimized in the regression procedure.
4. An analogous procedure is used to model the standard deviation of the observed values.
5. From the power–law regression model, we can estimate the error of the ensemble after $n+n'$ samples are presented and also some estimates of how the variation changes.

The results of this procedure are illustrated in Fig. 6.6. The five base classifiers listed above are combined using the Discounted Dempster–Shafer combination ensemble [32]. Or_n was measured for $n = \{100, 120, \ldots, 1{,}000\}$ samples. A constant regression was performed to model Or_n with a constant value and the obtained value is then used as an asymptote when modeling the ensemble errors. The errors the ensemble makes are again recorded for $n = \{100, 120, \ldots, 1{,}000\}$ samples and a (robust) regression model is built according to (6.12). The results of this procedure are illustrated in Fig. 6.6a for CD-Operator 4 and in Fig. 6.6b for Artificial 04. The values Or_n and the errors of the ensemble are shown for different n, as well as the regression models that are built for them, together with the estimated standard errors. Also the final error after evaluating the performance of the ensemble when it is trained on the entire data set is indicated, to show how accurately the errors are predicted for the ensemble when it would be trained using a larger number of training samples $(n+n')$.

First, in both cases the results show that the model of OR does as expected form a lower bound on the error. As can be seen from Fig. 6.6b, the use of the secondary robust regression method to predict the mean and standard deviation of the observed ensemble error (top set of curves) for the artificial data set, extrapolates well and the final observed error (large asterisk at $n+n' \approx 20{,}000$) falls inside these values. For the much smaller CD print data set the figure is less clear, and the estimated standard errors on the predicted asymptote Or_n (bottom set of curves) overlap those of the robust regression prediction. Nevertheless, again the observed final ensemble error lies within one standard deviation of the value predicted by the robust regression procedure.

6.6 Application to Nonstationary Environments

In this chapter, we have outlined two ways in which a decomposition of the observed error into bias and variance terms can lead to useful predictions of future behavior. In the last section, we also introduced a means of rapidly estimating a

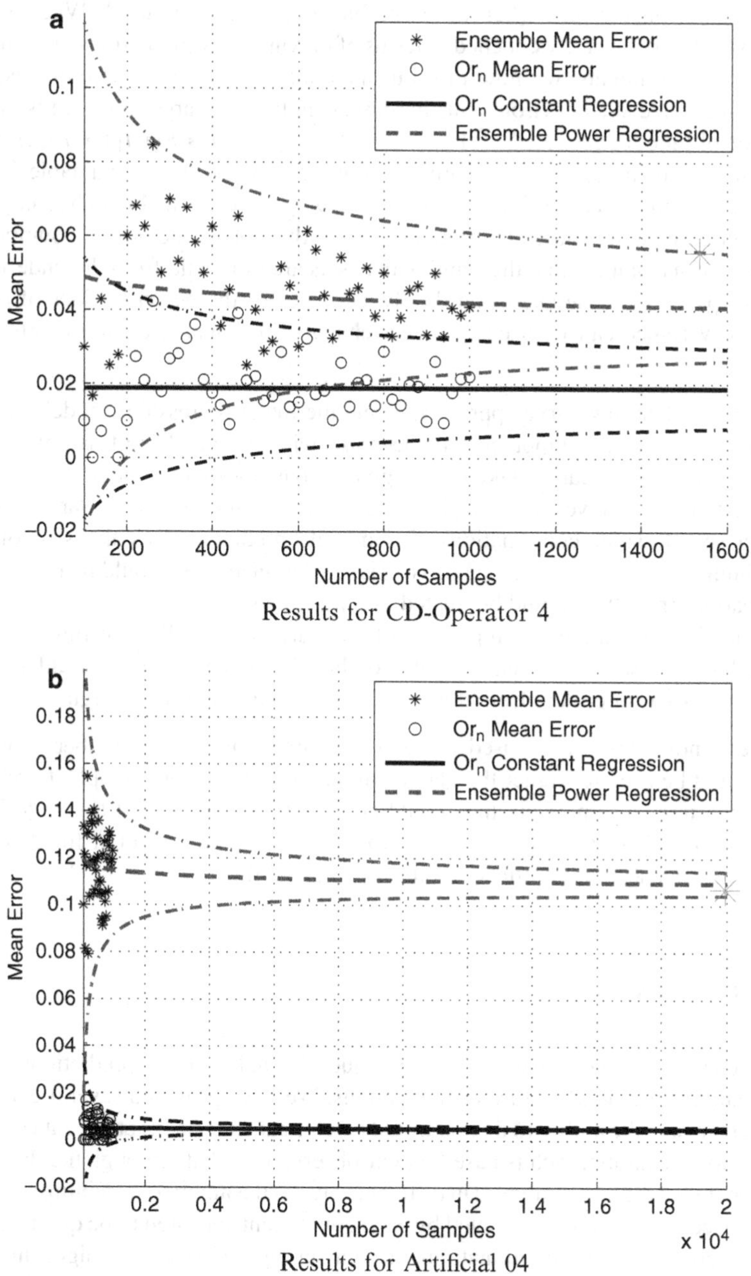

Fig. 6.6 Prediction of the errors of a (trainable) ensemble for $n+n'$ samples using a regression model which is built using n training samples. The mean values of Or_n and the mean errors of the ensemble are shown, together with their regression models (including standard errors). The error when training on the entire data set is depicted by the large $*$

lower bound on the bias when an ensemble of classifiers is used. Whereas the discussion has been partly couched in terms of making early predictions of accuracy in static environments, we now turn our attention to dynamic environments. We will characterize such environments by observing that the target function is now a function of time, i.e., $F : X,T \rightarrow Y$ and $\exists x \in X, t_1, t_2, t_1 < t_2 \bullet F(x,t_1) \neq F(x,t_2)$.

In this case the approaches outlined above can also act as a valuable "early warning system", because since for example $F(x,t_2)$ is not available to be sample at t_1 and vice versa, dynamic environments explicitly violate one of the assumptions of the analysis, namely that the training data sets are drawn iid from the underlying sample space. Thus the use of on-the-fly estimates of the *current* values of error, bias and variance can be used to detect changes in dynamic environments—for example:

- Both the "learning curve approaches" and the linear regressions models predict that the variance component will decrease with the number of samples. Any observed increase can be taken as a sign of a dynamic environment.
- Although the relative size of the bias and variance components will vary between data set–classifier combinations, our models predict that as the size of the training set increases then the ratio of bias to variance term should increase. Any departure from this should be treated as a warning sign.
- If the bias and variance components of error are periodically re-estimated from the last few samples, then the ratio of bias to variance term should remain constant. Any departure from this should be treated as a warning sign.

These indicators can be used as "early warning" signs in a number of ways. They could be used to trigger the classification algorithm—for example, to rebuild a decision tree. Alternatively they could be used to recognize that the underlying algorithm itself need changing to one which can explicitly account for dynamic situations, such as those outlined in other chapters of this book.

6.7 Conclusion

In this chapter, we have investigated techniques for making early predictions of the error rate achievable after further interactions. We have provided several example scenarios where the ability to do this would be of great value in practical data mining applications. Our approach is based on our observations that although the different components of the error progress in different ways as the number of training samples is increased, the behavior displayed by each component appeared to be qualitatively similar across different combinations of data set and classification algorithm. To investigate this finding, we have created a large set of results for many different combinations of data set, algorithm, and training set size (n) and applied statistical techniques to examine the relationship between the values observed after partial training (with n samples) and those after full training.

Perhaps surprisingly, the experimental results showed that in fact a simple linear model provided a highly accurately predictor for the subsequent behavior of different components. The results confirmed our hypothesis that these could be combined to produce highly accurate predictions of the total observed error. As there is no bias–variance(–covariance) decomposition available for 0/1 loss functions for ensembles of classifiers, it is not straightforward to apply the methodology used to accurately predict the performance of classifiers after further training to ensembles of classifiers. We have shown how a reformulation of the bias component can provide an estimate of the lower bound on the achievable error which may be more easily computed. This is especially important when the cost of training is high—for example, with trainable ensembles of classifiers. This bound is used as an asymptote in a power–law regression model to accurately predict the progression of the ensemble's error, independently for each data set.

For future work, we will focus in two directions. First, we will combine previous theoretical findings and the successful results from the two different approaches here. Taken together they suggest that for even more accurate predictions, it is worth combining the linear model for bias with an inverse power law model for variance using both the current estimates and the period over which to predict (n') as factors. This can be expected to prove particularly useful for classifiers where variance forms a major part of the observed error. Second, the work presented in this chapter used Kohavi and Wolpert's definition of bias and variance, and we will investigate whether using other definitions of bias and variance further improve the predicted accuracy.

Acknowledgements This work was supported by the European Commission (project Contract No. STRP016429, acronym DYNAVIS). This publication reflects only the authors' views.

References

1. P. L. Bartlett, S. Boucheron, and G. Lugosi. Model selection and error estimation. *Machine Learning*, **48**(1), 85–113 (2002)
2. S. Boucheron, O. Bousquet, and G. Lugosi. Theory of classification: A survey of some recent advances. *ESAIM: P&S*, **9**, 323–375 (2005)
3. O. Bousquet, S. Boucheron, and G. Lugosi. Introduction to statistical learning theory. *Advanced Lectures on Machine Learning*, pp. 169–207 (2004)
4. L. Breiman. Bagging predictors. *Machine Learning*, **24**(2), 123–140 (1996)
5. L. Breiman. Bias, variance, and arcing classifiers. Technical Report 460, Statistics Department, University of California, Berkeley, CA, 1996
6. L. Breiman. Random forests. *Machine Learning*, **45**(1), 5–32 (2001)
7. L. Breiman, J. H. Friedman, R. A. Olshen, and C. J. Stone. *Classification and Regression Trees*. Wadsworth International Group, Belmont, California. (1994)
8. D. Brian and G.I. Webb. On the effect of data set size on bias and variance in classification learning. In *Proceedings of the 4th Australian Knowledge Acquisition Workshop*, pp. 117–128 (1999)
9. G. Brown, J. Wyatt, R. Harris, and X. Yao. Diversity creation methods: A survey and categorisation. *Journal of Information Fusion*, **6**(1), 5–20 (2005)

10. C. Cortes, L.D. Jackel, S.A. Solla, V. Vapnik, and J.S. Denker. Learning curves: Asymptotic values and rate of convergence. In *Advances in Neural Information Processing Systems: 6*, pp. 327–334 (1994)
11. T.M. Cover and P.E. Hart. Nearest neighbor pattern classification. *IEEE Transactions on Information Theory*, **13**(1), 21–27 (1967)
12. Pedro Domingos. A unified bias–variance decomposition and its applications. In *Proceedings of the 17th International Conference on Machine Learning*, pp. 231–238. Morgan Kaufmann, San Francisco (2000)
13. R O. Duda, P. E. Hart, and D. G. Stork. *Pattern Classification*. Wiley Interscience, 2nd edition, New York (2000)
14. C. Eitzinger, W. Heidl, E. Lughofer, S. Raiser, J.E. Smith, M.A. Tahir, D. Sannen and H. Van Brussel. Assessment of the Influence of Adaptive Components in Trainable Surface Inspection Systems, *Machine Vision and Applications*, **21**(5), 613–626 (2010)
15. Yoav Freund and Robert E. Shapire. Experiments with a new boosting algorithm. In *Proceedings of 13th International Conference on Machine Learning*, pp. 148–156. (1996)
16. J. H. Friedman. On bias, variance, 0/1-loss, and the curse of dimensionality. *Data Mining and Knowledge Discovery*, **1**(1), 55–77 (2000)
17. S. Geman, E. Bienenstock, and R. Doursat. Neural networks and the bias/variance dilemma. *Neural Computation*, textbf4, 1–48 (1995)
18. T. Hastie, R. Tibshirani, and J. Friedman. *The Elements of Statistical Learning: Data Mining, Inference and Prediction.* Springer Verlag, New York, Heidelberg, London (2001)
19. G. James. Variance and bias for general loss functions. *Machine Learning*, **51**(2), 115–135 (2003)
20. R. Kohavi. The power of decision tables. In *Proceedings of the 8th European Conference on Machine Learning*, pp. 174–189. Springer Verlag, London, UK (1995)
21. R. Kohavi and D. H. Wolpert. Bias plus variance decomposition for zero-one loss functions. In *Proceedings of the 13th International Conference on Machine Learning.*, pp. 275–283 (1996)
22. B. E. Kong and T. G. Dietterich. Error-correcting output coding corrects bias and variance. In *Proceedings of the 12th International Conference on Machine Learning*, pp. 313–321, San Francisco, Morgan Kaufmann (1995)
23. R. Leite and P. Brazdil. Improving Progressive Sampling via Meta-learning on Learning Curves. In *Proceedings of the European Conference on machine Learning (ECML)* pp. 250–261 (2004)
24. E. Lughofer. Extensions of vector quantization for incremental clustering. *Pattern Recognition*, **41**(3), 995–1011 (2008)
25. E. Lughofer, J.E. Smith, M.A. Tahir, P. Caleb-Solly, C. Eitzinger, D. Sannen and M. Nuttin. Human–Machine Interaction Issues in Quality Control Based on On-Line Image Classification, IEEE Transactions on Systems, Man and Cybernetics, Part A: Systems and Humans, **39**(5), 960–971 (2009)
26. S. Mukherjee, P. Tamayo, S. Rogers, R.M. Rifkin, A. Engle, C. Campbell, T.R. Golub, and J.P. Mesirov. Estimating dataset size requirements for classifying DNA microarray data. *Journal of Computational Biology*, **10**(2), 119–142 (2003)
27. Bartlett P.L. and S Mendelson. Rademacher and gaussian complexities: Risk bounds and structural results. *Journal of Machine Learning Research*, **3**, 463–482 (2002)
28. J. Platt. Fast training of support vector machines using sequential minimal optimization. In B. Schoelkopf, C. Burges, and A. Smola, editors, *Advances in Kernel Methods—Support Vector Learning*. MIT Press, Mass (1998)
29. F.J. Provost, D. Jensen and T. Oates. Efficient Progressive Sampling. In *Proceedings of Knowledge Discovery in Databases (KDD)*, pp. 23–32 (1999)
30. J.R. Quinlan. *C4.5: Programs for Machine Learning*. Morgan Kaufmann, San Francisco (1993)
31. J. J. Rodriguez, C. J. Alonso, and O. J. Prieto. Bias and variance of rotation-based ensembles. In *Computational Intelligence and Bioinspired Systems*, number 3512 in Lecture Notes in Computer Science, pp. 779–786. Springer, Berlin, Heidelberg (2005)

32. D. Sannen, H. Van Brussel, and M. Nuttin. Classifier fusion using discounted Dempster–Shafer combination. In *Proceedings of the 5th International Conference on Machine Learning and Data Mining, Poster Proceedings*, pp. 216–230 (2007)
33. J. E. Smith and M. A. Tahir. Stop wasting time: On predicting the success or failure of learning for industrial applications. In *Proceedings of the 8th International Conference on Intelligent Data Engineering and Automated Learning (IDEAL'08)*, number 4881 in Lecture Notes in Computer Science, pp. 673–683. Springer Verlag, Berlin, Heidelberg (2007)
34. M. Stone. Cross-validatory choice and assessment of statistical predictions. *Journal of the Royal Statistical Society*, **36**, 111–147 (1974)
35. V. Vapnik. *The Nature of Statistical Learning Theory*. Springer, New York (2000)
36. V. Vapnik and O. Chapelle. Bounds on error expectation for support vector machines. *Neural Computation*, **12**(9), 2013–2036 (2000)
37. G. I. Webb. Multiboosting: A technique for combining boosting and wagging. *Machine Learning*, **40**(2), 159–196 (2000)
38. Geoffrey I. Webb and Paul Conilione. Estimating bias and variance from data. Technical report, Monash University, http://www.csse.monash.edu.au/webb/Files/WebbConilione03.pdf (2004)
39. I.H. Witten and E. Frank. *Data Mining: Practical machine learning tools and techniques*. Morgan Kaufmann, San Francisco, 2nd edition (2005)

Chapter 7
Incremental Classifier Fusion and Its Applications in Industrial Monitoring and Diagnostics

Davy Sannen, Jean-Michel Papy, Steve Vandenplas, Edwin Lughofer, and Hendrik Van Brussel

Abstract Pattern recognition techniques have shown their usefulness for monitoring and diagnosing many industrial applications. The increasing production rates and the growing databases generated by these applications require learning techniques that can adapt their models incrementally, without revisiting previously used data. Ensembles of classifiers have been shown to improve the predictive accuracy as well as the robustness of classification systems. In this work, several well-known classifier fusion methods (*Fuzzy Integral*, *Decision Templates*, *Dempster–Shafer Combination*, and *Discounted Dempster–Shafer Combination*) are extended to allow incremental adaptation. Additionally, an incremental classifier fusion method using an evolving clustering approach is introduced—named *Incremental Direct Cluster-based* ensemble. A framework for strict incremental learning is proposed in which the ensemble and its member classifiers are adapted concurrently. The proposed incremental classifier fusion methods are evaluated within this framework for two industrial applications: online visual quality inspection of CD imprints and prediction of maintenance actions for copiers from a large historical database.

D. Sannen (✉) • J.-M. Papy • S. Vandenplas
Flanders Mechatronics Technology Centre, Celestijnenlaan 300B,
B-3001 Leuven, Belgium
e-mail: davysannen@gmail.com; Jean-Michel.Papy@fmtc.be; Steve.Vandenplas@fmtc.be

E. Lughofer
Department of Knowledge-Based Mathematical Systems, Johannes Kepler
University Linz, Altenbergerstrasse 69, A-4040 Linz, Austria
e-mail: edwin.lughofer@jku.at

H. Van Brussel
Department of Mechanical Engineering, Katholieke Universiteit Leuven, Celestijnenlaan 300B,
B-3001 Leuven, Belgium
e-mail: hendrik.vanbrussel@mech.kuleuven.be

M. Sayed-Mouchaweh and E. Lughofer (eds.), *Learning in Non-Stationary Environments: Methods and Applications*, DOI 10.1007/978-1-4419-8020-5_7,
© Springer Science+Business Media New York 2012

7.1 Introduction

Automatic monitoring and diagnostics can be of great added value in industrial production and manufacturing environments. It increases the performance of the production processes and it increases the uptime of the production facilities. It ensures that the parts are produced according to the specifications and helps finding the root causes of problems. As the complexity of these machines increases, so does the need for integrated intelligence, including smart diagnostics, predictive maintenance, automatic quality inspection and decision support systems. In order to effectively use the large amounts of data the complex machinery is producing, pattern recognition [11,57] and data mining [21,40] have shown their usefulness in many applications, including (visual) quality inspection [51], fault detection [38], predictive maintenance [1], machine health prognostics [14], and decision support systems [13].

Common choices for constructing the classification models are *Support Vector Machines* [60], *rule-based classifiers* [47], *Decision Trees* [4, 45], *Neural Networks* [62] and *Gaussian Mixture Models* [54]. However, the "No Free Lunch" theorem [64] shows that no classification outperforms all other classification algorithms for all data sets, if no prior knowledge about the data exists. In practice, this means that often the appropriate classification algorithm is chosen based on a trial-and-error procedure in which many different classifiers are evaluated.

In order to alleviate the above-mentioned problem it can be beneficial to use multiple classifier systems (also called *ensembles* of classifiers) [30, 44], which produce their decision based on the decisions of a set of different classifiers. The combination algorithm used by the ensemble tries to exploit the diversity between the classifiers to produce a prediction which is more robust and more accurate than the ensemble's member classifiers. There are several reasons why ensembles can perform better than single classifiers, based on statistical, computational, and representational motivations [9]. Two necessary and sufficient conditions for ensembles of classifiers for achieving a higher accuracy than any of their individual members are [20, 26]:

1. The classifiers are accurate (they perform better than random guessing).
2. The classifiers are diverse (their errors are uncorrelated).

As the complexity and the degree of autonomy of industrial machines grows continuously, so does the need for intelligent analysis of their behavior. This means it is required for the machines to log sensor data, operating modes, operating conditions, etc. As this data is usually recorded online, while the system is fully operational, more and more industrial systems are producing large amounts of data. For these systems, there are two main scenarios in which incremental learning is required:

Computational resources Industrial systems can produce huge amounts of data, possibly even stored in Very Large DataBases (VLDBs).The learning algorithms should not iterate over the data repeatedly, as is done by many classification algorithms, because this would consume too much computational and memory resources.

Online availability When running online in a production environment, the measured data can be considered as a data stream [18]. This data stream can be thought of as an ordered sequence of data items which arrive more or less continuously as time progresses. Modeling such data streams requires a permanent updating of the classifiers in order to take into account the changing system dynamics. Updating the classification system (which can even be running on a Digital Signal Processor (DSP) or a simple microprocessor) should in this case not revisit data previously considered, as this would be too time consuming.

The settings in which the incremental learning is to be applied can be categorized mainly along two different axes. The first axis is the data itself which is to be learned: the target concept might remain fixed over time (referred to as *true incremental learning*) or might change over time (referred to as *concept drift*) [48, 63]. The second axis is the way the data becomes available to the classification system: the data might be presented to the classification system in (relatively small) batches (referred to as *incremental data mining*) or it might be presented one instance at a time [referred to as *incremental machine learning (IML)*] [24]. IML is in [56] further divided into *weakly incremental* and *strictly incremental* methods. Weakly incremental method require additional computation or memory when the number of data samples increases, in contrast to strictly incremental methods.

A number of incremental classification algorithms have been developed for various incremental learning settings, including incremental Support Vector Machines [8, 16, 48], incremental Decision Trees [6, 58], incremental discriminant analysis [42], and evolving fuzzy classifiers [37]. These classifiers are able to adapt their models and parameters based on the newly arriving data samples.

As accuracy, robustness, and reliability are highly important requirements for industrial monitoring and diagnostics systems, it would be interesting to also have ensemble methods at our disposal which can be applied in incremental learning settings. Fixed ensemble methods (such as *Voting* [32]) are a first candidate as they do not need any updating (they simply apply fixed rules without any optimization based on training data). These algorithms have interesting properties such as their simplicity and the good results they can produce in certain situations. However, they are usually suboptimal. Trainable ensembles, on the other hand, are optimized using the available training data and the outputs of their member classifiers for this data (see, e.g. [30, 44]). Of course, these algorithms also need incremental updating if they are to be applied in online classification systems, in contrast to the fixed ensemble algorithms.

In this work *true incremental machine learning*, as defined above, is considered. This means that the target concept is considered to be fixed over time (although the presented algorithms are also able to incorporate new target classes which might not be available during the initial training), and the newly arriving data samples can be presented to the learning algorithms whenever they become available and discarded immediately afterwards.

The remainder of this work is organized as follows. First, the proposed classification system architecture for the incremental adaptation of both the ensemble and its member classifiers is described in Sect. 7.2, together with an overview of related work. Section 7.3 introduces classifier fusion more formally and Sect. 7.4 describes the incremental extensions of several batch classifier fusion methods (*Fuzzy Integral* [5, 17], *Decision Templates* [27, 31], *Dempster–Shafer Combination* [46] and *Discounted Dempster–Shafer Combination* [52]), together with the proposed Incremental Direct Cluster-based ensemble method, in detail. Section 7.5 demonstrates the effectiveness of the proposed framework and the developed incremental classifier fusion methods therein for several industrial monitoring and diagnostics applications. Finally, the conclusions which can be drawn from this work are formulated in Sect. 7.6.

7.2 Proposed Architecture for Incremental Classifier Fusion and Related Work

Different strategies for incrementally updating an ensemble of classifiers have been explored [29], including:

Dynamic combiners The individual classifiers in the ensemble are trained off-line during the initial training; only the combination algorithm is adapted online. One example of this strategy is the *Mixture of Experts* methodology [23]. Another example is the use of trainable classifier fusion methods which can be updated incrementally (such as *Naive Bayes* [66] and *Behavior-Knowledge Space* (BKS) [22]), as is done in [34].

Updating the ensemble members Newly arriving data is used to adapt the individual classifiers in the ensemble online; the combination algorithm might or might not be adapted. Typically the data is sampled appropriately before using it to adapt the different classifiers. Algorithms using this strategy include the online *Bagging* and *Boosting* algorithms [41], the *Pasting Small Votes* method [3] and the *Learn++* algorithm [43].

Structural changes The classifiers are re-evaluated and can be removed or replaced by a newly trained classifier, based on this evaluation. An example of this strategy is presented in [61], in which the classifiers are evaluated on the most recent block of data.

The approach followed in this work is somewhat different from the strategies described above. A classification framework is presented in this section which allows a sample-wise, strictly incremental learning. For combining the decisions of the different individual classifiers, adaptive extensions are presented of a number of well-known batch classifier fusion algorithms [30]: *Fuzzy Integral* [5, 17], *Decision Templates* [27, 31] and ensembles based on *Dempster–Shafer theory* [46, 52]. In this sense, this work can be seen as an extension of the work presented in [34], in which *Naive Bayes* [66] and *BKS* [22] are used as the combination algorithm. However, in this work the ensemble's member classifiers are updated as well, concurrently with the combination algorithm (in this context we use "concurrently" to indicate that the classifiers as well as the combination algorithms are adapted continuously, sample by sample; neither is kept fixed at any stage of the incremental adaptation process). Besides the adaptive extensions of the batch classifier fusion algorithms mentioned above, a novel ensemble method based on an evolving clustering based-approach (*eVQ* [35]) is proposed which is capable of batch as well as incremental learning—named *Incremental Direct Cluster-based* ensemble.

The architecture of the proposed classification system which allows an incremental adaptation of the ensemble method as well as its member classifiers is visualized in Fig. 7.1. The top part of this flow diagram shows the initial (off-line) training phase in which the classifiers and ensemble are trained in batch mode. This modus operandi usually results in better performance compared to starting the incremental learning completely from scratch. During this phase good parameter settings can be determined for the classification algorithms, which can afterwards also be used during the online adaptation. In our case, these parameter settings are determined using a cross-validation (CV) procedure [55] coupled to a parameter grid search. After a set of (diverse) classifiers has been trained using this approach an ensemble algorithm is trained (also in batch mode) on the outputs of the individual classifiers together with the target labels (a detailed description is given in Sect. 7.4).

In the bottom left part of the flow diagram in Fig. 7.1, the incremental adaptation phase is shown. The adaptation, based on a newly arriving data sample and the corresponding feedback from a human operator (if available), is first done for the individual classifiers and then for the ensemble. All training data (in batch mode as well as during the incremental adaptation) is presented to both the classifiers and the ensemble algorithm. This ensures they are all up-to-date at any given time (so-called "any-time-learning") and they are all trained using the maximum amount of information.

In industrial applications, often certain target classes are strongly underrepresented in the data. In order to avoid an "unlearning" effect for the underrepresented classes the system is only adapted in the following cases (termed as *balanced learning*—a similar concept is proposed in [39] for updating individual classifiers based on the feedback of the operators):

1. Whenever the operator overrules a decision from the classifier: this increases the likelihood that the classifier will produce the correct decision when a similar image is presented again;
2. Whenever the relative proportion of the samples belonging to every class is equally balanced: in this case the refinement of the classifier always increases its classification accuracy; or
3. Whenever the relative proportion of the samples belonging to the current class is lower than the relative proportion of samples belonging to any other class: this enriches the classifier by balancing out the non-equal class distribution, which further increases the accuracy.

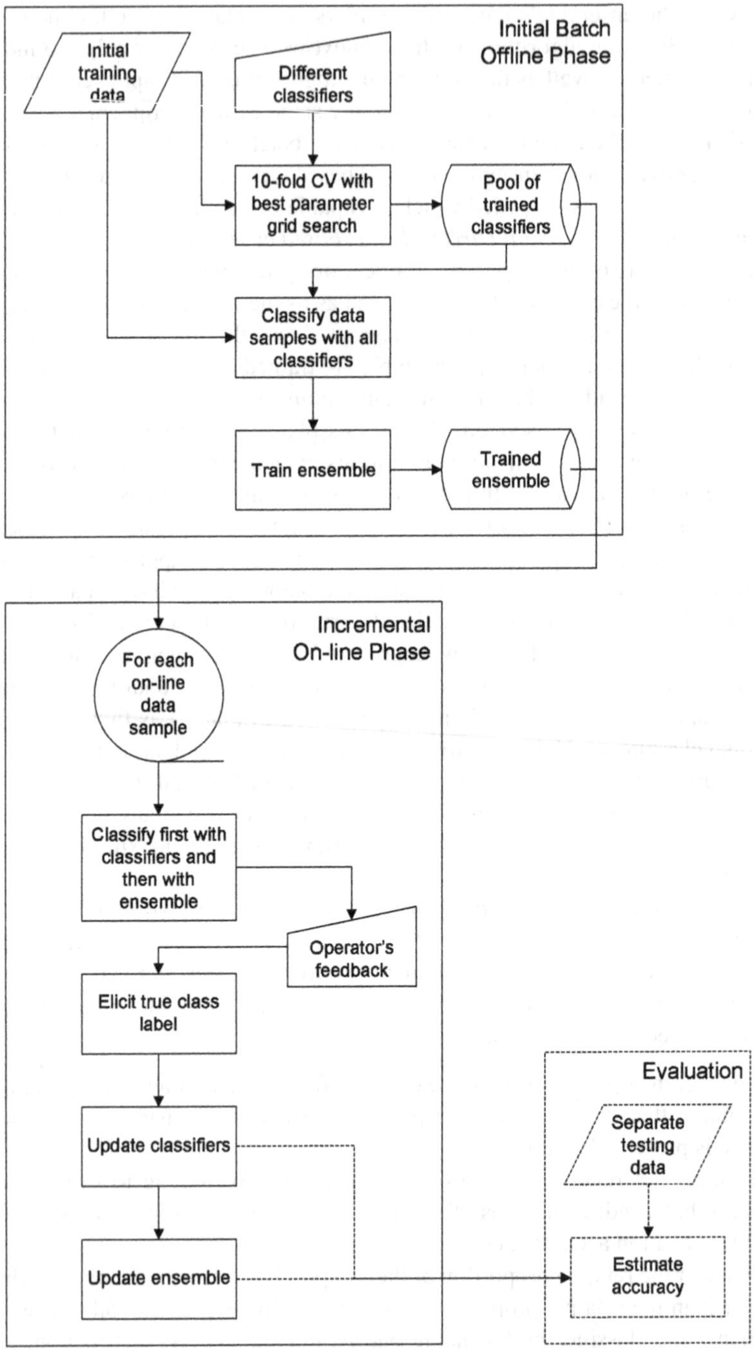

Fig. 7.1 Framework for the training and adaptation of the individual classifiers and the classifier fusion method.

In most production environments, time for providing feedback is very limited for the operators. It is therefore not mandatory for the operators to provide their feedback for every new data sample. In order to save the labeling workload, it is sufficient that they provide their input when they think the system makes an incorrect decision and they want to overrule this decision. *Active learning* concepts, where the feedback is only requested from the operators for data samples for which the classifier is uncertain in its predictions, may also be useful in this respect [36].

The bottom right part of Fig. 7.1 shows additional components (with dashed lines) used for evaluation and testing purposes (also used for obtaining the experimental results in Sect. 7.5). When data which is stored off-line is used in the evaluation (e.g., data captured online and stored in a database), the online operation can be simulated by incorporating the different data samples one by one into the learning system. This gives the operators more time to label the data (which is often not possible online).

7.3 Classifier Fusion: Basic Concepts and Notations

The combination methods in the ensemble that will be considered in this work are classifier fusion methods. In order to explain them in detail in Sect. 7.4 some basic concepts and definitions will be provided in this section, together with the notations that will be used, based on [30].

In general, two different ensemble strategies can be distinguished: generative and nongenerative ensembles [59]. *Generative* ensemble methods generate sets of classifiers, trying to actively improve their diversity and accuracy. Examples of algorithms using this strategy include the well-known *Bagging* [2] and *Boosting* [15] methods. In contrast, *non-generative* ensemble methods do not actively generate new classifiers but they try to combine the predictions of a set of different, existing classifiers in an appropriate way. Non-generative ensembles can be generally divided into two groups: *classifier selection* and *classifier fusion* [65]. The assumption of the former is that each classifier is an "expert" in some local part of the feature space. The latter assumes that all classifiers are trained over the entire feature space. In this work *classifier fusion* will be considered.

Let $\mathbf{x} \in \mathbb{R}^n$ be a feature vector and $\Omega = \{\omega_1, \ldots, \omega_c\}$ the set of labels for c mutually exclusive classes. $\tau(\mathbf{x}) \in \Omega$ denotes the class label assigned to $\mathbf{x}$. Suppose L classifiers $\{D_1, \ldots, D_L\}$ are available. In general, three types of information can be obtained for classification systems [30, 49, 66]:

Abstract level The classifier outputs a single class label; no information about uncertainty of the classifier's decision or possible alternatives is available.

Rank level The classifier ranks the possible classes in order of plausibility; this is especially suited for problems with large numbers of classes [30].

Measurement level The classifier outputs a vector $D_i(\mathbf{x}) = [d_{i,1}(\mathbf{x}), \ldots, d_{i,c}(\mathbf{x})]$. The values $d_{i,j}$ represent the support of classifier D_i for the hypothesis that vector

$\mathbf{x}$ belongs to class ω_j. Using a probabilistic interpretation, the $d_{i,j}(\mathbf{x})$ can be regarded as estimates of the posterior probabilities for the classes given $\mathbf{x}$ (i.e., $d_{i,j}(\mathbf{x}) = P(\omega_j|\mathbf{x})$). A more general interpretation of the classifier outputs is to consider them as the support (belief, possibility, etc.) for the classes [31]. Without loss of generality the condition $d_{i,j}(\mathbf{x}) \in [0,1]$ can be imposed.

The measurement level provides the most information, as quantitative information about alternative classifications is available. Hence, it is to be expected that combining this type of classifier outputs produces the most significant improvement in classification performance [49]. It should be noted that the other types can be easily represented as a measurement level output, so all further discussions will use the previous notation for classifiers producing measurement level outputs.

An intuitive and compact (matrix) representation of the outputs of all the classifiers in an ensemble for a given feature vector $\mathbf{x}$ is suggested in [27,31], named the *Decision Profile* (DP):

$$\mathrm{DP}(\mathbf{x}) = \begin{pmatrix} D_1(\mathbf{x}) \\ \vdots \\ D_L(\mathbf{x}) \end{pmatrix} = \begin{pmatrix} d_{1,1}(\mathbf{x}) & \dots & d_{1,c}(\mathbf{x}) \\ \vdots & & \vdots \\ d_{L,1}(\mathbf{x}) & \dots & d_{L,c}(\mathbf{x}) \end{pmatrix}. \tag{7.1}$$

The combined output of the L classifiers in the ensemble, μ, is then defined as

$$\mu(\mathbf{x}) = [\mu_1(\mathbf{x}), \dots, \mu_c(\mathbf{x})] = \mathscr{F}(D_1(\mathbf{x}), \dots, D_L(\mathbf{x})), \tag{7.2}$$

where $\mathscr{F}$ is called an *aggregation rule* [31].

As explained above, only the classifier outputs are considered by the classifier fusion methods. This provides an abstraction of the classification methods themselves and of whether they are being adapted incrementally or not. Hence, in general the batch classifier fusion methods (as well as their incremental extensions) can take the outputs of any set of batch and/or incremental classifiers as their input, as is possible as well in the framework proposed in Sect. 7.2.

The fusion of the outputs of the individual classifiers into one final decision can be done using fixed rules or trainable algorithms. *Fixed rules* are simple and can produce good results for some tasks, but they are usually suboptimal. To achieve better results *trainable fusion* methods can be used, which try to optimize the fused decision using some learning algorithm. These methods are discussed in detail in Sect. 7.4, together with their proposed adaptive extensions.

7.4 Classifier Fusion Algorithms and Their Incremental Extensions

Classifier fusion methods can be generally divided into two classes: class-conscious and class-indifferent methods [30, 31]. *Class-conscious* fusion methods use one column of the Decision Profile $\mathrm{DP}(\mathbf{x})$ at a time to classify a feature vector $\mathbf{x}$, taking into account that this column contains the individual classifiers' supports for class ω_j. *Class-indifferent* fusion methods ignore the structure of DP($\mathbf{x}$) and treat all values $d_{i,j}(\mathbf{x})$ as feature values in a new feature space. The final decision is made by another classifier that takes this new feature space as input and provides the final classification as its output.

In this section, different well-known batch classifier fusion methods are discussed in detail (based on [30]), together with extensions to make them capable of incremental learning. The remainder of this section is organized as follows. In Sect. 7.4.1 the batch algorithms for class-conscious classifier fusion are detailed, followed by their adaptive extensions in Sect. 7.4.2. In Sect. 7.4.3 the batch algorithms for class-indifferent classifier fusion are detailed, followed by their adaptive extensions (which are also able to incorporate new class labels) in Sect. 7.4.4. In Sect. 7.4.5, the Incremental Direct Cluster-based ensemble is presented. This method is actually a normal trainable class-indifferent classifier fusion method, but as this is a newly proposed method it is discussed in a bit more detail in a separate section.

7.4.1 Batch Algorithms for Class-Conscious Classifier Fusion

This section describes the batch learning and classification algorithms for a number of class-conscious classifier fusion methods. Different *fixed* classifier fusion methods and the *Fuzzy Integral* algorithm are described in Sects. 7.4.1.1 and 7.4.1.2, respectively.

7.4.1.1 Fixed Classifier Fusion

A popular and simple way to combine the outputs of classifiers is the use of a fixed combination rule. These simple rules can have a good performance in certain situations; however, they are usually suboptimal (as they are not adapted to the specific problem). A very well-known and widely used combination method is voting [32, 66]. Note that voting rules are applied to abstract level classifier outputs, so if the outputs of the classifiers in the ensemble are of the measurement level their outputs need to be made crisp first (e.g., by taking the maximum). Other rules use simple algebraic combiners [25, 28]. These algebraic combiners are used as the aggregation rule $\mathscr{F}$ in (7.2), applied separately to the classifier outputs for each of the classes (i.e., applied separately to each of the columns of DP($\mathbf{x}$)).

$$\mu_j(\mathbf{x}) = \mathscr{F}\left(d_{1,j}(\mathbf{x}),\ldots,d_{L,j}(\mathbf{x})\right), \quad j = 1,\ldots,c, \tag{7.3}$$

where $\mu_j(\mathbf{x})$ is the ensemble's output for class ω_j. Examples of such combiners are *maximum*, *minimum*, *product*, *mean*, and *median* [25, 28].

7.4.1.2 Fuzzy Integral

The fuzzy integral algorithm [5, 17] tries to find a compromise between the competence of the classifiers in the ensemble (represented by a fuzzy measure g) and the evidence for class ω_j (the support for ω_j by the different classifiers) [30].

Batch Training

In order to compute the fuzzy integral for the different classes, first the degrees of support for ω_j (column j of the DP) are sorted in descending order, resulting in $d_{*,j} = [d_{i_1,j}(\mathbf{x}),\ldots,d_{i_L,j}(\mathbf{x})]^{\mathsf{T}}$. For each $\alpha \in d_{*,j}$, the classifiers giving support for ω_j that is greater than or equal to α are identified. This subset of classifiers is called an *α-cut*, denoted as H_α. Second, *every subset* of classifiers in the ensemble must have a measure of competence assigned to it, indicating how good this *group of classifiers* is for the given input $\mathbf{x}$. This measure is called a *fuzzy measure, g.*

A problem is that usually a competence value is not available for each possible subset of classifiers. A solution is to calculate the so-called *λ-fuzzy measure*, which can provide confidence values for each of the classifier subsets, based on the competence values of the individual classifiers, $g^1,\ldots,g^L$ (called *fuzzy densities*). Typically the competence values for the individual classifiers are set to their (half of the) accuracies, estimated, e.g. on a separate data set or in a CV procedure.

The value of λ, needed to calculate the values of the λ-fuzzy measure g, is obtained as the unique real root of the polynomial

$$\lambda + 1 = \prod_{i=1}^{L} \left(1 + \lambda g^i\right), \quad 0 \neq \lambda > -1. \tag{7.4}$$

In order to compute the values of the fuzzy measure g, first the fuzzy densities $g^1,\ldots,g^L$ are arranged corresponding to the sorting used to obtain $d_{*,j}$, resulting in $g^{i_1},\ldots,g^{i_L}$. Then the following recursive calculation is computed [30]:

1. Set $g(1) = g^{i_1}$.
2. For $t = 2$ to L: $g(t) = g^{i_t} + g(t-1) + \lambda g^{i_t} g(t-1)$.

Here, $g(a)$ denotes the competence assigned to the subset of classifiers which are most confident in their prediction for class ω_j, as determined by the sorting of their supports for class ω_j to obtain $d_{*,j}$ (i.e., classifiers $D_{i_1},\ldots,D_{i_a}$).

Classification

To calculate the combined support for class ω_j of the ensemble, $\mu_j(\mathbf{x})$, the values of α and g are combined. The combined support for class ω_j calculated through the *Sugeno fuzzy integral* is given by

$$\mu_j(\mathbf{x}) = \max_{\alpha}\left[\min(\alpha, g(H_\alpha))\right], \quad j = 1,\ldots,c. \tag{7.5}$$

Filling in the value of α in (7.5), the final degree of support for class ω_j is calculated by

$$\mu_j(\mathbf{x}) = \max_{1\le k\le L}\left[\min\left(d_{i_k,j}(\mathbf{x}), g(k)\right)\right], \quad j = 1,\ldots,c. \tag{7.6}$$

Another popular type of fuzzy integral is the *Choquet fuzzy integral*. The calculations use the same λ-fuzzy measure g; the only difference with the calculation of the Sugeno fuzzy integral is (7.6), as detailed in, e.g., [30].

7.4.2 *Incremental Adaptation for Class-Conscious Classifier Fusion*

As the fixed classifier fusion algorithms are not optimized for the given data, they obviously need no adaptation. For the remaining class-conscious classifier fusion method discussed in Sect. 7.4.1, the Fuzzy Integral ensemble, it can be seen that the only part which is to be updated are the fuzzy densities, $g^1,\ldots,g^L$, after which the fuzzy integral is to be recomputed. As the fuzzy densities are usually represented by (half of) the accuracies of the classifiers, they need to be kept track of in an incremental way. This can be achieved by storing the number of samples that have been used for training so far, N^{tr}. When a new training sample becomes available, for each of the individual classifiers D_i, its updated accuracy, g^i_{new}, can be incrementally computed by

$$g^i_{\text{new}} = \begin{cases} \frac{g^i N^{\text{tr}}+1}{N^{\text{tr}}+1} & \text{If classifier } i \text{ correctly classified the new sample;} \\ \frac{g^i N^{\text{tr}}}{N^{\text{tr}}+1} & \text{If classifier } i \text{ incorrectly classified the new sample.} \end{cases} \tag{7.7}$$

After the accuracy estimates have been updated for each of the classifiers, N^{tr} is incremented by 1 (the new training sample has been used for updating the Fuzzy Integral ensemble).

Incorporation of New Classes

When a new classification target is added to the data set, this means that an additional column is added to the DP. As the class-conscious classifier fusion methods by definition use each of the columns of the DP independently, the new class can be incorporated in these methods without any problems; c simply needs to be replaced by $(c+1)$ in (7.3)–(7.6).

7.4.3 *Batch Algorithms for Class-Indifferent Classifier Fusion*

This section describes the batch learning and classification algorithms for a number of class-indifferent classifier fusion methods. The *Decision Templates*, *Dempster–Shafer combination*, and *Discounted Dempster–Shafer combination* algorithms are discussed in Sects. 7.4.3.1, 7.4.3.2 and 7.4.3.3, respectively.

7.4.3.1 Decision Templates

The idea of the *Decision Templates* combination method [27, 31] is to remember the most typical DP for each class ω_j, called the *Decision Template for class j* (DT_j). The combined decision for a feature vector $\mathbf{x}$ is determined by comparing its Decision Profile $\mathrm{DP}(\mathbf{x})$ with the Decision Templates for each of the classes, using some similarity measure $\mathcal{S}$. The closest match labels $\mathbf{x}$. This comes down to applying the *Nearest Mean* classifier [21] in the space of classifier outputs.

Batch Training

The decision template DT_j for class ω_j is calculated as the mean of the DPs of all the members of ω_j from the training data set $\mathbf{Z}$:

$$\mathrm{DT}_j = \frac{1}{N_j} \sum_{\substack{\mathbf{z}_k \in \mathbf{Z} \\ \tau(\mathbf{z}_k)=\omega_j}} \mathrm{DP}(\mathbf{z}_k), \tag{7.8}$$

where N_j is the number of elements of $\mathbf{Z}$ belonging to class ω_j.

Classification

In [27, 31] several measures of similarity are proposed, including the *Euclidean*, *Hamming*, and *Mahalanobis* distances. The combined classification using the Euclidean distance is, e.g., computed by

$$\mu_j(\mathbf{x}) = 1 - \frac{1}{L \times c} \sum_{i=1}^{L} \sum_{k=1}^{c} \left[(\mathrm{DT}_j)_{i,k} - d_{i,k}(\mathbf{x}) \right]^2, \; j = 1, \ldots, c, \tag{7.9}$$

where $(\mathrm{DT}_j)_{i,k}$ denotes the element at row i and column k of the matrix DT_j.

7.4.3.2 Dempster–Shafer Combination Ensemble

The *Dempster–Shafer theory of evidence* is a generalization of the Bayesian reasoning used to represent and combine evidences, introduced by Dempster [7] and Shafer [53]. Without going into the details of this theory, its application to classifier fusion will be explained in this section and Sect. 7.4.3.3.

Batch Training

In [46], a way to apply the Dempster–Shafer theory of evidence to the problem of classifier fusion is described. The Dempster–Shafer combination training is like the Decision Templates training: the c decision templates are calculated from the data set in the same way—see (7.8).

Classification

Although the Dempster–Shafer combination training is equal to the Decision Templates training, determining the combined output for a new data sample is different. Instead of calculating the similarity between the decision template and the DP, the "proximity" between the decision templates and the output of *each* classifier is calculated. Let DT_j^i denote row i of decision template DT_j and $D_i(\mathbf{x})$ the output of D_i (row i of the Decision Profile $DP(\mathbf{x})$), as defined in Sect. 7.3. The proximity Φ between DT_j^i and $D_i(\mathbf{x})$ is then calculated by

$$\Phi_j^i(\mathbf{x}) = \frac{\left(1+\left\|\mathrm{DT}_j^i - D_i(\mathbf{x})\right\|^2\right)^{-1}}{\sum_{k=1}^{c}\left(1+\left\|\mathrm{DT}_k^i - D_i(\mathbf{x})\right\|^2\right)^{-1}}, \tag{7.10}$$

where $||\cdot||$ can be any matrix norm (e.g. the Euclidean distance between two vectors). This results in L proximities for each of the c decision templates.

Using these proximities, for every class ω_j and for every classifier D_i the following degrees of *belief* are calculated (based on the orthogonal sum rule of the Dempster–Shafer theory of evidence—for a detailed discussion, see [46,52]):

$$b_j^i(\mathbf{x}) = \frac{\Phi_j^i(\mathbf{x})\prod_{k\neq j}\left(1-\Phi_k^i(\mathbf{x})\right)}{1-\Phi_j^i(\mathbf{x})\left[1-\prod_{k\neq j}\left(1-\Phi_k^i(\mathbf{x})\right)\right]}. \tag{7.11}$$

From these degrees of belief, the final combined degrees of support for each class ω_j are calculated as follows:

$$\mu_j(\mathbf{x}) = K\prod_{i=1}^{L} b_j^i(\mathbf{x}), \quad j = 1,\ldots,c, \tag{7.12}$$

where K is a normalizing constant.

7.4.3.3 Discounted Dempster–Shafer Combination Ensemble

An extension to the Dempster–Shafer combination ensemble [46] (see Sect. 7.4.3.2) is described in [52]. The idea is to augment the information used in the ensemble with the accuracies of the individual classifiers, next to the information about the most typical classifier outputs (represented by the Decision Templates) which is also used by the (standard) Dempster–Shafer combination method.

Batch Training

As with the Dempster–Shafer combination ensemble, the first part of the training of the Discounted Dempster–Shafer combination ensemble is the same as the Decision Templates training: the c decision templates are calculated from the data set (see (7.8)). Additionally, the accuracies of the classifiers are estimated, e.g. on a separate data set or in a CV procedure.

Classification

The so-called "proximities" between the decision templates and the output of *each* classifier is calculated in the same way as for the standard Dempster–Shafer combination method (see (7.10)). Suppose now that the estimated accuracies for the L classifiers are $\alpha_1, \ldots, \alpha_L$. Using the proximities, for every class ω_j and for every classifier D_i the degrees of *belief* are again calculated using the orthogonal sum rule of the Dempster–Shafer theory of evidence. However, also the estimated classifier accuracies are taken into account, in contrast to the standard Dempster–Shafer combination. This is done using the discounting operation. Equation (7.11) is then replaced by the following equation (for a detailed discussion see [52]):

$$b_j^i(\mathbf{x}) = \frac{\alpha_i \Phi_j^i(\mathbf{x}) \left[(1-\alpha_i) + \alpha_i \left(\prod_{k \neq j} \left(1 - \Phi_k^i(\mathbf{x})\right)\right)\right]}{1 - \alpha_i{}^2 \Phi_j^i(\mathbf{x}) \left[1 - \prod_{k \neq j} \left(1 - \Phi_k^i(\mathbf{x})\right)\right]}. \tag{7.13}$$

From these degrees of belief, the final degrees of support for each class ω_j can be calculated using again (7.12).

7.4.4 *Incremental Adaptation for Class-Indifferent Classifier Fusion*

The training of all three classifier fusion methods described in Sect. 7.4.3 is based on the computation of the Decision Templates. Hence, for these methods to be adapted incrementally, the Decision Templates need to be updated. In order to do so, the number of training samples *per class* in the training set $\mathbf{Z}$ needs to remembered. Let us denote these numbers as N_j^{tr}, for all classes $j = 1, \ldots, c$. When a new training sample $\mathbf{x}_{\text{adapt}}$, belonging to class $\omega_{j_{\text{adapt}}}$, becomes available, the new Decision Template for the corresponding class, $\text{DT}_{j_{\text{adapt}}}^{\text{new}}$, can be obtained as follows:

$$\text{DT}_{j_{\text{adapt}}}^{\text{new}} = \frac{N_{j_{\text{adapt}}}^{\text{tr}} * \text{DT}_{j_{\text{adapt}}} + DP\left(\mathbf{x}_{\text{adapt}}\right)}{N_{j_{\text{adapt}}}^{\text{tr}} + 1}, \tag{7.14}$$

where $*$ denotes the elementwise scalar matrix multiplication.

For the Discounted Dempster–Shafer combination method, also incremental estimates of the accuracies of the classifiers are required. These can be computed completely analogously to the Fuzzy Integral combination (see (7.7)).

Incorporation of New Classes

The class-indifferent classifier fusion methods can also be adapted so that they are able to incorporate new classes that might appear. Let us assume that at a certain point in the training process there are L classifiers in the ensemble, classifying the data into c classes, $\omega_1, \ldots, \omega_c$. When a new class, ω_{c+1}, is present in the data, the classifiers' output will contain an additional value (representing their support for the new class ω_{c+1}). The classifier fusion methods can then detect a new class was present in the data if the DP which is used as input to the ensemble is of size $L \times (c+1)$.

Note that the training of (and the classification by) the class-indifferent fusion methods depends on the Decision Templates. Before the new class is presented to the fusion methods, there are c Decision Templates, $\text{DT}_1, \ldots, \text{DT}_c$, of size $L \times c$. After the new class ω_{c+1} is introduced, the DPs produced by the classifiers are of size $L \times (c+1)$. Therefore, the outputs of the classifiers for the new class ω_{c+1} (i.e. column $c+1$ of the DP) should be ignored by the fusion algorithms when classifying new data, until at least one sample of class ω_j is presented in the training data. This can be achieved by simply removing the column $c+1$ from the DP that is presented to the fusion method.

As new training samples are coming in, the Decision Templates for the different classes can be updated and the new class ω_{c+1} can be incorporated. The first time a sample belongs to class ω_{c+1}, a new decision template DT_{c+1} for class ω_{c+1} is created and initialized to the DP of this sample. Obviously, column $c+1$ of the DP of new samples should also be used for comparison with DT_{c+1}. For updating DT_{c+1} with samples belonging to class ω_{c+1} coming in afterward, the procedure described in the beginning of this section can be used. The first time a sample belongs to class $\omega_j \neq \omega_{c+1}$, the Decision Template for class ω_j, DT_j, can be updated to incorporate class ω_{c+1}. This is done by adding the classifier supports for class ω_{c+1} into DT_j (i.e., copying column $c+1$ from the DP of the new training data to column $c+1$ of DT_j). From this point on, column $c+1$ of the DP of new samples should also be used for comparison with DT_j. For updating DT_j with samples belonging to class ω_j coming in afterward, the procedure described in the beginning of this section can again be used.

7.4.5 *Incremental Direct Cluster-Based Ensemble*

The motivation to apply clustering directly on the outputs of the base classifiers, defined by $o(\mathbf{x}) = (D_1(\mathbf{x}),\ldots,D_L(\mathbf{x}))$, is the assumption that the output vectors from different classes form different groups in the high-dimensional classifier output space. Even if this assumption does not (completely) hold, i.e., classes are overlapping in some regions, a reasonable classifier can be built by exploiting the relative frequencies of the classes in these local regions and the distances of new samples to the nearest decision boundary. Furthermore, as the method performs a clustering on the classifier output vectors and contains no inversion of covariance matrices, it does not suffer instabilities, caused by possibly correlated outputs of two or more classifiers (as pointed out in [31]).

Batch and Incremental Training

For deducing an incremental classifier fusion method an evolving vector quantization approach (*eVQ*) [35] is used, which adapts and evolves centers and surfaces of ellipsoidal clusters in a samplewise and single-pass manner. This method is inherently incremental; hence, there is no separate description of its batch training.

The update of the winning cluster center, $\mathbf{c}_{\text{win}}$ (i.e., the cluster center which is closest to the output of the classifiers for the current data sample, $\mathbf{o}^{\text{new}}$), is given by

$$\mathbf{c}_{\text{win}}^{\text{new}} = \mathbf{c}_{\text{win}}^{\text{old}} + \eta\left(\mathbf{o}^{\text{new}} - \mathbf{c}_{\text{win}}^{\text{old}}\right), \tag{7.15}$$

with the learning gain η monotonically decreasing with the number of samples forming each cluster, k_c (usually set to $\frac{0.5}{k_c}$).

The distance of a new classifier output vector $\mathbf{o}_{\text{new}}$ is not calculated to the cluster center (as done in conventional vector quantization [19]), but to the surface of the ellipsoids. The distance to cluster i, dist_i, is then calculated as follows (for a detailed discussion, see [35]):

$$\text{dist}_i = (1-t)\sqrt{\sum_{j=1}^{L\times c}\left(\mathbf{o}_j^{\text{new}} - \mathbf{c}_{ij}\right)^2}, \quad \text{with } t = \frac{1}{\sqrt{\sum_{j=1}^{L\times c}\frac{\left(\mathbf{o}_j^{\text{new}}-\mathbf{c}_{ij}\right)^2}{\sigma_{ij}^2}}}, \tag{7.16}$$

where σ_{ij}^2 denotes the spread of the data in dimension j for cluster i (as axis-parallel ellipsoids are used, this can be estimated by the variance of the data samples in the corresponding dimension).

The update of the ellipsoidal axes is done using the recursive variance estimation formula [33], i.e., by (here again for the winning cluster):

$$\forall j: k_{\text{win}}\sigma_{\text{win},j}^2(\text{new}) = (k_{\text{win}}-1)\,\sigma_{\text{win},j}^2(\text{old}) + k_i \Delta \mathbf{c}_{\text{win},j}^2 + \left(\mathbf{c}_{\text{win},j} - \mathbf{o}_j^{\text{new}}\right)^2, \quad (7.17)$$

with k_{win} the number of confidence vectors forming the winning cluster and $\Delta c_{\text{win},j}^2$ the quadratic difference between the old and the new (updated) cluster center. The evolution of a new cluster center is done by checking whether (7.16) is greater than a vigilance parameter ρ, which is defined in the range of $\left[0, \sqrt{L\times c}\right]$, with L the number of classifiers and c the number of classes. Note that the classifier outputs are already normalized, so $\sqrt{L\times c}$ represents the diagonal of the space and hence the largest distance. A good value for the vigilance parameter can be obtained with an offline best parameter grid search (see Sect. 7.2).

This algorithm is extended to classification problems by introducing a hit matrix, H, the rows of which represent the clusters and the columns of which represent the classes. The entry h_{ij} of this matrix simply contains the number of samples falling into cluster i and class j. This means each row of H contains the relative frequencies of each class falling into the corresponding cluster. Note that the incremental update of this matrix is straightforward by simply incrementing the entries h_{ij}, whenever a sample of class j is attached to cluster i.

Classification

When classifying a new data sample, a winner-takes-all approach is used where the nearest cluster is seen as the best representative to label the current sample. Similar to the training process it is simply checked whether the classifier outputs for the new data sample, $\mathbf{o}^{\text{new}}$, lies inside any evolved cluster by checking the following condition:

$$\exists i \in \{1,\ldots,C\}: \sum_{j=1}^{L\times c} \frac{\left(\mathbf{o}_j^{\text{new}} - \mathbf{c}_{ij}\right)^2}{\sigma_{ij}^2} \leq 1, \quad (7.18)$$

with C the number of clusters generated so far. If condition (7.18) is fulfilled, the distance of the classifier outputs for the data sample to be classified to all the clusters fulfilling the condition is calculated and the nearest cluster is elicited. If (7.18) is not fulfilled, the distance of the classifiers for the data sample to be classified to all evolved clusters is calculated using (7.16) and the nearest cluster is elicited. In this sense, the fused output for each of the classes, $\mu_j(\mathbf{x}),\ j = 1,\ldots,c$, is calculated by the relative frequencies of each class falling into the winning cluster:

$$\mu_j(\mathbf{x}) = \frac{h_{\mathrm{win},j}}{\sum_{k=1}^{c} h_{\mathrm{win},k}}. \tag{7.19}$$

The final class output, R, can then be computed as $R = \arg\max_{1\leq j\leq c}(\mu_j(\mathbf{x}))$.

7.5 Experimental Results

In this section, the proposed incremental classifier fusion techniques are evaluated against their batch counterparts as well as against the individual classifiers. The experimental setup is detailed in Sect. 7.5.1. For the evaluation two industrial monitoring and diagnostics applications are considered: online visual quality inspection of CD imprints and predicting maintenance actions for copiers from a large historical database. These applications are described in more detail in Sect. 7.5.2. The evaluation follows the workflow depicted in Fig. 7.1, which serves as the classification framework for all experiments reported in this section. The experimental results for the different applications considered here are shown and discussed in detail in Sect. 7.5.3.

7.5.1 Experimental Setup

For all the experiments, three inherently incremental classifiers are used as members of the ensemble: *k-Nearest Neighbors* (*k*-NN) [21], *eVQ-Class* [35], and *(Gaussian) Naive Bayes* (NB) [10]. These classifiers are based on very different principles for producing their classifications (distances, clusters and densities). Therefore, it can be expected that their decisions are to a certain degree uncorrelated, which is crucial for the ensemble to perform well. This is confirmed by the experimental results in Sect. 7.5.3, which show that the different classifiers achieve quite different accuracies for most of the data sets. Also the accuracies of the classifiers as more and more samples are used for adapting them show different evolutions for the different classifiers.

The relatively small ensemble size is motivated by the applications considered here (see Sect. 7.5.2). The first application in Sect. 7.5.2.1, visual inspection of CD imprints, is intended to be used in an online industrial production setting. Such systems are typically deployed on hardware with very limited computational resources (processing power and memory), on which also many other processes need to run (for controlling the production, logging, reporting, etc.). Therefore, the goal was to make the footprint of the classification system as small as possible, which resulted in the choice of a relatively small ensemble size. The same architecture was chosen for the second application in Sect. 7.5.2.2, the prediction of maintenance actions for copiers from a large historical database, such that its results can be compared to the first application.

The batch and incremental classifier fusion methods which are evaluated are the methods described in Sect. 7.4: as fixed classifier fusion methods the *Plurality Voting* rule (Vote) [32, 66] and the *Mean rule* (Mean) [25, 28] are used; as trainable classifier fusion methods the *Fuzzy Integral* (FI) [5, 17], *Decision Templates* (DT) [27, 31], *Dempster–Shafer Combination* (DS) [46], *Discounted Dempster–Shafer Combination* (DDS) [52], and the proposed *Direct Cluster-Based ensemble* (DC) are used. The proposed incremental variants of these trainable classifier fusion methods are abbreviated by prefixing an 'I' to the abbreviations of their batch counterparts (e.g., "IFI" represents the incremental version of the Fuzzy Integral (FI) method).

The performance of both the individual classifiers and their combinations using the different classifier fusion methods is estimated at two stages in the learning process: after the initial batch learning and after the incremental adaptation. This provides two evaluation criteria: (1) comparison of the performance of the ensembles against the performance of the individual classifiers, and (2) the performance increase by using the incremental adaptation of the system. For comparison the performance of the ensembles after the incremental adaptation is compared to their performance when no incremental adaptation is applied. For comparing the performance of the incremental classifiers and ensembles with their batch counterparts, they are also trained in batch mode on the complete training data (the data used for the initial batch learning and the data used for the incremental adaptation). Besides the prediction accuracies also the computation times are compared between the batch and incremental algorithms.

7.5.2 Applications

For the evaluation two industrial monitoring and diagnostics applications are considered: online visual quality inspection of CD imprints and predicting maintenance actions for copiers from a large historical database. For the former application, incremental learning is important as the system needs to be adapted while the production system is fully operational; for the latter, it is important because the database is too large for repeated training in batch mode, whenever new maintenance data becomes available. These applications are described in Sects. 7.5.2.1 and 7.5.2.2, respectively.

7.5.2.1 Online Visual Quality Inspection

For evaluating the proposed methodology a first application is the online visual quality inspection of CD imprints. The data used for this application was recorded on-line in an industrial production line. The potential defects in the recorded images,

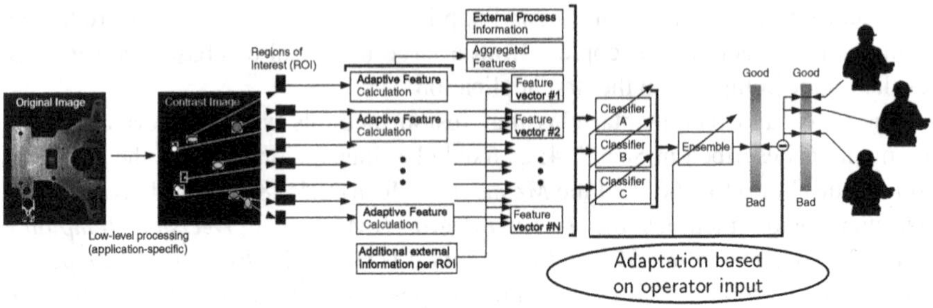

Fig. 7.2 Framework for on-line visual quality inspection

caused by, e.g., wrong palettes, color weaknesses, etc., should be automatically detected and classified into "good" and "bad" according to the quality decision made by the quality control operators. This should happen while the production system is fully operational; hence, the requirement for incremental adaptation of the classification system.

The complete system architecture of the online, self-adaptive visual inspection system is shown in Fig. 7.2. As depicted, the individual classifiers as well as the ensembles are adapted according to the operators' decisions (if available, as discussed in Sect. 7.2). After removing the application-dependent elements (e.g., by comparing newly recorded images with a master), encoded in the contrast image, Regions Of Interest (ROIs) are recognized by clustering and object recognition approaches, marking the potential defects. For these ROIs a set features is calculated, which are used for the initial (offline) training and further (online) adaptation of the classifiers and ensemble. This feature set includes 57 so-called *object features* (characterizing a single ROI, e.g., the shape of the circumscribing ellipse, the brightness, etc.), and 17 so-called *aggregated features* (characterizing the image as a whole, e.g., the number of objects in the image, the maximal local density of the objects, etc.), resulting in 74 features in total per image. For some of the features there is an implicit parameter (e.g., above which gray-level value are the pixels taken into account for computing the object's area). These feature parameters can be optimized with respect to the classification problem, which we refer to as "adaptive feature calculation" [12]; also refer to Chap. 13. After the initial training, when the system is fully operational, the classifiers and ensemble are updated based on the decisions of the operators (by overruling the classification system's decisions); refer to Chap. 14, where evolving fuzzy classifiers are used for online quality control in a wider range of inspection systems (eggs, rotor, bearings). For more detailed information about this image classification framework and its individual components, see [50, 51].

A set of 1,534 images was recorded from the production line and labeled independently by five different quality control operators into two classes ("good" and "bad"), resulting in 5 different data sets (referred to as CD1–CD5). In the experiments for these data sets the performance evaluation is always done using the last third of the data. The initial batch training is done using the first half of the data not used for evaluation (i.e., the first 33% of the data), during which also the optimal parameter settings of the classifiers are found; the incremental adaptation is done using the second half of the data not used for evaluation (i.e., the 33% of the data after the data used for the initial training).

As the data was recorded online as the different CDs were produced, the true behavior of the classification system can be estimated. If the data was shuffled this would likely lead to overly high estimates of the predictive accuracy of the algorithms, as more different types of failures would be present in the initial batch training set. In practice, however, this is not the case, as specific failures might develop while the system is fully functional.

7.5.2.2 Prediction of Maintenance Actions

The second application for evaluating the proposed methodology is the prediction of maintenance actions for copiers. The data used for this application was recorded in a central database, containing information about the replacement of different components by service technicians. 17570 maintenance actions for different types of copiers, operational at different locations, are available. The replacement of different components of the copiers are logged, together with the time the replacement has taken place. Concurrently, the maintenance action for the Toner Transfer Fusing (TTF) belt cleaner is logged. The TTF belt cleaner replacement is the most frequently occurring replacement and hence it can be important to predict whether it needs maintenance, so that the maintenance policy can be optimized. This maintenance action, which is to be predicted by the classification system, can take three states: "No maintenance," "Predictive maintenance," and "Corrective maintenance." Using such a classification system, it becomes possible to predict whether the TTF belt cleaner needs maintenance or not, based on the replacements of other components. As the database is too large for repetitively retraining the classifiers and ensembles in batch mode, incremental learning techniques are also required for this application. A detailed description of this database can be found in [1].

In the experiments for these data sets, the performance evaluation is always done using the last third of the data. The initial batch training is done using the first 20% of the data not used for evaluation (i.e., the first 13.2% of the data), during which also the optimal parameter settings of the classifiers are found; the incremental adaptation is done using the second half of the data not used for evaluation (i.e., the 52.8% of the data after the data used for the initial training). As the data was recorded while the copiers were fully functional, the true behavior of the classification system can be estimated, as the last data recorded is the most relevant to estimate the predictive performance (as explained in Sect. 7.5.2.1).

Table 7.1 Classifier and ensemble accuracies after the initial batch training (in %)

	CD1	CD2	CD3	CD4	CD5	Copier
Classifiers						
NB	82.55	82.35	**81.18**	**78.63**	69.22	27.35
k-NN	68.04	79.61	79.41	77.06	64.31	**84.22**
eVQ-Class	**85.88**	**89.80**	75.29	75.69	**70.00**	75.86
Ensembles						
Vote	**85.88**	**90.59**	**85.29**	**86.08**	**72.35**	75.40
Mean	**85.88**	**90.59**	**85.29**	**86.08**	**72.35**	75.52
FI	**85.88**	**90.59**	**85.29**	**86.08**	**72.35**	75.41
DT	78.63	82.35	80.39	78.43	64.71	75.02
DS	82.55	82.35	80.39	78.63	64.31	75.07
DDS	78.63	82.35	80.39	78.63	64.31	**84.22**
DC	**85.88**	**90.59**	84.51	85.88	64.31	**84.22**

7.5.3 Results and Discussion

The classifiers and ensembles are evaluated based on two criteria: the accuracy and the robustness of their predictions against badly performing classifiers, and the computation time needed to train, adapt, and evaluate them. The former is discussed in Sect. 7.5.3.1; the latter is discussed in Sect. 7.5.3.2.

7.5.3.1 Evaluation of the Classification Performances

For all six data sets (five from the CD imprint inspection application and one from the maintenance prediction), the predictive accuracies of the different classifiers and ensembles for the initial batch training (on the first part of the training data) are shown in Table 7.1; the accuracies after incremental adaptation of the classifiers as well as the ensembles using the second part of the training data are shown in Table 7.2; the accuracies after incremental adaptation of the classifiers using the second part of the training data while keeping the ensembles fixed are shown in Table 7.3; and the accuracies after training the classifiers and ensembles in batch mode on all training data are shown in Table 7.4. The classifiers and ensembles achieving the highest accuracies are highlighted in boldface for each of the different data sets. These results can be summarized as follows.

The Need for Incremental Adaptation For some of the data sets, it is required to use an incremental adaptation of the classifiers and ensembles. By comparing Tables 7.1 and 7.2, one can see that the accuracies of the static classifiers and ensembles (i.e., classifiers and ensembles which are not updated after being trained on the first part of the training data) may be significantly worse than the dynamic classifiers and ensembles (i.e., adapting both classifiers and ensembles

Table 7.2 Classifier and ensemble accuracies after the incremental adaptation of the initial models (in %)

	CD1	CD2	CD3	CD4	CD5	Copier
Classifiers						
NB	78.63	80.98	81.76	80.00	61.96	78.64
k-NN	81.76	**90.98**	84.12	88.43	**83.33**	84.05
eVQ-Class	**82.94**	90.78	**88.82**	**89.22**	65.88	**85.68**
Ensembles						
Vote	**85.69**	**90.78**	89.02	**89.41**	70.00	85.91
Mean	**85.69**	**90.78**	89.02	**89.41**	70.00	85.78
IFI	**85.69**	**90.78**	89.02	**89.41**	70.00	86.02
IDT	**85.69**	**90.78**	89.02	**89.41**	82.94	76.32
IDS	**85.69**	**90.78**	89.02	**89.41**	76.27	80.54
IDDS	**85.69**	**90.78**	**89.80**	**89.41**	82.94	83.32
IDC	82.35	**90.78**	84.12	**89.41**	**83.33**	**86.92**

Table 7.3 Classifier and ensemble accuracies after the incremental adaptation of only the base classifiers; the ensembles are not adapted (in %)

	CD1	CD2	CD3	CD4	CD5	Copier
Classifiers						
NB	78.63	80.98	81.76	80.00	61.96	78.64
k-NN	81.76	**90.98**	84.12	88.43	**83.33**	84.05
eVQ-Class	**82.94**	90.78	**88.82**	**89.22**	65.88	**85.68**
Ensembles						
Vote	**85.69**	**90.78**	**89.02**	**89.41**	70.00	**85.91**
Mean	**85.69**	**90.78**	**89.02**	**89.41**	70.00	85.78
IFI	**85.69**	**90.78**	**89.02**	**89.41**	70.00	85.62
IDT	75.29	80.98	76.08	80.00	82.94	84.09
IDS	78.63	80.98	80.00	80.00	82.94	84.77
IDDS	76.08	80.98	76.08	80.00	82.94	83.99
IDC	**85.69**	**90.78**	83.33	**89.41**	**83.33**	84.05

using the second part of the training data after the initial training). For example for the ensembles the largest increase in performance is up to more than 19% for the CD5 data set (computed as the difference between Tables 7.1 and 7.2). Obviously, this is due to the limited number of samples presented to the classifiers and ensembles during the initial training (especially for the CD data sets), resulting in undertrained classifiers and ensembles.

By comparing Tables 7.2 and 7.3, one can see that when the (trainable) ensembles are kept fixed during the incremental adaptation phase while adapting the classifiers incrementally, the accuracies of IDT, IDS and IDDS are significantly lower than when they are adapted. The only exceptions are data set CD5 (the accuracies are approximately the same for IDT and IDDS and the accuracy of IDS is even better when it is not adapted) and Copier (the accuracy of IDT is better when it is not adapted).

Table 7.4 Classifier and ensemble accuracies after batch training using all training data (in %)

	CD1	CD2	CD3	CD4	CD5	Copier
Classifiers						
NB	**83.53**	87.65	86.86	87.06	64.71	76.44
k-NN	82.15	**91.18**	84.31	**88.43**	**82.94**	84.50
eVQ-Class	81.76	90.39	**87.25**	87.25	64.90	**84.51**
Ensembles						
Vote	**84.71**	90.98	**87.84**	88.24	69.41	85.71
Mean	**84.71**	90.98	**87.84**	88.24	69.41	**85.83**
FI	**84.71**	90.98	**87.84**	88.24	69.41	85.47
DT	80.78	**91.18**	82.55	**88.63**	81.57	84.38
DS	**84.71**	90.98	**87.84**	**88.63**	74.71	84.75
DDS	82.16	90.98	82.55	**88.63**	82.55	84.39
DC	82.16	**91.18**	82.16	88.04	**82.94**	84.43

The Need for Ensembles Inspecting all resulting accuracies (Tables 7.1–7.4), one can see that at any stage of the learning process the performance of the best ensemble is better than or equal to the performance of the best classifier. The largest improvement (7.45%) is achieved for CD4, when classifiers as well as ensembles are not adapted incrementally. The only exception is the CD2 data set, for which the result of the best classifier is 0.2% better than the best ensemble when the classifiers are adapted incrementally (see Tables 7.2 and 7.3). It is thus safe to state that ensembles of classifiers should definitely be considered in order to achieve a higher predictive performance during the online classification if enough computing power is available during operation.

The Effectiveness of the Incremental Adaptation By comparing Tables 7.2 and 7.4, one can see that the predictive accuracy of the incremental classifier fusion methods (trained in batch mode on the first part of the training data and afterward incrementally adapted using the second part of the training data) is in general approximately equal to the accuracy of the classifier fusion methods trained in batch mode on the entire training data. For most data sets (except CD2), the incrementally adapted classifier fusion methods even perform better than their batch counterparts, ranging up to 2.49% for the Copier data set. Note that in both cases the algorithms have been presented with exactly the same training data.

This result is quite important, and it even opens the question whether a retraining of the system in batch mode is useful at all, as this puts severe constraints on the computational requirements when the system should be kept up-to-date at a high rate (see also the results of the computation times required by the classifiers and ensembles discussed below).

The Robustness of the Ensembles Even though one of the ensembles' base classifiers, Naive Bayes, is for some data sets performing much worse than the others (even below 50% after the initial training for the Copier data set), the ensembles' accuracies are mostly unaffected by this. The ensembles (especially the trainable ones) show a robust performance in all of the experiments. This underlines the fact that the diversity between the classifiers helps building good ensembles. In all results (Tables 7.1–7.4), it can be seen that different classifiers are better for different data sets. However, after further adaptation it can be seen that yet different classifiers become the best performers (by comparing Tables 7.1 and 7.2). This makes it very difficult to select the most appropriate classifier for a given application. Using an ensemble, this selection problem is eliminated: the ensemble will in general perform at least as good as its best member. When the performances of the classifiers change over time, the ensembles will also adjust to this new situation.

The robustness of the ensembles is underlined by Figs. 7.3 and 7.4, showing the evolution of the accuracies of the incremental classifiers and ensembles for the CD5 and Copier data sets. Even though some classifiers show a poor performance for these data sets, most of the ensembles remain unaffected by this and show a quite stable performance.

The Need for Incrementally Adaptable Ensembles Inspecting the results of the incremental ensembles in Table 7.2, one can see that the trainable ensembles are the best performers, when they are incrementally adapted, for all of the experiments. For CD5, the largest improvement between the best fixed ensemble and the best trainable one is achieved (13.33%). This again underlines the importance of the development of incrementally adaptable ensembles.

Comparison of the Ensembles From the results presented above, no clear winner can be identified between the different incremental classifier fusion methods. A slight preference could be given to the IDDS and the proposed IDC methods, as they are usually among the top performers.

7.5.3.2 Evaluation of the Computation Times

The computational complexity of the algorithms is evaluated empirically by measuring the computation times on a conventional PC. As the CD1–CD5 data sets contain the same number of instances and features, the computation times are very similar and thus only the computation times for one of these data sets are reported. The resulting computation times for the CD data sets and the Copier data set are reported in Table 7.5, comparing the time required for both the incremental algorithms as their batch counterparts (retrained every 10 samples for the CD data sets and every 100 samples for the Copier data set).

The importance of the incremental adaptation of the classifiers as well as the ensembles is immediately apparent in Table 7.5. For the NB and k-NN classifiers the reduction in computation time is not so large (a factor 1.2–1.5). For the eVQ-Class classifier, however, the reduction is huge (a factor 10–26). In fact, retraining eVQ-Class every 100 samples for the Copier data set takes over 3 h, while this is less than 10 min for k-NN and approximately 0.5 min for NB. In this case, the incremental version of eVQ-Class reduces the computation time to approximately the level of

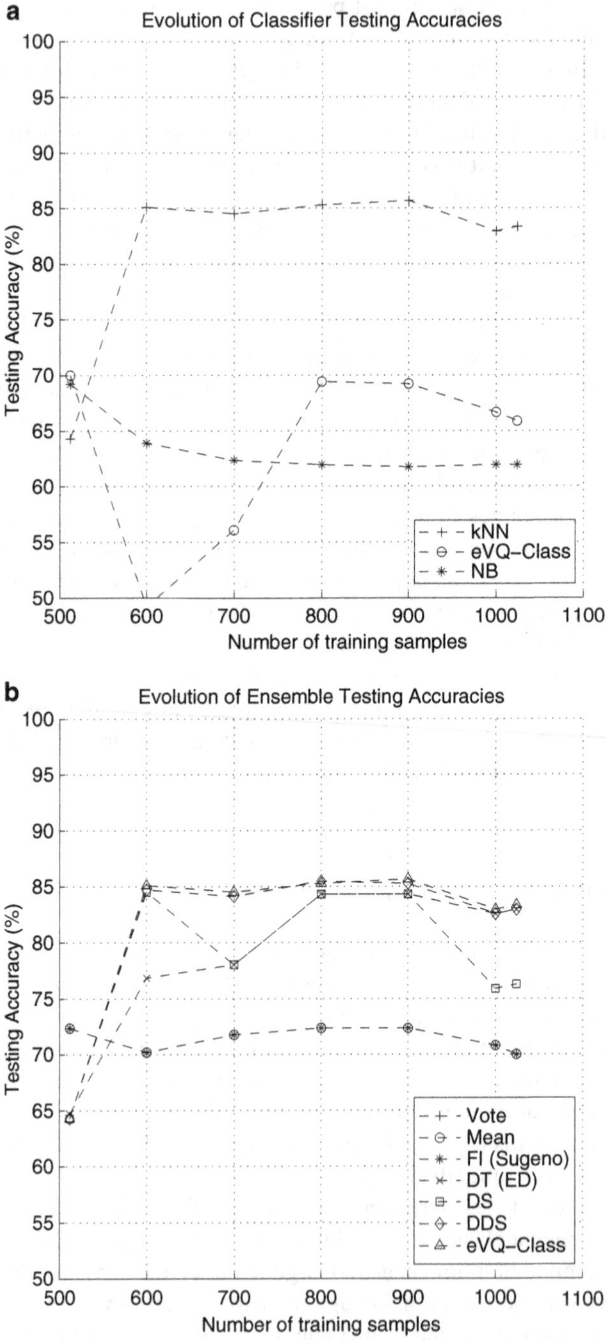

Fig. 7.3 Evolution of the accuracies for the CD5 data set: (**a**) Incremental classifiers; (**b**) Incremental ensembles

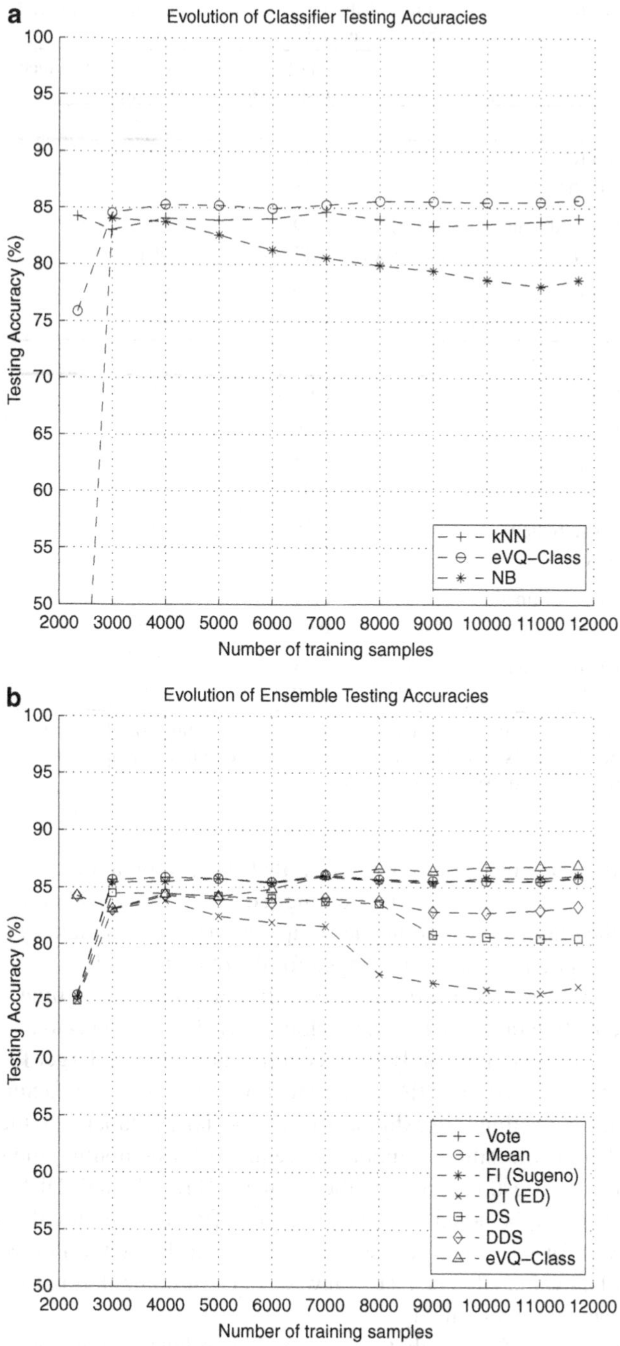

Fig. 7.4 Evolution of the accuracies for the Copier maintenance data set: (**a**) Incremental classifiers; (**b**) Incremental ensembles

Table 7.5 Training times of the batch retrained and the incremental classifiers and ensembles (in seconds)[a]

	CD imprint inspection	Copier maintenance prediction
Classifiers		
NB retrained	0.34	34
NB incremental	0.28	33
k-NN (update of database)	0.69	549
k-NN (no update of database)	0.47	391
eVQ-Class retrained	1.50	10,909
eVQ-Class incremental	0.13	414
Ensembles		
FI retrained	0.06	8
IFI	0.02	6
DT retrained	0.55	12
IDT	0.05	6
DS retrained	0.55	12
IDS	0.05	6
DDS retrained	0.59	12
IDDS	0.06	6
DC retrained	1.13	36
IDC	0.08	7

[a]For *k*-NN the classification time (i.e. searching for the nearest neighbors) is stated as the training time (i.e. adding samples to the reference database) is negligible.

k-NN. Important to note here, however, is that the computation time of *k*-NN is determined by its *evaluation*, not its training. Hence, if more evaluations would be done during the training/adaptation, the computation time consumed by *k*-NN keeps on increasing, while it remains unchanged for the other methods.

For the ensembles also, a significant reduction in computation time is achieved. The lowest reduction (a factor 1.3–3) is achieved by IFI (which is already the fastest trainable classifier fusion method). The reduction is somewhat larger for IDT, IDS, and IDDS (a factor 2–10). For IDC, the reduction is the largest (a factor 5–14). All of the ensemble algorithms are extremely fast, especially compared to the classifiers (of course, also the classifiers' outputs are required for computing the ensemble's output). IFI is the fastest, followed by the group of IDT, IDS, and IDDS, and IDC is the slowest. It should be noted, however, that the differences are not that large. This shows that even for large data sets such as the Copier data set, all classifier fusion methods can be evaluated very fast (a few seconds of computation time) and thus the best one can be found quickly.

Putting the results of the predictive accuracy and robustness and the computation times together, for online modeling tasks and repeatedly learning in large data sets a clear preference can be given to an *incremental* learning framework as the one presented in this work. The main benefits are that the system is up-to-date at all time and that the computational complexity is greatly reduced, while the accuracies remain at the same level. A clear preference can also be given to ensembles of classifiers as they provide the robustness necessary for industrial environments, and in some cases they further improve the predictive accuracy.

7.6 Conclusions

In this work extensions of various classifier fusion algorithms are presented which allow a sample-wise, strictly incremental adaptation based on newly arriving data samples. Additionally, a novel classifier fusion method is proposed, based on an evolving clustering approach, which can be trained in batch mode as well as adapted incrementally. These algorithms only require a small amount of computation time and memory, especially compared to the individual classifiers within the ensemble. This makes these techniques suitable for learning in fast on-line applications and for learning from large databases. The classifiers and classifier fusion methods are integrated in an on-line classification framework in which they are adapted concurrently, ensuring that all classifiers and classifier fusion methods are up-to-date at any given time.

The framework and the classifier fusion methods therein are evaluated for two industrial monitoring and diagnostics tasks: online visual quality inspection for a CD imprint process and predicting maintenance actions for copiers. The experimental results show that the performance of the incremental classifier fusion methods effectively increases when more data is presented to them, to a level which is even higher than the best incremental classifier within the ensemble. The predictive accuracy after the incremental adaptation is approximately the same as the accuracy of the classifier fusion methods' batch counterparts when trained in batch mode on the same data, but the computation times for training in batch mode are much larger compared to the incremental adaptation. An additional advantage of the use of the (batch and incremental) classifier fusion methods is that they are robust with respect to changes in the performance of their member classifiers, which ensures that their performance is at least at the same level as the best individual member classifier. This alleviates the typically very time-consuming task of having to select the most appropriate classifier for a given classification problem. The robustness increases the usefulness of the framework for industrial applications even further as reliability is one of the most important characteristics if these techniques are to be applied in real-world settings.

Acknowledgments This work was partly supported by the European Commission (project Contract No. STRP016429, acronym DYNAVIS) and partly carried out within the framework of the technological advice project td-diagmon (grant no. 070522), which is financially supported by the Institute for the Promotion of Innovation through Science and Technology in Flanders (IWT-Vlaanderen). This publication reflects only the authors' views.

References

1. Bey-Temsamani, A., Engels, M., Motten, A., Vandenplas, S., Ompusunggu, A.: A practical approach to combine data mining and prognostics for improved predictive maintenance. In: Proceedings of the Data Mining Case Studies Workshop, 15th ACM SIGKDD Conference on Knowledge Discovery and Data Mining, pp. 37–44 (2009)
2. Breiman, L.: Bagging predictors. Machine Learning **24**(2), 123–140 (1996)
3. Breiman, L.: Pasting small votes for classification in large databases and on-line. Mach. Learn. **36**(1–2), 85–103 (1999)
4. Breiman, L., Friedman, J., Olshen, R., Stone, C.: Classification and Regression Trees. Wadsworth International Group, Belmont, CA, USA (1984)
5. Cho, S., Kim, J.: Combining multiple neural networks by fuzzy integral for robust classification. IEEE Trans. Syst. Man Cybern. **25**(2), 380–384 (1995)
6. Crawford, S.: Extensions to the CART algorithm. Int. J. Man Mach. Stud. **31**(2), 197–217 (1989)
7. Dempster, A.: A generalization of Bayesian inference. J. Roy. Stat. Soc. B **30**, 205–247 (1968)
8. Diehl, C., Cauwenberghs, G.: SVM incremental learning, adaptation and optimization. In: Proceedings of the 2003 International Joint Conference on Neural Networks, pp. 2685–2690 (2003)
9. Dietterich, T.: Ensemble methods in machine learning. In: J. Kittler, F. Roli (eds.) Multiple Classifier Systems, *Lecture Notes in Computer Science*, vol. 1857, pp. 1–15. Springer, Berlin/Heidelberg, Germany (2000)
10. Duda, R., Hart, P.: Pattern Classification and Scene Analysis. John Wiley & Sons, New York (1973)
11. Duda, R., Hart, P., Stork, D.: Pattern Classification, 2nd edn. John Wiley & Sons, New York, USA (2000)
12. Eitzinger, C., Gmainer, M., Heidl, W., Lughofer, E.: Increasing classification robustness with adaptive features. In: A. Gasteratos, M. Vincze, J. Tsotsos (eds.) Computer Vision Systems, *Lecture Notes in Computer Science*, vol. 5008, pp. 445–453. Springer, Berlin/Heidelberg (2008)
13. Feng, J.: An intelligent decision support system based on machine learning and dynamic track of psychological evaluation criterion. In: J. Kacpzryk (ed.) Intelligent Decision and Policy Making Support Systems, pp. 141–157. Springer, Berlin/Heidelberg (2008)
14. Filev, D., Tseng, F.: Novelty detection based machine health prognostics. In: Proceedings of the 2006 International Symposium on Evolving Fuzzy Systems, pp. 193–199 (2006)
15. Freund, Y., Schapire, R.: A decision-theoretic generalization of on-line learning and an application to boosting. J. Comput. Syst. Sci. Int. **55**(1), 119–139 (1997)
16. Fung, G., Mangasarian, O.: Incremental support vector machine classification. In: Second SIAM International Conference on Data Mining, pp. 247–260 (2002)
17. Gader, P., Mohamed, M., Keller, J.: Fusion of handwritten word classifiers. Pattern Recognit. Lett. **17**(6), 577–584 (1996)
18. Gama, J.: Knowledge Discovery from Data Streams. Chapman & Hall / CRC, Boca Raton, Florida, USA (2010)
19. Gray, R.M.: Vector quantization. IEEE ASSP Mag. **1**(2), 4–29 (1984)
20. Hansen, L., Salamon, P.: Neural network ensembles. IEEE Trans. Pattern Anal. Mach. **12**(10), 993–1001 (1990)
21. Hastie, T., Tibshirani, R., Friedman, J.: The Elements of Statistical Learning: Data Mining, Inference, and Prediction. Springer Verlag, New York, USA (2001)
22. Huang, Y., Suen, C.: A method of combining multiple experts for the recognition of unconstrained handwritten numerals. IEEE Trans. Pattern Anal. Mach. **17**(1), 90–94 (1995)
23. Jacobs, R., Jordan, M., Nowlan, S., Hinton, G.: Adaptive mixtures of local experts. Neural Comput. **3**, 79–87 (1991)

24. Kerdprasop, N., Kerdprasop, K.: Data partitioning for incremental data mining. In: Proceedings of the First International Forum on Information and Computer Science, pp. 114–118. Hamamatsu, Japan (2003)
25. Kittler, J., Hatef, M., Duin, R., Matas, J.: On combining classifiers. IEEE Trans. Pattern Anal. Mach. **20**(3), 226–239 (1998)
26. Krogh, A., Vedelsby, J.: Neural network ensembles, cross validation, and active learning. In: G. Tesauro, D. Touretzky, T. Leen (eds.) Proceedings of the 7th Conference on Advances in Neural Information Processing Systems, vol. 7, pp. 231–238. MIT Press, Cambridge, MA, USA (1995)
27. Kuncheva, L.: Using measures of similarity and inclusion for multiple classifier fusion by decision templates. Fuzzy Sets Syst. **122**(3), 401–407 (2001)
28. Kuncheva, L.: A theoretical study on six classifier fusion strategies. IEEE Trans. Pattern Anal. Mach. **24**(2), 281–286 (2002)
29. Kuncheva, L.: Classifier ensembles for changing environments. In: F. Roli, J. Kittler, T. Windeatt (eds.) Multiple Classifier Systems, *Lecture Notes in Computer Science*, vol. 3077, pp. 1–15. Springer, Berlin/Heidelberg, Germany (2004)
30. Kuncheva, L.: Combining Pattern Classifiers: Methods and Algorithms. Wiley, New Jersey (2004)
31. Kuncheva, L., Bezdek, J., Duin, R.: Decision templates for multiple classifier fusion: An experimental comparison. Pattern Recognit. **34**(2), 299–314 (2001)
32. Kuncheva, L., Whitaker, C., Shipp, C., Duin, R.: Limits on the majority vote accuracy in classifier fusion. Pattern Anal. Appl. **6**(1), 22–31 (2003)
33. Li, W., Yue, H., Valle-Cervantes, S., Qin, S.: Recursive PCA for adaptive process monitoring. J. Process Control **10**(5), 471–486 (2000)
34. Lim, C., Harrison, R.: Online pattern classification with multiple neural network systems: An experimental study. IEEE Trans. Syst. Man Cybern. Part C Appl. Rev. **33**, 235–247 (2003)
35. Lughofer, E.: Extensions of vector quantization for incremental clustering. Pattern Recognition, **41**(3), 995–1011 (2008)
36. Lughofer, E.: Hybrid active learning for reducing the annotation efforts of operators in classification systems. Pattern Recognition, **45**(2), 884–896 (2012)
37. Lughofer, E.: Evolving Fuzzy Systems: Methodologies, Advanced Concepts and Applications. Springer, Berlin/Heidelberg (2011)
38. Lughofer, E., Guardiola, C.: On-line fault detection with data-driven evolving fuzzy models. J. Contr. Intell. Syst. **36**(4), 307–317 (2008)
39. Lughofer, E., Smith, J., Caleb-Solly, P., Tahir, M., Eitzinger, C., Sannen, D., Nuttin, M.: Human-machine interaction issues in quality control based on online image classification. IEEE Trans. Syst. Man Cybern. Part A Syst. Humans **39**(5), 960–971 (2009)
40. Nisbet, R., Elder, J., Miner, G.: Handbook of Statistical Analysis and Data Mining Applications. Academic Press, Burlington, MA, USA (2009)
41. Oza, N.: Online ensemble learning. Ph.D. thesis, University of California, USA (2001)
42. Pang, S., Ozawa, S., Kasabov, N.: Incremental linear discriminant analysis for classification of data streams. IEEE Trans. Syst. Man Cybern. Part B Cybern. **35**(5), 905–914 (2005)
43. Polikar, R., Upda, L., Upda, S., Honavar, V.: Learn++: An incremental learning algorithm for supervised neural networks. IEEE Trans. Syst. Man Cybern. Part C Appl. Rev. **31**(4), 497–508 (2001)
44. Polikar, R.: Ensemble based systems in decision making. IEEE Circuits Syst. Mag. **6**(3), 21–45 (2006)
45. Quinlan, J.: C4.5: Programs for Machine Learning. Morgan Kaufmann, San Mateo, CA, USA (1993)
46. Rogova, G.: Combining the results of several neural network classifiers. Neural Networks **7**(5), 777–781 (1994)
47. Roubos, J., Setnes, M., Abonyi, J.: Learning fuzzy classification rules from data. Inform. Sciences **150**, 77–93 (2003)

48. Rüping, S.: Incremental learning with support vector machines. Ph.D. thesis, Universität Dortmund (2002)
49. Ruta, D., Gabrys, B.: An overview of classifier fusion methods. Comput. and Inform. Syst. J. **7**(1), 1–10 (2000)
50. Sannen, D.: A flexible framework for learning visual quality inspection: Information fusion approaches. Ph.D. thesis, Katholieke Universiteit Leuven, Belgium (2010)
51. Sannen, D., Nuttin, M., Smith, J., Tahir, M., Caleb-Solly, P., Lughofer, E., Eitzinger, C.: An on-line interactive self-adaptive image classification framework. In: A. Gasteratos, M. Vincze, J. Tsotsos (eds.) Computer Vision Systems, *Lecture Notes in Computer Science*, vol. 5008, pp. 171–180. Springer, Berlin/Heidelberg, Germany (2008)
52. Sannen, D., Van Brussel, H., Nuttin, M.: Classifier fusion using discounted Dempster–Shafer combination. In: P. Perner (ed.) Machine Learning and Data Mining in Pattern Recognition, Poster Proceedings, pp. 216–230. IBal Publishing, Leipzig, Germany (2007)
53. Shafer, G.: A Mathematical Theory of Evidence. Princeton University Press, Princeton, NJ, USA (1976)
54. Song, M., Wang, H.: Incremental estimation of gaussian mixture models for online data stream clustering. In: Proceedings of International Conference on Bioinformatics and its Applications (2004)
55. Stone, M.: Cross-validatory choice and assessment of statistical predictions (with discussion). J. Roy. Stat. Soc. B **36**, 111–147 (1974)
56. Sutton, R., Whitehead, S.: Online learning with random representations. In: Proceedings of the 10th International Conference on Machine Learning, pp. 314–321. Morgan Kaufmann, San Mateo, CA, USA (1993)
57. Theodoridis, S., Koutroumbas, K.: Pattern Recognition, 4th edn. Academic Press, Burlington, MA, USA (2009)
58. Utgoff, P.: Incremental induction of decision trees. Mach. Learn. **4**, 161–186 (1989)
59. Valentini, G., Masulli, F.: Ensembles of learning machines. In: M. Marinaro, R. Tagliaferri (eds.) Neural Nets: 13th Italian Workshop on Neural Nets, Vietri sul Mare, Italy, 2002, Revised Papers, *Lecture Notes in Computer Science*, vol. 2486, pp. 3–22. Springer, Berlin/Heidelberg, Germany (2002)
60. Vapnik, V., Golowich, S., Smola, A.: Support vector method for function approximation, regression estimation and signal processing. Adv. Neural Inf. Process. Syst. **9**, 281–287 (1997)
61. Wang, H., Fan, W., Yu, P., Han, J.: Mining concept-drifting data streams using ensemble classifiers. In: Proceedings of the 9th ACM SIGKDD International Conference on Knowledge Discovery and Data Mining, pp. 226–235. ACM, New York, NY, USA (2003)
62. Wasserman, P.: Advanced Methods in Neural Computing. Van Nostrand Reinhold, New York, USA (1993)
63. Widmer, G., Kubat, M.: Learning in the presence of concept drift and hidden contexts. Mach. Learn. **23**(1), 69–101 (1996)
64. Wolpert, D., Macready, W.: No free lunch theorems for optimization. IEEE Trans. Evol. Comput. **1**(1), 67–82 (1997)
65. Woods, K., Kegelmeyer, W., Bowyer, K.: Combination of multiple classifiers using local accuracy estimates. IEEE Trans. Pattern Anal. Mach. **19**(4), 405–410 (1997)
66. Xu, L., Krzyźak, A., Suen, C.: Methods of combining multiple classifiers and their applications to handwriting recognition. IEEE Trans. Syst. Man Cybern. **22**(3), 418–435 (1992)

Chapter 8
Instance-Based Classification and Regression on Data Streams

Ammar Shaker and Eyke Hüllermeier

Abstract In order to be useful and effectively applicable in dynamically evolving environments, machine learning methods have to meet several requirements, including the ability to analyze incoming data in an online, incremental manner, to observe tight time and memory constraints, and to appropriately respond to changes of the data characteristics and underlying distributions. This paper advocates an instance-based learning algorithm for that purpose, both for classification and regression problems. This algorithm has a number of desirable properties that are not, at least not as a whole, shared by currently existing alternatives. Notably, our method is very flexible and thus able to adapt to an evolving environment quickly, a point of utmost importance in the data stream context. At the same time, the algorithm is relatively robust and thus applicable to streams with different characteristics.

8.1 Introduction

The idea of adaptive learning in dynamical environments has recently received increasing attention in different research communities, for example, in the database and data mining community under the slogan of "learning from data streams" [17, 18], and in the computational intelligence community under the notion of "evolving fuzzy systems" [4, 5, 24, 25]. Despite small differences regarding the basic assumptions and the technical setting, the emphasis of goals and performance criteria, and the focus on specific types of applications, the key motivation of these and related fields is the idea of a system that learns incrementally, and maybe even in real-time, on a continuous stream of data, and which is able to properly adapt itself to changes of environmental conditions or properties of the data-generating process.

A. Shaker • E. Hüllermeier (✉)
Department of Mathematics and Computer Science, Philipps-Universität Marburg, D-35032 Marburg, Germany
e-mail: shaker@Mathematik.Uni-Marburg.de; eyke@mathematik.uni-marburg.de

M. Sayed-Mouchaweh and E. Lughofer (eds.), *Learning in Non-Stationary Environments: Methods and Applications*, DOI 10.1007/978-1-4419-8020-5_8,
© Springer Science+Business Media New York 2012

Systems with these properties have been developed for different machine learning and data mining problems, such as clustering [1], classification [22], and frequent pattern mining [10].

Domingos and Hulten [15] list a number of properties that an ideal stream mining system should possess, and suggest corresponding design decisions: the system uses only a limited amount of memory; the time to process a single record is short and ideally constant; the data is volatile and a single data record accessed only once; the model produced in an incremental way is equivalent to the model that would have been obtained through common batch learning (on all data records so far); the learning algorithm should react to concept drift [32] (i.e., any change of the underlying data-generating process) in a proper way and maintain a model that always reflects the current concept.

Given the existence of a number of sophisticated and partly quite complicated methods for learning on data streams, it is surprising that one of the simplest approaches to machine learning, namely the instance-based (case-based) learning paradigm, has only received very little attention so far—all the more since the nearest neighbor estimation principle, the core of this paradigm, is a standard method in machine learning, pattern recognition, and related fields. In this chapter, we elaborate on the potential of the instance-based approach to supervised learning within the context of data streams and propose an efficient instance-based learning algorithm for classification and regression. To this end, we build on [6], in which our approach to classification was introduced.

The remainder of the paper is organized as follows: The next section recalls the basic ideas of instance-based learning, along with a short discussion of its possible advantages and disadvantages in a streaming context. Our approach to instance-based learning on data streams, IBL-DS, is introduced in Sect. 8.3. In Sect. 8.4, we provide some information about the MOA (Massive Online Analysis) framework for mining data streams, in which IBL-DS is implemented. Experimental results are presented in Sect. 8.5.

8.2 Instance-Based Learning

The term instance-based learning (IBL) stands for a family of machine learning algorithms, including well-known variants such as memory-based learning, exemplar-based learning and case-based learning [23, 27, 28]. As the term suggests, in instance-based algorithms special importance is attached to the concept of an *instance* [3]. An instance or exemplar can be thought of as a single experience, such as a pattern (along with its classification) in pattern recognition or a problem (along with a solution) in case-based reasoning.

As opposed to model-based machine learning methods which induce a general model (theory) from the data and use that model for further reasoning, IBL algorithms simply store the data itself. They defer the processing of the data until a prediction (or some other type of query) is actually requested, a property which qualifies them as a *lazy* learning method [2]. Predictions are then derived by combining the information provided by the stored examples.

Such a combination is typically accomplished by means of the *nearest neighbor* (NN) estimation principle [11]. Consider the following setting: Let $\mathscr{X}$ denote the instance space, where an instance corresponds to the description x of an object (usually although not necessarily in attribute-value form). $\mathscr{X}$ is endowed with a distance measure $\Delta(\cdot)$, i.e., $\Delta(x,x')$ is the distance between instances $x,x' \in \mathscr{X}$. $\mathscr{Y}$ is the output space and $\langle x,y\rangle \in \mathscr{X} \times \mathscr{Y}$ is called a labeled instance, a case, or an example. In classification, $\mathscr{Y}$ is a finite (usually small) set comprised of m classes $\{\lambda_1,\ldots,\lambda_m\}$, whereas $\mathscr{Y} = \mathbb{R}$ in regression.

The current experience of the learning system is represented in terms of a set $\mathscr{D}$ of examples $\langle x_i, y_i\rangle$, $1 \le i \le n = |\mathscr{D}|$. From a machine learning point of view, $\mathscr{D}$ plays the role of the *training set* of the learner. More precisely, since not all examples will necessarily be stored by an instance-based learner, $\mathscr{D}$ is only a subset of the training set. In case-based reasoning, it is also referred to as the *case base*.

Finally, suppose a novel instance $x_0 \in \mathscr{X}$ (a query) to be given. The NN principle prescribes to estimate the corresponding output y_0 by the output of the nearest (most similar) sample instance. The *k-nearest neighbor* (*k*-NN) approach is a slight generalization, which takes the $k \ge 1$ nearest neighbors of x_0 into account. That is, an estimation y_0^{est} of y_0 is derived from the set $\mathscr{N}_k(x_0)$ of the k nearest neighbors of x_0. In classification, this is usually done by means of a *majority vote*, i.e.,

$$y_0^{\text{est}} = \arg\max_{\lambda \in \mathscr{L}} \#\{x_i \in \mathscr{N}_k(x_0) \,|\, y_i = \lambda\}, \tag{8.1}$$

with $\mathscr{L}$ the set of class labels, whereas in regression, a weighted average of the outputs of the neighbors is predicted:

$$y_0^{\text{est}} = \sum_{x_i \in \mathscr{N}_k(x_0)} w(x_i) \cdot y_i, \tag{8.2}$$

with

$$w(x_i) = \frac{f(\Delta(x_i, x_0))}{\sum_{x_j \in \mathscr{N}_k(x_0)} f(\Delta(x_f, x_0))}.$$

Here, $f(\cdot)$ is a decreasing function $\mathbb{R}_+ \to \mathbb{R}_+$, which means that the smaller $\Delta(x_i, x_0)$, the stronger the weight of y_i.

Recall the aforementioned key requirements for learning and data mining algorithms on data streams: Above all, such algorithms must be incremental, highly adaptive, and they must be able to deal with concepts that may change over time. Is lazy, instance-based learning preferable to eager, model-based learning under these conditions? Unfortunately, this question cannot be answered unequivocally.

Obviously, IBL algorithms are inherently incremental, since adaptation basically comes down to adding or removing observed cases. Thus, incremental learning and model adaptation is simple and cheap in the case of IBL. As opposed to this, incremental learning is much more difficult to realize for most model-based approaches. Even though incremental versions do exist for a number of well-known learning methods, such as decision tree induction [30], the incremental update of a model is often quite complex and in many cases assumes the storage of a considerable amount of additional information.

The training efficiency of lazy learners does not come for free, however. Compared with model-based approaches, IBL has higher computational costs when it comes to answering new queries. In fact, the latter requires finding the k nearest neighbors of the query, and even though this retrieval step can be supported by efficient data and indexing structures, it remains costly in comparison with deriving a model-based prediction.

Consequently, IBL might be preferable in a data stream application if the number of incoming data is large compared with the number of queries to be answered, i.e., if model updating is the dominant factor. On the other hand, if queries must be answered frequently and under tight time constraints, whereas a need for updating the model due to newly observed examples rarely occurs, a model-based method might be the better choice.

Regarding the handling of concept drift, a definite answer cannot be given either. Appropriately reacting to concept drift requires, apart from its discovery, flexible updating, and adaptation strategies. In instance-based learning, model adaptation basically comes down to editing the case base, that is, adding new and/or deleting old examples. Whether or not this can be done more efficiently than adapting an other type of model, such as a classification tree or a neural network, does of course strongly depend on the particular model at hand. In any case, maintaining an implicit concept description by storing observations, as done by IBL, facilitates "forgetting" examples that seem to be outdated. In fact, such examples can simply be removed, while retracting the influence of outdated examples is usually more difficult in model-based approaches. In a neural network, for example, a new observation causes an update of the network weights, and this influence on the network cannot simply be cancelled later on.

8.3 Instance-Based Learning on Data Streams

This section introduces our approach to instance-based learning on data streams, referred to as IBL-DS. Our learning scenario consists of a data stream that permanently produces examples, potentially with a very high arrival rate, and a second stream producing query instances to be classified. The key problem for our learning system is to maintain an implicit concept description in the form of a case base (memory). Before presenting details of IBL-DS, some general aspects and requirements of concept adaptation (case-base maintenance) in a streaming context will be discussed.

8.3.1 Concept Adaptation

The simplest adaptive learners are those using sliding windows of fixed size. Since the update is very simple, these learners are also very fast. On the other hand, the assumption that the data which is currently relevant forms a fixed-sized window, i.e., that it consists of a *fixed* number of *consecutive* observations, is quite restrictive. In fact, by fixing the number of examples in advance, it is impossible to optimally adapt the size of the case base to the complexity of the concept to be learned, and to react to changes of this concept appropriately. Moreover, being restricted to selecting a subset of successive observations in the form of a window, it is impossible to disregard a portion of observations in the middle (e.g., outliers) while retaining preceding and succeeding blocks of data.

To avoid both of the aforementioned drawbacks, nonwindow-based approaches are needed that do not only adapt the size of the training data but also have the liberty to select an arbitrary *subset* of examples from the data seen so far. Needless to say, such flexibility does not come for free. Apart from higher computational costs, additional problems such as avoiding an unlimited growth of the training set and, more generally, trading off accuracy against efficiency, have to be solved.

Instance-based learning seems to be attractive in light of the above requirements, mainly because of its inherently incremental nature and the simplicity of model adaptation. In particular, since in IBL an example has only local influence, the update triggered by a new example can be restricted to a local region around that observation.

Regarding the updating (editing) of the case base in IBL, an example should in principle be retained if it improves the predictive performance (classification accuracy) of the classifier; otherwise, it should better be removed.[1] Unfortunately, this criterion cannot be used directly, since the (future) usefulness of an example in this sense is simply not known. Instead, existing approaches fall back on suitable *indicators* of usefulness:

- Temporal relevance: According to this indicator, recent observations are considered as potentially more useful and, hence, are preferred to older examples.
- Spatial relevance: The relevance of an example can also depend on its position in the instance space. This is the case, for example, if a concept drift only affects a part of the instance space. Besides, a more or less uniform coverage of the instance space is usually desirable, especially for local learning methods. In IBL, examples can be redundant in the sense that they do not change the nearest neighbor classification of any query. More generally (and less stringently), one might consider a set of examples redundant if they are closely neighbored in the instance space and, hence, have a similar region of influence. In other words, a new example in a region of the instance space already occupied by many other examples is considered less relevant than a new example in a sparsely covered region.

[1] Of course, this maxim disregards other criteria, such as the complexity of the method.

- Consistency: An example should be removed if it seems to be inconsistent with the current concept, e.g., if its own output (strongly) differs from those in its neighborhood.

Many algorithms use only one indicator, either temporal relevance (e.g., window-based approaches), spatial relevance (e.g., Lightweight Frequency Counting, LWF), or consistency (e.g., Instance-Based learning algorithm 3, IB3). A few methods also use a second indicator, e.g., the approach of Klinkenberg (temporal relevance and consistency), but only the window-based system FLORA4 (Floating Rough Approximation) uses all three aspects.

8.3.2 IBL-DS

In this section, we describe the main ideas of IBL-DS, our approach to IBL on data streams that not only takes all of the aforementioned three indicators into account but also meets the efficiency requirements of the data stream setting.

IBL-DS optimizes the composition and size of the case base autonomously. On arrival of a new example $\langle x_0, y_0 \rangle$, this example is first added to the case base. Moreover, it is checked whether other examples might be removed, either since they have become redundant or since they are outliers (noisy data). To this end, a set C of examples within a neighborhood of x_0 are considered as candidates. This neighborhood is given by the k_{cand} nearest neighbors of x_0, determined according a distance measure Δ (see Sect. 8.7), and the candidate set C consists of the examples within that neighborhood. The most recent examples are excluded from removal due to the difficulty to distinguish potentially noisy data from the beginning of a concept change. Even though unexpected observations will be made in both cases, noise and concept change, these observations should be removed only in the former but not in the latter case.

In the classification scenario, the most frequent class among the k_{cand} youngest examples in a larger test environment of size[2] $k_{\mathrm{test}} = (k_{\mathrm{cand}})^2 + k_{\mathrm{cand}}$ is determined. If this class corresponds to the current class y_0, those candidates in C are removed that have a different class label and do not belong to the k_{cand} youngest examples in the larger test environment. Furthermore, to guarantee an upper bound on the size of the case base, the oldest element of the similarity environment is deleted, regardless of its class, whenever the upper bound would be exceeded by adding the new example. The similarity environment constitutes the set of instances in the vicinity of the query instance, while the test environment can be seen as the union of the similarity environments of the neighbored instances.

---

[2]This choice of k_{test} aims at including in the test environment the similarity environments of all examples in the similarity environment of x_0; of course, it does not guarantee to do so.

In the regression scenario, the k_{cand} youngest examples in the neighborhood set C determines a confidence interval $\left[\bar{y} - Z_{\frac{\alpha}{2}} \frac{\sigma}{\sqrt{k_{\text{cand}}}}, \bar{y} + Z_{\frac{\alpha}{2}} \frac{\sigma}{\sqrt{k_{\text{cand}}}}\right]$, where $\bar{y}$ is the average target value for the considered examples and σ is the standard deviation. A class values y_0 outside this interval indicates an unexpected change in the neighborhood when this instance was generated. In this case, instances not belonging to the confidence interval are removed from the larger test environment.

Using this strategy, the algorithm is able to adapt to concept drift but will also have a high accuracy for nondrifting data streams. Still, these two situations—drifting and stable concept—are to some extent conflicting with regard to the size of the case base: If the concept to be learned is stable, classification accuracy will increase with the size of the case base. On the other hand, a large case base turns out to be disadvantageous in situations where concept drift occurs, and even more in the case of concept shift. In fact, the larger the case base is, the more outdated examples will have to be removed and, hence, the more sluggish the adaptation process will be.

For this reason, we try to detect an abrupt change of the concept using a statistical test as in [19, 20]. If a corresponding change has been detected, a large number of examples will be removed instantaneously from the case base. In the classification scenario, the test is performed as follows: We maintain the prediction error p and standard deviation $s = \sqrt{\frac{p(1-p)}{100}}$ for the last 100 training instances. Let p_{min} denote the smallest among these errors and s_{min} the associated standard deviation. A change is detected if the current value of p is significantly higher than p_{min}. Here, statistical significance is determined by testing the null hypothesis $H_0 : p \leq p_{\text{min}}$ against the alternative hypothesis $H_1 : p > p_{\text{min}}$. This is accomplished by using a standard (one-sided) z-test, i.e., the condition to be tested is $p + s > p_{\text{min}} + z_\alpha s_{\text{min}}$, where α is the level of confidence (we use $\alpha = 0.999$).

Finally, in case a change has been detected, we try to estimate its extent in order to determine the number of examples that need to be removed. More specifically, we delete p_{dif} percent of the current examples, where p_{dif} is the difference between p_{min} and the classification error for the last 20 instances; the latter serves as an estimation of the current classification error.[3] Examples to be removed are chosen at random according to a distribution which is spatially uniform but temporally skewed; see [6] for details.

In the regression scenario, the above test is conducted with the mean absolute error instead of the classification rate, and the percentage of examples to be removed is determined by the relative increase of this error.

[3]Note that, if this error, p, is estimated from the last k instances, the variance of this estimation is $\approx p(1-p)/k$. Moreover, the estimate is unbiased, provided that the error remained constant during the last k time steps. The value $k = 20$ provides a good trade-off between bias and precision.

8.4 MOA

IBL-DS is implemented under the MOA (Massive Online Analysis) framework, an open source software for mining and analyzing large data sets in a stream-like manner. MOA is written in Java and is closely related to WEKA [31], the Waikato Environment for Knowledge Analysis, which is presently the most commonly used machine learning software.

MOA supports the development of classifiers that can learn either in a purely incremental mode, or in batch mode first (on an initial part of a data stream) and incrementally afterward. The implementation of an evolving classifier is supported by a Java interface called UpdateableClassifier. This operation simulates the case of online learning, which means that each instance is accessed only once. A few incremental classifiers are already included in MOA, notably the Hoeffding tree [22], a state-of-the-art classifier often used as a baseline in experimental studies. Some meta learning techniques are implemented, too, such as online bagging and boosting both for static [26] and evolving streams [8].

8.4.1 Stream Generators

MOA supports the simulation of data streams by means of synthetic stream generators. An example is the Hyperplane generator that was originally used in [22]. It generates data for a binary classification problem, taking a random hyperplane in d-dimensional Euclidean space as a decision boundary; a certain percentage of instances is corrupted with noise.

Another important stream generator is the RandomTree generator. Its underlying model is a decision tree for a desired number of attributes and classes. The tree is built by splitting on randomly chosen attributes and then giving random class labels to the leaf nodes. Instances are generated with uniformly distributed values in the attributes while the class label is determined by the tree.

MOA offers the ConceptDriftStream procedure for simulating concept drift. The idea underlying this procedure is to mix two pure distributions in a probabilistic way, smoothly varying the corresponding probability degrees. In the beginning, examples are taken from the first pure stream with probability 1, and this probability is decreased in favor of the second stream in the course of time. More specifically, the probability is controlled by means of the sigmoid function

$$f(t) = \left(1 + e^{-4(t-t_0)/w}\right)^{-1}.$$

This function has two parameters: t_0 is the mid point of the change process, while w is the length of this process.

8.4.2 Model Evaluation

The evaluation of an evolving classifier is clearly a nontrivial issue. In fact, compared to standard batch learning, simple one-dimensional performance measures such as classification accuracy are not immediately applicable, or at least not able to capture the time-varying behavior of a classifier in a proper way. MOA offers different solutions for this problem.

The *holdout procedure* is a generalization of the cross-validation procedure commonly used in batch learning. Here, the training and the testing phase of a classifier are interleaved as follows: the classifier is trained incrementally on a block of M instances and then evaluated (but no longer adapted) on the next N instances, then again trained on the next M and tested on the subsequent N instances, and so forth. Thus, it becomes possible to monitor the performance of the model as time progresses; this information can also be used as an indicator of possible changes of the underlying concept [7, 9].

While the holdout procedure uses an instance either for training or for testing, each instance is used for both in the *prequential* approach [12]: First, the model is evaluated on the instance, and then a single incremental learning step is carried out. The prequential error is advocated in [21], where it is also shown to converge to the holdout measure when using a sliding window or a fading factor (exponential weighting).

8.5 Experiments

In this section, we compare IBL-DS with state-of-the-art learners in terms of performance and handling of concept drift, namely Hoeffding trees for classification [22] and the FLEXFIS approach for regression [24]. Hoeffding trees is a decision tree approach suitable for learning on data streams, whereas FLEXFIS constructs and maintains a specific kind of fuzzy rule-based model, namely a model of the Takagi–Sugeno type [29]. Our study is not meant as an extensive empirical evaluation that supports statistically valid conclusions. Instead, it is only supposed to serve an illustration purpose. We refer to [6] for more experiments with classification problems.

We use IBL-DS in its default setting unless otherwise stated (in some binary classification problems, we try different values for the maximum size of the instance base). Experiments are not only conducted with real data sets, but also with synthetic data. As an important advantage of synthetic data, let us note that it allows for conducting experiments in a *controlled* way and, therefore, to investigate the performance of a method under specific conditions. In particular, synthetic data is useful for simulating a concept drift.

The experiments are performed in the MOA framework, using the holdout procedure for measuring predictive accuracy. The parameters M and N vary

Table 8.1 Summary of the data sets used in the experiments

Data Set	Instances	Attributes	Holdout evaluation
Statlog (shuttle)	58,000	9	$M = 5{,}000$ and $N = 1{,}000$
Red wine	1,599	11	$M = 100$ and $N = 25$
White wine	4,889	11	$M = 200$ and $N = 50$
YearPredictionMSD	515,345	90	$M = 200$ and $N = 50$

depending on the size of the data set (we take $M = 5{,}000$ and $N = 1{,}000$ in the first two experiments with synthetic data). For the experiments with real data, these parameters are adapted to the size of the respective data set; see Table 8.1 for an overview of the main characteristics of these data sets. The real data sets are standard benchmarks taken from the Statlib archive[4] and the UCI repository [16]. Since they do not have an inherent temporal order, we average the performance curves over 100 randomly shuffled versions of these data sets.

8.5.1 Classification

8.5.1.1 Synthetic Data

The first two experiments are based on synthetic data with different characteristics (i.e., different types of decision boundaries). The first experiment uses data taken from the hyperplane generator. The ConceptDriftStream procedure mixing streams produced by two different hyperplanes simulates a rotating hyperplane. Using this procedure, we generated 12,000,000 examples connecting two hyperplanes in four-dimensional space, with $t_0 = 500{,}000$ and $w = 100{,}000$.

We compare the performance of two different settings of IBL-DS, one with a value of 400 for the maximum size of the instance base and the other one with 5,000. Figure 8.1 shows that both versions of IBL-DS initially outperform the Hoeffding tree. The Hoeffding tree is also more affected by the concept drift, showing a more pronounced "valley" in the performance curve, and also taking more time to recover. IBL-DS recognizes and adapts to the concept drift quite early, recovering its original performance as soon as the drift is over.

In a second experiment, we use the random tree generator to produce examples. Obviously, this generator is favorable for the Hoeffding tree. Again, the same ConceptDriftStream is used, but this time mixing two random tree generators. As can be seen in Fig. 8.2, the Hoeffding tree is now able to outperform IBL-DS in the first phase of the learning process; in fact, reaching an accuracy of close to 100%, which is not unexpected given that the Hoeffding tree is ideally tailored for this kind of data. Once again, however, the Hoeffding tree is much more affected by the

[4] http://lib.stat.cmu.edu/.

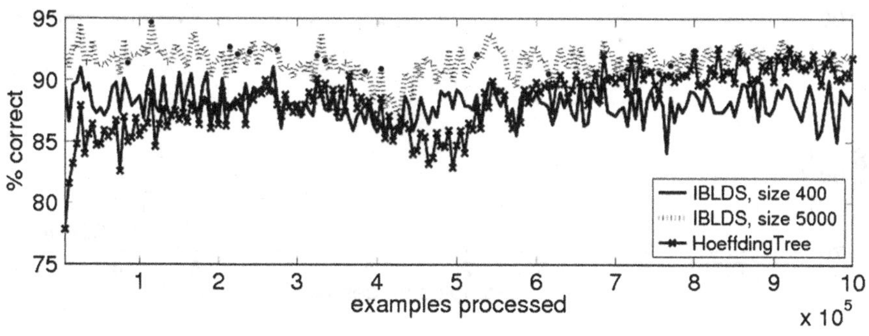

Fig. 8.1 Classification rate on the hyperplane data (binary)

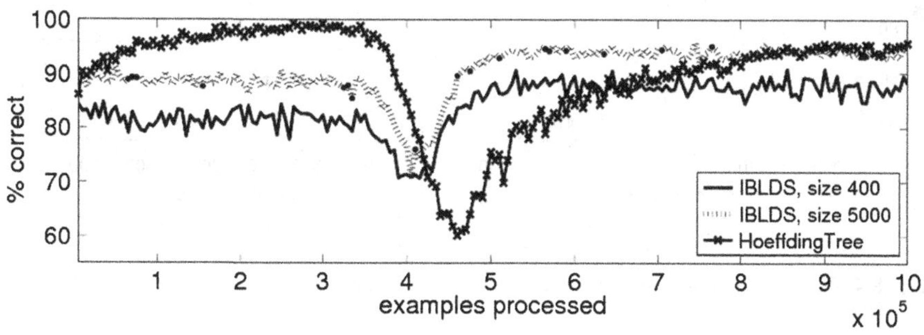

Fig. 8.2 Classification rate on the RandomTree data (binary)

concept drift than the IBL-DS. Both variants of IBL-DS suffer from a drop of about 15% in terms of classification rate, and recover quickly during the phase of the drift, whereas the Hoeffding tree loses about 40% of its accuracy.

8.5.1.2 Real Data

In this experiment, we used the Shuttle data from the Statlog repository, for which the task is to predict the class of a shuttle. The data set is highly imbalanced, with 80% of the instances belonging to one class and the remaining 20% distributed among six other classes; in order to obtain a binary problem, we grouped these six classes into a single one. The new problem thus consists of predicting whether a shuttle belongs to the majority class or not. Both algorithms were initially trained on 300 instances in batch mode; for the holdout evaluation, we used $M = 200$ and $N = 50$. Figure 8.3 shows the results averaged over 100 randomly shuffled versions of the data set. As can be seen, IBL-DS starts with a very strong performance, close to 99% accuracy; the Hoeffding tree reaches this accuracy, too, but not before observing three quarters of the whole stream.

The wine quality data is an ordinal classification problem, in which a wine (characterized by several chemical properties) is put into a discrete category ranging

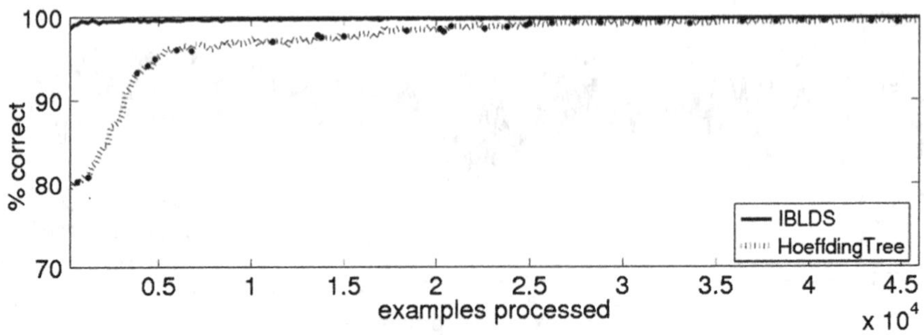

Fig. 8.3 Classification rate on the Shuttle data (binary)

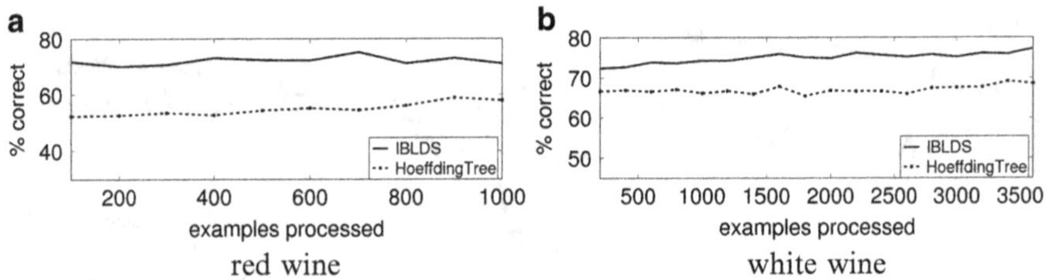

Fig. 8.4 Classification rate on the wine quality data set (binary)

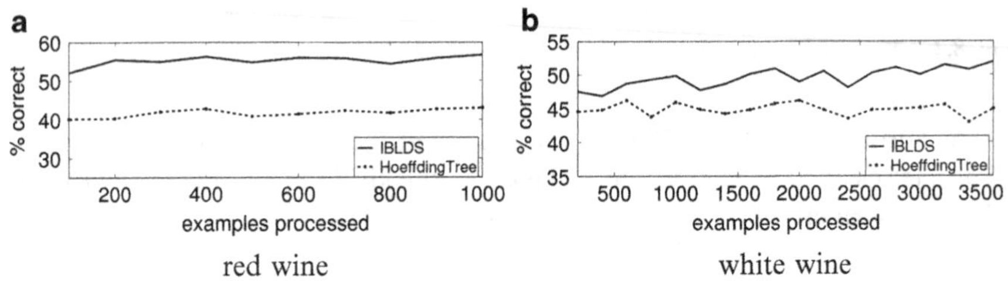

Fig. 8.5 Classification rate on the wine quality data set (multiclass)

from 10 (best) to 0 (worst). We turned this problem into a binary classification task by grouping the top-5 and bottom-6 classes. Actually, the data set consists of two subsets, one for white wine and one for red wine. For both data sets, the initial learning is done on 300 instances. In all our experiments on the wine quality data, we average the results over 100 randomly shuffled versions. For the evaluation on the red wine data, we used $M = 100$ and $N = 25$, because this data set is relatively small (about 1,600 examples); for white wine, we used $M = 200$ and $N = 50$. Figure 8.4 shows the results of both experiments. As can be seen, IBL-DS is clearly superior to Hoeffding trees on these data sets.

For evaluating the muticlass case, we used the same real data sets as above, but without grouping the output categories. As can be seen from Fig. 8.5, the performance of both IBL-DS and Hoeffding trees on the wine data is lower than

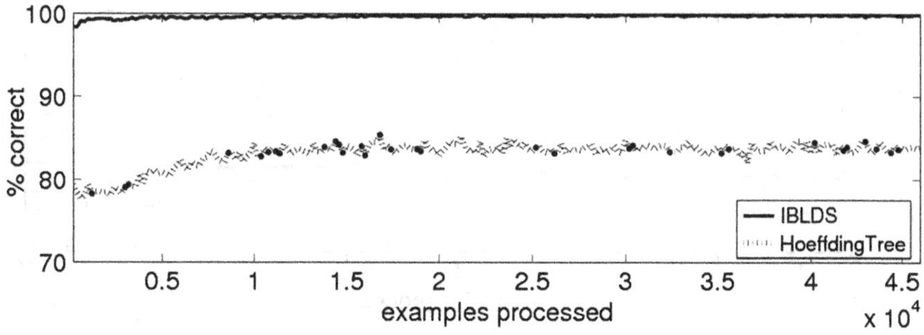

Fig. 8.6 Classification rate on the Shuttle data (multiclass)

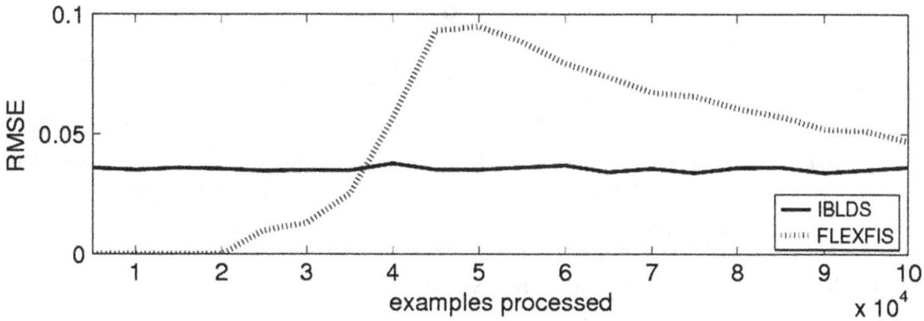

Fig. 8.7 RMSE for the hyperplane data (regression, linear case)

that for the binary case, an observation that is clearly expected. Still, IBD-DS remains superior on the whole stream. For the Shuttle data, Fig. 8.6 shows that the performance of IBL-DS remains almost the same, compared to the binary case, whereas the Hoeffding tree again starts with low classification rate and never exceeds the 85% limit.

8.5.2 Regression

For the case of regression, we modified the hyperplane generator in MOA as follows: The output for an instance x is not determined by the sign of $w^{\mathrm{T}}x$, where w is the normal vector of the hyperplane, but by the absolute value $|w^{\mathrm{T}}x|$. In other words, the problem is to predict the distance to the hyperplane. As an alternative, we also tried $\left(w^{\mathrm{T}}x\right)^2$, i.e., the squared distance. Again, ConceptDriftStream was used for simulating a concept drift by mixing two streams.

Figures 8.7 and 8.8 show the performance of IBL-DS and FLEXFIS, in terms of the root mean squared error (RMSE), for the (piecewise) linear and the quadratic case (and dimension $d = 4$), respectively. As can be seen, FLEXFIS performs quite

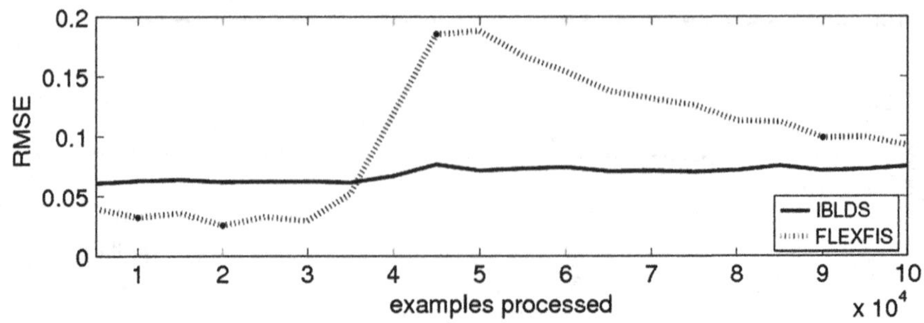

Fig. 8.8 RMSE for the hyperplane data (regression, quadratic case)

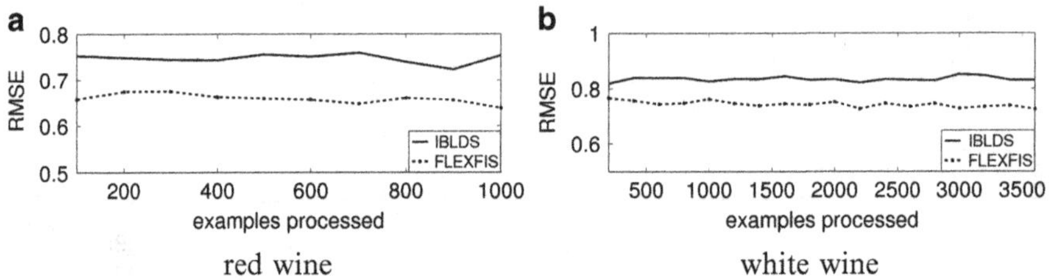

Fig. 8.9 RMSE for wine quality data set (regression)

well in the linear case. This behavior is expected and can easily be explained by its model structure (FLEXFIS uses fuzzy rules with linear functions as consequent parts). What is more interesting, however, is the observation that IBL-DS is much less affected by the concept drift, both in the linear and the quadratic case. In fact, while FLEXFIS deteriorates significantly and needs quite some time to recover, the performance of IBL-DS remains almost unchanged.

As a real data set, we again used the wine data, this time treating the quality level as a numerical value. Figure 8.9 shows that IBL-DS is slightly worse than FLEXFIS [24] on these two data sets.

8.6 Summary

We have presented an instance-based algorithm for classification and regression on data streams. This algorithm, called IBL-DS, has a number of desirable properties that are not, at least not as a whole, shared by existing alternative methods. The experiments presented in [6], complemented by those in this paper, suggest that IBL-DS is very flexible and thus able to adapt to an evolving environment quickly, a point of utmost importance in the data stream context. In particular, two specially designed editing strategies are used in combination in order to successfully deal with both gradual concept drift and abrupt concept shift. Besides, IBL-DS is relatively robust and produces good results when being used in a default setting for its parameters. An implementation of IBL-DS under the MOA framework, along with a documentation, can be downloaded under the following address: http://www.uni-marburg.de/fb12/kebi/research/software/iblstreams/.

8.7 Distance Function

The distance function used in IBL-DS is an incremental variant of SVDM (Simple Value Difference Metric) which is a simplified version of the VDM (Value Difference Metric) distance measure [28] and was successfully used in the classification algorithm RISE [13, 14]. Let an instance x be specified in terms of ℓ features $F_1,\dots,F_\ell$, i.e., as a vector $x=(f_1,\dots,f_\ell)\in D_1\times\cdots\times D_\ell$.

Numerical features F_i with domain $D_i=\mathbb{R}$ are first normalized by the mapping $f_i \mapsto f_i/(\max-\min)$, where max and min denote, respectively, the largest and smallest value for F_i observed so far; these values are permanently updated.[5] Then, $\delta_i(f_i,f_i')$ is defined by the Euclidean distance between the normalized values of f_i and f_i'.

For a discrete attribute F_j, the distance between two values f_j and f_j' is defined by the following measure:

$$\delta_i(f_j,f_j') = \sum_{k=1}^{m} \left\| P(\lambda_k \,|\, F_j = f_j) - P(\lambda_k \,|\, F_j = f_j') \right\|,$$

where m is the number of classes and $P(\lambda\,|\,F=f)$ is the probability of the class λ given the value f for attribute F. Finally, the distance between two instances x and x' is given by the mean squared distance

$$\Delta(x,x') = \frac{1}{\ell}\sum_{i=1}^{\ell} \delta_i(f_i,f_i')^2 .$$

References

1. Aggarwal, C.C., Han, J., Wang, J., Yu, P.S.: A framework for clustering evolving data streams. In: Proceedings of VLDB 2003, the 29th International Conference on Very Large Data Bases. Berlin, Germany (2003)
2. Aha, D.W. (ed.): Lazy Learning. Kluwer Academic Publ., Dordrecht, Netherlands (1997)
3. Aha, D.W., Kibler, D.F., Albert, M.K.: Instance-based learning algorithms. Machine Learning **6**(1), 37–66 (1991)

[5] To make the transformation more robust toward outliers, it makes sense to replace max and min by appropriate percentiles of the empirical distribution.

4. Angelov, P.P., Filev, D.P., Kasabov, N.: Evolving Intelligent Systems. John Wiley and Sons, New York (2010)
5. Angelov, P.P., Lughofer, E., Zhou, X.: Evolving fuzzy classifiers using different model architectures. Fuzzy Sets and Systems **159**(23), 3160–3182 (2008)
6. Beringer, J., Hüllermeier, E.: Efficient instance-based learning on data streams. Intelligent Data Analysis **11**(6), 627–650 (2007)
7. Bifet, A., Holmes, G., Kirkby, R., Pfahringer, B.: MOA: massive online analysis. Journal of Machine Learning Research **11**, 1601–1604 (2010)
8. Bifet, A., Holmes, G., Pfahringer, B., Kirkby, R., Gavaldà, R.: New ensemble methods for evolving data streams. In: Proceedings of the 15th ACM SIGKDD International Conference on Knowledge Discovery and Data Mining, pp. 139–148. Paris, France (2009)
9. Bifet, A., Kirkby, R.: Massive Online Analysis Manual (2009)
10. Cormode, G., Muthukrishnan, S.: What's hot and what's not: Tracking most frequent items dynamically. In: ACM Symposium on Principles of Database Systems (PODS). San Diego, California (2003)
11. Dasarathy, B.V. (ed.): Nearest Neighbor (NN) Norms: NN Pattern Classification Techniques. IEEE Computer Society Press, Los Alamitos, California (1991)
12. Dawid, A.P.: Statistical theory: The prequential approach. In: Journal of the Royal Statistical Society-A, pp. 147:278–292 (1984)
13. Domingos, P.: Rule induction and instance-based learning: A unified approach. In: C. Mellish (ed.) Proceedings of the 14th International Joint Conference on Artificial Intelligence, IJCAI 95, vol. 2, pp. 1226–1232. Morgan Kaufmann, Montral, Qubec, Canada (1995)
14. Domingos, P.: Unifying instance-based and rule-based induction. Machine Learning **24**, 141–168 (1996)
15. Domingos, P., Hulten, G.: A general framework for mining massive data streams. Journal of Computational and Graphical Statistics **12** (2003)
16. Frank, A., Asuncion, A.: UCI machine learning repository (2010). URL http://archive.ics.uci.edu/ml
17. Gaber, M.M., Zaslavsky, A., Krishnaswamy, S.: Mining data streams: A review. ACM SIGMOD Record, ACM Special Interest Group on Management of Data **34**(1) (2005)
18. Gama, J., Gaber, M.M.: Learning from Data Streams. Springer-Verlag, Berlin, New York (2007)
19. Gama, J., Medas, P., Castillo, G., Rodrigues, P.: Learning with drift detection. In: Proceedings SBIA 2004, the 17th Brazilian Symposium on Artificial Intelligence, pp. 286–295. São Luis, Maranhão, Brazil (2004)
20. Gama, J., Medas, P., Rodrigues, P.: Learning decision trees from dynamic data streams. In: SAC '05: Proceedings of the 2005 ACM symposium on Applied computing, pp. 573–577. ACM Press, New York, NY, USA (2005). DOI http://doi.acm.org/10.1145/1066677.1066809
21. Gama, J., Sebastião, R., Rodrigues, P.P.: Issues in evaluation of stream learning algorithms. In: Proceedings of 15th ACM SIGKDD International Conference on Knowledge Discovery and Data Mining. Paris, France (2009)
22. Hulten, G., Spencer, L., Domingos, P.: Mining timechanging data streams. In: Proceedings 7th ACM SIGKDD International Conference on Knowledge Discovery and Data Mining, pp. 97–106. San Francisco, CA, USA (2001)
23. Kolodner, J.L.: Case-based Reasoning. Morgan Kaufmann, San Mateo (1993)
24. Lughofer, E.: FLEXFIS: A robust incremental learning approach for evolving takagi-sugeno fuzzy models. IEEE Transactions on Fuzzy Systems **16**(6), 1393–1410 (2008)
25. Lughofer, E.: Evolving Fuzzy Systems: Methodologies, Advanced Concepts and Applications. Springer-Verlag, Berlin, Heidelberg (2011)
26. Oza, N.C., Russell, S.: Online bagging and boosting. Artificial Intelligence and Statistics pp. 105–112 (2001)
27. Salzberg, S.: A nearest hyperrectangle learning method. Machine Learning **6**, 251–276 (1991)

28. Stanfill, C., Waltz, D.: Toward memory-based reasoning. Communications of the ACM **29**, 1213–1228 (1986)
29. Takagi, T., Sugeno, M.: Fuzzy identification of systems and its applications to modeling and control. IEEE Transactions on Systems, Man, and Cybernetics **15**(1), 116–132 (1985)
30. Utgoff, P.E.: Incremental induction of decision trees. Machine Learning **4**, 161–186 (1989)
31. Witten, I.H., Frank, E.: Data Mining: Practical machine learning tools and techniques, 2 edn. Morgan Kaufmann, San Francisco (2005)
32. Widmer, G. and Kubat, M.: Learning in the Presence of Concept Drift and Hidden Contexts. Machine Learning **23**, 69–101 (1996)

Part III
Dynamic Methods for Supervised Regression Problems

Chapter 9
Flexible Evolving Fuzzy Inference Systems from Data Streams (FLEXFIS++)

Edwin Lughofer

Abstract Data streams are usually characterized by an ordered sequence of samples recorded and loaded on-line with a certain frequency arriving continuously over time. Extracting models from such type of data within a reasonable on-line computational performance can be only achieved by a training procedure which is able to incrementally build up the models, ideally in a single-pass fashion (not using any prior samples). This chapter deals with data-driven design of fuzzy systems which are able to handle sample-wise loaded data within a streaming context. These are called *flexible evolving fuzzy inference systems (FLEXFIS)* as they may permanently change their structures and parameters with newly recorded data, achieving maximal flexibility according to new operating conditions, dynamic system behaviors, or exceptional occurrences. We are explaining how to deal with parameter adaptation and structure evolution on demand for regression as well as classification problems. In the second part of the chapter, several key extensions of the *FLEXFIS* family will be described (leading to the *FLEXFIS++* and *FLEXFIS-Class++* variants), including concepts for on-line rule merging, dealing with drifts, dynamically reducing the curse of dimensionality, as well as interpretability considerations and reliability in model predictions. Successful applications of the *FLEXFIS* family are summarized in a separate section. An extensive evaluation of the proposed methods and techniques will be demonstrated in a separate chapter (Chap. 14), when dealing with the application of flexible fuzzy systems in on-line quality-control systems.

E. Lughofer (✉)
Department of Knowledge-Based Mathematical Systems, Johannes Kepler University Linz, Altenbergerstr. 69, A-4040 Linz, Austria
e-mail: edwin.lughofer@jku.at

M. Sayed-Mouchaweh and E. Lughofer (eds.), *Learning in Non-Stationary Environments: Methods and Applications*, DOI 10.1007/978-1-4419-8020-5_9,
© Springer Science+Business Media New York 2012

9.1 Introduction

9.1.1 Motivation

In today's industrial real-world applications, there is a significant growth of sensor networks [23], TCP/IP traffic, GPS data, and parallel production lines [20]. The amount of data which needs to be synchronously processed can become so huge that storing it as persistent tables or (feature) matrices and building process models from these in a batch off-line manner is simply not feasible as taking too much computation time or virtual memory requirements. Furthermore, the processes may become very non-stationary, meaning that changing system states, varying environmental conditions, or human behaviors may arise which have to be integrated into the models. Under these considerations, the automatic learning and permanent adaptation of models from on-line and massive data streams [8] are playing a more and more central role in data-driven model design. A data stream can be seen as a stochastic process in which events occur continuously and independently from each other. In particular, it is characterized by [22]:

- The data samples or data blocks are continuously arriving on-line over time. The frequency depends on the frequency of the measurement recording process.
- The data samples are arriving in a specific order, over which the system has no control.
- Data streams are usually not bounded in a size; that is, a data stream is alive as long as some interfaces, devices, or components at the system are switched on and are collecting data.
- Once a data sample/block is processed, it is usually discarded immediately, afterwards.

Furthermore, a data stream model may also be applied in case of very large data bases (VLDB)[1] (e.g., consider the storage of customers' attitudes in a supermarket or long-term metrological data): in this case, the data needs to be treated as a kind of pseudo stream as have to be loaded step-wise (in blocks or single samples) into virtual memory in order to avoid memory over-flows.

In order to build models from data streams within a reasonable time frame (ideally in real-time manner), the incremental (step-wise) learning concept plays an important role, as re-training steps on all or partial blocks of data seen so far often becomes too slow [2]. Single-pass incremental training techniques where one sample/block of data is loaded, the model updated based on this data, and then the data discarded require minimal virtual memory and computation time. In order to account for changing system dynamics, two learning concepts are essential: (1) dynamic adaptation of model parameters, ideally guided by some

[1] http://en.wikipedia.org/wiki/Very_large_database.

optimization procedure and (2) evolving new structural components on demand to extend the model to unexplored regions in the feature space and to prevent (precarious) extrapolation situations at an early stage.

9.1.2 Our Contribution

In literature, various methods exist which can handle the above mentioned circumstances appropriately. Some can be found under the umbrella of incremental learning methods (e.g., incremental classifiers, clustering approaches), which are purchasing their motivation from the machine learning community and their basic concepts. The massive on-line analysis tool MOA[2] developed by the Department of Computer Science at the Waikato University [8] integrates some important incremental learning methods (including Hoeffding trees [16], their extension with naive Bayes leaves, on-line OZA bagging and boosting [58], and ultra fast forest of trees (UFFT) [24]). Alternative approaches have emerged during the last decade within the soft computing community under the scope of evolving (intelligent) systems [2], evolving connectionist systems including dynamic neural networks approaches and beyond [31], as well as evolving fuzzy systems (EFS) [44]. The later are enjoying a great attraction for several years due to achieving a reasonable trade-off between universal approximation capability [13] and interpretability [11] during the extraction of models from data (streams). In this sense, pure black box models are prevented (as achieved by neural networks of support vector machines) and some insights in the process can be offered. In the subsequent section, we will present the core methodologies, basic strategies, and concepts of one of the widely used approaches within the field of EFS, namely, the *FLEXFIS* family (for a comprehensive survey on various EFS methods, see [44]), which is coming with a regression variant (*FLEXFIS*) [41] and a classification variant (*FLEXFIS-Class*) [4], together with a pure clustering-based spin-off classifier called *eVQ-Class* [42]. Furthermore, we will demonstrate important extensions of the *FLEXFIS* family (leading to *FLEXFIS++* and *FLEXFIS-Class++* variants) to guide the evolved models/classifiers to higher predictive performance and transparency (Sect. 9.3). This will include a rule merging strategy in order to reduce unnecessary complexity, an attempt on how to reduce curse of dimensionality in flexible fuzzy classifiers in a smooth and incremental manner, concepts on how to deal with drifts in data streams, some considerations regarding interpretability of the evolved models, and concepts dealing with reliability (self-awareness) in model predictions.

[2]http://moa.cs.waikato.ac.nz/.

9.2 The *FLEXFIS* Family: Core Learning Engines

The *FLEXFIS* family contains a bunch of algorithms, which were developed during the last years and are summarized as family tree in Fig. 9.1. The incremental evolving clustering engine *eVQ* is used in all regression and classification variants for rule evolution and updating the antecedent parameters. The methods for improved performance which are placed more to the left-hand side ("complexity reduction" and "drift handling") are actually integrated in the regression resp.

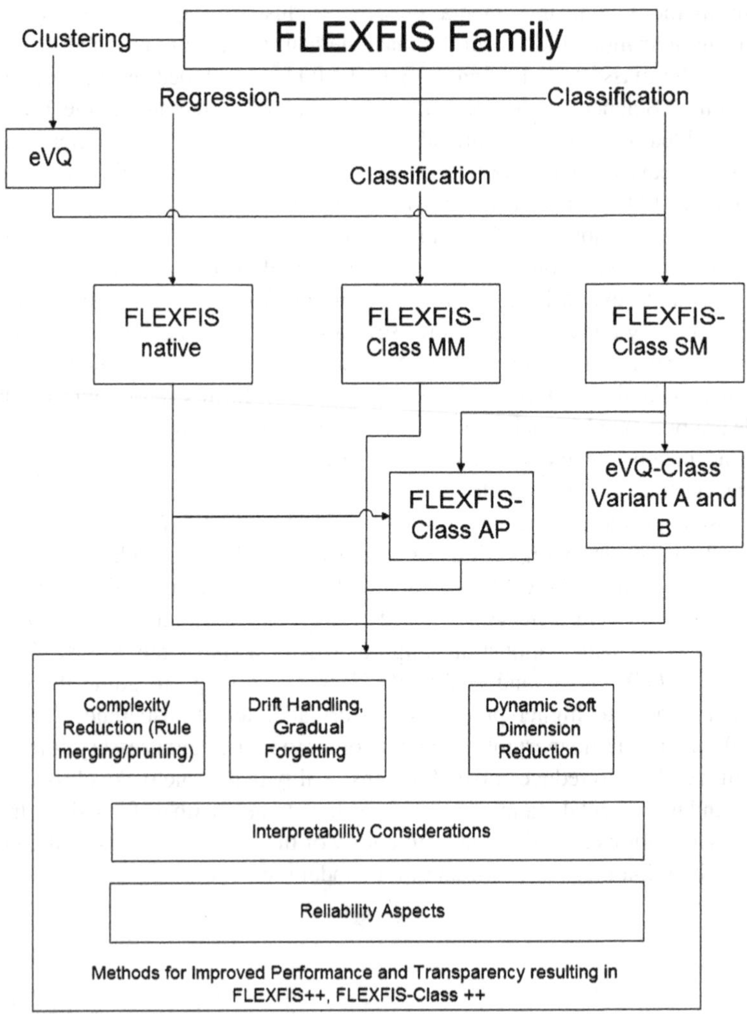

Fig. 9.1 Family tree of *FLEXFIS* including extensions for improved performance and transparency resulting in FLEXFIS++ and FLEXFIS-Class++ approaches

MM classification variants, that which is placed more to the right ("dynamic soft dimension reduction") is actually integrated in all classification variants. Interpretability considerations and reliability aspects refer to all regression and classification variants. In the following sections and subsections, we summarize the most important facets in each of these components.

9.2.1 Flexible Evolving Fuzzy Regression Models

In a regression context, a data stream is defined by an (theoretically) infinite sequence of samples $(\mathbf{x}(1),\mathbf{y}(1)),(\mathbf{x}(2),\mathbf{y}(2)),(\mathbf{x}(3),\mathbf{y}(3)),\ldots$ or by an infinite sequence of data blocks $(\mathbf{x}(1:N),\mathbf{y}(1:N)),(\mathbf{x}(N+1:2N),\mathbf{y}(N+1:2N)),(\mathbf{x}(2N+1:3N),\mathbf{y}(2N+1:3N)),\ldots$ (each containing N samples), where the output vector $\mathbf{y}(k)$ contains continuous values recorded at time instance k. In the first case, we speak about sample-wise evolving data streams, and in the second case, about block-wise evolving data streams. For the sake of simplicity and without loss of generality, we are considering only single-output systems, reducing $\mathbf{y}(k)$ for all outputs to single values denoted as $y(k)$. The task of regression is now to build up models which perform a mapping of (a subset of) the input space $\mathbf{x}$ onto the target concept y, explaining the functional coherence between inputs and output as close as possible. In literature, often least squares error is applied as underlying optimization criterion, which expresses the minimization of the average squared deviation between estimated $\hat{y}$ and measured y output values:

$$J=\frac{1}{N}\sum_{i=1}^{N}(y(i)-\hat{y}(i))^2=\min! \tag{9.1}$$

In general, non-linearities are implicitly contained in the data streams; hence, linear models are usually not able to resolve the coherences with sufficient accuracy.

The Takagi–Sugeno model architecture [65] is a widely used fuzzy model architecture, which fulfills the universal approximation theorem [13] (i.e., can approximate any non-linear functional behavior with sufficient accuracy) and also allows some sort of interpretability, as (1) the antecedent parts consist of linguistic conditions and terms and (2) the consequents may indicate some local trends (see also Sect. 9.3.4). The ith rule of a Takagi–Sugeno fuzzy model is defined by:

$$\begin{aligned}&\text{Rule}_i:\ \text{IF } x_1 \text{ IS } \mu_{i1} \text{ AND...AND } x_p \text{ IS } \mu_{ip} \text{ THEN}\\ &\qquad l_i=w_{i0}+w_{i1}x_1+w_{i2}x_2+\ldots+w_{ip}x_p,\end{aligned} \tag{9.2}$$

with μ_{ij} the fuzzy set in the jth antecedent part and l_i a hyper-plane defining the consequent of the ith rule. In functional form, the Takagi–Sugeno fuzzy model containing C rules is defined by:

$$\hat{f}(\mathbf{x})=\hat{y}=\sum_{i=1}^{C}l_i\Psi_i(\mathbf{x})\quad \Psi_i(\mathbf{x})=\frac{\mu_i(\mathbf{x})}{\sum_{j=1}^{C}\mu_j(\mathbf{x})}\quad \mu_i(\mathbf{x})=\mathop{\mathrm{T}}_{j=1}^{p}\mu_{ij}(x_j), \tag{9.3}$$

where p the dimensionality of the input space and x_j the jth component in the data vector, hence reflecting the value of the jth variable. The symbol T denotes a t-norm in general. The most common choice in the data-driven design of Takagi–Sugeno fuzzy models and also used in the *FLEXFIS* approach, is the product t-norm in connection with Gaussian fuzzy sets (called *fuzzy basis function networks* going back to Wang and Mendel [68]). This yields smooth approximation surfaces and always guarantees well-defined input states. Furthermore, the basis function networks have a close synergy to normalized radial basis function (RBF) networks [33,57], which means that most of the data-driven learning algorithms designed for fuzzy basis function networks can be easily adopted to RBF networks.

In an evolving data stream context, the fuzzy models have to be learned in an incremental manner, ideally in a single-pass fashion (see Sect. 9.1.1). This requires:

1. A permanent adaptation of previously learned parameters with new incoming samples.
2. An evolution of new structural elements on demand due to changing system dynamics, new environmental influences, and operating conditions.

The first issue accounts for a model refinement and intensification of the learned relationship; the second expands the model to new regions and prevents long-term extrapolation effects, which may cause significant uncertainties and bad quality in its predictions.

9.2.1.1 Consequent Learning

In *FLEXFIS* [41], the adaptation of linear consequent parameters is achieved by using the least squares optimization function (9.1) as optimization setting and applying the *recursive least squares estimator (RLS)* [38] in weighted form in order to achieve locality effects of the consequent hyper-planes and handling the learning of each rule consequent separately. In particular, the hyper-planes are then snuggling along the real local trend of the functional dependency/approximation to be learned and can be seen as piecewise local linear approximators [71]. This yields nice interpretation capabilities (trend analysis, feature weights), see Sect. 9.3.4. The weights are given by the normalized membership degree of single samples to the current evolved rules, defined by $\Psi_i(\mathbf{x}(N))$ (9.3) for the Nth data sample in the ith rule. This reduces the influence of rules lying far away from actual samples in the recursive least squares estimator, which is called *recursive fuzzily weighted least squares estimator (RFWLS)* [4] and for the ith rule defined by:

$$\hat{\mathbf{w}}_{\mathbf{i}}(N+1) = \hat{\mathbf{w}}_{\mathbf{i}}(N) + \gamma(N)(y(N+1) - \mathbf{r}^T(N+1)\hat{\mathbf{w}}_{\mathbf{i}}(N)) \tag{9.4}$$

$$\gamma(N) = \frac{P_i(N)\mathbf{r}(N+1)}{\frac{1}{\Psi_i(\mathbf{x}(N+1))} + \mathbf{r}^T(N+1)P_i(N)\mathbf{r}(N+1)} \tag{9.5}$$

$$P_i(N+1) = (I - \gamma(N)\mathbf{r}^T(N+1))P_i(N), \tag{9.6}$$

with $P_i = (R_i^T Q_i R_i)^{-1}$ the weighted inverse Hessian matrix and synchronously updated to the linear parameter vector $\mathbf{w_i}$ (omitting time-intensive matrix inversion steps), and $\mathbf{r}(N+1) = [1 \;\; x_1(N+1) \;\; x_2(N+1) \;\; \ldots \;\; x_p(N+1)]^T$ the regressor values of the $(N+1)$th data sample and $y(N+1)$ the target value.

The major important characteristics of these adaptive update scheme for consequent parameters can be summarized as follows [41]:

- In order to achieve maximal convergence to the optimum at the beginning of a data stream, reasonable start parameters are as follows: $\mathbf{w_i} = \mathbf{0}$ and $P_i = \alpha I$ with α a big integer number, for example, 1,000. After having some rules evolved, for a new rule i, a more stable initialization can be achieved by setting $\mathbf{w_i} = \mathbf{w_j}$ with j the nearest rule center to the center of the newly evolved rule i.
- It is a recursive adaptation scheme, meaning that, once an optimum is found, it converges to the exact optimum within each iteration step for each newly loaded data sample [5]. This is due to the hyper-parabolic shape of the optimization function and the gradient-based Newton step in the update (9.4) being able to find the global minimum of the hyper-parabola.
- For each rule, a separate recursive estimation scheme according to (9.4)–(9.6) is conducted. This yields maximal flexibility in terms of evolving new rules and pruning/merging old rules, as parameters from the other, non-concerned rules are not affected, "disturbed", therefore staying optimal in the least squares sense.
- The computation time for an update of the parameters with a single sample is of complexity $O(C(p+1)^2)$ with C the current number of rules and p the dimensionality of the input space.

9.2.1.2 Antecedent Learning and Rule Evolution

The second part in *FLEXFIS* is dedicated to antecedent learning and rule evolution. The update of already available rules accounts for *plasticity* in the learning scheme, while the generation of new rules on demand accounts for *stability* (older learned relations are not attached, changed). Finally, it is an essential matter to find a reasonable trade-off between stability and plasticity, also known as the *plasticity-stability dilemma* [1]; according to the current nature of the data distribution: if for instance a drift in the underlying data distribution arises (see also Sect. 9.3.1.2), the update of already available rules should be stronger than in non-drift cases (plasticity), triggering an out-dating effect of older learned relationships; if, on the other hand, a new previously unseen system state arises, new rules should be evolved in order to include these in the model within a life-long learning concept (stability), not changing any previously learned rules. This enriches and expands the knowledge and memory of the model, achieving some sort of computational intelligence [3].

Rule evolution and update is performed in the high-dimensional feature space by exploiting an incremental clustering algorithm, which partitions the feature space in local regions in order to account for non-linearity in the regression problem. For each local region (cluster), a rule is defined. Intuitively, it is clear, that the higher the non-linearity of the concept to be learned is, the more clusters should be generated. In this sense, *FLEXFIS* uses the concept of input–output space clustering in order to account for non-linearity in the target concept, too.

For incremental clustering, an evolving version of vector quantization (*eVQ*) [40] is applied, which is deduced from conventional vector quantization [25] and moves the cluster centers through the feature space with new incoming samples in a single-pass manner, updates the ranges of influence of the clusters in a recursive manner and evolves new rules on demand. Cluster center movement accounts for plasticity in the learning algorithm and is defined by:

$$\mathbf{c}_{\text{win}}^{(\text{new})} = \mathbf{c}_{\text{win}}^{(\text{old})} + \eta_{\text{win}} \left(\mathbf{x} - \mathbf{c}_{\text{win}}^{(\text{old})} \right), \tag{9.7}$$

where $\mathbf{c}_{\text{win}}$ denotes the cluster center of the winning cluster = that cluster which is nearest to the current data sample $\mathbf{x}$ with respect to a distance metric:

$$\text{win} = \text{argmin}_{i=1,\ldots,C} \left(\|\mathbf{x} - \mathbf{c}_i\|_A \right). \tag{9.8}$$

If using Euclidean distance for A, ellipsoidal clusters in main position, and if using Mahalanobis distance [55], ellipsoidal clusters in arbitrary position are triggered. The learning gain η_{win} controls the movement degree of the center, hence the plasticity-stability trade-off of the incremental clustering process; it monotonically decreases with the number of samples belonging to the ith cluster, that is, for which the ith cluster was the winning cluster: this finally guarantees convergence of the cluster (centers and ranges of influence) over time (stability). In Sect. 9.3.1.2, we will demonstrate how stronger center movements are enforced in drift cases in order to react on changing data distributions. An evolution of a new rule accounts for stability and is triggered whenever the current sample does not fit in the already generated cluster partition, that is, when the following condition is fulfilled:

$$\|\mathbf{x} - \mathbf{c}_{\text{win}}\|_A \geq \rho, \tag{9.9}$$

where A is an arbitrary distance metric (should be the same as used in (9.8)) and ρ the so-called vigilance parameter controlling the trade-off between stability and plasticity. It is the essential parameter in the *FLEXFIS* algorithm, controlling the number of generated rules. Assuming normalized data streams in the hyper-cube $[0,1]^{p+1}$ with p the dimensionality of the input/output space, ρ is set to a fraction of the unit space diagonal:

$$\rho = \text{fac} * \frac{\sqrt{p+1}}{\sqrt{2}}, \tag{9.10}$$

with fac $\in [0,1]$. The default setting of fac is 0.3, usually, to our best knowledge from various real-world data streams; values lying in $[0.2, 0.4]$ produce best results. fac can be tuned in an initial modeling phase (step 1a in Algorithm 1) on first block(s) of samples from the stream, stepwise by changing from 0.1 to 0.9 in steps of 0.1, to elicit the optimal setting in terms of validation error or a combined criterion

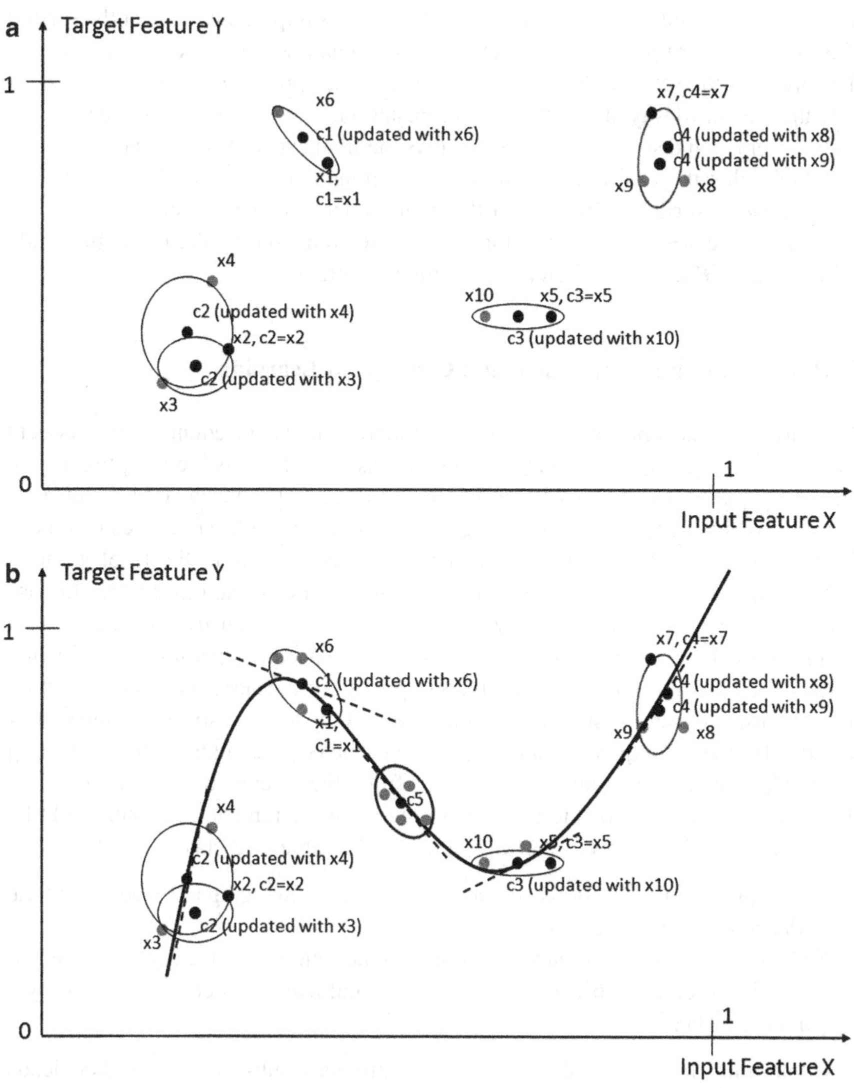

Fig. 9.2 (**a**) Update progress of clusters in the product space through eVQ using the first ten samples $x_1, x_2, \ldots, x_{10}$ from a stream. (**b**) Final achieved model (*solid line*) and consequent functions (*dotted straight lines*)

including also the model complexity. Normalization of data streams to $[0,1]$ is achieved by dividing the feature values through their ranges after subtracting their minimal values seen so far.

Figure 9.2a visualizes an example of the update progress of *eVQ* for the first ten samples of a data stream (numerated as $x_1, x_2, \ldots, x_{10}$): the first center c_1 is set to the first data sample x_1, and the second data sample x_2 is far away from c_1; therefore, a new cluster is evolved, and its center c_2 is set to x_2; sample x_3 lying near the second cluster is used for updating the center c_2, moving half-way between x_2 and x_3, and the spread of the second cluster; x_4 triggers a similar procedure, while center c_2 is only updated third way along its old position and x_4; x_5 again opens up a new cluster with center c_3; and so on. Figure 9.2b shows the final evolved cluster partitions and also highlights the estimated consequent functions (straight dotted lines) and the approximation surface (function) of the final model. The consequent functions are serving a piece-wise linear predictors, which are combined by the normalized rule firing degrees $\Psi_.$ to a smooth non-linear function after (9.3).

9.2.1.3 Connecting Antecedent and Consequent Learning

The purpose of an appropriate connection of incremental antecedent and consequent learning is to keep the consequent parameters as close as possible to optimality in the least squares sense. This is automatically assured whenever a new cluster is evolved, as this triggers an opening up of a new recursive least squares estimator for the new rule (which is independent from the others due to the local learning approach), not "disturbing" the consequent parameters for the other rules. In case when a cluster center and its range of influence is updated, it triggers a structural change in the RFWLS estimator which makes the consequent parameters estimated based on the previous cluster position non-optimal for the current position. In order to compensate this non-optimal situation, a correction vector resp. correction matrix are added to the consequent parameters resp. inverse Hessian matrix before updating them with the current data block/sample from the stream through (9.4)–(9.6). An explicit calculation of these correction terms in incremental manner could be not achieved so far; however, partial results are available (see [44]):

1. The sequence of corrections vectors/matrices are converging to 0 over time (due to the convergence effect of cluster centers).
2. This convergence takes place in a quasi-monotonic fashion; hence, the sum of correction vectors/matrices over the incremental learning steps is bounded by a (small) constant.

Combining the aspects discussed in the previous subsections together, leads to the *FLEXFIS* algorithm for evolving TS fuzzy systems from data streams as summarized in Algorithm 1.

Algorithm 1 FLEXFIS+ (FLEXible Fuzzy Inference Systems from Data Streams)

1a **Option 1 (most commonly used)**: Estimate the (initial) ranges of all variables; perform an initial training in batch mode (on pre-collected training samples or using first on-line samples) and obtain the optimal value for the vigilance parameter ρ in a step-wise evaluation process (e.g., testing values of $fac =$ 0.1,0.2...,0.9 in a CV process). Initial training is done by sending the initial data set into *eVQ* first (steps 3 to 7 for each sample in the set) and then perform consequent estimation on all data using steps 8 to 11.

1b **Option 2**: Alternatively, (in case when no off-line recordings are available) incremental learning from scratch can be carried out, by estimating the (initial) ranges of all variables from a few dozens of on-line data stream samples and using $fac = 0.3$ as default parameter; set $C = 0$.
2 For each new incoming data sample $\mathbf{x}$, do (steps 3 to 12).
3 If rule base is empty, set the center of the first rule to the current sample, $\mathbf{c}_1 = \mathbf{x}$; its spread to 0, $\sigma_1 = 0$; and its consequent parameters $\mathbf{w_1} = \mathbf{0}$ and $P_1 = \alpha I$ with α a big integer number, for example, 1,000; go to 11.
4 Normalize $\mathbf{x}$ to $[0,1]$ and the cluster centers according to the estimated feature ranges in step 1a or 1b.
5 Elicit winning cluster according to (9.8).
6 **If** condition (9.9) is fulfilled, then evolve a new rule by increasing the number of rules $C = C + 1$ and setting its center to the current data sample $\mathbf{c}_C = \mathbf{x}$ and its range of influence in each direction by $\sigma_C = 0$. Set consequent parameters $\mathbf{w}_C = \mathbf{w}_j$ and inverse Hessian matrix $P_C = P_j$, where j is the rule nearest center to $\mathbf{c}_C$.
7 **Else** update the center of the nearest cluster $\mathbf{c}_{win}$ by moving it towards the current sample as in (9.7) and update its range of influence (see below).
8 Transfer the clusters back to original feature space, according to the ranges of the features.
9 Project modified/evolved cluster, whose support k_{win} exceeds a threshold (default value = 3), to the axes in order to update/evolve the fuzzy set partition in each dimension (feature): the centers of the fuzzy sets are associated with the corresponding center coordinates of the clusters, and the widths of the fuzzy sets are set to $\max(\sigma_{.}, \varepsilon)$ with ε a small positive constant, in order to avoid numerical instabilities.
10 Add correction vectors to the linear consequent parameters and correction matrices to the inverse Hessian matrices estimated in the previous step.
11 Perform recursive fuzzily weighted least squares (for local learning) using (9.4) to (9.6) for all C rules.
12 Update the ranges of all features.

The algorithm contains two options for setting up initial models: (1a) uses an initial batch training cycle, where the vigilance parameter may be optimized within a CV process; this also opens the possibility to perform an a priori selection of the most important variables in order to crisply reduce the curse of dimensionality and to continue the incremental phase on the selected variables only. (1b) supports the option of incremental train, the model from scratch, using the default setting of the factor fac $= 0.3$ in the vigilance formula (9.10).

The concrete update formula for the range of influence of the winning cluster in each incremental learning step (step 7 in Algorithm 1) depends on the applied distance metric A: in case of Euclidean distance, axis-parallel ellipsoids are triggered, which requires an update of the data stream variance along each dimension; this can be achieved by the recursive variance formula with rank-one modification for achieving fast convergence to the batch variance, see [61]. In case of Mahalanobis distance, ellipsoids in arbitrary position are triggered which requires an update of the covariance matrix (each cluster/rule is defined by the inverse of this matrix also called *shape matrix* [36]), on a new sample $\mathbf{x}(N+1)$ achieved by [60]:

$$\Sigma(\text{new}) = \frac{1}{N+1}(N\Sigma(\text{old}) + \frac{N}{N+1}(\bar{X}(N) - \mathbf{x}(N+1))^T(\bar{X}(N) - \mathbf{x}(N+1))), \tag{9.11}$$

where $\Sigma(\text{old})$, the old covariance matrix and $\bar{X}(N)$, the mean values of all input features, which are represented by the updated cluster centers $\mathbf{c}$. In order to speed up the incremental learning process, the inverse covariance matrix $\Sigma^{-1}(\text{old})$ can be directly updated in sample-wise manner by [6]:

$$\Sigma^{-1}(\text{new}) = \frac{\Sigma^{-1}(\text{old})}{1-\alpha} - \frac{\alpha}{1-\alpha}\frac{(\Sigma^{-1}(\text{old})(\mathbf{x}(N+1)-\mathbf{c}))(\Sigma^{-1}(\text{old})(\mathbf{x}(N+1)-\mathbf{c}))^T}{1+\alpha((\mathbf{x}(N+1)-\mathbf{c})^T\Sigma^{-1}(\text{old})(\mathbf{x}(N+1)-\mathbf{c}))}, \tag{9.12}$$

with $\alpha = \frac{1}{N+1}$.

9.2.2 *Flexible Evolving Fuzzy Classifiers*

In a classification context, a data stream is defined by an (theoretically) infinite sequence of samples $(\mathbf{x}(1),\mathbf{y}(1)),(\mathbf{x}(2),\mathbf{y}(2)),(\mathbf{x}(3),\mathbf{y}(3)),\ldots$ or by an infinite sequence of data blocks $(\mathbf{x}(1:N),\mathbf{y}(1:N)),(\mathbf{x}(N+1:2N),\mathbf{y}(N+1:2N)),(\mathbf{x}(2N+1:3N),\mathbf{y}(2N+1:3N)),\ldots$ (each containing N samples), where the output vectors $\mathbf{y}(k)$ contain integer values as class labels $L=1,\ldots,K$ recorded at time instance k. The task of learning classifiers from data is now to establish decision boundaries which are able to discriminate between the classes in form of decision or classification models. In the fuzzy case, these classification models are again representing a mapping between input and output space, explaining the coherence between local input regions and their preferred classification responses.

9.2.2.1 Single Model Architecture

The classical and widely applied fuzzy model architecture [35, 62], which is also used in the single model (SM) variant of *FLEXFIS-Class*, is that one based on singleton consequent class labels. The ith rule is defined by:

$$\text{Rule}_i: \text{IF } x_1 \text{ IS } \mu_{i1} \text{ AND...AND } x_p \text{ IS } \mu_{ip} \text{ THEN } l_i = L_i, \tag{9.13}$$

with $L_i \in \{1,\ldots,K\}$ the output class label of the ith rule.

Rule extraction from data is performed by applying *eVQ* in the same manner as described in Sect. 9.2.1.2 and its update process through stream samples visualized in Fig. 9.2. The only difference is that for the classification case, clustering is conducted in the input space only and a separate hit matrix H carries the information about the number of samples attached to each cluster: the entry h_{ik} denotes the number of samples belonging to cluster i and falling into class k. The clusters are projected to the input feature space to form the fuzzy sets and antecedent parts of the rules; the consequent class labels are elicited by:

$$L_i = \text{argmax}_{k=1}^{K} h_{ik}, \tag{9.14}$$

that is, the most frequent class in the ith rule = cluster is taken as response class for the ith rule.

Classifying a new instance in the native *FLEXFIS-Class SM* is conducted through a winner-takes-it-all classification scheme:

$$L = L_{i^*} \text{ with } i^* = \text{argmax}_{1 \le i \le C}\ \mu_i, \tag{9.15}$$

that is, that rule with the highest membership degree in the current sample is sought and its consequent class label used as final response. Furthermore, an important measure is also the confidence in the classifier decisions as they may be of great help for a better interpretation of the model's final output. This is obtained by:

$$\text{conf}_L = \frac{\text{conf}_{i^*L}}{\text{conf}_{i^*L} + \text{conf}_{i^*L_2}}, \tag{9.16}$$

with i^* as defined in (9.15), L_2 the second most supported class in rule $i*$, and conf_{ik} the confidence of the ith rule in the kth class (obtained by the relative frequency of the kth class in the ith rule). An extended calculation of confidence levels is treated in Sect. 9.3.5, dealing with a more appropriate handling of conflict cases between classes.

9.2.2.2 Multimodel Architecture

Although the single model architecture offers some nice features such as a transparent rule structure (a user can easily recognize in which local regions which class dominates), it often has some short-comings in terms of classification accuracy [42] (compare also with Chap. 14). This is especially the case in multi-class classification problems with a high number of classes ($K \ge 8$), as then the decision boundaries are usually getting quite complex to learn when applying a direct multi-class mapping [21]. Therefore, a multi-model architecture is exploited in *FLEXFIS-Class* [48], which uses one Takagi–Sugeno fuzzy model (as defined in (9.3)) per class and performs a one-versus-rest classification over the regression-based single model outputs:

$$L = \text{class}(\mathbf{x}) = \text{argmax}_{m=1,\dots,K} \hat{f}_m(\mathbf{x}), \tag{9.17}$$

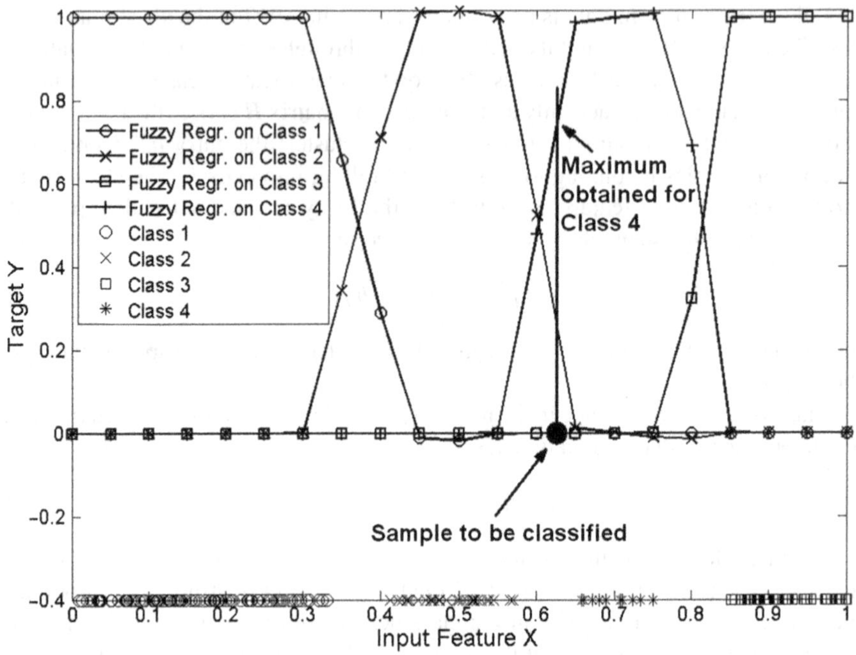

Fig. 9.3 Fuzzy regression functions approaching 1 in regions where the corresponding class is located, and a sample to be classified are shown as *big dot* →; the circular class takes the maximum and is therefore outputted by (9.17)

with $\hat{f}_m$ the estimated regression value in $[0,1]$ for the mth model. This means that a decision boundary is made easier by being reduced from a direct multi-class boundary to a boundary discriminating between one class versus the others. The training of each TS fuzzy model is conducted in the same manner as in regression-based *FLEXFIS* after Algorithm 1 by using indicator target entries in the data stream vectors for $y(k)$: $y(k) = 1$ is used when updating the mth model in the kth time instance when the current data sample belongs to class $m, y(k) = 0$, is used otherwise. This is also called *fuzzy regression by indicator matrix* in accordance to the *linear regression by indicator matrix* technique. The latter suffers from the masking problem in case of $K > 2$ classes [27] which can be solved in the fuzzy case. An example of how regression functions are spanned and how a new sample is classified according to (9.17) for a four-class problem in the one-dimensional input case is visualized in Fig. 9.3.

The confidence $\text{con}f_L$ of the overall output value $L = m \in \{1, \dots, K\}$ is elicited by normalizing the maximal output value with the sum of the two maximal output values from all K models:

$$\text{con}f_L = \frac{\max_{m=1,\dots,K} \hat{g}_m(\mathbf{x})}{\max_{m=1,\dots,K} \hat{g}_m(\mathbf{x}) + \hat{g}_n(\mathbf{x})} \tag{9.18}$$

where $\hat{g}_m(\mathbf{x}) = \hat{f}_m(\mathbf{x}) + |\min(0, \min_{j=1,...,K} \hat{f}_j(\mathbf{x}))|$ and $\hat{g}_n(\mathbf{x})$ defined as $\hat{g}_m(\mathbf{x})$ corresponding to the second maximal output value $\hat{f}_n(\mathbf{x})$.

The *FLEXFIS-Class MM* learning method is summarized in Algorithm 2 [42].

Algorithm 2 FLEXFIS-Class MM

1. Setup phase with two options as in Algorithm 1, and training K TS fuzzy models $TS_i, i = \{1,...,K\}$ for the K classes by estimating K cluster partitions $CL_i, i = \{1,...,K\}$ with *eVQ* separately and independently and project clusters from each partition on the axes.
2. Take the next incoming data sample $\mathbf{x}$.
3. Elicit its class label, say L.
4. **If** $L \leq K$ (no new class is introduced).
5. Update the Lth TS fuzzy model by:

   a. Taking $y = 1$ as response (target) value,
   b. Performing steps 3 to 11 of Algorithm 1

6. **Else** (a new class label is introduced):

   a. $K = K + 1$.
   b. A new TS fuzzy system is initiated with a single rule whose center is set to the current data sample, using the same value for ρ as in the other models.

7. Update all other TS fuzzy models by taking $y = 0$ as response (target) value and performing steps 3 to 11 of Algorithm 1.
8. Update ranges of all features.
9. If new incoming samples are still available then go to step 2; otherwise *stop*.

One central issue in above algorithm is how to elicit the class label in a data streaming context: usually, the class labels are not defined a priori or are given at hand; therefore, an operator has to provide them during on-line operation mode. This causes significant supervision effort, which can be reduced with so-called active learning approaches [15, 37], where operator's feedback is only required in specific circumstances, that is, when the classifier is quite uncertain in its prediction. An approach which tackles this problem in an unsupervised off-line setting coupled with a supervised on-line setting is the so-called *hybrid active learning (HAL)* approach [45], which uses the same multi model evolving fuzzy classifiers concept as defined in this section. There, it could be empirically verified that a certainty threshold of down to 0.7 (only those samples for which the classification certainty is below this value are used for further adaptation of the models) hardly influences the final classification rate compared to when using all samples for model adaptation.

9.2.2.3 All-Pairs Strategy for Multiclass Classification Problems: Outline

Recently, for multi-class classification problems ($K > 2$), a new architecture within the context of evolving fuzzy classifiers was introduced, the so-called all-pairs classification technique, leading to the *FLEXFIS-Class-AP* or shortly to the *EFC-AP = evolving fuzzy classifiers using all-pairs technique* approach [43]. For each single pair of classes, a binary classifier is trained and evolved either based on a singleton consequent class label architecture (as used in *FLEXFIS-Class SM*) or based on TS-type fuzzy models (as used in *FLEXFIS*). Regression is conducted on $[0,1]$, using target values of 0 for all samples belonging to the first class and target values of 1 for all samples belonging to the other class. Thus, each binary classifier is only trained on a small sub-set of samples, achieving some sort of a collection of weak classifiers. In the classification phase, each binary classifier provides a preference degree of one class over the other. These degrees are stored into a preference relation matrix based on which a final classification response is provided with the help of (weighted) voting strategies. Weights may point to conflict and ignorance cases (see also Sect. 9.3.5.1) and are obtained based on enhanced reliability models. Results in [43] show that *EFC-AP* can out-perform EFC using single-model and multi-model architecture (*EFC-SM* and *EFC-MM*) with statistical significance (based on Wilcoxon test) on seven high-dimensional real-world data sets. Furthermore, it is analytically underlined that *EFC-AP* requires less computational complexity than *EFC-MM* for performing model updates with new incoming streaming samples.

9.2.3 eVQ-Class as Spin-Off

A spin-off from the *FLEXFIS-Class SM* method is proposed in [39], compared with *FLEXFIS-Class MM* in [42] and acting directly in the high-dimensional feature space, that is, no projection of fuzzy sets is carried out, but the classification takes place directly according to the position of the current sample w.r.t. to the extracted clusters. The main advantage of this approach is that it is able to take into account:

1. The distance of a new sample to be classified to the decision boundary, that is, to the border between the two nearest clusters having majority support in two different classes.
2. The relative support of the most frequent classes in the two nearest clusters.

This results in a weighted classification scheme, representing a better model for conflict situations between classes (see also Sect. 9.3.5 for underlining this), that is, the final confidence values of the $k = 1,\dots,K$ classes are obtained by summing the confidence values over the two nearest clusters $i1$ and $i2$:

$$\mathrm{con}f_k = w_{i1}\frac{h_{i1,k}}{\sum_{j=1}^{K} h_{i1,j}} + w_{i2}\frac{h_{i2,k}}{\sum_{j=1}^{K} h_{i2,j}}, \tag{9.19}$$

with

$$w_{i1} = 0.5 + \frac{I}{2} \text{ and } \quad w_{i2} = 0.5 - \frac{I}{2} \tag{9.20}$$

the weight factors of the nearest and second nearest clusters, and $I = 1 - \frac{\text{dist}_{c_{i1}}}{\text{dist}_{c_{i2}}}$, where $\text{dist}_{c_{i1}}$ the distance to the nearest cluster and $\text{dist}_{c_{i2}}$ the distance to the second nearest cluster. This incorporates a kind of "gravity" of clusters, pulling the decision boundary more to the less clean cluster, such that the majority class in the more clean cluster has a larger range of influence.

9.3 FLEXFIS++: Extensions for Improved Performance and Transparency

The second part of this chapter summarizes important extensions of the *FLEXFIS family* (highlighted in the bottom box in Fig. 9.1), which were developed mainly for the purpose to achieve higher predictive quality and transparency of the models.

9.3.1 *Drift Handling*

The common denominator in the aforementioned incremental learning methods is that they are life-long learning approaches [26], that is, treating all samples from a data stream as equally important. This means that the models expand their memory over time by compressing all knowledge extracted so far from the stream in their structures (rule-base) and parameters. In a lot of learning scenarios, this kind of modus operandi is the best choice, as older learned behaviors, relations, and dependencies may get activated again at a later stage and then a model of this partial local region (e.g., in form of a rule) is already available at hand. However, in some cases, the underlying data distribution may change over time, such that older learned behaviors are not valid any longer and should therefore be out-dated. Such cases are referred as *concept drifts* [69] or shortly *drifts* [32] in literature. For instance, consider an on-line production system, where (parts of) items are permanently produced and should be supervised whether they contain any failures, etc.; there, a drift could be in form of some changing characteristics of the items or a new production type. In [7], several examples of real problems are presented where change detection is necessary, including user modeling as well as monitoring in bio-medicine and industrial processes.

In principle, three different types of drifts may occur [22]:

- Drifts in the mean of the (local) data distribution.
- Drifts in the variance of the (local) data distribution.
- Drifts in the target concept, also referred as drifts in the correlation between inputs and outputs.

The effect of these three cases on the cluster/rule models as appearing in the inputs of the fuzzy models are visualized in Fig. 9.4a, b: a stronger movement of either the centers (in (a)) or the range of influence (in (b)) is required in order to model the new

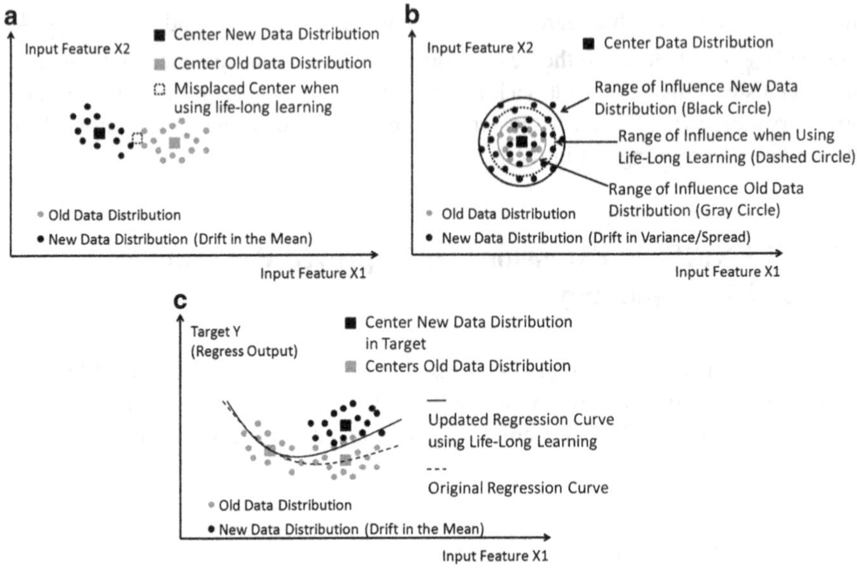

Fig. 9.4 Three drift concepts: (**a**) drift in the mean of a local region (cluster, rule), (**b**) drift in the variance, and (**c**) drift in the target (correlation)

distributions appropriately; otherwise, the rules (clusters) are somehow misplaced and usually cause a severe drop in predictive accuracy during the on-line process. Figure 9.4c shows the effect of a drifting target concept, where the y-axis denotes the output = target variable. There, also the consequents (parameters) of the rules have to be enforced to a stronger movement in order to model the new correlation in that local part appropriately.

For all these types, the drifts can be either abrupt or gradual, the latter showing a smooth transition from the older distribution to the newer one.

9.3.1.1 Drift Detection

In literature, there exist many methods for detecting drifts in data streams fully automatically. Some are based on the evolution of performance indicators over time, and some others, on monitoring distributions on two different time windows. Widely known methods for the first approach (which is the most commonly used one), are the *cumulative sum* [59] and the *Page-Hinkley (PH)* test [56]. The latter considers a cumulative variable $m_{i,T}$ defined as the cumulative difference between the observed values and their mean till the current moment (here for the ith input variable):

$$m_{i,T} = \left| \sum_{t=1}^{T} (x_i(t) - \bar{x} - \varepsilon) \right|, \tag{9.21}$$

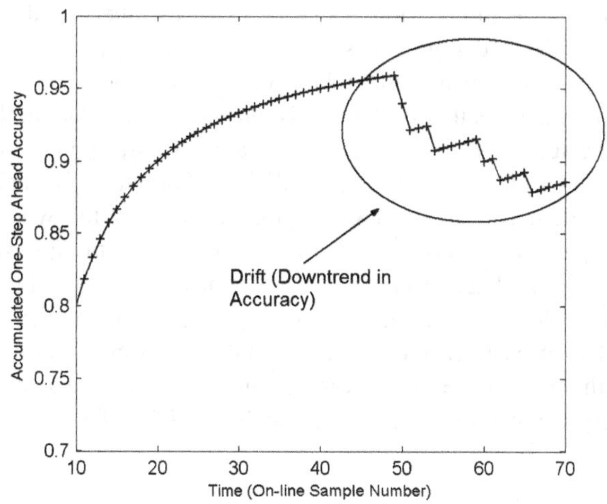

Fig. 9.5 Evolution of accumulated one-step-ahead accuracy over time for a specific on-line data set; note the abrupt decrease of accuracy towards the end of the data stream

where ε corresponds to the magnitude of changes that are allowed. The PH test is then simply to monitor the difference:

$$PH_{i,T} = m_{i,T} - M_{i,T} \tag{9.22}$$

for all variables with $M_{i,T}$ the minimal value of all $m_{i,T}$ seen so far, and provide a drift warning whenever this difference exceeds a certain threshold. Similar checks could be applied on the variances of single variables (instead of on original samples) in order to detect variance drifts.

Instead of examining the single variables separately, another possibility is to look at the joint development of all the variables by observing the evolution of the predictive accuracy. Especially when using the block-wise holdout test [9], but also in an interleaved test-and-then-train scenario, drifts can be recognized by sudden downtrends of the accuracy curves or abrupt increase of the error curves. An example is presented in Fig. 9.5 which shows the evolution of accuracies for a visual inspection system handling metal rotor parts (see also Chap. 14). Such significant downtrends in accuracies as shown at the end of the on-line process indicate a drift in the data stream and can be detected with the Page-Hinkley test (suddenly decreasing accuracies increase $m_{i,T}$ as measuring the absolute differences to the mean over all accuracies). Downtrends in accuracies usually indicate drifts somewhere in the model and does neither specify the type nor the location of the drift.

Therefore, an alternative concept for drift detection embedded in the incremental learning cycle of evolving fuzzy systems is presented in [47], where the concept of rule ages is used for tracking the points of time when the evolved rules are attached and to which extent (here for the ith rule):

$$\text{age}^i = k - \frac{\sum_{l=1}^{n_i} I_l \Psi_{i,l}}{n_i}, \tag{9.23}$$

where i is the rule index and $\Psi_{i,l}$ is the membership degree of the lth sample in the ith rule, n_i denotes the support of rule i, I_l denotes the time instance at which the data sample was read, and k is the current time instance. If this attachment changes significantly, that is, the rule is out-dated faster than in the past resp. the gradient of the age curves (= the second derivative) suddenly increases, a drift is detected in that local region where the rule is defined. Forgetting mechanisms can then be triggered particularly for that local region and not for the whole model. A possibility for separately eliciting whether a drift in the target/output variable occurs, would be to apply the Page-Hinkley test, as defined in (9.22), onto the output variable for each rule (local region) separately: for each sample, the rule with highest membership degree (nearest rule) is elicited and its drift indicator $m_{i,T}(j)$ updated, where x_i the ith value of the output variable in the stream and $\bar{x}$ the mean of the output values corresponding to rule j (for which rule j was the nearest one).

9.3.1.2 Drift Reaction

Once a drift is detected, an appropriate reaction strategy during the learning process has to be integrated into the flexible fuzzy systems approach. In particular, older learned relations, distributions, and behaviors should be out-dated over time, ideally in a smooth manner. Thereby, the intensity of forgetting should depend on the intensity of the drift. Furthermore, we suggest to integrate such a strategy always in both parts: antecedent and consequent space. In fact, in case of a drift occurring in the target (see Fig. 9.4c for an example), for accuracy reasons, it may be sufficient to react appropriately only in the consequent part (bending the model); however, for interpretability reasons, it is also recommended to shift the rule/cluster center (in the product space) to the distribution it actually should represent (after the drift).

In the Antecedent Space

Learning and evolution of rules' antecedents are conducted with the support of *eVQ* clustering algorithm (see Sect. 9.2.1.2), where two basic operations are guiding the learning process:

1. Rule evolution due to checking whether a new sample fits into the current cluster partition and
2. Rule movement where the nearest rule is moved by a fraction towards the current data sample.

The first case is usually triggered when new system states, operating conditions arise which are completely falling out of the scope of previous ones. It may also be triggered in case of extreme drifts in the mean or correlation (also referred as *shifts*). Forgetting of older rules in such cases is not required as a new rule representing the new distribution will emerge, anyway. An older rule (before the shift happened) stays in the model, however will not contribute with significant membership degree to the final model output any longer. Hence, in case of *shifts*, the deletion of older rules only would account for a complexity reduction step. The second case is triggered in normal operation modes already seen so far, but also in (conventional) *drift* cases (see e.g., Fig. 9.4a), where still samples from the new distribution would be attached to an old rule being the nearest one. However, in drift cases, the center as well as the range of influence should be moved stronger than in non-drift cases in order to represent the new distribution appropriately. In *eVQ*, this is achieved by re-setting the learning gain in (9.7) (see also [47]):

$$n_{\text{win}} = n_{\text{win}} - n_{\text{win}} * \lambda\_trans, \tag{9.24}$$

where n_{win}, the number of samples attached to the winning cluster so far and

$$\lambda\_trans = -9.9\lambda + 9.9 \tag{9.25}$$

and where λ, a forgetting factor which is set according to the intensity of the drift (e.g., how strong (9.22) is violated). This reactivates the movement of (already converged) cluster centers and furthermore also the ranges of influence according to recursive variance formula or incremental covariance matrix (9.11), both including the cluster centers. With new attachments to $\mathbf{c}_{\text{win}}, n_{\text{win}}$ again increases, achieving a decrease of the learning gain η_{win} in a monotonic fashion and finally a convergence within the new (drifted) distribution.

In the Consequents

Reacting on drifts in the consequents goes hand in hand with a gradual forgetting of the linear consequent parameters in the hyper-planes. This can be achieved by including an exponential forgetting directly in the least squares optimization problem:

$$J_i = \sum_{k=1}^{N} \lambda^{N-k} \Psi_i(\mathbf{x}(k)) e_i^2(k) \longrightarrow \min_{\mathbf{w_i}}, \tag{9.26}$$

with $e_i(k) = y(k) - \hat{y}(k)$ the error of the ith rule in sample k and λ a forgetting factor. Deducing the recursive least squares estimator for this modified optimization functions leads to the following consequent update [47] (compare also with formulas (9.4)–(9.6)):

$$\hat{\mathbf{w}}_\mathbf{i}(N+1) = \hat{\mathbf{w}}_\mathbf{i}(N) + \gamma(k)(y(N+1) - \mathbf{r}^T(N+1)\hat{\mathbf{w}}_\mathbf{i}(N)) \tag{9.27}$$

$$\gamma(N) = \frac{P_i(N)\mathbf{r}(N+1)}{\frac{\lambda}{\Psi_i(\mathbf{x}(N+1))} + \mathbf{r}^T(N+1)P_i(N)\mathbf{r}(N+1)} \tag{9.28}$$

$$P_i(N+1) = (I - \gamma(N)\mathbf{r}^T(N+1))P_i(N)\frac{1}{\lambda}. \tag{9.29}$$

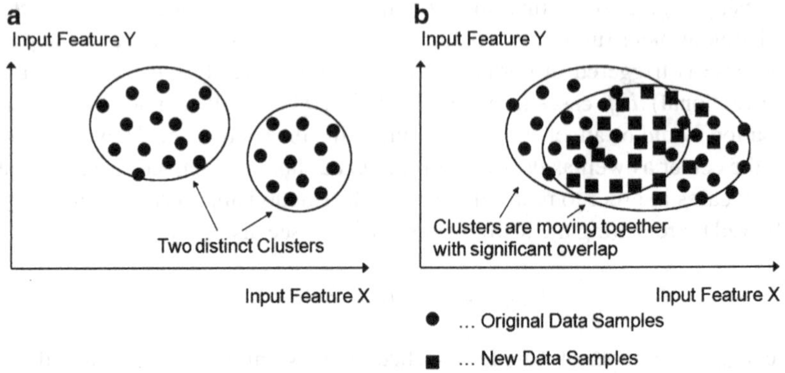

Fig. 9.6 (**a**) Two distinct clusters from original data. (**b**) Samples are filling up the gap between the two original clusters which get overlapping due to movements of their centers and expansion of their ranges of influence

Again, here, the forgetting factor λ plays an important role keeping a reasonable trade-off between fast update ($\lambda = 0.9$) and no update ($\lambda = 1.0$) and should depend on the intensity of the drift (i.e., on the intensity of accuracy downtrend).

9.3.2 Reducing Complexity, Enhancing Transparency

This section deals with the reduction of unnecessary complexity as it may come into being due to the nature of the incremental learning process from data streams, as Algorithm 1 always sees only a small snapshot of the whole data at once. This means that rules may be evolved at a former stage which turn out to be superfluous at a later stage. This especially comes true in *shift* cases (see previous section) or when two rules are moving together and finally may end up with significant overlap. An example of the latter case, which is also denoted as *rule redundancy*, is visualized in Fig. 9.6, (a) denotes the initial partition, (b) the updated one according to new samples filling up the gap in-between the original clusters.

9.3.2.1 Rule Merging in the Feature Space

An elimination strategy of occasions shown in Fig. 9.6 consists of two stages:

- The first stage detects significantly overlapping rules based on a high-dimensional similarity measure.
- The second stage performs a merging of two overlapping rules.

For tackling the first stage, we apply a post-processing step after each incremental learning step, where we check whether the latest rule movement leads to significantly overlapping clusters. In particular, the latest moved rule A is checked whether it has become redundant to any of the other rules B in the cluster partition. This can be achieved for arbitrary fuzzy sets used in the evolved fuzzy system by using an inclusion measure defined as:

$$\mathrm{inc}(A,B) = \mathop{\mathrm{T}}_{j=1}^{p} \mathrm{inc}(A_p, B_p) \tag{9.30}$$

with A_p the fuzzy set in the pth antecedent part of rule A and

$$\mathrm{inc}(A_i, B_i) = \frac{\int \min(A_i(x), B_i(x)) \mathrm{d}x}{\int B_i(x) \mathrm{d}x}, \tag{9.31}$$

denoting the inclusion degree of the fuzzy set A_i in B_i, and in discrete (computational) form, it becomes:

$$\mathrm{inc}(A_i, B_i) = \frac{\sum_{j=1}^{N} \min(A_i(x_j), B_i(x_j))}{\sum_{j=1}^{N} B_i(x_j)}. \tag{9.32}$$

Then, the overlap (similarity) measure can be defined as a two-sided inclusion measure:

$$\mathrm{overlap}(A,B) = S(\mathrm{inc}(A,B), \mathrm{inc}(B,A)). \tag{9.33}$$

The justification of using a t-norm in (9.30) (e.g., minimum) is underlined by the fact that a strong non-overlap along one single dimension i (achieving a low inclusion degree of A_i in B_i and vice versa) is sufficient that the clusters do not overlap at all (as torn apart). A feasible choice for the t-conorm S in (9.33) is the maximum operator, as it points to the maximal inclusion of A in B and B in A.

In case of Gaussian membership functions, a significantly faster version of (9.33) can be achieved by using the membership values at the intersection points of dimension-wise over-lapping Gaussian fuzzy sets [49]:

$$\mathrm{overlap}(A,B) = \mathrm{Agg}_{j=1}^{p} \mathrm{overlap}_{A,B}(j) \tag{9.34}$$

with

$$\mathrm{overlap}_{A,B}(j) = \max(\mu(\mathrm{inter}_x(1)), \mu(\mathrm{inter}_x(2))), \tag{9.35}$$

$\mu(x)$ the membership degree to the univariate Gaussian, and

$$\mathrm{inter}_x(1) = -\frac{c_{kj}\sigma_{ij}^2 - c_{ij}\sigma_{kj}^2}{\sigma_{kj}^2 - \sigma_{ij}^2} + \sqrt{\left(\frac{c_{kj}\sigma_{ij}^2 - c_{ij}\sigma_{kj}^2}{\sigma_{kj}^2 - \sigma_{ij}^2}\right)^2 - \frac{c_{ij}^2\sigma_{kj}^2 - c_{kj}^2\sigma_{ij}^2}{\sigma_{kj}^2 - \sigma_{ij}^2}} \quad \text{and}$$

$$\text{inter}_x(2) = -\frac{c_{kj}\sigma_{ij}^2 - c_{ij}\sigma_{kj}^2}{\sigma_{kj}^2 - \sigma_{ij}^2} - \sqrt{\left(\frac{c_{kj}\sigma_{ij}^2 - c_{ij}\sigma_{kj}^2}{\sigma_{kj}^2 - \sigma_{ij}^2}\right)^2 - \frac{c_{ij}^2\sigma_{kj}^2 - c_{kj}^2\sigma_{ij}^2}{\sigma_{kj}^2 - \sigma_{ij}^2}}, \quad (9.36)$$

the two intersections points of the two Gaussians obtained by projecting the two rules A and B onto the jth dimension (for the sake of equation length, we use $i = A$ and $k = B$). Taking the maximum in (9.35), has the positive side effect that "cluster crosses" can be eliminated as well.

Once two significantly overlapping clusters A and B are found (i.e., (9.33) or (9.34) is higher than a pre-defined threshold), the two clusters have to be merged to one single rule. This is achieved in two stages:

- Merging of the clusters in the produce space and projecting the new merged clusters to the input axes to form the new fuzzy sets and rule's antecedent part.
- Merging of the consequents of the two redundant rules or deleting the consequent of the less supported rule subject to a consistency check.

The merging of clusters in the product space is achieved by (1) calculating a weighted average of the two cluster centers where the weights represent the support of the two clusters (hence, the new center will lie in-between and closer to the center of the more supported cluster), (2) adaptively estimating the range of influence of the new merged cluster by updating the range of influence of the more supported cluster A with the range of influence of the less supported cluster B using recursive variance formula: the less supported cluster represents a collection of data samples with which the range of influence of the more supported cluster is updated, and (3) summing up the support of the two clusters:

$$\begin{aligned} c_j^{\text{new}} &= \frac{c_j^A k_A + c_j^B k_B}{k_A + k_B} \\ \sigma_j^{\text{new}} &= \sqrt{\frac{k_A(\sigma_j^A)^2}{k_A + k_B} + (c_j^A - c_j^{\text{new}})^2 + \frac{(c_j^{\text{new}} - c_j^B)^2}{k_A + k_B}} \\ &\quad + \frac{k_B \sigma_j^B}{k_A + k_B} \\ k_{\text{new}} &= k_A + k_B, \end{aligned} \quad (9.37)$$

where k_A denotes the number of samples falling into rule A and k_B the number of samples falling into rule $B, k_A > k_B$. The last term in the second formula of (9.37) expands the range of influence by a fraction of the variance of samples belonging to the less significant cluster in order to achieve a good coverage of the data cloud covered by the merged cluster, see [49].

The merging strategy of consequents includes a consistency check of the two over-lapping rules according to the theory of propositions in fuzzy logic. In particular, whenever the similarity of the rules antecedent parts is higher than the similarity of their consequents, two contradictory rules (in the fuzzy sense) are present in the rule base and the consequent of one of these (that one with lower support) is deleted; otherwise, the consequents are merged by a weighted average according to the support of the rules. This leads us to the following combination rule (inspired by Yager's idea of *participatory learning* [70]):

$$\mathbf{w}_{\text{new}} = \mathbf{w}_A + \alpha \cdot \rho(\mathbf{w}_A, \mathbf{w}_B) \cdot (\mathbf{w}_B - \mathbf{w}_A), \quad (9.38)$$

where $\alpha = k_B/(k_A + k_B)$ and $\rho(\mathbf{w}_A, \mathbf{w}_B)$ is a measure of consistency of the two rule consequents. This measure can be defined in different ways, for example, "smoothly" by $\rho(\mathbf{w}_A, \mathbf{w}_B) = S_{\text{cons}}(y_A, y_B)$ or, more drastically, by:

$$\rho(\mathbf{w}_A, \mathbf{w}_B) = \begin{cases} 1 & \text{if } S_{\text{cons}}(y_A, y_B) \geq \text{overlap}(A,B) \\ 0 & \text{if } S_{\text{cons}}(y_A, y_B) < \text{overlap}(A,B) \end{cases},$$

with S_{cons} the similarity degree of the two consequents belonging to rules A and B. For $\rho = 0$, indicating an inconsistency in the rule base, we obtain $\mathbf{w}_{\text{new}} = \mathbf{w}_A$, that is, the consequent of the more relevant rule. For $\rho = 1$, on the other hand, we obtain the weighted average of the two consequent functions according to their support. The similarity of the consequents is elicited based on the angle spanned between the two hyper-planes y_A and y_B, denoting the consequent function of the two rules:

$$S_{\text{cons}}(y_A, y_B) = \begin{cases} 1 - \frac{2}{\pi} * \phi & \phi \in [0, \frac{\pi}{2}], \\ \frac{2}{\pi} * (\phi - \frac{\pi}{2}) & \phi \in [\frac{\pi}{2}, \pi] \end{cases} \quad (9.39)$$

where

$$\phi = \arccos\left(\left| \frac{\mathbf{a}^T \mathbf{b}}{|\mathbf{a}||\mathbf{b}|} \right| \right) \quad (9.40)$$

and a and b denoting the normal vectors of the hyper-planes defined by $a = (w_{m1}\ w_{m2}\ \dots\ w_{mp}\ \ -1)^T$ and $b = (w_{n1}\ w_{n2}\ \dots\ w_{np}\ \ -1)^T$. Figure 9.7 visualizes two examples of redundant rules, where in one case merging is conducted and in the other, deletion.

Summarizing the aspects above, the elimination of redundancy can be integrated in flexible fuzzy systems by using the following algorithmic steps (right after step 12 in Algorithm 1):

13. **If** a new cluster was evolved, do nothing.
14. **Else**, perform the following steps:
15. Check if similarity of moved/updated rule A with any other rule $R \in \mathscr{R} \setminus \{A\}$ is higher than a pre-defined threshold *sim_thr* (default setting is 0.35 when using (9.33), 0.8 when using (9.34))
16. **If** yes:

    a. Perform rule merging of rule A with rule $B = \arg\max_{B \in \mathscr{R} \setminus \{A\}} OL(A,B)$ according to (9.37).
    b. Perform merging of corresponding rule consequent functions according to (9.38).

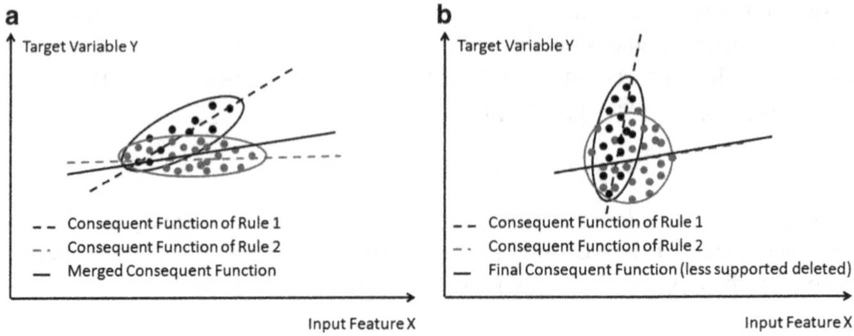

Fig. 9.7 (**a**) Merging of two rule consequent function by weighted average where weights are identical to cluster support (number of samples belonging to a cluster): the lower rule has much more support; hence, the consequent function of the merged rule is closer to its consequent function than to that one of the other rule and (**b**) deleting of the consequent function of the rule with lower weight, as the consequent functions are very dissimilar (as spanning an angle of almost 90°)

c. Overwrite parameters $(\mathbf{c}^A, \sigma^A)$ of rule A with the parameters of the merged rule $(\mathbf{c}^{new}, \sigma^{new})$; overwrite consequents of rule $A, \mathbf{w}_A$, with $\mathbf{w}_new$.
d. Delete rule B.
e. Decrease number of rules: $C = |\mathscr{R}| = C - 1$.

Finally, we want to mention that a further complexity reduction concept in EFS which takes into account the statistical influence of fuzzy rules over time will be handled in Chap. 10, Sect. 10.2.

9.3.2.2 Fuzzy Set Merging in the Partition Space

Redundancy on fuzzy set level arises not only when (partially) overlapping rules are present, but also due to the projection concept of a high-dimensional space to low-dimensional partitions—consider one cluster laying over the other with respect to the second dimension where in the first dimension they are covering is nearly the same range.

Detection of similarity among fuzzy sets A and B (S_{set}) in the single partitions can be achieved by the same measure as in as defined in (9.33) on one-dimensional fuzzy set level, again using (9.32) and applying the minimum operator over the inclusion of fuzzy set A in B and B in A. A faster kernel-based similarity measures (based on distances between two centers and spreads in the exponent) can be defined for Gaussians, see [49]:

$$S_{\text{ker}}(A,B) = e^{-|c_A - c_B| - |\sigma_A - \sigma_B|}. \tag{9.41}$$

In order to be scale invariant, the centers and widths of the Gaussians should be normalized beforehand according to the ranges of the variables.

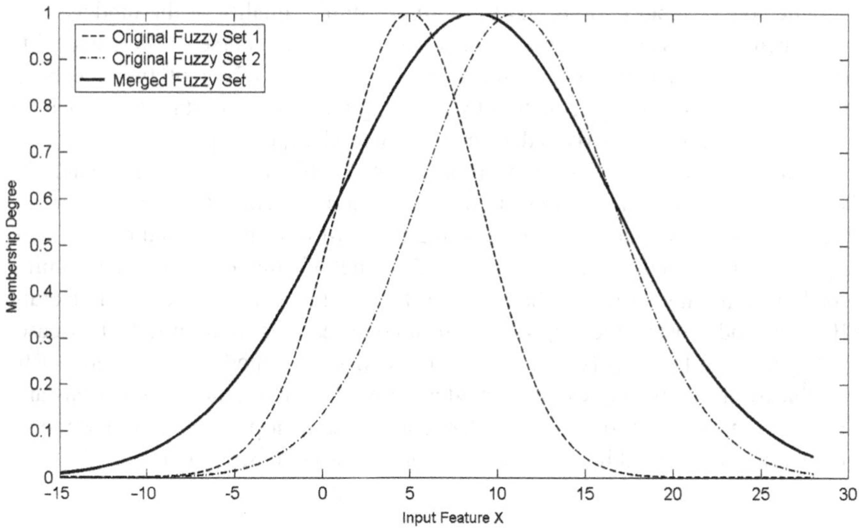

Fig. 9.8 Merging of two Gaussian fuzzy sets (*dashed* and *dotted dashed lines*) to a new fuzzy set according to (9.42)

Merging is caused whenever S_{set} exceeds a certain threshold (default 0.35 and 0.8 when using kernel-based metric). In case of Gaussian membership functions with μ as center and σ as spread, two Gaussian fuzzy sets are merged into a new Gaussian kernel with the following parameters:

$$\mu_{new} = (\max(U) + \min(U))/2 \text{ and}$$
$$\sigma_{new} = (\max(U) - \min(U))/2, \tag{9.42}$$

where $U = \{\mu_A \pm \sigma_A, \mu_B \pm \sigma_B\}$. The idea underlying this definition is to reduce the *approximate* merging of two Gaussian kernels to the *exact* merging of two of their α-cuts, for a specific value of α. Here, we choose $\alpha = \exp(-1/2) \approx 0.6$, which is the membership degree of the inflection points $\mu \pm \sigma$ of a Gaussian kernel with parameters μ and σ. A merging example is presented in Fig. 9.8. Similar merging considerations can be made for trapezoidal or bell-shaped membership functions.

9.3.3 Soft Dynamic Dimension Reduction

In case of high-dimensional feature spaces, curse of dimensionality may become a significant pitfall when evolving non-linear models from data. This is because input feature spaces are inherently sparse [10], that is, with an increasing number of features, the number of data samples should increase exponentially in order to guarantee dense input regions and furthermore similar quality of the models as in low-dimensional cases [28]. In case of fuzzy systems, the curse of dimensionality may become even more severe due to the localization effect of the rule base. Therefore, dimensionality reduction by selecting the most important features is an essential step in the data-driven design of fuzzy models [12, 63].

During the incremental learning process, the problem is how to exchange the list of most important features used in the model architecture, whenever the characteristics of the data stream regarding the most informative features w.r.t. the target concept changes. For instance, based on the information of a new incoming data block, it may turn out that feature #3 should be exchanged with feature #10 in a model using the five most informative features as inputs. In this case, a straightforward switch (feature #3 is erased from the model while feature #10 is included in the model) would immediately produce many false predictions and classifications as the current rule centers and consequent parameters in the model were learned for the old input space based on the past data stream blocks.

9.3.3.1 Feature Weight Integration Concept

Therefore, we pursue a *soft dynamic dimension reduction* approach, which uses the concept of *adaptive feature weights* denoting the importance of features and their dynamic change behavior over time. Each feature is associated with a weight λ_i in $[0,1]$, whereas a value near 0 would indicate a feature with low importance and a value near 1 a feature with high importance. This also accounts for a kind of complexity reduction and interpretability enhancement step (see subsequent section), as unimportant features may be completely discarded in the antecedent and consequent parts of the final evolved models (or of the transferred model shown to the operator during the on-line process) as not contributing to the fuzzy inference process for producing predictions/classifications. In particular, in the classification phase, the integration of the feature weights $\lambda_i, i = 1, \ldots, p$ for the p input variables in an EFS is achieved through inclusion in the calculation of the membership degrees of the single rules (here stated for the ith rule):

$$\mu_i(\mathbf{x}) = \prod_{j=1}^{p}((\mu_{ij} - 1)\lambda_j + 1) \; i = 1, \ldots, C. \tag{9.43}$$

Due to the use of the product t-norm as conjunction operator, a value of $\lambda_j = 0$ would deliver a product term of 1, meaning that the antecedent part corresponding to the jth input feature serves as do not care part; hence, it depends on the other features whether this rule fires with a significant degree or not, that is, the jth feature is ignored. A value of $\lambda_j = 1$ delivers a product term of μ_{ij} which means that the real activation level of the jth antecedent part (belonging to feature j) is used in the final rule membership value.

The inclusion of feature weights in the training process may decrease the complexity of the models as distances or rule movements along unnecessary

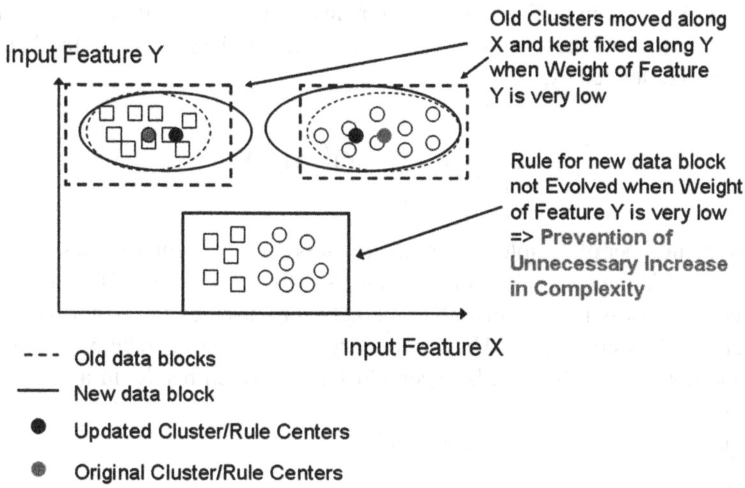

Fig. 9.9 Impact of including feature weights when during incremental learning phase—the new data block (circumvented by the solid box) does not trigger a new rule when Feature Y is unimportant (out-weighted) → avoidance of unnecessary clusters; instead, the original two clusters are expanded by extending their ranges of influences towards the middle area of Feature X

dimensions may be down-weighted. Thereby, consider the example demonstrated in Fig. 9.9, the integration into the training is included in the cluster movement:

$$\mathbf{c}_{\text{win}}^{(\text{new})} = \mathbf{c}_{\text{win}}^{(\text{old})} + \eta_{\text{win}} \lambda I \left(\mathbf{x} - \mathbf{c}_{\text{win}}^{(\text{old})} \right), \tag{9.44}$$

with I the identity matrix and η_{win} the decreasing learning, and in the distance calculation (here when using Mahalanobis distance measure for A):

$$\text{mahal} = \sqrt{(\lambda . * (\mathbf{x} - \mathbf{c}))\Sigma^{-1}(\lambda . * (\mathbf{x} - \mathbf{c}))}, \tag{9.45}$$

with Σ^{-1} the inverse of the covariance matrix and $.*$ the component-wise product of two vectors.

9.3.3.2 Incremental Feature Weight Calculation

In a classification setting, the feature weights are extracted based on Dy-Brodley's interclass separability criterion [19], which is a modified, more stable version of Fisher's interclass separability criterion [18]. The criterion is defined by:

$$J = \text{trace}(S_w^{-1} S_b), \tag{9.46}$$

with trace(A) the sum of the diagonal elements in $A; S_w$ the within scatter matrix; which is the sum of the covariance matrices over all classes and S_b the between scatter matrix, defined by:

$$S_b = \sum_{j=1}^{K} N_j(\bar{X}_j - \bar{X})^T(\bar{X}_j - \bar{X}), \tag{9.47}$$

with N_j the number of samples belonging to class $j, \bar{X}_j$ the center of class j (for all features), and $\bar{X}$ the mean over all data samples (for all features). The term within the trace operator is nothing else than the between-class scatter S_b normalized by the average class covariance. Hence, the larger the value of trace$(S_w^{-1}S_b)$ is, the larger the normalized distance between clusters is, which results in a better class discrimination.

Then, feature weights are defined by [46]:

$$\lambda_j = 1 - \frac{J_j - \min_{1,\ldots,p}(J_j)}{\max_{j=1,\ldots,p}(J_j) - \min_{1,\ldots,p}(J_j)}, \tag{9.48}$$

with J_j the separability criterion (9.46) calculated for the reduced input feature space by neglecting the jth feature (hence p J_j criteria are obtained). Thus, a low value of J_j means that feature j is very important, as it is discarded from the complete set of features resulting in a significant drop in the separability criterion. This triggers a high value of λ_j after (9.48). In particular, for the most important feature, the value of (9.48) will get 1, according to the normalization term.

The incremental adaptation of (9.46) for each j is achieved by updating the single components in S_b (means and number of samples falling into each class), the single covariance matrices for each class after (9.11) and re-calculating (9.48) through Dy-Brodley's criterion.

In Chap. 14, it will be shown how incremental feature weights integrated into evolving fuzzy classifiers in fact significantly improve the accumulated one-step-ahead prediction accuracies over time for visual inspection data.

9.3.4 Interpretability Considerations

Interpretability is one of the most important key drivers for choosing fuzzy system architecture in the data-driven and evolving model design. This property is a valuable characteristic of this kind of model architecture, providing readable rules in linguistic (IF-THEN) form (compare with (9.2)) and linguistic terms in form of fuzzy sets (as part of the antecedents) with a clear semantic meaning. This valuable insight can be not achieved with other model architectures such as neural networks [29] or support vector machines [64] which usually provide complete black-box models. There are several benefits drawn from interpretable (fuzzy) models such as:

- Providing (additional) insight for operators into the nature of the process: there may be some unknown or hidden dependencies between certain process parameters or variables within the system from which the operators may get a deeper understanding, etc.
- Enhanced user interaction in the on-line modeling process, where the user may interact with the model on a structural level.
- Easiness of on-line supervision purposes; for instance, the user wants to find reasons for certain model decisions.
- Increasing the motivation and consistency of operators' feedback to the modeling process (e.g., overruling model decisions).

When extracting fuzzy systems from data, the interpretability is usually significantly worsened compared to knowledge-based systems [34] which were built upon experiences, views, and opinions from experts. This is simply because most of the modeling methods including all conventional EFS approaches and in particular the *FLEXFIS* family are precise modeling approaches, that is, they are trying to model the basic trend of the natural data distribution at hand as accurately as possible without taking care of interpretability aspects (see below for a list of those). For batch modeling methods, a large collection of interpretability improvement techniques exist [11], ranging from constrained-based optimization techniques via orthogonal rule parameter learning to post-processing techniques in which fuzzy systems are post-"beautified".

There are several important criteria for achieving interpretable models which can be divided into high-level (on fuzzy rule level) and low-level interpretability [72], that is, optimizing the membership functions in terms of semantic criteria on fuzzy set/fuzzy partition level. Important facets at the high-level interpretability stage include:

- Distinguishability
- Consistency
- Simplicity (low number of rules)
- Rule length
- Feature information
- Interpretable consequents

In the flexible EFS concept, the first two issues are tackled by the strategies described in Sect. 9.3.2, as consistency corresponds to omittance of contradictions during the learning process which is guaranteed by the consistency check in Sect. 9.3.2.1 through comparing the similarity of rule consequent parts with the similarity of the rule antecedents. Simplicity is partially achieved through merging/deletion of redundant rules; some further improvements may be expected when merging clusters grown together. The rule length depends on the number of antecedent parts in the single rules, that is, as using a flat model architecture, on the number of inputs. In case of classification problems, the dynamic assignment and adaptation of feature weights provides a possibility on how to reduce the number of antecedent parts, as features with low weights can be neglected when interpreting the models. Regarding interpretation of the consequents, it is essential to use local learning by applying recursive fuzzily weighted least squares estimator (9.4)–(9.6). This guarantees a snuggling of the partial rule hyper-planes along the real trend of the functional behavior [51]. This also opens up the possibility of a local feature weighting/selection step in regression problems, as being achieved for batch-trained TS fuzzy systems, see, for example, [52]. In case of classification problems, the consequents contain singleton class labels (corresponding to the majority class) and are therefore interpretable per se.

9.3.5 Reliability Aspects

Apart from the predictive behavior of a model in terms of its accuracy, another prerequisite for the user acceptance is the *reliability* of a model. Reliability accounts for the importance to offer, apart from the prediction itself, information about how reliable this prediction is. Ideally, a learning algorithm is "self-aware" in the sense of knowing what it knows, and what it does not. Based on the reliability degree of a prediction (due to noise variance, high bias of the model, or conflicting situations), the user is able to decide whether the prediction has to be treated with caution. Therefore, reliability also can be seen as a sort of interpretability of the model responses/outputs.

9.3.5.1 For Classification Problems

For classification problems, the uncertainty of models for a given query sample (=sample to be classified) can be expressed by two concepts [30]: conflict and ignorance. Conflict refers to the degree to which two classes are in conflict with each other, as the classifier supports both of them almost equally as potential classifications. Thus, conflict is usually caused by a query instance which lies near the decision boundary of the classes. Ignorance represents that part of the uncertainty which may arise due to a query point lying in a region of the feature space which was not covered by any training samples. Thus, ignorance is somehow related to extrapolation cases of query points. Figure 9.10(a) visualizes an example of a conflict case and (b) shows an ignorance case. The latter arises due to an uncertainty in the position of the decision boundary (straight lines in our case): obviously, a lot of decision boundaries are possible, and not all may classify the query point to the same class. This circumstance diminishes when the query point lies closer to one of the clusters.

In the conflict case, it is not really clear whether the sample should be classified to the first or second class; hence, someone may expect a value of 0.5 for the

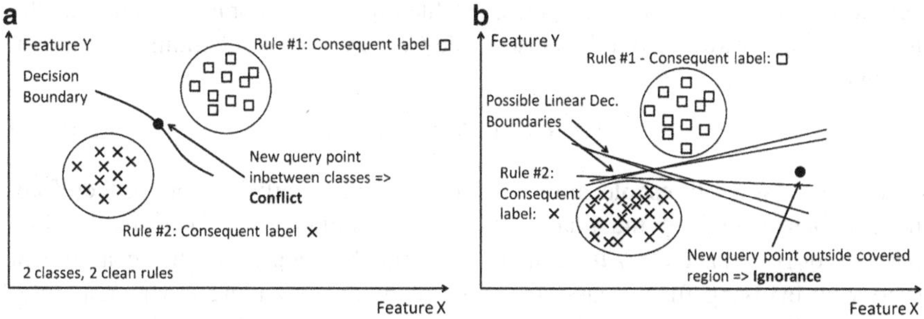

Fig. 9.10 (**a**) Conflict case for new query point to be classified as lying in-between both classes. (**b**) Ignorance case for a new query point to be classified as lying significantly away from both rules (marked as clusters)

confidence calculated by *FLEXFIS-Class* resp. *eVQ-Class*. However, when using (9.16) as confidence value for the final class label according to the winner-takes-it-all classification concept, such conflict occurrences as shown in Fig. 9.10a cannot be modeled appropriately: the confidence value would be 1 (indicating no conflict) for that class lying a bit nearer to the sample to be classified, as the cluster is completely clean, that is, represented by a single class. Therefore, we extend (9.16) and use a weighted combination of relative frequencies in order to achieve a final confidence:

$$\mathrm{con}f_{k,l} = \frac{\mu_1 h*_{1,k} + \mu_2 h*_{2,k}}{\mu_1 + \mu_2}, \tag{9.49}$$

with

$$h*_{1,k} = \frac{h_{1,k}}{h_{1,k} + h_{1,l}} \quad \text{and} \quad h*_{2,k} = \frac{h_{2,k}}{h_{2,k} + h_{2,l}} \tag{9.50}$$

the relative frequency (weight) of class k in the two nearest rules supporting two different classes k and l ($h*_1$ belongs to the nearest rule supporting class k and $h*_2$ belongs to the nearest rule supporting *another* class l), and μ_1 the membership degree of the current sample to the nearest rule supporting class k and μ_2 the membership degree of the current sample to the nearest rule supporting another class l. If a sample falls exactly into one rule (which supports a specific class k) and all other rules are far away, then $\mu_2 \approx 0$ which reduces (9.49) to (9.16), reflecting the degree of conflict within the nearest rule. In *eVQ-Class*, a similar weighted confidence scheme is applied, compare with (9.19), also delivering a good model for conflict cases. In *FLEXFIS-Class MM*, a conflict occurs when the regression output value of two classes are nearly identical, which would deliver a value around 0.5 according to (9.18).

Fuzzy rule bases are able to resolve ignorance in a natural way, as they deliver (fuzzy) activation degrees of rules. Then, it is quite obvious that if no rule fires significantly in the actual query point, the likelihood of ignorance is high, as the query is lying far away from the closest rule (representing some training samples). This means whenever

$$\mu_i < \varepsilon \text{ and } \forall i = 1, \ldots, C, \tag{9.51}$$

with $\varepsilon > 0$ a small threshold, no class label is significantly supported; therefore, the confidence level of the final output class should be set to a small value, that is, $conf_L = max_{i=1,\ldots,C}\mu_i$. Finally, this can be seen as a strong argument as to why evolving fuzzy classifiers may be preferred among other crisp rule-based classifiers (here rule activation degrees are either 0 or 1) and also among other machine learning classifiers such as incremental SVMs, tree-based methods (such as Hoeffding trees), or incremental statistical approaches (e.g., naive Bayes rules), based on which it is much harder to resolve the problem of ignorance [30]. Latest results show that the integration of ignorance levels as down-weighting multiplicators of single predictions from binary models in multi-model classifiers (such as *FLEXFIS-Class MM* or *AP*) mostly increases classification accuracy significantly.

9.3.5.2 For Regression Problems

In case of regression problems, the uncertainty about predictions arises either due to sufficient noise in the data, extrapolation regions/holes in the data clouds, or the inflexibility of a model to follow the non-linear trend of the approximation surface. All these can be tracked by so-called adaptive local error bars [50], which change their behavior over the input space locally with respect to the natural characteristics of the data distribution and the model behavior.

The local error bars for Takagi–Sugeno fuzzy systems can be calculated as an extension of the formula for error bars for linear regression models [17], as each consequent part denotes a linear hyper-plane and furthermore represents a linear approximation of the local dependency in that region where the corresponding rule antecedent parts is defined (achieved by local learning). Then, the single hyper-planes of the single rules just need to be connected to form an over-all error bar for the prediction on a current data sample x_k using the membership fulfillment degrees μ of all rules:

$$\hat{y}_{\text{fuz}} \pm \sqrt{\text{cov}\{\hat{y}_{\text{fuz}}\}} = \hat{y}_{\text{fuz}} \pm \frac{\sum_{i=1}^{C} \mu_i(\mathbf{x}_{\text{act}}) \sqrt{\text{cov}\{\hat{y}_i\}}}{\sum_{i=1}^{C} \mu_i(\mathbf{x}_{\text{act}})} \tag{9.52}$$

where $\hat{y}_i$ the estimated value of the ith rule consequent function, for which cov is calculated by:

$$\text{cov}\{\hat{y}\} = X_{\text{act}} \sigma^2 (X_i^T Q_i X_i)^{-1} X_{\text{act}}^T, \tag{9.53}$$

with X_{act} the current sample for which a prediction is queried and its corresponding certainty level sought (thus, $\mathrm{cov}\{\hat{y}\}$ reduces to a single value). Q_i is the weighting matrix as used in the local learning approach, $(X_i^T Q_i X_i)^{-1}$ is the inverse weighted Hessian matrix, and σ is the noise variance estimated by:

$$\hat{\sigma}^2 = \frac{2\sum_{j=1}^{N}(y(j) - \hat{y}(j))^2}{N - \deg}, \tag{9.54}$$

with N the number of samples and deg the degrees of freedom (estimated by $\deg = 2Cp + kC$) with C the number of fuzzy rules and p the number of inputs and k the number of parameters in the rule consequents. The update of the error bars is automatically given as $P_i = (X_i^T Q_i X_i)^{-1}$ is updated through (9.4)–(9.6), and the noise variance can be updated by:

$$\hat{\sigma}^2(\text{new}) = \frac{(N - \deg(\text{old}))\hat{\sigma}^2(\text{old}) + 2\sum_{j=N}^{N+m}(y(j) - \hat{y}_{\text{fuz}}(j))^2}{N + m - \deg(\text{new})}, \tag{9.55}$$

where deg(new), the new number of parameters, if changed in the model due to an evolution of the structure. In Chap. 14, the adaptive local error bars will be used for producing reliable fault indicators in an on-line multi-channel fault detection system, improving the performance of constant error bands.

9.4 Applications of the *FLEXFIS* Family: Summary

In this section, we provide a short summary about successful applications of the *FLEXFIS* family, including references to publications (if any):

- On-line system identification at engine test benches ([44], Chap. 7): the task was to identify unknown interrelationships and dependencies between measurement channels in order to characterize newly developed engines during the on-line test phase, and to gain some important and interesting insights.
- The identified models plus confidence regions (previous section) served as reference oracles for an extended plausibility analysis of measurements, based on which faults and system failures can be elicited—see Chap. 14.
- Prediction of NOx emissions of diesel and petrol engines [53]: the aim was to substitute expensive hardware sensors with a soft sensor, which is able to predict the prospective NOx content (up to 10 s ahead); time offsets and changing engine operating conditions had to be incorporated in the models.
- On-line prediction of resistance values at cold rolling mills ([44], Chap. 7): the task was to predict the yield strength of a steel plate in order to guarantee a smooth rolling process, that is, finally a steel plate with the intended a priori defined thickness. The predictions have to be carried out during the on-line rolling process in order to compensate deviations from intended values quickly.

FLEXFIS could out-perform state-of-the-art physical models, especially when applying time-delayed inputs and triggering forgetting mechanisms as soon as new stitches appeared.

- Prediction of residential premise prices [54]: based on some main drivers such as usable area of premises, age of a building, number of rooms in a flat, floor on which a flat is located, number of storeys in a building, as well as the distance from the city center, the aim was to predict the prices of houses and premises for future years; re-setting the models with new data recordings should be avoided; instead, the dynamic changes in the real-estate market in the course of time should be included on the fly. Rule merging option could significantly reduce the complexity of the evolved models and in some cases even improve accuracy by reducing over-fitting; batch off-line expert-based estimation methods could be significantly out-performed.
- Evolving chemometric models for viscose production process (publication submitted [14]): the aim is to set up calibration models based on NIR spectra with a very high # of wavelengths (=dimensionality) for automatically quantifying the contents of chemical substances H_2SO_4,Na_2SO_4, and $ZnSO_4$. As the process is very dynamic according to a permanently changing state of the spin bath, models need to be updated persistently to guarantee quantifications with high quality; *FLEXFIS* could significantly outperform state-of-the-art (statistical-oriented) chemometric methods [67], which, in large parts, produced high unacceptable errors; gradual forgetting played an important role in order to properly react onto the highly dynamic changes.
- On-line image classifiers as part of an on-line quality-control framework based on machine vision technology (part of an European project)—see Chap. 14.
- Perception-based texture models (part of an European project) [66]: the aim was to build models for associating perceptions and emotions with visual textures; six core adjectives served as final output of a MIMO model (warm, rough, complex, elegant, natural, and like); as inputs, we used selected low-level features extracted from the textures and characterizing best emotional states of humans when inspecting the textures. Models with max. 20% error performance over all human beings could be established, even showing some synergies to psychological-oriented models.
- Application of the *FLEXFIS* family onto several data sets from the UCI repository and widely used dynamic data sets (Mackey-Glass, Box-Jenkins, etc.), showing similar performance to batch modeling methods, and justifying its convergence to batch solutions. Specific tests were made on a hyper-plane data set including a huge data stream of 1.2 million samples and a drift case: rule merging option was indispensable in order to obtain results within a reasonable time frame (automatic reduction from about 4,000 to 8–10 rules resulted in a reduction of computation times from hours to minutes); forgetting was essential to increase *FLEXFIS's* accuracy during the occurrence of a drift, which could out-perform the accuracy of Hoeffding trees as alternative incremental classification method (from the MOA framework) [49].

A thorough discussion on the usage of the *FLEXFIS* family and a detailed evaluation and performance analysis within on-line quality control systems (measurement and image based) will follow in Chap. 14.

9.5 Conclusion

This chapter provides an overview of the flexible evolving fuzzy systems family, including the most important algorithms and techniques for guiding the incremental learning process of parameters to optimal solutions and the model structures (rules) to an appropriate representation of local data distributions and clouds. The flexible fuzzy systems and classifiers are successively (step-wise) built up, evolved, and further expanded on the basis of data streams which are arriving on-line in a block- or sample-wise fashion and may represent changing systems and varying environmental influences. Therefore, they can be seen as a substantial contribution to *dynamic learning in non-stationary environments*. The chapter also embraces advanced concepts for achieving higher predictive accuracy of the evolved models, including an appropriate handling of concept drift in data streams; a dynamic soft dimension reduction approach, where unimportant features are smoothly out-weighted over time; and an incremental concept for merging and pruning unnecessary rules in order to guarantee distinguishability of the rules and fuzzy sets. Finally, this chapter raises considerations and aspects about interpretability of the evolved models and the reliability of their predictions. For the latter, conflict and ignorance models for predictions in classification settings are represented, as well as adaptive local error bars for tracking the uncertainty due to noise, bias, and extrapolation regions in regression problems. Therefore, a major strength of the *FLEXFIS* family is the manifold of extensions which can cope with different situations in resp. characteristics of data streams, namely, drifts, rule fusion over time, dynamically changing curse of dimensionality, and varying noise levels. Additional central strengths are its robustness in terms of achieving convergence of parameters with respect to an optimization criterion and the possibility to reduce unnecessary complexity in the evolved models, achieving more transparent fuzzy systems. A weakness may be that an initial fuzzy model (regression type or classifier) is needed, or at least an initial data set for estimating the ranges of features. The evaluation and performance of the various methods (*FLEXFIS*, *FLEXFIS-Class SM, MM, AP*, and *eVQ-Class*) as well as most of the enhanced concepts will be thoroughly discussed based on several on-line quality-control applications in Chap. 14.

Finally, we want to point out that other variants of EFS approaches, in particular, using TS fuzzy systems architecture, emerged during the last decade, and a comprehensive survey; and study on these can be found in [44]; a small overview of approaches is also mentioned at the beginning of the next chapter.

Acknowledgments This work was funded by the Austrian fund for promoting scientific research (FWF, contract number I328-N23, acronym IREFS). It reflects only the authors' views.

References

1. Abraham, W., Robins, A.: Memory retention - the synaptic stability versus plasticity dilemma. Trends in Neurosciences **28**(2), 73–78 (2005)
2. Angelov, P., Filev, D., Kasabov, N.: Evolving Intelligent Systems—Methodology and Applications. John Wiley & Sons, New York (2010)
3. Angelov, P., Kasabov, N.: Evolving computational intelligence systems. In: Proceedings of the 1st International Workshop on Genetic Fuzzy Systems, pp. 76–82. Granada, Spain (2005)
4. Angelov, P., Lughofer, E., Zhou, X.: Evolving fuzzy classifiers using different model architectures. Fuzzy Sets and Systems **159**(23), 3160–3182 (2008)
5. Aström, K., Wittenmark, B.: Adaptive Control - Second Edition. Addison-Wesley Longman Publishing Co., Inc., Boston, MA, USA (1994)
6. Backer, S.D., Scheunders, P.: Texture segmentation by frequency-sensitive elliptical competitive learning. Image and Vision Computing **19**(9–10), 639–648 (2001)
7. Basseville, M., Nikiforov, I.: Detection of Abrupt Changes. Prentice Hall Inc. (1993)
8. Bifet, A., Holmes, G., Kirkby, R., Pfahringer, B.: MOA: Massive online analysis. Journal of Machine Learning Research **11**, 1601–1604 (2010)
9. Bifet, A., Kirkby, R.: Data stream mining — a practical approach. Tech. rep., Department of Computer Sciences, University of Waikato, Japan (2009)
10. Carreira-Perpinan, M.: A review of dimension reduction techniques. Tech. Rep. CS-96-09, Dept. of Computer Science, University of Sheffield, Sheffield, U.K. (1997)
11. Casillas, J., Cordon, O., Herrera, F., Magdalena, L.: Interpretability Issues in Fuzzy Modeling. Springer Verlag, Berlin Heidelberg (2003)
12. Casillas, J., Cordon, O., Jesus, M.D., Herrera, F.: Genetic feature selection in a fuzzy rule-based classification system learning process for high-dimensional problems. Information Sciences **136**(1–4), 135–157 (2001)
13. Castro, J., Delgado, M.: Fuzzy systems with defuzzification are universal approximators. IEEE Transactions on Systems, Man and Cybernetics, part B: Cybernetics **26**(1), 149–152 (1996)
14. Cernuda, C., Lughofer, E., Suppan, L., Röder, T., Schmuck, R., Hintenaus, P., Märzinger, W., Kasberger, J.: Evolving Chemometric Models for Predicting Dynamic Process Parameters in Viscose Production. Analytica Chimica Acta online and in press, (2012), doi:10.1016/j.aca.2012.03.012
15. Cohn, D., Atlas, L., Ladner, R.: Improving generalization with active learning. Machine Learning **15**(2), 201–221 (1994)
16. Domingos, P., Hulten, G.: Mining high-speed data streams. In: Proceedings of the Sixth ACM SIGKDD International Conference on Knowledge Discovery and Data Mining, pp. 71–80. Boston, MA (2000)
17. Draper, N., Smith, H.: Applied Regression Analysis. Probability and Mathematical Statistics. John Wiley & Sons, New York (1981)
18. Duda, R., Hart, P., Stork, D.: Pattern Classification - Second Edition. Wiley-Interscience (John Wiley & Sons), Southern Gate, Chichester, West Sussex, England (2000)
19. Dy, J., Brodley, C.: Feature selection for unsupervised learning. Journal of Machine Learning Research **5**, 845–889 (2004)
20. Freiheit, T., Koren, Y., Hu, S.: Productivity of parallel production lines with unreliable machines and material handling. IEEE Transactions on Automation Sciences and Engineering **1**(1), 98–103 (2004)
21. Fürnkranz, J.: Round robin classification. Journal of Machine Learning Research **2**, 721–747 (2002)

22. Gama, J.: Knowledge Discovery from Data Streams. Chapman & Hall/CRC, Boca Raton, Florida (2010)
23. Gama, J., Gaber, M.M. (eds.): Learning from Data Streams: Processing Techniques in Sensor Networks. Springer-Verlag (2007)
24. Gama, J., Medas, P., Rocha, R.: Forest trees for on-line data. In: Proceedings of the 2004 ACM symposium on Applied computing, pp. 632–636. New York (2004)
25. Gray, R.: Vector quantization. IEEE ASSP Magazine **1**(2), 4–29 (1984)
26. Hamker, F.: RBF learning in a non-stationary environment: the stability-plasticity dilemma. In: R. Howlett, L. Jain (eds.) Radial basis function networks 1: recent developments in theory and applications, pp. 219–251. Physica Verlag, Heidelberg, New York (2001)
27. Hastie, T., Tibshirani, R., Friedman, J.: The Elements of Statistical Learning: Data Mining, Inference and Prediction. Springer Verlag, New York, Berlin, Heidelberg, Germany (2001)
28. Hastie, T., Tibshirani, R., Friedman, J.: The Elements of Statistical Learning: Data Mining, Inference and Prediction - Second Edition. Springer, New York Berlin Heidelberg (2009)
29. Haykin, S.: Neural Networks: A Comprehensive Foundation (2nd Edition). Prentice Hall Inc., Upper Saddle River, New Jersey (1999)
30. Hühn, J., Hüllermeier, E.: FR3: A fuzzy rule learner for inducing reliable classifiers. IEEE Transactions on Fuzzy Systems **17**(1), 138–149 (2009)
31. Kasabov, N.: Evolving Connectionist Systems: The Knowledge Engineering Approach - Second Edition. Springer Verlag, London (2007)
32. Klinkenberg, R.: Learning drifting concepts: example selection vs. example weighting. Intelligent Data Analysis **8**(3), 281–300 (2004)
33. Koczy, L., Tikk, D., Gedeon, T.: On functional equivalence of certain fuzzy controllers and RBF type approximation schemes. International Journal of Fuzzy Systems **2**(3), 164–175 (2000)
34. Kruse, R., Gebhardt, J., Klawonn, F.: Foundations of Fuzzy Systems. John Wiley & Sons, New York (1994)
35. Kuncheva, L.: Fuzzy Classifier Design. Physica-Verlag, Heidelberg (2000)
36. Kurzhanskiy, A.A., Varaiya, P.: Ellipsoidal toolbox. Tech. rep. (2006)
37. Li, X., Wang, L., Sung, E.: Multilabel SVM active learning for image classification. In: Proceedings of the International Conference on Image Processing (ICIP) vol. 4, pp. 2207–2010. Singapore (2004)
38. Ljung, L.: System Identification: Theory for the User. Prentice Hall PTR, Prentic Hall Inc., Upper Saddle River, New Jersey (1999)
39. Lughofer, E.: Evolving vector quantization for classification of on-line data streams. In: Proc. of the Conference on Computational Intelligence for Modelling, Control and Automation (CIMCA 2008), pp. 780–786. Vienna, Austria (2008)
40. Lughofer, E.: Extensions of vector quantization for incremental clustering. Pattern Recognition **41**(3), 995–1011 (2008)
41. Lughofer, E.: FLEXFIS: A robust incremental learning approach for evolving TS fuzzy models. IEEE Transactions on Fuzzy Systems **16**(6), 1393–1410 (2008)
42. Lughofer, E.: On-line evolving image classifiers and their application to surface inspection. Image and Vision Computing **28**(7), 1065–1079 (2010)
43. Lughofer, E.: All-pairs evolving fuzzy classifiers for on-line multi-class classification problems. In: Proceedings of the EUSFLAT 2011 Conference, pp. 372–379. Elsevier, Aix-Les-Bains, France (2011)
44. Lughofer, E.: Evolving Fuzzy Systems — Methodologies, Advanced Concepts and Applications. Springer, Berlin Heidelberg (2011)
45. Lughofer, E.: Hybrid active learning (HAL) for reducing the annotation efforts of operators in classification systems. Pattern Recognition **45**(2), pp. 884–896 (2012)
46. Lughofer, E.: On-line incremental feature weighting in evolving fuzzy classifiers. Fuzzy Sets and Systems **163**(1), 1–23 (2011)
47. Lughofer, E., Angelov, P.: Handling drifts and shifts in on-line data streams with evolving fuzzy systems. Applied Soft Computing **11**(2), 2057–2068 (2011)

48. Lughofer, E., Angelov, P., Zhou, X.: Evolving single- and multi-model fuzzy classifiers with FLEXFIS-Class. In: Proceedings of FUZZ-IEEE 2007, pp. 363–368. London, UK (2007)
49. Lughofer, E., Bouchot, J.L., Shaker, A.: On-line elimination of local redundancies in evolving fuzzy systems. Evolving Systems **2**(3), 165–187 (2011)
50. Lughofer, E., Guardiola, C.: Applying evolving fuzzy models with adaptive local error bars to on-line fault detection. In: Proceedings of Genetic and Evolving Fuzzy Systems 2008, pp. 35–40. Witten-Bommerholz, Germany (2008)
51. Lughofer, E., Hüllermeier, E., Klement, E.: Improving the interpretability of data-driven evolving fuzzy systems. In: Proceedings of EUSFLAT 2005, pp. 28–33. Barcelona, Spain (2005)
52. Lughofer, E., Kindermann, S.: SparseFIS: Data-driven learning of fuzzy systems with sparsity constraints. IEEE Transactions on Fuzzy Systems **18**(2), 396–411 (2010)
53. Lughofer, E., Macian, V., Guardiola, C., Klement, E.: Identifying static and dynamic prediction models for nox emissions with evolving fuzzy systems. Applied Soft Computing **11**(2), 2487–2500 (2011)
54. Lughofer, E., Trawinski, B., Trawinski, K., Kempa, O., Lasota, T.: On employing fuzzy modeling algorithms for the valuation of residential premises. Information Sciences, **181**(23), 5123–5142 (2011)
55. Mahalanobis, P.C.: On the generalised distance in statistics. In: Proceedings of the National Institute of Sciences of India, vol. 2 (1), pp. 49–55 (1936)
56. Mouss, H., Mouss, D., Mouss, N., Sefouhi, L.: Test of Page-Hinkley, an approach for fault detection in an agro-alimentary production system. In: Proceedings of the Asian Control Conference, Volume 2, pp. 815–818 (2004)
57. Nelles, O.: Nonlinear System Identification. Springer, Berlin (2001)
58. Oza, N.C., Russell, S.: Online bagging and boosting. Artificial Intelligence and Statistics pp. 105–112 (2001)
59. Page, E.: Continuous inspection schemes. Biometrika **41**(1-2), 100–115 (1954)
60. Pang, S., Ozawa, S., Kasabov, N.: Incremental linear discriminant analysis for classification of data streams. IEEE Transaction on Systems, Men and Cybernetics, part B: Cybernetics **35**(5), 905–914 (2005)
61. Qin, S., Li, W., Yue, H.: Recursive PCA for adaptive process monitoring. Journal of Process Control **10**(5), 471–486 (2000)
62. Roubos, J., Setnes, M., Abonyi, J.: Learning fuzzy classification rules from data. Information Sciences **150**(1–2), 77–93 (2003)
63. Sanchez, L., Suarez, M., Villar, J., Couso, I.: Mutual information-based feature selection and partition design in fuzzy rule-based classifiers from vague data. International Journal of Approximate Reasoning **49**(3), 607–622 (2008)
64. Schölkopf, B., Smola, A.: Learning with Kernels - Support Vector Machines, Regularization, Optimization and Beyond. MIT Press, London, England (2002)
65. Takagi, T., Sugeno, M.: Fuzzy identification of systems and its applications to modeling and control. IEEE Transactions on Systems, Man and Cybernetics **15**(1), 116–132 (1985)
66. Thumfart, S., Jacobs, R., Lughofer, E., Cornelissen, F., Maak, H., Groissboeck, W., Richter, R.: Modelling human aesthetic perception of visual textures. ACM Transactions on Applied Perception, **8**(4), (2011)
67. Varmuza, K., Filzmoser, P.: Introduction to Multivariate Statistical Analysis in Chemometrics. CRC Press, Boca Raton (2009)
68. Wang, L., Mendel, J.: Fuzzy basis functions, universal approximation and orthogonal least-squares learning. IEEE Transactions on Neural Networks **3**(5), 807–814 (1992)
69. Widmer, G., Kubat, M.: Learning in the presence of concept drift and hidden contexts. Machine Learning **23**(1), 69–101 (1996)
70. Yager, R.R.: A model of participatory learning. IEEE Transactions on Systems, Man and Cybernetics **20**(5), 1229–1234 (1990)

71. Yen, J., Wang, L., Gillespie, C.: Improving the interpretability of TSK fuzzy models by combining global learning and local learning. IEEE Transactions on Fuzzy Systems **6**(4), 530–537 (1998)
72. Zhou, S., Gan, J.: Low-level interpretability and high-level interpretability: a unified view of data-driven interpretable fuzzy systems modelling. Fuzzy Sets and Systems **159**(23), 3091–3131 (2008)

Chapter 10
Sequential Adaptive Fuzzy Inference System for Function Approximation Problems

Hai-Jun Rong

Abstract In the classic approaches to design a fuzzy inference system, the fuzzy rules are determined by a domain expert a priori and then they are maintained unchanged during the learning. These fixed fuzzy rules may not be appropriate in real-time applications where the environment or model often meets unpredicted disturbances or damages. Hence, poor performance may be observed. In comparison to the conventional methods, fuzzy inference systems based on neural networks, called fuzzy-neural systems, have begun to exhibit great potential for adapting to the changes by utilizing the learning ability and adaptive capability of neural networks. Thus, a fuzzy inference system can be built using the standard structure of neural networks. Nevertheless, the determination of the number of fuzzy rules and the adjustment of the parameters in the if-then fuzzy rules are still open issues. A sequential adaptive fuzzy inference system (SAFIS) is developed to determine the number of fuzzy rules during learning and modify the parameters in fuzzy rules simultaneously. SAFIS uses the concept of influence of a fuzzy rule for adding and removing rules during learning. The influence of a fuzzy rule is defined as its contribution to the system output in a statistical sense when the input data is uniformly distributed. When there is no addition of fuzzy rules, only the parameters of the "closest" (in a Euclidean sense) rule are updated using an extended Kalman filter (EKF) scheme. The performance of SAFIS is evaluated based on some function approximation problems, via, nonlinear system identification problems and a chaotic time-series prediction problem. Results indicate that SAFIS produces similar or better accuracies with lesser number of rules compared to other algorithms.

H.-J. Rong (✉)
State Key Laboratory of Strength and Vibration, School of Aerospace,
Xi'an Jiaotong University, No.28 Xianning West Road, Xi'an, ShaanXi, China
e-mail: hjrong@mail.xjtu.edu.cn

M. Sayed-Mouchaweh and E. Lughofer (eds.), *Learning in Non-Stationary Environments: Methods and Applications*, DOI 10.1007/978-1-4419-8020-5_10,
© Springer Science+Business Media New York 2012

10.1 Introduction

A fuzzy inference system can model the qualitative aspects of human knowledge and reasoning processes using fuzzy if-then rules. Based on this, many ill-defined and uncertain systems in some disciplines such as engineering, economics, and other areas [6, 10, 16, 18, 20, 23, 24, 29, 31] can be handled without employing precise quantitative analysis. The experiment results have demonstrated that the fuzzy inference systems are very useful to solve some practical problems involving a high level of uncertainty, complexity, or nonlinearity compared with the conventional modeling methods. However, the superior performance of fuzzy inference systems mainly depends on the fuzzy rules. If the fuzzy rules are not appropriate and deviate from the requirement of the system itself, this may result in poor performance. On the other hand, although the rules are correct, it is hard to determine the appropriate parameters for the fuzzy rules. Inappropriate parameters also may result in poor performance.

To solve these problems, many researchers have built fuzzy-neural systems by incorporating the fuzzy inference process in the structure of neural networks and then the learning ability of neural networks are used to adjust the fuzzy rules. Except for some special fuzzy-neural systems which made use of fuzzy neurons and fuzzy weights [7, 25], most of the recent fuzzy-neural systems [1, 3, 5, 11, 13, 17, 19, 32] have been built based on the standard feed-forward network with local fields to approximate the fuzzy inference systems with local properties. In these fuzzy-neural systems, the neurons with local fields correspond to the fuzzy rules and the proposed algorithms for designing the fuzzy-neural systems have considered two issues, that is, the structure identification and the parameter adjustment. Structure identification determines the input–output space partition, antecedent and consequent variables of if-then rules, number of such rules, and initial positions of membership functions. The task of parameter adjustment involves realizing the parameters for the fuzzy system structure determined in the previous step [21].

The researchers [5, 13, 15, 19, 27, 28, 30, 32] have tried to develop many efficient approaches for solving the two issues. These methods can be broadly divided into two classes, namely, batch learning schemes and sequential learning schemes. In batch learning, it is assumed that the complete training data is available before the training commences. The training usually involves cycling the data over a number of epochs. In sequential learning, the data arrives one by one, and after the learning of each data, it is discarded and the notion of epoch does not exist. In practical applications, new training data arrives sequentially, and to handle this using batch learning, one has to retrain the network all over again, resulting in large training time. Hence, in these cases, sequential learning algorithms are generally preferred over batch learning algorithms as they do not require retraining whenever a new data is received. The sequential fuzzy-neural scheme, which is discussed in this chapter, has the following distinguishing features:

1. All the training observations are *sequentially* (one by one) presented to the system.
2. At any time, *only one* training observation is seen and learned.

3. A training observation is *discarded* as soon as the learning procedure for that particular observation is completed.
4. The learning system has no *prior* knowledge as to how many total training observations will be presented.

Thus, if one strictly applies the above features of the sequential algorithms, many of the existing algorithms are not sequential. One major bottleneck seems to be that they need the entire training data ready for training before the training procedure starts and thus they are not really sequential. This point is highlighted in a brief review of the existing algorithms given below.

Jang [11] has developed an adaptive-network-based fuzzy inference system (ANFIS) where a hybrid learning method was utilized to identify the system parameters. The parameters in the membership functions were updated by a gradient descent method, and the parameters in the consequent parts were adjusted by means of a least-square error method. The number of fuzzy rules was determined according to a grid-type partition which resulted in the exponential increase of the number of fuzzy rules as the input variables increased. Chiu [4] solved this problem by selecting some significant input variables from all the input variables as the input of the fuzzy systems. However, these algorithms require cycling the whole training data over a number of learning cycles (epochs). Thus, they are batch learning algorithms. Besides, in these algorithms, the number of fuzzy rules are determined beforehand and cannot be varied according to learning process.

Many approaches [5, 13, 19, 32] have been proposed based on the functional equivalence between a radial basis function (RBF) neural network and a fuzzy inference system to achieve the determination of the number of fuzzy rules and parameter adjustment simultaneously during learning. These schemes utilize the learning capabilities of the RBF for changing the rules as well as adjusting the parameters since the hidden neurons of the RBF networks are related to the fuzzy rules [12]. A significant contribution to sequential learning in RBF network was made by Platt [26] through the development of resource allocation network (RAN). In RAN, the network starts with no hidden neurons but adds hidden neurons based on the novelty of the input data. Most of the recent algorithms for adaptively creating fuzzy systems are based on the ideas of RAN. These algorithms claim to be "on-line" algorithms, and if one looks closely at them, they are not sequential as per the above distinguishing features.

A hierarchically self-organizing approach proposed by Cho and Wang [5] automatically generated fuzzy rules without predefining the number of fuzzy rules based on the error and distance criterion of fuzzy basis functions. The parameters in the fuzzy rules were modified by the gradient descent algorithm. However, the algorithm requires cycling the whole training data over a number of learning cycles (epochs), and hence, it is not a truly sequential learning scheme.

Juang and Lin [13] have proposed a self-constructing neural fuzzy inference network (SONFIN) in which the fuzzy rules were extracted online from the training data together with the parameter update for all existing fuzzy rules using the gradient descent method. For adding a new fuzzy rule, SONFIN utilized the distance criterion between the new input data and the center of the Gaussian membership function in the existing fuzzy rules. Although this algorithm is sequential in nature, it does not remove the fuzzy rules once created even though that rule is not effective. This may result in a structure where the number of rules may be large.

In most of the real applications, not all fuzzy rules contribute significantly to the system performance during the entire time period. A fuzzy rule may be active initially, but may later contribute little to the system output. For this reason, the insignificant fuzzy rules have to be removed during learning to realize a compact fuzzy system structure. Using the ideas of adding and pruning hidden neurons to form a minimal RBF network in [33], a hierarchical on-line self-organizing learning algorithm for dynamic fuzzy neural networks (DFNN) has been proposed in [32]. Another on-line self-organizing fuzzy neural network (SOFNN) proposed by Leng et al. [19] also included a pruning method. The pruning method utilized the optimal brain surgeon (OBS) approach to determine the importance of each rule. In the two algorithms, the least-square error method was utilized to update the parameters for all the existing fuzzy rules. However, in these two algorithms the pruning criteria need all the past data received so far. Hence, they are not strictly sequential and further requires increased memory for storing all the past data.

A dynamic evolving neural-fuzzy inference system (DENFIS) was proposed by Kasabov and Song [17] where the fuzzy rules were created depending on the position of the input vector in the input space and the output was dynamically calculated based on m-most active fuzzy rules which have been created during the past learning process. Angelov and Filev [3] proposed an evolving Takagi–Sugeno model (eTS) that recursively updated TS model structure based on the *potential* of the input data (defined based on its distances to all other data points received so far). In this algorithm, a new rule was added when the potential of the new data was higher than the potential of the existing rules, or a new rule was modified when the potential of the new data was higher than the potential of the existing rules and the new data was close to an old rule. These two algorithms are truly sequential learning algorithms. However, the algorithms cannot simplify the rule base during learning by ignoring the rules which may become irrelevant with the future data samples when the data sample sequentially arrives. A simplified version of the eTS learning algorithm that simplified the rule base, called the simpl_eTS, was proposed by Angelov and Filev [1]. The algorithm utilized the concept of the scatter which was similar to the notion of potential but computationally more efficient. The algorithm could simplify the rule base to make the rules representative based on the population of each rule determined by the number of the data samples that belonged to a particular cluster. If the population of a rule was less than 1% of the total data at the moment of appearance of a rule, the rule was ignored from the rule base by setting its firing strength to zero. Besides, these algorithms employed the least-square error method to modify the parameters of the existing fuzzy rules.

In this chapter, a sequential adaptive fuzzy inference system (SAFIS) is developed to realize a compact fuzzy system with lesser number of rules. SAFIS uses the idea of functional equivalence between a RBF neural network and a fuzzy inference system. Here, SAFIS uses the growing and pruning RBF (GAP-RBF) neural network proposed by Huang et al. [8]. The SAFIS algorithm consists of two aspects: determination of the fuzzy rules and adjustment of the premise and consequent parameters in fuzzy rules.

SAFIS uses the concept of *influence* of a fuzzy rule to add and remove rules during learning. SAFIS starts with no fuzzy rules, and based on the data, builds up a compact rule base. During the learning, only the current data is made use of, and there is no need to store all the past data. The *influence* of a fuzzy rule is defined as its contribution to the system output in a statistical sense. Here, we have derived an expression for this for the case where the input data is uniformly distributed. The parameter adjustment is done using a winner rule strategy where the winner rule is defined as the one closest to the latest input data, and the parameter update is done using an extended Kalman filter (EKF) mechanism.

10.2 Architecture of SAFIS

A function approximation problem can be described as follows. Suppose the sample data, $\{(\mathbf{x}_n, \mathbf{y}_n) : n = 1,2,\ldots\}$, are observed, where $\mathbf{x}_n$ is a N_x-dimensional features of observation n and $\mathbf{y}_n$ is its target output of dimension N_y. It is assumed that the observation data are free of noise, and an underlying function f exists between the target output $\mathbf{y}_n$ and feature space $\mathbf{x}_n$ from the known set of data:,

$$\mathbf{y}_n = f(\mathbf{x}_n). \tag{10.1}$$

The aim of the SAFIS algorithm is to approximate f such that:

$$\hat{\mathbf{y}}_n = \hat{f}(\mathbf{x}_n), \tag{10.2}$$

where $\hat{\mathbf{y}}_n$ is the output of SAFIS. This means that the objective is to minimize the error between the system output and the output of SAFIS, $\|\mathbf{y}_n - \hat{\mathbf{y}}_n\|$. Before describing the details of the algorithm, the structure of SAFIS network is first described below.

The structure of SAFIS illustrated by Fig. 10.1 consists of five layers to realize the following fuzzy rule model:

Rule k : if (x_1 is A_{1k}) ... (x_{N_x} is A_{N_xk}), then ($\hat{y}_1$ is a_{1k}) ... ($\hat{y}_{N_y}$ is a_{N_yk}), where $a_{jk}(j = 1,2,\ldots,N_y; k = 1,2,\ldots,N_h)$ is a constant consequent parameter in rule k, $A_{ik}(i = 1,2,\ldots,N_x)$ is the membership value of the ith input variable x_i in rule k, N_x is the dimension of the input vector $\mathbf{x}(\mathbf{x} = [x_1,\ldots,x_{N_x}]^T)$, N_h is the number of fuzzy rules, and N_y is the dimension of the output vector $\hat{\mathbf{y}}(\hat{\mathbf{y}} = [\hat{y}_1,\ldots,\hat{y}_{N_y}]^T)$. In SAFIS, the number of fuzzy rules N_h varies. Initially, there is no fuzzy rule and then during learning, fuzzy rules are added and removed.

Layer 1: In layer 1, each node represents an input variable and directly transmits the input signal to layer 2.

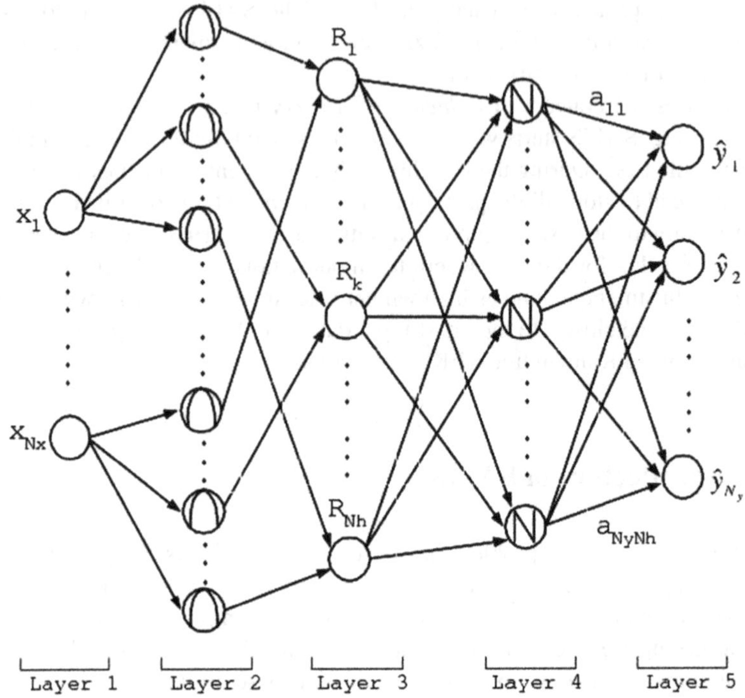

Fig. 10.1 Structure of SAFIS

Layer 2: In this layer, each node represents the membership value of each input variable. SAFIS utilizes the function equivalence between a RBF network and a FIS, and thus, its antecedent part (if part) in fuzzy rules is achieved by Gaussian functions of the RBF network. The membership value $A_{ik}(x_i)$ of the ith input variable x_i in the kth Gaussian function is given by:

$$A_{ik}(x_i) = \exp\left(-\frac{(x_i - \mu_{ik})^2}{\sigma_k^2}\right), k = 1, 2, \ldots, N_h, \tag{10.3}$$

where N_h is the number of the Gaussian functions, μ_{ik} is the center of the kth Gaussian function for the ith input variable, and σ_k is the width of the kth Gaussian function. In SAFIS, the width of all the input variables in the kth Gaussian function is the same.

Layer 3: Each node in the layer represents the if part of if-then rules obtained by the sum-product composition, and the total number of such rules is N_h. The firing strength (if part) of the kth rule is given by:

$$R_k(\mathbf{x}) = \prod_{i=1}^{N_x} A_{ik}(x_i) = \exp\left(-\sum_{i=1}^{N_x} \frac{(x_i - \mu_{ik})^2}{\sigma_k^2}\right) = \exp\left(-\frac{\|\mathbf{x} - \mu_k\|^2}{\sigma_k^2}\right). \tag{10.4}$$

Layer 4: The nodes in the layer are named as normalized nodes whose number is equal to the number of the nodes in third layer. The kth-normalized node is given by:

$$\bar{R}_k = \frac{R_k(\mathbf{x})}{\sum_{k=1}^{N_h} R_k(\mathbf{x})}. \tag{10.5}$$

Layer 5: Each node in this layer corresponds to an output variable, which is given by the weighted sum of the output of each normalized rule. The system output is calculated by:

$$\hat{\mathbf{y}} = \frac{\sum_{k=1}^{N_h} R_k(\mathbf{x})\mathbf{a}_k}{\sum_{k=1}^{N_h} R_k(\mathbf{x})}, \tag{10.6}$$

where $\hat{\mathbf{y}} = [\hat{y}_1, \hat{y}_2, \ldots, \hat{y}_{N_y}]^T, \mathbf{a}_k = [a_{k1}, a_{k2}, \ldots, a_{kN_y}]^T$.

Similar to the significance concept of a neuron in GAP-RBF [8], the SAFIS algorithm uses the concept of "influence" of a rule to realize the growing and pruning of fuzzy rules. It is described below.

10.2.1 "Influence" of a Fuzzy Rule

As per (10.6), the contribution of the kth rule to the overall output for an input observation $\mathbf{x}_l$ is given by:

$$E(k,l) = \|\mathbf{a}_k\| \frac{R_k(\mathbf{x}_l)}{\sum_{k=1}^{N_h} R_k(\mathbf{x}_l)}. \tag{10.7}$$

Then the contribution of the kth rule to the overall output based on all input data N received so far is obtained by:

$$E(k) = \|\mathbf{a}_k\| \frac{\sum_{l=1}^{N} R_k(\mathbf{x}_l)}{\sum_{k=1}^{N_h}\sum_{l=1}^{N} R_k(\mathbf{x}_l)}. \tag{10.8}$$

Dividing both the numerator and denominator by N in (10.8), the equation becomes:

$$E(k) = \|\mathbf{a}_k\| \frac{\sum_{l=1}^{N} R_k(\mathbf{x}_l)/N}{\sum_{k=1}^{N_h}\sum_{l=1}^{N} R_k(\mathbf{x}_l)/N}. \tag{10.9}$$

Using the significance concept of GAP-RBF [8], the *influence* of the kth fuzzy rule is defined as its statistical contribution to the overall output of SAFIS. When $N \to \infty$, the influence of the kth rule is given by:

$$E_{\text{inf}}(k) = \lim_{N\to\infty} E(k) = \lim_{N\to\infty} \|\mathbf{a}_k\| \frac{\sum_{l=1}^{N} R_k(\mathbf{x}_l)/N}{\sum_{k=1}^{N_h}\sum_{l=1}^{N} R_k(\mathbf{x}_l)/N}. \tag{10.10}$$

Calculation of $E_{\text{inf}}(k)$ using the above equation requires the knowledge of $(\mathbf{x}_l, \mathbf{y}_l)$, $l = 1,\ldots,N$. In the truly sequential learning scheme, this is not possible. An alternate way of calculating $E_{\text{inf}}(k)$ is by using the distribution of the inputs and follows the same approach as introduced in [9]. In order to compute $E_{\text{inf}}(k)$, one has to compute first E_k defined by:

$$E_k = \lim_{N\to\infty} \frac{\sum_{l=1}^{N} R_k(\mathbf{x}_l)}{N}. \tag{10.11}$$

Assume that the observations, $(\mathbf{x}_l, \mathbf{y}_l)$, $l = 1,\ldots$, are drawn from a sampling range X with a sampling density function $p(\mathbf{x})$. Consider a situation where N observations have been learned by the sequential learning scheme. Let the sampling range X be divided into M small spaces $\Delta_j, j = 1,\ldots,M$. The size of Δ_j is represented by $S(\Delta_j)$. Since the sampling density function is $p(\mathbf{x})$, there are around $N \cdot p(\mathbf{x_j}) \cdot S(\Delta_j)$ samples in each Δ_j, where $\mathbf{x}_j$ is any point chosen in Δ_j. When the number of input observations N is large and Δ_j is small from (10.11), we have:

$$\begin{aligned} E_k &\approx \lim_{M\to\infty} \frac{\sum_{j=1}^{M} R_k(\mathbf{x}_j)\cdot Np(\mathbf{x}_j)\cdot S(\Delta_j)}{N} \\ &= \lim_{M\to\infty} \sum_{j=1}^{M} R_k(\mathbf{x}_j)\cdot p(\mathbf{x}_j)\cdot S(\Delta_j) \\ &= \int_X R_k(\mathbf{x})p(\mathbf{x})d\mathbf{x} \\ &= \int_X \exp\left(-\frac{\|\mathbf{x}-\mu_k\|^2}{\sigma_k^2}\right) p(\mathbf{x})d\mathbf{x}. \end{aligned} \tag{10.12}$$

If the distribution of the N_x attributes $(x_1,\ldots,x_i,\ldots,x_{N_x})^T$ of observations $\mathbf{x}$'s are independent from each other, the density function $p(\mathbf{x})$ of $\mathbf{x}$ can be written as follows: $p(\mathbf{x}) = \prod_{i=1}^{N_x} p_i(x_i)$, where $p_i(x)$ is the density function of the i-th attribute x_i of observations. In this case, (10.12) can be re-written as:

$$\begin{aligned} E_k &= \int \cdots \int_X \exp\left(-\frac{\|\mathbf{x}-\mu_k\|^2}{\sigma_k^2}\right) p(\mathbf{x})\mathrm{d}\mathbf{x} \\ &= \prod_{i=1}^{N_x}\left(\int_{a_i}^{b_i} \exp\left(-\frac{\|x-\mu_{k,i}\|^2}{\sigma_k^2}\right) p_i(x)\mathrm{d}x\right), \end{aligned} \tag{10.13}$$

where N_x is the dimension of the input space X and (a_i, b_i) is the interval of the i-th attribute x_i of observations, $i = 1, \ldots, N_x$.

Equation (10.13) involves the integration of the probability density function $p(\mathbf{x})$ in the sampling range X. When the input samples are uniformly drawn from a range X, the sampling density function $p(\mathbf{x})$ is given by $p(\mathbf{x}) = \frac{1}{S(X)}$, where $S(X)$ is the size of the range X given by $S(X) = \int_{\mathbf{x}} 1 d\mathbf{x}$. Substituting for $p(\mathbf{x})$ in (10.12), we get:

$$E_k = \int_X \exp\left(-\frac{\|\mathbf{x} - \mu_k\|^2}{\sigma_k^2}\right) \frac{1}{S(X)} d\mathbf{x}. \tag{10.14}$$

Note that, in general, the width σ_k of a rule k is much less than the size of range X, the above equation can be approximated as:

$$\begin{aligned} E_k &\approx \frac{1}{S(X)} \left(2 \int_0^{+\infty} \exp\left(-\frac{x^2}{\sigma_k^2}\right) dx\right)^{N_x} \\ &= \frac{\pi^{N_x/2} \sigma_k^{N_x}}{S(X)} \\ &= \frac{(1.8\sigma_k)^{N_x}}{S(X)}. \end{aligned} \tag{10.15}$$

Thus, based on (10.15), the influence of the kth rule is given by:

$$E_{\text{inf}}(k) = \|\mathbf{a}_k\| \frac{(1.8\sigma_k)^{N_x}}{\sum_{k=1}^{N_h} (1.8\sigma_k)^{N_x}}. \tag{10.16}$$

It is noteworthy that the significance of a neuron proposed in GAP-RBF [8] is defined based on the average contribution of an individual neuron to the output of the RBF network. Under this definition, one may need to estimate the input distribution range S(X). However, the influence of a rule introduced here is different from the significance of a neuron proposed in GAP-RBF [8]. In fact, the influence of a neuron is defined as the relevant significance of the neuron compared to summation of significance of all the existing RBF neurons. Seen from (10.16), with the introduction of influence, one need not estimate the input distribution range S(X) and the implementation has been simplified.

Influence of a rule is utilized for the addition and deletion of a fuzzy rule in SAFIS algorithm as indicated below.

10.2.2 SAFIS Algorithm

The learning algorithm of SAFIS consists of two aspects: determination of fuzzy rules and adjustment of the premise and consequent parameters in fuzzy rules.

SAFIS can automatically add and remove fuzzy rules using ideas similar to GAP-RBF [8] for hidden neurons. A description of dynamically adding and removing the fuzzy rules along with the details of parameter adjustment when there are no addition of rules is given below.

10.2.2.1 Adding of Fuzzy Rules

SAFIS begins with no fuzzy rules. When the first input $\mathbf{x}_1, \mathbf{y}_1$ is received, it is translated into the first rule whose parameters are given as, $\mu_1 = \mathbf{x}_1, \mathbf{a}_1 = \mathbf{y}_1, \sigma_1 = \kappa\|\mathbf{x}_1\|$. Then, as inputs $\mathbf{x}_n, \mathbf{y}_n$ ($n > 1$ is the time index) are received sequentially during learning, growing of fuzzy rules is based on the following two criteria which are distance criterion and the influence of the new added fuzzy rule $N_h + 1$:

$$\begin{aligned} &\|\mathbf{x}_n - \mu_{nr}\| > \varepsilon_n \\ &E_{\text{inf}}(N_h+1) = \|\mathbf{e}_n\| \frac{(1.8\kappa\|\mathbf{x}_n - \mu_{nr}\|)^{N_x}}{\sum_{k=1}^{N_h+1} (1.8\sigma_k)^{N_x}} > e_g, \end{aligned} \tag{10.17}$$

where ε_n, e_g are thresholds to be selected appropriately, $\mathbf{x}_n$ is the latest input data, μ_{nr} is the center of the fuzzy rule nearest to $\mathbf{x}_n$, and e_g is the growing threshold and is chosen according to the desired accuracy of SAFIS. $\mathbf{e}_n = \mathbf{y}_n - \hat{\mathbf{y}}_n$, $\mathbf{y}_n$ is the true value, $\hat{\mathbf{y}}_n$ is the approximated value, κ is an overlap factor that determines the overlap of fuzzy rules in the input space, and ε_n is the distance threshold which decays exponentially and is given by:

$$\varepsilon_n = \max\{\varepsilon_{\max} \times \gamma^n, \varepsilon_{\min}\}, \tag{10.18}$$

where $\varepsilon_{\max}, \varepsilon_{\min}$ are the largest and smallest length of interest and γ is the decay constant. The equation shows that initially it is the largest length of interest in the input space which allows fewer fuzzy rules to coarsely learn the system and then it decreases exponentially to the smallest length of interest in the input space which allows more fuzzy rules to finely learn the system.

10.2.2.2 Allocation of Antecedent and Consequent Parameters

When the new fuzzy rule $N_h + 1$ is added, its corresponding antecedent and consequent parameters are allocated as follows:

$$\begin{cases} \mathbf{a}_{N_h+1} &= \mathbf{e}_n \\ \mu_{N_h+1} &= \mathbf{x}_n \\ \sigma_{N_h+1} &= \kappa\|\mathbf{x}_n - \mu_{nr}\| \end{cases} \tag{10.19}$$

10.2.2.3 Parameter Adjustment

In parameter modification, SAFIS utilizes a winner rule strategy similar to the work done by Huang et al. [8]. The key idea of the winner rule strategy is that only the parameters related to the selected winner rule are updated by the EKF algorithm in every step. The "winner rule" is defined as the rule that is closest (in the Euclidean distance sense) to the current input data as in [8]. As a result, a fast computation is achieved in SAFIS.

The parameter vector existing in all the fuzzy rules is given by:

$$\begin{aligned}\theta_n &= \left[\,\theta_1\,\ldots\,\theta_{nr}\,\ldots\,\theta_{N_h}\,\right]^T\\ &= \left[\,\mathbf{a}_1,\mu_1,\sigma_1,\,\ldots,\,\mathbf{a}_{nr},\mu_{nr},\sigma_{nr},\,\ldots,\,\mathbf{a}_{N_h},\mu_{N_h},\sigma_{N_h}\,\right]^T,\end{aligned} \tag{10.20}$$

where $\theta_{nr} = [\mathbf{a}_{nr},\mu_{nr},\sigma_{nr}]$ is the parameter vector of the nearest fuzzy rule and its gradient is derived as follows:

$$\begin{aligned}\dot{\mathbf{a}}_{nr} &= \frac{\partial\hat{\mathbf{y}}_n}{\partial\mathbf{a}_{nr}} = \frac{\partial\hat{\mathbf{y}}_n}{\partial R_{nr}}\frac{\partial R_{nr}}{\partial\mathbf{a}_{nr}} = \frac{R_{nr}}{\sum_{k=1}^{N_h} R_k}\\ \dot{\mu}_{nr} &= \frac{\partial\hat{\mathbf{y}}_n}{\partial\mu_{nr}} = \frac{\partial\hat{\mathbf{y}}_n}{\partial R_{nr}}\frac{\partial R_{nr}}{\partial\mu_{nr}} = \frac{\mathbf{a}_{nr}-\hat{\mathbf{y}}_n}{\sum_{k=1}^{N_h} R_k}\frac{\partial R_{nr}}{\partial\mu_{nr}}\\ \dot{\sigma}_{nr} &= \frac{\partial\hat{\mathbf{y}}_n}{\partial\sigma_{nr}} = \frac{\partial\hat{\mathbf{y}}_n}{\partial R_{nr}}\frac{\partial R_{nr}}{\partial\sigma_{nr}} = \frac{\mathbf{a}_{nr}-\hat{\mathbf{y}}_n}{\sum_{k=1}^{N_h} R_k}\frac{\partial R_{nr}}{\partial\sigma_{nr}}\\ \frac{\partial R_{nr}}{\partial\mu_{nr}} &= 2R_{nr}\frac{\mathbf{x}_n-\mu_{nr}}{\sigma_{nr}^2}\\ \frac{\partial R_{nr}}{\partial\sigma_{nr}} &= 2R_{nr}\frac{||\mathbf{x}_n-\mu_{nr}||^2}{\sigma_{nr}^3}.\end{aligned} \tag{10.21}$$

After obtaining the gradient vector of the nearest fuzzy rule, that is, $\mathbf{B}_{nr} = [\dot{\mathbf{a}}_{nr},\dot{\mu}_{nr},\dot{\sigma}_{nr}]^T$, EKF is used to update its parameters as follows:

$$\begin{aligned}\mathbf{K}_n &= \mathbf{P}_{n-1}\mathbf{B}_n[\mathbf{R}_n+\mathbf{B}_n^T\mathbf{P}_{n-1}\mathbf{B}_n]^{-1}\\ \theta_n &= \theta_{n-1}+\mathbf{K}_n\mathbf{e}_n\\ \mathbf{P}_n &= [\mathbf{I}_{Z\times Z}-\mathbf{K}_n\mathbf{B}_n^T]\mathbf{P}_{n-1}+q\mathbf{I}_{Z\times Z},\end{aligned} \tag{10.22}$$

where q is a scalar that determines the allowed step in the direction of the gradient vector and Z is the dimension of parameters to be adjusted. When a new rule is added, the dimension of P_n increases to:

$$\begin{pmatrix} \mathbf{P}_{n-1} & 0 \\ 0 & p_0 \mathbf{I}_{Z_1 \times Z_1} \end{pmatrix}, \tag{10.23}$$

where Z_1 is the dimension of the parameters introduced by the newly added rule and p_0 is an initial value of the uncertainty assigned to the newly allocated rule.

10.2.2.4 Removing of a Fuzzy Rule

If the influence of rule k is less than a certain pruning threshold e_p, the rule k is insignificant to the output and should be removed. The pruning threshold e_p is chosen a priori. Given the pruning threshold e_p, rule k will be removed if:

$$E_{\mathrm{inf}}(k) = \|\mathbf{a}_k\| \frac{(1.8\sigma_k)^{N_x}}{\sum\limits_{k=1}^{N_h} (1.8\sigma_k)^{N_x}} < e_\mathrm{p}. \tag{10.24}$$

In SAFIS, only the nearest rule instead of all the existing rules will be considered for removing. This is explained as follows. Considering the Gaussian function $R(x) = \exp(-\frac{x^2}{\sigma^2})$, its first and second derivatives will approach zero much faster when x moves away from zero. Thus, in EKF, the gradient vector of the parameters for all the rules except the nearest rule will approach zero more quickly than those of the nearest rule that are given by:

$$\left(\frac{R_{nr}}{\sum\limits_{k=1}^{N_h} R_k}, \frac{2(\mathbf{a}_{nr} - \hat{\mathbf{y}}_n) R_{nr}}{\sum\limits_{k=1}^{N_h} R_k} \frac{\mathbf{x}_n - \mu_{nr}}{\sigma_{nr}^2}, \frac{2(\mathbf{a}_{nr} - \hat{\mathbf{y}}_n) R_{nr}}{\sum\limits_{k=1}^{N_h} R_k} \frac{\|\mathbf{x}_n - \mu_{nr}\|^2}{\sigma_{nr}^3} \right).$$

In this case, one may only need to adjust parameters of the nearest rule without adjusting the parameters of all rules when a new observation enters and a new rule needs not be added. At the same time, all rules need not be checked for possible pruning. If a new observation arrives and the growing criteria (10.17) is satisfied, a new rule will be added. The existing rules will maintain their influence because their parameters remain unchanged after learning the new observation. Simultaneously, the newly added rule is also influencing, and therefore it is not necessary to check for pruning after a new rule is added. If the growing criteria (10.17) is not satisfied after a new observation arrives, a new rule will not be added and only the parameters of the nearest rule will be modified. As such, only the nearest rule needs to be checked for pruning.

The SAFIS algorithm is summarized below:

Given the growing and pruning thresholds e_g, e_p, for each observation $(\mathbf{x}_n, \mathbf{y}_n)$, where $\mathbf{x}_n \in R^{N_x}$, $\mathbf{y}_n \in R^{N_y}$ and $n = 1, 2, \ldots$, do

1. **compute** the overall system output:

$$\hat{\mathbf{y}}_n = \frac{\sum_{k=1}^{N_h} \mathbf{a}_k R_k(\mathbf{x}_n)}{\sum_{k=1}^{N_h} R_k(\mathbf{x}_n)}$$

$$R_k(\mathbf{x}_n) = \exp\left(-\frac{1}{\sigma_k^2}\|\mathbf{x}_n - \mu_k\|^2\right) \tag{10.25}$$

where N_h is the number of fuzzy rules.

2. **calculate** the parameters required in the growth criterion:

$$\varepsilon_n = \max\{\varepsilon_{\max}\gamma^n, \varepsilon_{\min}\}, (0 < \gamma < 1)$$

$$\mathbf{e}_n = \mathbf{y}_n - \hat{\mathbf{y}}_n \tag{10.26}$$

3. **apply** the criterion for adding rules:

If $\|\mathbf{x}_n - \mu_{nr}\| > \varepsilon_n$ and $E_{\inf}(N_h+1) = \|\mathbf{e}_n\| \frac{(1.8\kappa\|\mathbf{x}_n - \mu_{nr}\|)^{N_x}}{\sum_{k=1}^{N_h+1}(1.8\sigma_k)^{N_x}} > e_g$

allocate a new rule $N_h + 1$ with

$$\mathbf{a}_{N_h+1} = \mathbf{e}_n$$

$$\mu_{N_h+1} = \mathbf{x}_n$$

$$\sigma_{N_h+1} = \kappa\|\mathbf{x}_n - \mu_{nr}\| \tag{10.27}$$

Else

adjust the system parameters $\mathbf{a}_{nr}$, μ_{nr}, σ_{nr} for the nearest rule only by using the EKF method.

check the criterion for pruning the rule:

If $E_{\inf}(nr) = \|\mathbf{a}_{nr}\| \frac{(1.8\sigma_{nr})^{N_x}}{\sum_{k=1}^{N_h}(1.8\sigma_k)^{N_x}} < e_p$

remove the nr-th rule

reduce the dimensionality of EKF

Endif

Endif

10.2.3 Selecting of Predefined Parameters

In SAFIS, some parameters need to be decided in advance according to the problems considered. They include the distance thresholds ($\varepsilon_{\max}$, $\varepsilon_{\min}$, γ), the overlap factor (κ) for determining the width of the newly added rule, the growing threshold (e_g)

Table 10.1 Effects of parameter e_g on system performance (number of rules and the testing RMS error) under different ε_{max} values and $\kappa = 1.0$

$\varepsilon_{max} \backslash e_g$	0.001	0.005	0.01	0.05
1.0	(61, 0.0198)	(17, 0.0385)	(11, 0.0535)	(2, 0.0912)
5.0	(45, 0.0233)	(17, 0.0385)	(11, 0.0535)	(2, 0.0912)
10.0	(41, 0.0249)	(14, 0.0386)	(9, 0.0461)	(2, 0.0912)

Table 10.2 Effects of parameter e_g on system performance (number of rules and the testing RMS error) under different κ values and $\varepsilon_{max} = 10.0$

$\kappa \backslash e_g$	0.001	0.005	0.01	0.05
1.0	(41, 0.0249)	(14, 0.0386)	(9, 0.0461)	(2, 0.0912)
1.5	(50, 0.0350)	(18, 0.0586)	(15, 0.0598)	(3, 0.1382)
2.0	(52, 0.0557)	(25, 0.0902)	(15, 0.1384)	(3, 0.1391)

for a new rule, and the pruning threshold (e_p) for removing an insignificant rule. Based on the observation from many experiments, a general selection procedure for the predefined parameters is given as follows: ε_{max} is set to around the upper bound of input variables; ε_{min} is set to around 10% of ε_{max}; γ is set to around 0.99; and e_p is set to around 10% of e_g. The overlap factor (κ) is utilized to initialize the width of the newly added rule and chosen according to different problems. κ is suggested to be chosen in the range $[1.0, 2.0]$. The growing threshold e_g is chosen according to the system performance. The smaller e_g, the better the system performance, but the resulting system structure is more complex.

An example is given to illustrate the effects of the parameters (e_g, κ, ε_{max}) on the system structure and performance. Consider the following two-dimension sinc function:

$$z = \text{sinc}(x, y) = \frac{\sin(x)\sin(y)}{xy}. \tag{10.28}$$

In the simulation, 2,500 training data pairs (x, y) are drawn from the input range $[-10, 10] \times [-10, 10]$. At the same time, 100 testing data pairs (x, y) are drawn from the same input range.

The general rule for choosing the parameters ($\varepsilon_{min}, \gamma, e_p$) are obeyed. ε_{min} is set to 10% of ε_{max}; γ is set to 0.997; and e_p is set to the 10% of e_g. The parameters e_g, ε_{max}, and κ are observed in the range $[0.001, 0.05]$, $[1.0, 10.0]$, and $[1.0, 2.0]$, respectively, to illustrate their effect on the resulting system structure and testing accuracy. Tables 10.1 and 10.2 give the effects of parameter e_g on system performance in terms of number of rules and the testing RMS error under different κ or ε_{max} values. From the two tables, it is easy to find that with the increase of e_g the number of rules is decreased and also system performance (testing RMS error) becomes worse with the same κ or ε_{max} value. Furthermore, it can be found from the two tables that the resulting system structure and testing accuracy have no very big change when the parameter κ or $\varepsilon_{\max}$ appears different values. However, these parameters are problem dependent and need to be determined according to the problem considered. Besides the above guidelines for setting the parameters, the optimal parameters can be determined using search techniques like GA for some complex problems in the future work.

10.3 Performance Evaluation of SAFIS

In this section, the performance of SAFIS is evaluated based on two nonlinear system identification problems and one chaotic time-series (Mackey-Glass) prediction problem. For the first system identification problem, performance of SAFIS is compared with other well-known sequential algorithms such as MRAN [33], RANEKF [14], eTS [3], Simpl_eTS [1], and hybrid algorithm (HA) [30]. For the second system identification problem performance of SAFIS is compared with MRAN [33], RANEKF [14], eTS [3], Simpl_eTS [1], and SONFIN [13]. For the chaotic time-series prediction problem, the comparison is done with MRAN [33], RANEKF [14], eTS [3], and Simpl_eTS [1]. In all the studies, the parameters (r, Ω) for eTS and Simpl_eTS where r is the distance and Ω is the least-square error parameter [1, 3] are tuned to obtain the best performance.

Performance comparison is done in terms of accuracy and the complexity (the number of rules) of the fuzzy system. For these problems, the SAFIS algorithm goes through the training data sequentially in a single pass and builds up the fuzzy inference system by adding and removing the rules along with their parameters. Then, its performance is evaluated on the unseen test data.

10.3.1 Nonlinear Dynamic System Identification

Generally, a wide class of MIMO nonlinear dynamic systems can be represented by the nonlinear discrete model with an input–output description form:

$$\mathbf{y}(n) = \mathbf{f}[\mathbf{y}(n-1), \mathbf{y}(n-2), \ldots, \mathbf{y}(n-k+1); \mathbf{u}(n), \mathbf{u}(n-1), \ldots, \mathbf{u}(n-p+1)], \tag{10.29}$$

where $\mathbf{y}$ is a vector containing N_y system outputs, $\mathbf{u}$ is a vector for N_u system inputs, $\mathbf{f}$ is a nonlinear vector function, representing N_y hypersurfaces of the system, and k and p are the maximum lags of the output and input, respectively.

Selecting $[\mathbf{y}(n-1), \ldots, \mathbf{y}(n-k+1); \mathbf{u}(n), \mathbf{u}(n-1), \ldots, \mathbf{u}(n-p+1)], \mathbf{y}(n)$ as the fuzzy system's input–output $\mathbf{x}_n, \mathbf{y}_n$ at time n, the above equation can be put as:

$$\mathbf{y}_n = \mathbf{f}(\mathbf{x}_n). \tag{10.30}$$

The SAFIS algorithm is used to approximate **f** such that:

$$\hat{\mathbf{y}}_n = \hat{\mathbf{f}}(\mathbf{x}_n), \tag{10.31}$$

and the error between the system output $\mathbf{y}_n$ and the output of SAFIS $\hat{\mathbf{y}}_n$, $\|\mathbf{y}_n - \hat{\mathbf{y}}_n\|$ is minimized.

Narendra and Parthasarathy [22] have suggested two special forms of the nonlinear system model given in (10.32) and (10.33).

Model I:

$$y(n+1) = f[y(n), y(n-1), \ldots, y(n-k+1)] + \sum_{i=0}^{p-1} \beta_i u(n-i), \tag{10.32}$$

where β_i is the constant unknown parameter.

Model II:

$$y(n+1) = f[y(n), y(n-1), \ldots, y(n-k+1)] + g[u(n), u(n-1), \ldots, u(n-p+1)]. \tag{10.33}$$

These two models of nonlinear systems have been used here for performance comparison.

Selecting $[y(n), y(n-1), \ldots, y(n-k+1), u(n), u(n-1), \ldots, u(n-p+1)]$, and $y(n+1)$ as the input–output of SAFIS, the identified model is given by this equation:

$$\hat{y}(n+1) = \hat{f}(y(n), y(n-1), \ldots, y(n-k+1), u(n), u(n-1), \ldots, u(n-p+1)), \tag{10.34}$$

where $\hat{f}$ is the SAFIS approximation and $\hat{y}(n+1)$ is the output of the SAFIS.

10.3.1.1 Identification Problem 1

The first nonlinear dynamic system to be identified represents model I and is described by Wang and Yen [30]:

$$y(n) = \frac{y(n-1)y(n-2)(y(n-1)-0.5)}{1+y^2(n-1)+y^2(n-2)} + u(n-1). \tag{10.35}$$

The equilibrium state of the unforced system given by (10.35) is $(0,0)$. As in [30], the input $u(n)$ is uniformly selected in the range $[-1.5, 1.5]$ and the test input $u(n)$ is given by $u(n) = \sin(2\pi n/25)$; 5,000 and 200 observation data are produced for the purpose of training and testing. The different parameter values for SAFIS are chosen as follows: $\gamma = 0.997, \varepsilon_{\max} = 1.0, \varepsilon_{\min} = 0.1, \kappa = 1.0, e_g = 0.05$, and $e_p = 0.005$.

The average performance comparison of SAFIS with MRAN, RANEKF, eTS, Simpl_eTS, and HA is shown in Table 10.3 based on 50 experimental trials. From the table, it can be seen that SAFIS obtains similar testing accuracy compared

Table 10.3 Results of nonlinear identification problem 1

Methods	No. of rules	Training RMSE	Testing RMSE
SAFIS	17	0.0539	0.0221
MRAN	22	0.0371	0.0271
RANEKF	35	0.0273	0.0297
Simpl_eTS (r = 2.0, $\Omega = 10^6$)	22	0.0528	0.0225
eTS (r = 1.8, $\Omega = 10^6$)	49	0.0292	0.0212
HA [30]	28	0.0182	0.0244

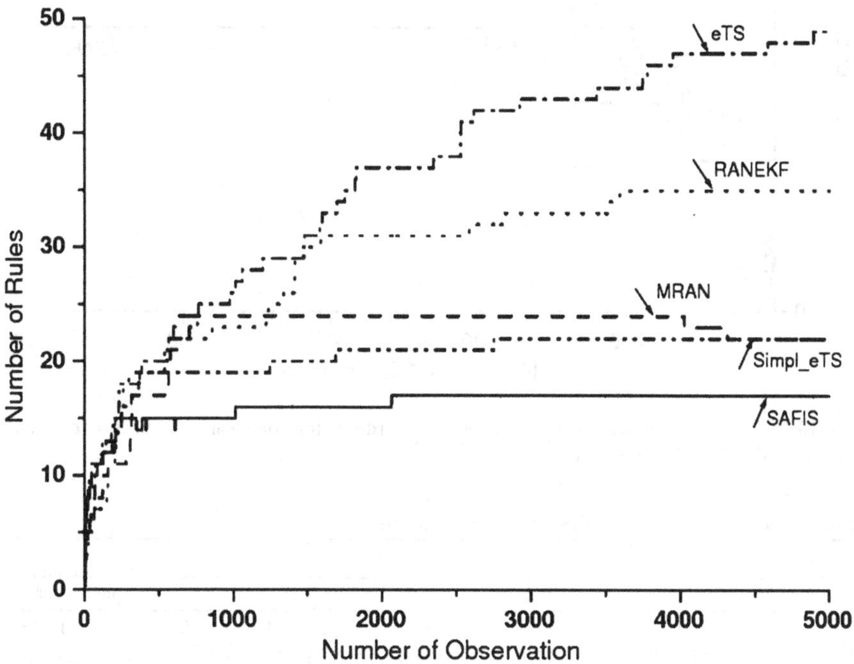

Fig. 10.2 Rule update process between different algorithms for nonlinear identification problem 1 during the whole observation

to MRAN, RANEKF, eTS, Simpl_eTS, and HA. However, SAFIS achieves this accuracy with smallest number of rules. It is worth noting that HA is based on GA iterative learning and is not sequential. The evolution of the fuzzy rules for SAFIS, MRAN, RANEKF, eTS, and Simpl_eTS for a typical run is shown by Fig. 10.2. It can be seen from the figure that SAFIS produces least number of rules. Besides, Fig. 10.3 gives a clear illustration for the rule evolution tendency between 0 and 1,000 observation and shows that SAFIS can automatically add and delete a rule during learning, which is manifested by increasing and reducing the number of rules by one. The fuzzy rules for the typical run are listed in Table 10.4, where G(.) represents the Gaussian membership function. The first and second values in G(.) indicate the center and the width of the Gaussian function, respectively.

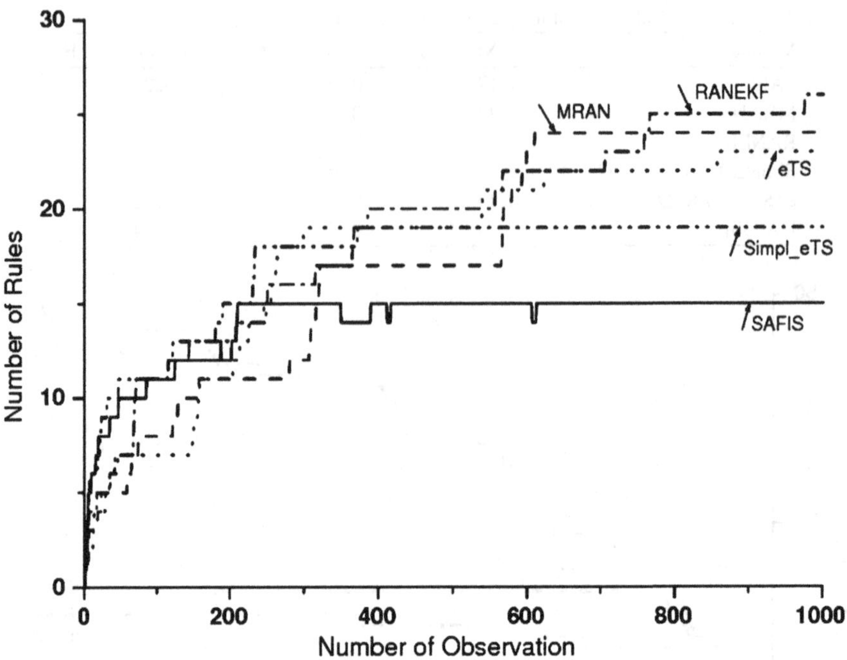

Fig. 10.3 Rule update process between different algorithms for nonlinear identification problem 1 between 0 and 1,000 observations

Table 10.4 Fuzzy rules of SAFIS for nonlinear identification problem 1

No. of rules	Antecedent parameters			Consequent parameters
	y(n-1)	y(n-2)	u(n-1)	
1	G(0.6883,0.8504)	G(0.3062,0.8504)	G(−2.0115,0.8504)	a = −1.6137
2	G(−1.6117,0.9709)	G(1.0091,0.9709)	G(1.7480,0.9709)	a = 2.5292
3	G(−0.2325,1.1461)	G(−1.5054,1.1461)	G(0.2722,1.1461)	a = 0.5625
4	G(−0.1653,1.1633)	G(−1.0712,1.1633)	G(1.9589,1.1633)	a = 1.5430
5	G(1.8338,1.2341)	G(0.3378,1.2341)	G(1.2441,1.2341)	a = 2.0280
6	G(1.5042,1.3481)	G(−0.5239,1.3481)	G(0.0087,1.3481)	a = −0.4277
7	G(0.3110,0.9829)	G(−1.0793,0.9829)	G(1.7363,0.9829)	a = 1.9069
8	G(−0.8126,0.6423)	G(−1.2611,0.6423)	G(−0.5928,0.6423)	a = −1.1355
9	G(−0.6152,0.9283)	G(−2.0362,0.9283)	G(−1.4239,0.9283)	a = −1.4374
10	G(−1.3413,0.7751)	G(−1.0834,0.7751)	G(−1.8843,0.7751)	a = −2.4472
11	G(1.8475,1.1035)	G(−1.0128,1.1035)	G(−1.4383,1.1035)	a = −2.1617
12	G(0.7468,0.7356)	G(2.2865,0.7356)	G(1.4947,0.7356)	a = 1.8152
13	G(−2.3833,1.9263)	G(−1.8191,1.9263)	G(−1.0221,1.9263)	a = −2.9007
14	G(−0.4007,1.7921)	G(1.9986,1.7921)	G(−1.6721,1.7921)	a = −2.0982
15	G(−1.6354,1.8484)	G(1.8161,1.8484)	G(0.1940,1.8484)	a = 1.6630
16	G(−2.8360,1.9440)	G(−2.0148,1.9440)	G(1.0852,1.9440)	a = 0.9242
17	G(1.7938,0.9549)	G(1.4088,0.9549)	G(−0.3515,0.9549)	a = 0.2905

Table 10.5 Results of nonlinear identification problem 2

Methods	No. of rules	Testing RMSE
SAFIS	8	0.0116
MRAN	10	0.0129
RANEKF	11	0.0184
Simpl_eTS (r = 0.075, $\Omega = 10^6$)	18	0.0122
eTS (r = 1.0, $\Omega = 10^6$)	19	0.0082
SONFIN [13]	10	0.0130

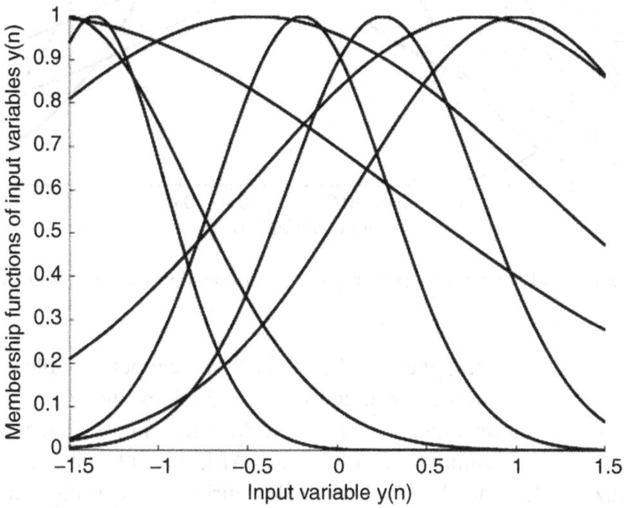

Fig. 10.4 Membership functions of input variable $y(n)$ for nonlinear identification problem 2

10.3.1.2 Identification Problem 2

The second nonlinear dynamic system to be identified represents model II and is described by Juang and Lin [13]:

$$y(n+1) = \frac{y(n)}{1+y^2(n)} + u^3(n). \tag{10.36}$$

In accordance with [13], the input signal $u(n)$ is given by $\sin(2\pi n/100)$; 50,000 and 200 observation data are produced for the purpose of training and testing. The SAFIS parameter values chosen are as follows: $\gamma = 0.997, \varepsilon_{\max} = 2.0, \varepsilon_{\min} = 0.2, \kappa = 2.0, e_g = 0.03$, and $e_p = 0.003$. The input variables $y(n), u(n)$, respectively, follow the uniform sample distribution in the range $[-1.5, 1.5]$ and $[-1.0, 1.0]$.

Table 10.5 shows the performance comparison of SAFIS with MRAN, RANEKF, eTS, Simpl_eTS, and SONFIN [13]. It can be seen from the table that SAFIS achieves similar accuracy with a lesser number of rules. Figures 10.4 and 10.5 show the final membership functions of input variables $y(n), u(n)$ achieved by

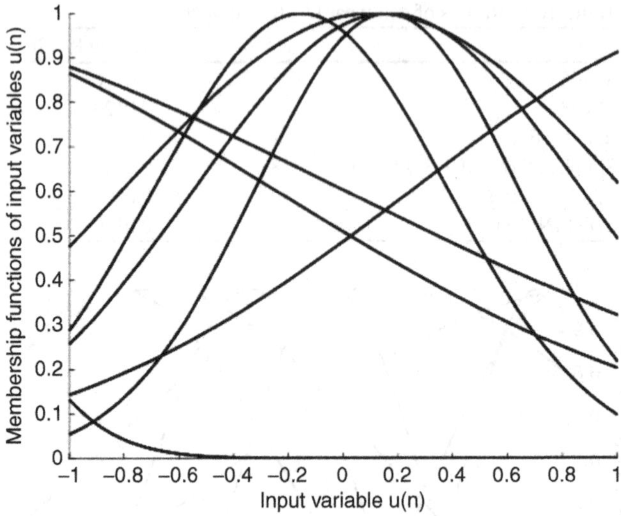

Fig. 10.5 Membership functions of input variable $u(n)$ for nonlinear identification problem 2

SAFIS. From the two figures, one can clearly see that the input variable membership functions are distributed in their own entire range. Besides, the testing accuracy of SAFIS is slightly better than those of MRAN, RANEKF, Simpl_eTS, and SONFIN, which verifies that the learning performance of SAFIS is not lost by only modifying the nearest fuzzy rule instead of all fuzzy rules during the learning. The evolution of the fuzzy rules for SAFIS, MRAN, RANEKF, eTS, and Simpl_eTS is shown by Fig. 10.6. It can be seen from the figure that SAFIS is able to add and delete rules during learning and produces least number of rules. The details of the fuzzy rules are depicted in Table 10.6.

10.3.2 Mackey-Glass Time-Series Prediction

In this example, the SAFIS is applied to predict complex time series, a special function approximation problem. The time-series prediction is very important in solving real-world problems such as the detection of arrhythmia in heartbeats. The chaotic Mackey-Glass time series is recognized as one of the time series benchmark problems, which is generated from the following differential equation [2]:

$$\frac{dx(t)}{dt} = \frac{0.2x(t-\tau)}{1+x^{10}(t-\tau)} - 0.1x(t), \tag{10.37}$$

where $\tau = 17$ and $x(0) = 1.2$. For the purpose of training and testing, 6,000 samples are produced by means of the fourth-order Runge-Kutta method with the step size 0.1. The prediction task is to predict the value $x(t+85)$ from the input

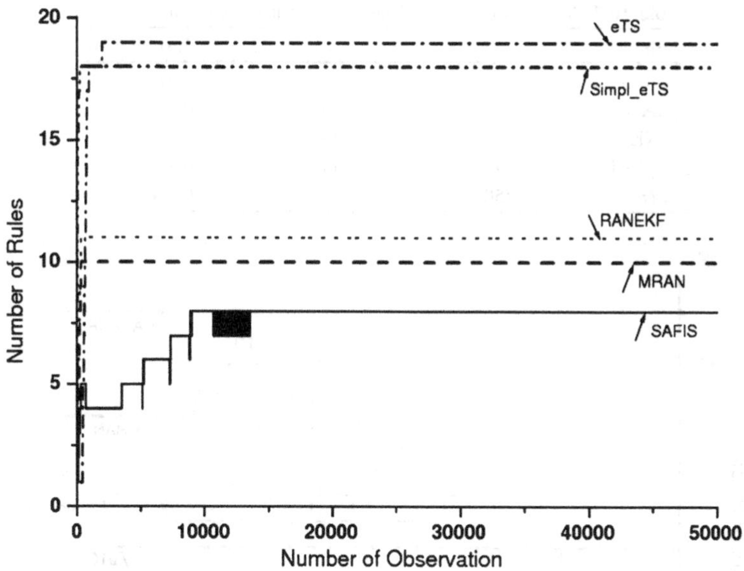

Fig. 10.6 Rule update process between different algorithms for nonlinear identification problem 2

Table 10.6 Fuzzy rules of SAFIS for nonlinear identification problem 2

No. of rules	Antecedent parameters		Consequent parameters
	y(n)	u(n)	
1	G(−0.9729, 1.0670)	G(−1.2625, 1.0670)	a = −6.3086
2	G(0.2731,1.4623)	G(1.9572,1.4623)	a = 4.5472
3	G(0.3706,0.8085)	G(−0.0884, 0.8085)	a = 3.1725
4	G(−0.1623, 1.1271)	G(0.1100,1.1271)	a = −6.4255
5	G(−0.8399, 1.1393)	G(−0.5162, 1.1393)	a = 4.6764
6	G(1.9988,2.1081)	G(1.8151,2.1081)	a = 1.8266
7	G(1.4377,2.3607)	G(1.2834,2.3607)	a = 2.2549
8	G(1.2992,0.8117)	G(0.3417,0.8117)	a = −1.8183

vector $[x(t-18)\ \ x(t-12)\ \ x(t-6)\ \ x(t)]$ for any value of the time t. As in [2], the observations between $t = 201$ and $t = 3{,}200$ and the observations between $t = 5{,}001$ and $t = 5{,}500$ are extracted from the series and used as training and testing data. For this problem, the parameters for SAFIS are selected as follows: $\gamma = 0.98, \varepsilon_{\max} = 1.6, \varepsilon_{\min} = 0.16, \kappa = 1.68, e_g = 0.0005$, and $e_p = 0.00005$. The data follow a uniform sample distribution in the range $[0.4, 1.4]$.

Table 10.7 shows the prediction accuracies and the number of rules obtained by SAFIS, MRAN, RANEKF, eTS, and Simpl_eTS. For comparison purposes, the prediction accuracy is based on the non-dimensional error index (NDEI) defined as the RMSE divided by the standard deviation of the true output values. As observed

Table 10.7 Results of Mackey-Glass time-series prediction

Methods	No. of rules	Testing NDEI
SAFIS	6	0.376
MRAN	14	0.375
RANEKF	18	0.378
Simpl_eTS($r = 0.25$, $\Omega = 750$) [1]	11	0.394
eTS($r = 0.25$, $\Omega = 750$) [2]	9	0.380

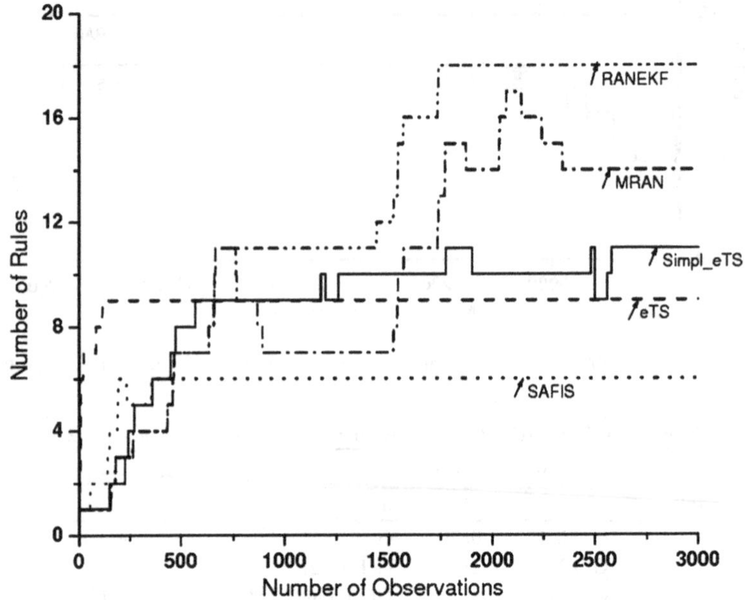

Fig. 10.7 Rule update process between different algorithms for Mackey-Glass time-series prediction

from Table 10.7, all the algorithms produce similar accuracies; however, SAFIS obtains the smallest number of fuzzy rules. The evolution of the fuzzy rules for SAFIS, MRAN, RANEKF, eTS, and Simpl_eTS is shown in Fig. 10.7.

10.4 Summary

In this chapter, a sequential fuzzy inference system called SAFIS is presented to automatically construct a fuzzy inference system using the training data during the learning process. Specifically, SAFIS algorithm implements the structure identification and parameter adjustment for a fuzzy inference system using the ideas from GAP-RBF algorithm. SAFIS algorithm utilizes the influence of a fuzzy rule to add and remove the fuzzy rules during learning. At the same time, the SAFIS algorithm utilizes the EKF to update the parameters of the nearest rule instead of all the rules without losing the approximation performance. Its performance has been evaluated by some function approximation benchmark problems including two nonlinear system identification problems and the Mackey-Glass time-series prediction problem. The simulation results from these benchmark problems show that, compared with other algorithms, SAFIS produces similar or better testing accuracies with lesser number of rules.

However, for large systems, EKF algorithm used in the parameter update equation increases the computation burden. Also, the calculation of rule influence requires uniform distribution of the input data, and this may degrade the performance. Further studies in these directions are required in the future.

References

1. Angelov, P., Filev, D.: Simpl_eTS: A simplified method for learning evolving Takagi-Sugeno fuzzy models. In: The 14th IEEE International Conference on Fuzzy Systems, pp. 1068–1073 (2005)
2. Angelov, P., Victor, J., Dourado, A., Filev, D.: On-line evolution of Takagi-Sugeno fuzzy models. In: Proceedings of the 2nd IFAC Workshop on Advanced Fuzzy/Neural Control, pp. 67–72. Oulu, Finland (2004)
3. Angelov, P.P., Filev, D.P.: An approach to online identification of Takagi-Sugeno fuzzy models. IEEE Transactions on Systems, Man, and Cybernetics-Part B: Cybernetics **34**(1), 484–498 (2004)
4. Chiu, S.L.: Selecting input variables for fuzzy models. Journal of Intelligent and Fuzzy Systems **4**, 243–256 (1996)
5. Cho, K.B., Wang, B.H.: Radial basis function based adaptive fuzzy systems and their applications to system identification and prediction. Fuzzy Sets and Systems **83**, 325–339 (1996)
6. Gopal, S., Karthikeyan, B., Kavitha, D.: Partial discharge pattern classification using fuzzy expert system. In: Proceedings of the 2004 IEEE International Conference on Solid Dielectrics, pp. 653–656. Toulouse, France (2004)
7. Gupta, M.M., Rao, D.H.: On the principles of fuzzy neural networks. Fuzzy Sets and Systems **61**(1), 1–18 (1994)
8. Huang, G.B., Saratchandran, P., Sundararajan, N.: An efficient sequential learning algorithm for growing and pruning RBF (GAP-RBF) networks. IEEE Transactions on Systems, Man, Cybernetics-Part B: Cybernetics **34**(6), 2284–2292 (2004)
9. Huang, G.B., Saratchandran, P., Sundararajan, N.: A generalized growing and pruning RBF (GGAP-RBF) neural network for function approximation. IEEE Transactions on Neural Networks **16**(1), 57–67 (2005)
10. Iqdour, R., Zeroual, A.: A rule based fuzzy model for the prediction of daily solar radiation. In: 2004 IEEE International Conference on Industrial Technology (ICIT), pp. 1482–1487. Hammamet, Tunisia (2004)
11. Jang, J.S.R.: ANFIS: Adaptive-network-based fuzzy inference system. IEEE Transactions on Systems, Man, and Cybernetics **23**(3), 665–685 (1993)
12. Jang, J.S.R., Sun, C.T.: Functional equivalence between radial basis function networks and fuzzy inference systems. IEEE Transactions on Neural Networks **4**(1), 156–159 (1993)
13. Juang, C.F., Lin, C.T.: An on-line self-constructing neural fuzzy inference network and its applications. IEEE Transactions on Fuzzy Systems **10**(2), 144–154 (2002)

14. Kadirkamanathan, V., Niranjan, M.: A function estimation approach to sequential learning with neural networks. Neural Computation **5**(6), 954–975 (1993)
15. Kalhor, A., Araabi, B.N., Lucas, C.: An online predictor model as adaptive habitually linear and transiently nonlinear model. Evolving Systems **1**, 29–41 (2010)
16. Kandel, A.: Fuzzy expert systems. Boca Raton, FL CRC Press (1992)
17. Kasaov, N.K., Song, Q.: DENFIS: Dynamic evolving neural-fuzzy inference system and its application for time series prediction. IEEE Transactions on Fuzzy Systems **10**(2), 144–154 (2002)
18. Konjic, T., Miranda, V., Kapetanovic, I.: Prediction of LV substation load curves with fuzzy inference systems. In: Proceedings of the 5th International Conference on Probabilistic Methods Applied to Power Systems, pp. 129–134. Ames, Iowa (2004)
19. Leng, G., McGinnity, T.M., Prasad, G.: An approach for on-line extraction of fuzzy rules using a self-organising fuzzy neural network. Fuzzy Sets and Systems **150**(2), 211–243 (2005)
20. Mamdani, E.H., Assilian, S.: An experiment in linguistic synthesis with a fuzzy logic controller. International Journal of Man-Machine Studies **7**(1), 1–13 (1975)
21. Mitra, S., Hayashi, Y.: Neuro-fuzzy rule generation: Survey in soft computing framework. IEEE Transactions on Neural Networks **11**(3), 748–768 (2000)
22. Narendra, K.S., Parthasarathy, K.: Identification and control of dynamical systems using neural networks. IEEE Transactions on Neural Networks **1**(1), 4–27 (1990)
23. Olej, V., Krupka, J.: Prediction of gross domestic product development by Takagi-Sugeno fuzzy inference systems. In: Proceedings of the 5th International Conference on Intelligent Systems Design and Applications (ISDA'05), pp. 186–191. Wroclaw, Poland (2005)
24. Pedrycz, W.: Fuzzy Control and Fuzzy Systems. New York: Wiley (1993)
25. Pedrycz, W., Rocha, A.F.: Fuzzy-set based models of neurons and knowledge-based networks. IEEE Transactions on Fuzzy Systems **1**(4), 254–266 (1993)
26. Platt, J.: A resource allocating network for function interpolation. Neural Computation **3**(2), 213–225 (1991)
27. Rubio, J.J.: SOFMLS: Online self-organizing fuzzy modified least-squares network. IEEE Transactions on Fuzzy Systems **17**(6), 1296–1309 (2009)
28. Soleimani, H., Lucas, C., Araabi, B.N.: Recursive Gath-Geva clustering as a basis for evolving neuro fuzzy modeling. Evolving Systems **1**, 59–71 (2010)
29. Tagaki, T., Sugeno, M.: Fuzzy identification of systems and its application to modelling and control. IEEE Transactions on Systems, Man, and Cybernetic **15**(1), 116–132 (1985)
30. Wang, L., Yen, J.: Extracting fuzzy rules for system modeling using a hybrid of genetic algorithm and Kalman Filter. Fuzzy Sets and Systems **101**, 353–362 (1999)
31. Wei, W., Mendel, J.M.: A fuzzy logic method for modulation classification in nonideal environments. IEEE Transactions on Fuzzy Systems **7**(3), 333–344 (1999)
32. Wu, S., Er, M.J.: Dynamic fuzzy neural networks - a novel approach to function approximation. IEEE Transactions on Systems, Man, and Cybernetics-Part B: Cybernetics **30**(2), 358–364 (2000)
33. Yingwei, L., Sundararajan, N., Saratchandran, P.: A sequential learning scheme for function approximation using minimal radial basis neural networks. Neural Computation **9**(2), 461–478 (1997)

Chapter 11
Interval Approach for Evolving Granular System Modeling

Daniel Leite, Pyramo Costa, and Fernando Gomide

Abstract Physical systems change over time and usually produce considerable amount of nonstationary data. Evolving modeling of time-varying systems requires adaptive and flexible procedures to deal with heterogeneous data. Granular computing provides a rich framework for modeling time-varying systems using nonstationary granular data streams. This work considers interval granular objects to accommodate essential information from data streams and simplify complex real-world problems. We briefly discuss a new class of problems emerging in data stream mining where data may be either singular or granular. Particularly, we emphasize interval data and interval modeling framework. Interval-based evolving modeling (IBeM) approach recursively adapts both parameters and structure of rule-based models. IBeM uses ∪-closure granular structures to approximate functions. In general, approximand functions can be time series, decision boundaries between classes, control, or regression functions. Essentially, IBeM accesses data sequentially and discards previous examples; incoming data may trigger structural adaptation of models. The IBeM learning algorithm evolves and updates rules quickly to track system and environment changes. Experiments using heterogeneous streams of meteorological and financial data are performed to show the usefulness of the IBeM approach in actual scenarios.

D. Leite (✉) • F. Gomide
University of Campinas, School of Electrical and Computer Engineering, Sao Paulo, Brazil
e-mail: danfl7@dca.fee.unicamp.br; gomide@dca.fee.unicamp.br

P. Costa
Pontifical Catholic University of Minas Gerais, Graduate Program in Electrical Engineering, Belo Horizonte, Brazil
e-mail: pyramo@pucminas.br

M. Sayed-Mouchaweh and E. Lughofer (eds.), *Learning in Non-Stationary Environments: Methods and Applications*, DOI 10.1007/978-1-4419-8020-5_11,
© Springer Science+Business Media New York 2012

11.1 Introduction

Measurements and expert estimates are never exact [1]. Novel technologies have created problems in which uncertainty, nonlinearity, nonstationarity, and complexity are crucial. Adaptive modeling from data streams, with minimal or no supervision, maximally exploits the information flow in dynamic environments.

Data stream modeling for knowledge discovery has recently become an important topic in various research areas. Modeling efforts are driven to processing continuously incoming examples from quickly changing, heterogeneous, nonstationary, and endless flows of data. Data-stream-oriented adaptive models receive examples one at a time and are constrained by the impossibility of storing previous examples. Fundamentally, recursive algorithm scans unbounded streaming dataset only once and should deliver models and results on demand; algorithmic procedures must account for the fact that the unknown is likely to matter. Classification, clustering, prediction, frequent pattern mining, regression, and control are examples of problems addressed in context. Essentially, neither the time complexity of structural adaptation of models, nor memory usage should scale with the number of streaming examples.

Recent research on evolving granular systems [2–8] relies on the concepts of granular view, information granule, and granular mapping in the process of modeling streaming data. Emphasis is on the tasks of data granulation and computing with granules [9–12]. The granularity of information explicitly embedded into granular systems offers key features in dynamic modeling, for example, transparency and flexibility. Concept change, missing and noisy values, and superfluous and outlier instances are common in online environment and require automatic intervention. Particularly, structured representation of data flows is a key contribution. By structured representation, we mean a collection of rules that tells the very essence of the data.

Information granulation splits a problem into simpler subproblems. In this work, the quotient structure of such a granulation process is a granular model of an evolving system built from a repertoire of data mining and machine learning procedures. Constructing granular models of large spatiotemporal data sets requires choosing a computational framework to return a proper granulation and draw conclusions useful for practical purposes. Evolvable granular models may be expressed in the framework of interval mathematics, statistics, fuzzy sets, rough sets, shadow sets, cluster analysis, decision trees, neighborhood systems, or hybrids. This concedes ample freedom in electing representative granular objects and handling tools. Regardless of the framework chosen, granulation aims to retain the essence of original streaming data and reveal local models. Computing with granules aims at looking to the data under different resolutions (shift back and forth between simpler and more detailed views of data) and extract from it features of interest to attain efficient and practical solution.

This work suggests multidimensional intervals (axis-aligned hyper-boxes) as formal granular object to wrap uncertainty in data stream. Features that make interval representation attractive include: (1) easiness of acquiring parameters. Only two parameters related with the real features (upper and lower bounds) need to be captured. These are not cognitively complex and appear straight from data flows; (2) adaptation of intervals demands basic fully formalized operations of interval arithmetic; (3) intervals make no specific assumption about the content of an information granule. This means that intervals do not require assumptions on probability distributions, membership functions, belief intervals, or possibility values—intervals are everything we wish to know from large amounts of data; (4) interval data is common in practice. Moreover, interval model has a great deal of appeal to represent counterpart interval data. Intervals may also rise after preprocessing numeric data, that is, by comprising it into a smaller set of granular data; (5) intervals can be translated quite easily to linguistic propositions. Interval precision facilitates comprehension when supported by a context.

Interval-based evolving modeling (IBeM) [7, 8] considers heterogeneous (singular and granular parts) streaming data, one-pass recursive learning algorithm, and monotonic interval inclusion functions associated with hyperrectangle-like forms of information granules to provide singular and granular approximations of nonstationary functions. Essentially, an interval evolving model self-adapts its structure when new concepts appear in data streams. Here, model structure means interval-type information granules, IF-THEN rules, and a concept. IBeM algorithms accumulate values associated to granules and rules. Granulation eases incremental updating and discovering of the essence of the structure of the data with modest storage and processing costs. Experts usually prefer models that approximate physical system outputs and provide estimates of the approximation bounds. Building intervals in bounded-error context is the IBeM approach for enveloping uncertainty.

The contribution of this work over [7] and [8] is twofold. First, we extend the IBeM approach to deal with augmented nonnumeric type of data. Second, we employ a preprocessing time-granulation step in the algorithm. Time granulation aims at synchronizing concurrent data flows, possibly from different sources, incoming at random time intervals. We examine both, spatial and temporal aspects of data stream processing, from a granular perspective.

Next section introduces a granulation approach for time and space events. Granulation of time and space leads to temporal and spatial granularities. We argue that granular framework better supports modeling of manifold heterogeneous data. Sections 11.3 and 11.4 address the formalization of concepts of interval mathematics, and data stream modeling using interval representation. A data-stream-driven recursive learning algorithm capable of operating in online environment and dealing with nonsynchronized heterogeneous data is suggested in Sect. 11.5. Section 11.6 provides detailed analysis of the behavior of the IBeM granular approach in different application domains, particularly, meteorology and finance. Section 11.7 concludes the paper and lists research issues for further investigation.

11.2 Granular Data and Systems

Information granules are conceptual objects that catch the essence of the overall data in a concise and explainable manner [13, 14]. Granules may be interpreted from two points of view. From the uncertainty theory, a granule is a unit lacking precise knowledge. From the knowledge engineering, a granule is a unit of elementary knowledge. Granular computing is intended to identifying manifestations of granules through moving back and forth among granularities to afford more or less differentiation. Too much detail is wasteful, and too little renders a system useless.

Data and systems can be granular. Sometimes, data are not realized with full precision but are subjectively noticed as linguistic terms, fuzzy numbers, and intervals. Sometimes, it is hard to discriminate numeric (singular) data precisely, and we are compelled to consider granules. Systems are better supported by granular framework to suit granular data. Singular data is a particular case in which a granule degenerates into a singleton. The necessity of building systems in finer granularities, close to the singularity, justifies only when there are clear benefits on doing so.

Streaming data in online environment can be granular from different perspectives. A more intuitive perspective concerns data that is granular by itself. To elaborate on this approach, consider a simple example of predicting variable y from the last available observation x. This leads us to search for an approximand p to describe the process function f based on pairs (x, y). In this example, instances x and y and function f are both singular. Singular data does not restrain models to be singular but rather a granular system may use granular models whose size and placement reflect the information carried by singular data. A hypothesis is that granular representation helps to assess the structure of detailed singular data and organizes the data into an interpretable quotient structure.

Consider $x = [\underline{x}, \overline{x}]$ and $y = [\underline{y}, \overline{y}]$ as instances of granular data stream and intervals in this case. To exemplify, $\underline{x}$ and $\overline{x}$ may denote the minimum and maximum price of an economical index during a day, and $\underline{y}$ and $\overline{y}$, the range of fluctuation of the price in the next day. In this example, data is originally granular, the process function $f = [\underline{f}, \overline{f}]$ is also originally granular, and models $p = [\underline{p}, \overline{p}]$ must be granular to support granular data and granular approximation of f. Figure 11.1 illustrates the granular modeling approach for function approximation.

Figures 11.1a and 11.1b show that granular models outer approximate singular and granular functions, respectively. Outer approximations of functions can always be obtained, for example, at the top level; the coarsest possible granular approximation is the problem domain. Merely enclosing a solution may sound at first shallower than finding the solution itself. We should reflect that the degree of satisfaction involved in embracing a solution depends strongly on the compactness of the enclosure obtained [15]. Moreover, when processing streaming data, we rarely have idea about the error and uncertainty associated to the data. By contrary, if we can compute with granules containing a solution, then we can take, for example, the midpoint as a numeric approximation. Hence, we obtain both an

Fig. 11.1 Granular models: (**a**) singular functionand (**b**) granular function

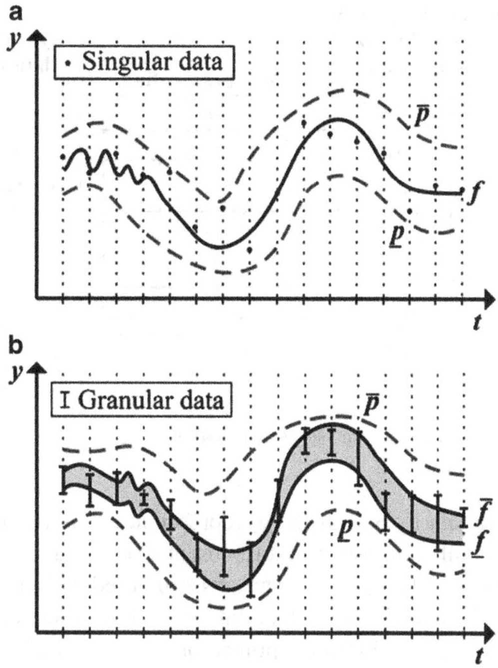

approximate numeric solution and tolerance bounds on the approximation. The key task of approximating with granules is seeking for the tightest envelope for the approximand.

Another perspective for the materialization of granules in data streams concerns the uncertainty introduced during preprocessing and analysis. Incomplete data makes precise discrimination of examples difficult. Missing values are usually predicted through imputation methods [16,17]. Imputed data is uncertain by the very nature of the prediction and motivates granules. Additionally, noise and disturbances of bounded-error dynamic context demand granular treatment of the information. Uncertainty in data representation may be useful to improve the quality of the results. For example, an instance with greater uncertainty may not be as important as one with smaller uncertainty. This incites incremental granular feature selection.

Time and space domains benefit from data granulation. Approaches for granules building regard temporal granulation earlier than spatial granulation, as illustrated in Fig. 11.2. This order is maintained due to several reasons. Occasionally, instances are recorded at different time intervals, for example, as in events stream. The need of synchronized analysis of manifold data streams and search for time correlated structures plead us to consider temporal granulation firstly. Temporal granulation tends to slow down the data flow once several streaming instances can be encapsulated by a granular object and further computations be based on granules. Time granules grant synchronism and smaller amount of data for subsequent spatial analysis. Spatial correlation among heterogeneous data with multiple levels of

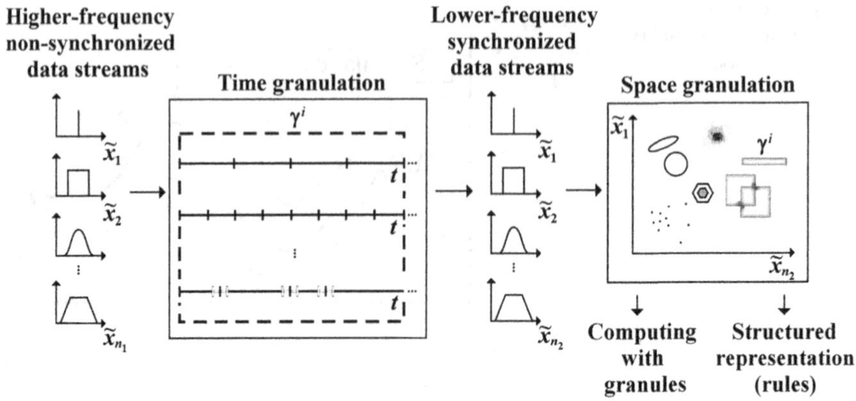

Fig. 11.2 Time and space granulation

granularity and different representations is captured during the process of spatial granulation. Structured representation of data is preserved over time as a synopsis of the data stream; it warrants structured problem solving at the practical level.

The flexibility of handling data using granular framework enables us to describe granules in different application domains without deep knowledge about the problem. Tight time constraints of online environment and interpretability requirements inspire granulated views of detailed data and computing at coarser granularities.

11.2.1 Time-Domain Granulation

Streaming data values are ideally recorded at equal time intervals. Exception happens either when instances arise at random time intervals or when concurrent data flows (usually from different sources or tasks) income at different sampling frequencies. The necessity of a synchronized analysis of concurrent data streams demands forming time granules. A time granule describes the data for a certain period.

Appropriate arrangement plays a key role in the definition of a time granule. If the borders of a time granule are aligned with significant changes of the function behavior, the resulting granulation provides a good abstraction of the data and the function. If alignment is poorly done, models may return inadequate results. Manifold granularities require temporal reasoning and formalizations. At this point, it is worth to distinguish time granule from time window.

Time window [18, 19] stands for a prespecified or adaptive duration interval within which data assembles a representation. Generally, a fixed number of samplings or error values define the size of the window. Windowing the time domain attempts to produce as few segments as possible to avoid data overfitting. Few time segments may hide information if the concept changes. Nonstationarity modifies "ideal" window lengths by its own dynamic. Approaches for testing window lengths are computationally costly and, hence, infeasible in environments with narrow time constraints. Essentially, there may exist several information granules in a time window. Data chunk analysis belongs to window-based approaches for information extraction and analysis.

A time granule groups data according to their indistinguishability in time. Since a time granule conveys similar data indexed in time, its bounds are naturally aligned with substantial changes in the function. The result of dynamic time granulation is a unique granule per segment. Time granules assume manifold levels of data abstraction and are aware of the pace of concept changes.

Event streams usually come at different time granularities. They require analysis of time-domain granules for commonalities extraction prior to space-domain analysis. Information evoked from time granules can be bounds of intervals, statistical or membership functions, and associated features, for example, frequency of a certain event, correlation between events, regular patterns, and the like. The internal structure of a granule and associated data provide full description and characterization of the granule. A granule may have complex structure itself, but it does not come for free.

Particularly, whenever manifold data streams mismatch each other at finer time granularities, we resort to a granulated view of the time-domain and data-mining approach. Resulting granulation should be at least as coarse as the coarsest individual stream to agree with the notion of outer approximation and guaranteed solution.

11.2.2 Space-Domain Granulation

Data granulation over the space domain is a process of organization for comprehension. Data flow triggers a mechanism to collect similar examples. Basically, granulation enables us to view different examples as being the same if low level details are neglected. Granulating of the domain space is fundamental in methods of clustering and information integration. Resulting granules may compose antecedent and consequent of rules in rule-based systems.

Whenever variables are recorded simultaneously and the sampling frequency is not extremely high so that we have enough time to step recursive algorithms, the time granulation stage can be ignored and efforts fully put on spatial granulation. In fact, time and space granulation somewhat relate to each other. For instance, (1) with the minimal and maximal values occurring in a time granule, we may form an interval granular object; (2) taking a representative mean or median of instances resting into a time granule and the confidence interval around it, we may form a statistical granular object; and (3) capturing the core and the uncertainty of instances falling in a same time granule may give rise to a fuzzy granular object. Granular objects of any precedence may be taken into consideration as input to the stage of spatial granulation.

Spatial location of a granule and its size play a role in the process of granulation. Original streaming data is compressed to few granules whose location and granularity reflect the structure of the data. There are many granulated views of the same problem. When evolving granular structures, granules are created as instances of the current knowledge. Next, granules may expand and occupy the space wherever new instances arrive. Operations on granules combine granules to form a coarser granule or decompose a granule into finer granules. Operations on granules should be consistent with the size of the granules and relations between granules; they provide the basic ingredients for the granular computing.

While concept drift and shift are terms related to the joint time-space domain, the descriptions of data density and information specificity [20] concern to the space domain and are choices to guide spatial granulation. Bargiela and Pedrycz [14] state that granules should embrace as many data as possible while maintain certain specificity in what they called principle of the maximization of the information density. Next, the authors suggest a principle for a balanced information granularity. This principle gives preference to the design of granules balanced along all dimensions rather than granules with unbalanced geometry. Hyperbox-based spatial granulation results in a description fully compatible with the intervals description. With intervals, the pursuit of a balanced granularity and refining and coarsening of granules are reduced to arithmetic of intervals.

11.3 Interval Analysis

Interval analysis is a branch of mathematics that provides reliable numerical tools for problem solving; it treats an interval both as a set and as a number [15, 21–26]. While arithmetic carries operations on numbers, interval arithmetic carries operations on intervals. Generally speaking, intervals are instances of granules. Granular computing materializes in the framework of interval analysis and provides features for interpretability.

Interval analysis is a theory oriented for computational implementation because it supports the development of interval-based algorithms. These algorithms are mainly designed to automatically provide rigorous bounds on approximation errors, rounding errors, and propagated uncertainties in initial data. This is of utmost importance because modeling of complex systems must compromise complexity and precision. Operations involving imprecise objects must consider the nature of the imprecision.

The main concern of the interval analysis is to provide a guaranteed approximation of the set of solutions of the underlying problem. "Guaranteed" in this context means that outer approximations of intervals can always be obtained and, moreover, be made as precise as desired. Intervals acknowledge limited precision by associating with a variable of the model under investigation a set of reals as possible values. For ease of storage and computation, these sets are restricted to intervals [26]. Essentials of the interval theory, which form a background of fundamentals for our investigations, are summarized below.

11.3.1 Interval Vectors

An interval I is a closed bounded set of real numbers

$$[l,L] = \{x : l \leq x \leq L\},$$

where l and L denote its endpoints. An n-dimensional interval vector is an ordered n-tuple of intervals $(I_1, \ldots, I_j, \ldots, I_n)$. If I is a, for example, two-dimensional interval vector, then $I = (I_1, I_2)$ for some, $I_1 = [l_1, L_1]$, and $I_2 = [l_2, L_2]$.

Set-theoretic operations of intersection, $\cap$, and union, $\cup$, are applicable to intervals. The intersection of two intervals, I^1 and I^2, is empty, $I^1 \cap I^2 = \emptyset$, if either $l^1 > L^2$ or $L^1 < l^2$. This indicates that I^1 and I^2 have no common points. Otherwise, the intersection of I^1 and I^2 is again an interval:

$$I^1 \cap I^2 = \left[\max\left(l^1, l^2\right),\ \min\left(L^1, L^2\right)\right].$$

The intersection of interval vectors is empty if the intersection of any of their items is empty. Otherwise, for $I^1 = \left(I_1^1, \ldots, I_j^1, \ldots, I_n^1\right)$ and $I^2 = \left(I_1^2, \ldots, I_j^2, \ldots, I_n^2\right)$, we have

$$I^1 \cap I^2 = \left(I_1^1 \cap I_1^2, \ldots, I_j^1 \cap I_j^2, \ldots, I_n^1 \cap I_n^2\right).$$

If two intervals have nonempty intersection, then their union,

$$I^1 \cup I^2 = \left[\min\left(l^1, l^2\right),\ \max\left(L^1, L^2\right)\right]$$

is an interval. Disconnected sets must not be expressed as a single interval.

The convex hull of two interval vectors, I^1 and I^2, namely, $\mathrm{ch}\left(I^1, I^2\right)$, is the smallest interval vector containing all their elements. Then,

$$\mathrm{ch}\left(I_j^1, I_j^2\right) = \left[\min\left(l_j^1, l_j^2\right),\ \max\left(L_j^1, L_j^2\right)\right],\ j = 1, \ldots, n.$$

Hull computation is an efficient procedure to combine sets independently of their connection. It follows that $I^1 \cup I^2 \subseteq \mathrm{ch}\left(I^1, I^2\right)$ for any I^1 and I^2.

If $I^1 = \left(I_1^1, \ldots, I_j^1, \ldots, I_n^1\right)$ and $I^2 = \left(I_1^2, \ldots, I_j^2, \ldots, I_n^2\right)$ are interval vectors, then,

$$I^1 \subseteq I^2 \quad \text{if and only if} \quad I_j^1 \subseteq I_j^2,\ j = 1, \ldots, n.$$

We denote the width of an interval vector, namely, $\mathrm{wdt}(I)$, as the length of its largest side:

$$\mathrm{wdt}(I) = \max(\mathrm{wdt}(I_1), \ldots, \mathrm{wdt}(I_j), \ldots, \mathrm{wdt}(I_n)).$$

The absolute value (magnitude) of an interval I is

$$|I| = \max(|l|, |L|).$$

It follows that $|x| \leq |I| \ \forall\, x \in I$. For the interval vector $I = (I_1, \ldots, I_j, \ldots, I_n)$, we use the vector norm:

$$||I|| = \max(|I_1|, \ldots, |I_j|, \ldots, |I_n|).$$

Finally, it is worth defining the midpoint of an interval I:

$$\mathrm{mp}(I) = (l + L)/2.$$

Analogously, if $I = (I_1, \ldots, I_j, \ldots, I_n)$ is an interval vector, then

$$\mathrm{mp}(I) = (\mathrm{mp}(I_1), \ldots, \mathrm{mp}(I_j), \ldots, \mathrm{mp}(I_n)).$$

11.3.2 Interval Arithmetic

Operations on real numbers can be extended to intervals. Interval arithmetic treats intervals as numbers: adding, subtracting, multiplying, and dividing them.

The rules for interval addition and subtraction are

$$I^1 + I^2 = [l^1, L^1] + [l^2, L^2] = [l^1 + l^2, L^1 + L^2],$$
$$I^1 - I^2 = [l^1, L^1] - [l^2, L^2] = [l^1 - L^2, L^1 - l^2].$$

Operations of addition and subtraction hold for interval vectors. For two interval vectors, $I^1 = (I^1_1, \ldots, I^1_j, \ldots, I^1_n)$ and $I^2 = (I^2_1, \ldots, I^2_j, \ldots, I^2_n)$, we have

$$I^1 + I^2 = (I^1_1 + I^2_1,\ \ldots,\ I^1_j + I^2_j,\ \ldots,\ I^1_n + I^2_n),$$
$$I^1 - I^2 = (I^1_1 - I^2_1,\ \ldots,\ I^1_j - I^2_j,\ \ldots,\ I^1_n - I^2_n).$$

For the product of two intervals, I^1 and I^2, we get

$$I^1 I^2 = \{x^1 x^2 : x^1 \in I^1, x^2 \in I^2\}.$$

Clearly, the result is again an interval, say I^3, whose endpoints are

$$[l^3, L^3] = \left[\min\left(l^1 l^2,\ l^1 L^2,\ L^1 l^2,\ L^1 L^2\right),\ \max\left(l^1 l^2,\ l^1 L^2,\ L^1 l^2,\ L^1 L^2\right)\right].$$

The reciprocal of an interval I yields

$$1/I = \{1/x : x \in I\}.$$

If I is an interval not containing the number 0, then $1/I = [1/L, 1/l]$ if $l > 0$ or $1/I = [1/l, 1/L]$ if $L < 0$. In case I contains 0 so that $l \leq 0 \leq L$, then the set is unbounded and cannot be represented as an interval whose endpoints are real numbers. For the quotient of two intervals, we have

$$I^1/I^2 = I^1\left(1/I^2\right) = \{x^1/x^2 : x^1 \in I^1, x^2 \in I^2\}.$$

If 0 is not contained in I^2, then I^1/I^2 is again an interval.

The product and quotient operations for interval numbers hold for interval vectors. For two interval vectors, $I^1 = (I_1^1, \ldots, I_j^1, \ldots, I_n^1)$ and $I^2 = (I_1^2, \ldots, I_j^2, \ldots, I_n^2)$, it follows that

$$I^1 I^2 = \left(I_1^1 I_1^2,\ \ldots,\ I_j^1 I_j^2,\ \ldots,\ I_n^1 I_n^2\right),$$
$$I^1/I^2 = \left(I_1^1/I_1^2,\ \ldots,\ I_j^1/I_j^2,\ \ldots,\ I_n^1/I_n^2\right).$$

11.3.3 Distance Between Intervals

A suitable metric to measure the distance between two intervals, I^1 and I^2, is

$$d\left(I^1, I^2\right) = \max\left(|l^1 - l^2|, |L^1 - L^2|\right).$$

With this metric, the correspondence between the interval number system and the real number system, $[x,x] \leftrightarrow x$, holds. The metric $d(.)$ preserves the distance between the corresponding items. We have that

$$d\left([x^1,x^1],[x^2,x^2]\right) = \max\left(|x^1 - x^2|, |x^1 - x^2|\right) = |x^1 - x^2|$$

for any x^1 and x^2. The real line is isometrically embedded into the metric space of intervals [27].

The distance between two interval vectors, $I^1 = (I_1^1, \ldots, I_n^1)$ and $I^2 = (I_1^2, \ldots, I_n^2)$,

$$d\left(I^1, I^2\right) = \left(\max\left(|l_1^1 - l_1^2|, |L_1^1 - L_1^2|\right),\ \ldots,\ \max\left(|l_n^1 - l_n^2|, |L_n^1 - L_n^2|\right)\right)$$

is an interval vector. Sometimes, we are more interested in a number to represent the overall distance between interval vectors. A measure for the overall distance between two interval vectors, I^1 and I^2, is

$$D(I^1, I^2) = \max\left(d(I^1, I^2)\right).$$

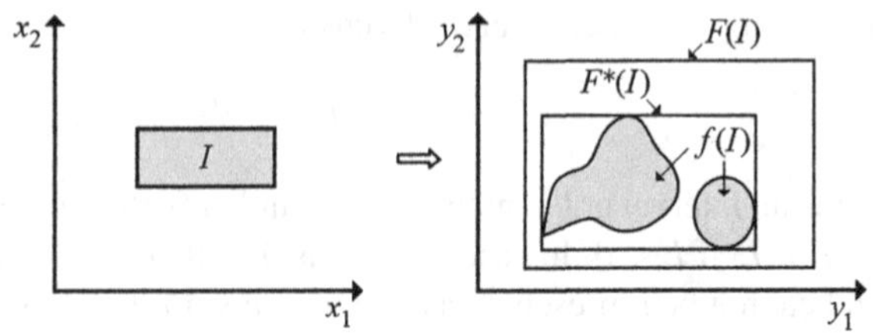

Fig. 11.3 Image f of box I and inclusion functions F and F^*

11.3.4 Interval Functions

Consider a real-valued function $f(x)$ and a corresponding interval-valued function $f(I)$. $f(I)$ is an interval extension of $f(x)$ if $f(I) = f(x)$ for any value of $x \in I$. If the parameters of $f(I)$ are degenerated, then $f(I)$ is a degenerated interval equal to $f(x)$. Formally, the image of an interval I under a real mapping f is

$$f(I) = \{f(x) : x \in I\}.$$

More generally, the image of a specified n-dimensional vector I admitting a multi-variable real function f is:

$$f(I_1,\ldots,I_j,\ldots,I_n) = \{f(x_1,\ldots,x_j,\ldots,x_n) : x_j \in I_j \ \forall j\}.$$

Generally, the image of an interval through f is not a box (see Fig. 11.3), and it may be difficult to obtain in closed form. In practice, $f(I)$ can be approximated by an inclusion function $F(I)$, which is a box in the range of f.

An interval function F from $\mathbb{IR}^n$ to $\mathbb{IR}^m$ is called inclusion function of f if

$$f(I) \subseteq F(I) \ \forall I \in \mathbb{IR}^n.$$

Inclusion functions are not unique, and they depend on how we choose F. An inclusion function is optimal if $F(I)$ is the interval hull of $f(I)$. In other words, the optimal inclusion function for $f(I)$ is the smallest box $F^*(I)$ that contains $f(I)$. Figure 11.3 illustrates the idea. $F^*(I)$ is unique.

In particular, for degenerated intervals I, it follows that

$$F(I) = f(I) = F^*(I).$$

Assume f is monotonically increasing in $I = [l, L]$. Then we can obtain $f(I)$ using

$$f(I) = [f(l), f(L)].$$

Consequently,

$$f(x) \subseteq [f(l), f(L)] \ \forall x \in I.$$

With monotonic decreasing functions, we have to order the resulting endpoints correctly. In these cases, $f(I) = [f(L), f(l)]$, that is, strict inclusion relationship holds.

Nonmonotonic functions could be monotonic under endpoint constraint. For example, $f(I) = \sin(I)$ is not monotonic in general but defining $I = [-\Pi/2, \Pi/2]$, then $f(I)$ is monotonic and $f(I) = \sin(I) = [\sin(l), \sin(L)]$.

An interval function $f(I)$ is inclusion isotonic, when for any interval vectors, I^1 and I^2,

$$\text{if } I^1 \subset I^2, \text{ then } f(I^1) \subset f(I^2).$$

Finite interval arithmetic [23] is inclusion isotonic. Consider that $\bullet$ denotes the operations of addition, subtraction, multiplication, and division, thus

$$I^1 \bullet I^2 \subset I^3 \bullet I^4$$

holds whenever $I^1 \subset I^3$ and $I^2 \subset I^4$. In this work, all interval enclosures are inclusion isotonic interval extensions of real-valued continuous functions.

An interval function $f(I) \in \mathbb{IR}$ is called *thin* when it involves only degenerate interval parameters or, equivalently, singular parameters. For instance, the interval function

$$f(I) = a_0 + \sum_{j=1}^{n} a_j I_j$$

is *thin* for $(a_0, \ldots, a_n)$ degenerated intervals. When an interval function involves at least one interval parameter of nonzero width, it is called *thick*. In this work, we consider *thin* interval functions only.

Interval analysis goes far beyond what has been covered in this section. We do not address both interval integration [27], complex interval arithmetic [28], interval statistics [29], and intervals in fuzzy set theory [30], but the essentials to pave the IBeM framework. Moving beyond the essentials and toward the development of effective approach to handle real-world problems is subject of the following section.

11.4 Interval-Based Evolving Modeling

The mathematical formalism of the interval analysis provides a robust framework for the analysis of granular structures. Interval mathematics supports the core of the IBeM learning algorithm and gives simplicity, correctness, totality, closeness, optimality, and efficiency [26].

IBeM originated from recent research on modeling nonstationary streaming data. IBeM models process data streams using recursive one-pass algorithm. It starts learning from scratch and dispenses knowledge about the properties of the data. Models developed by IBeM are interpretable via rules. Online learning algorithm casts the IBeM structure to learn new concepts, detect concept change, cope with uncertainty, learn forever, and to provide nonlinear approximation.

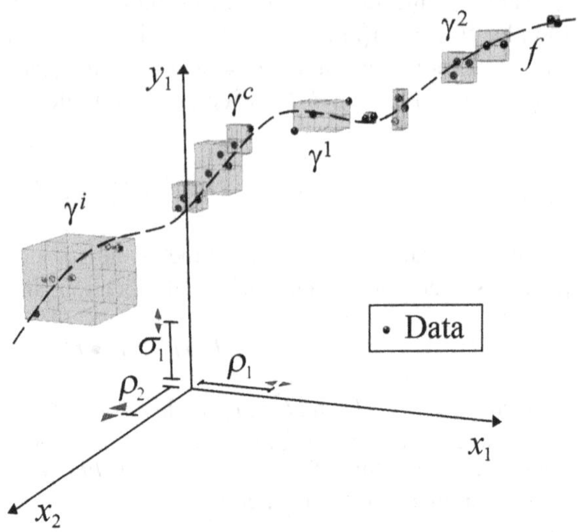

Fig. 11.4 Approximating a singular function with granules

IBeM exploits bottom–up procedures to form higher level granules from raw data. A ∪-closure granular structure ensues from more specific local granules. The internal representation of an IBeM granule, γ^i, in respect to antecedent variables, x_j, is empty. This means that bounds of intervals, $[l^i_j, L^i_j]$, are all IBeM records from input data stream. Consequent variables, y_k, are identically granulated whenever output data become available. The content of an output granule conveys additional information in respect of a rule, for example, inclusion monotonic functions p^i_k. Bounds of consequent variables, $\left[u^i_k, U^i_k\right]$, are determined by granulating the output data stream, processing the inclusion function using bounds of the antecedent variables, and performing the AND operation. The result is a granular approximation of a function. Computing p^i_k using x_j gives a singular approximation. Rules R^i associated with granules γ^i are of the type:

$$
\begin{aligned}
R^i\text{: IF } & (l^i_1 \le x_1 \le L^i_1) \text{ AND } (l^i_2 \le x_2 \le L^i_2) \text{ AND } \dots \text{ AND } (l^i_n \le x_n \le L^i_n) \\
\text{THEN } & (u^i_1 \le y_1 \le U^i_1) \text{ AND } p^i_1 = a^i_{01} + \Sigma^n_{j=1} a^i_{j1} [l^i_j, L^i_j] \text{ AND} \\
& \dots \\
& (u^i_k \le y_k \le U^i_k) \text{ AND } p^i_k = a^i_{0k} + \Sigma^n_{j=1} a^i_{jk} [l^i_j, L^i_j] \text{ AND} \\
& \dots \\
& (u^i_m \le y_m \le U^i_m) \text{ AND } p^i_m = a^i_{0m} + \Sigma^n_{j=1} a^i_{jm} [l^i_j, L^i_j].
\end{aligned}
$$

Functions p^i_k are thin and of first order in this case. In general, each p^i_k can be of different type and does not need to be linear. The recursive least mean square (RLMS) algorithm is used to determine the coefficients a^i_{jk} of p^i_k.

Assume that ρ_j and σ_k are the maximum width that intervals may take in the input and output spaces, respectively. Values of ρ and σ allow different representations of the same problem in different levels of detail. Figure 11.4 illustrates the idea in the input/output space. The case shown in the figure refers to a collection of granules

$\gamma^i, i = 1, \ldots, c$, of different sizes and geometric forms constructed in light of singular data being available. The learning approach relies on the information conveyed by the data to create and foster granules and to set the granularity. Granules are not allowed to grow beyond ρ and σ. If so, they are immediately required to be split.

11.5 Learning Algorithm

This section details the working principle of the IBeM learning algorithm. IBeM system grants important characteristics for evolving modeling. Its incremental learning approach spends a small and constant processing time, that is, processing time does not scale with the number of instances. Continuous processing on an instance-per-instance recursive basis enables IBeM to deal with concept drift within online environment. Nonstationarity requires detecting and tracking changes in the joint time-space structure of the underlying data. The IBeM approach for data flow mining and knowledge discovery relies predominantly on constructive bottom–up modeling procedures, but allows decomposition-based top–down procedures.

Formally, IBeM learns online from a sequence $(x,y)^{[h]}, h = 1, \ldots$, where $y^{[h]}$ is known given $x^{[h]}$ or will be revealed some steps latter. Each pair (x,y) is an observation of the target function f. When f is nonstationary, IBeM should track time varying function $f^{[h]}$. IBeM systems evolve whenever new information appears in the data. When new instances do not fit current knowledge, procedures create new information granules and rules managing the granules. Conversely, when instances fit current knowledge, procedures adapt existing granules and rules if necessary. Eventually, the quotient granular structure may be optimized, refined, or coarsed, agreeing with intergranule relationships.

11.5.1 Time Granulation

From a data stream $(x,y)^{[h]}, h = 1, \ldots$, time granulation groups successive instances $(x,y)^{[h]}, h = h_b, \ldots, h_{e-1}$, into a time granule. Indices h_b and h_{e-1} denote the beginning and the end of a time granule; h_e is a break point value. Strict relationship $h_e > h_b$ holds. The set of instances streaming during $[h_b, h_{e-1}]$ is considered indistinguishable and the inequalities

$$\mathrm{wdt}\left(\mathrm{ch}\left(x_j^{[h_b]}, \ldots, x_j^{[h_{e-1}]}\right)\right) \leq \rho_j,\; j = 1, \ldots, n, \text{ and}$$

$$\mathrm{wdt}\left(\mathrm{ch}\left(y_k^{[h_b]}, \ldots, y_k^{[h_{e-1}]}\right)\right) \leq \sigma_k,\; k = 1, \ldots, m,$$

are satisfied. The instance indexed by h_e conveys at least one contrasting value.

The collection $(x,y)^{[h]}, h = h_b, \ldots, h_{e-1}$, produces a unique interval granule with lower and upper endpoints determined by

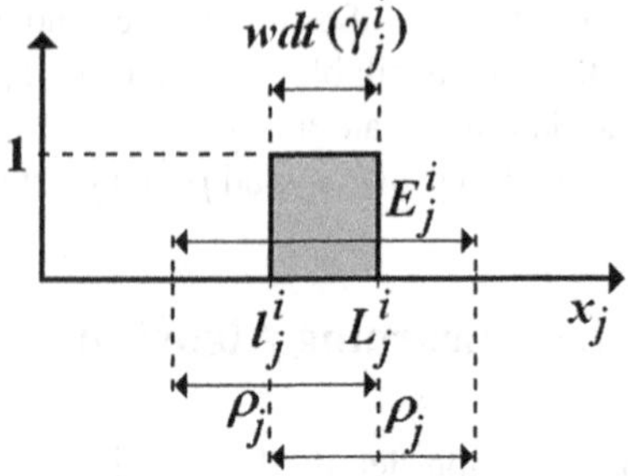

Fig. 11.5 Expansion of a granule

$$\left[\min\left(x_j^{[h]}\right), \max\left(x_j^{[h]}\right)\right],\ h = h_b, \ldots, h_{e-1};\ j = 1, \ldots, n, \text{ and}$$

$$\left[\min\left(y_k^{[h]}\right), \max\left(y_k^{[h]}\right)\right],\ h = h_b, \ldots, h_{e-1};\ k = 1, \ldots, m,$$

respectively, for input and output variables.

In the IBeM framework, time granulation is used as a preprocessing step especially on occasions where instances from different sources or tasks arise at random time intervals. Multiple time granularities allow synchronized analysis of concurrent data streams. Thereafter, learning within the space domain is based on time granule intervals, rather than on original data. IBeM is not exposed to all original data, which are far more numerous than time granules.

11.5.2 Birth and Growth of Space Granules

IBeM systems start learning from scratch. No granules and rules need to be preconceived nor needs the amount of granules to be set in advance. Granules and rules are created and adapted on demand, dynamically, steered by the behavior of the process function and information mirrored in the measured data. Whenever stream pairs $(x, y)^{[h]}$ arrive, a decision mechanism is trigged and granules and rules can be inserted into or adapt the IBeM structure.

Key questions to be answered for effective implementation of IBeM refer to when and how to create or adapt granules and rules recursively to consider new never-seen-before instances. Let E^i be the expansion region of a granule γ^i. Thus,

$$E^i_j = [L^i_j - \rho_j,\ l^i_j + \rho_j],\ j = 1, \ldots, n.$$

Bounds of expansion regions E^i_j help to derive criteria for deciding whether or not two objects should be put into the same granule. Figure 11.5 illustrates the expansion of an interval granule γ^i.

An information granule is born either when an input variable, say x_j, does not fit E^i_j for all i and some j or an output variable, say y_k, does not fit E^i_k for all i and some k. This means that existing granules must not expand their bounds beyond the limits dictated by ρ and σ to include the current input. Connective AND operators of IBeM rules suggest that both, E_j^i for all j and E_k^i for all k, suit (x, y) for the corresponding granule be considered. The new granule γ^{c+1} extends the current collection of granules $\gamma = \{\gamma^1, \ldots, \gamma^c\}$.

Adaptation of existing rules R^i expands the width of rules antecedent $[l_j^i,\ L_j^i]$ and consequent $[u_k^i,\ U_k^i]$ to accommodate new data and simultaneously adjusts the coefficients of local interval functions p_k^i. A rule R^i is adapted whenever an instance (x, y) falls into the region E^i of γ^i. This means, geometrically, that the instance lays inside the hyperrectangle of γ^i or close enough so that the granule is allowed to expand to include (x, y). Figure 11.6 summarizes nine situations that may happen depending on where the instance is confined and associated procedures.

In Fig. 11.6, recently arrived interval data, x, can be either outside, partially inside, or inside of a generic granule γ^i. Depending on the location of x, IBeM creates a new granule γ^{c+1} and/or adapts the bounds of an existing granule γ^i. Expansion is mainly based on union and convex hull operations. All uncertainty in the data is enveloped by some granule to guarantee outer approximation of the solution. Although data and granule may have some level of overlap, two granules are forbidden to overlap as result of these adaptation procedures.

Adaptation of consequent intervals $[u_k^i,\ U_k^i]$ uses outcome data y_k. Thin polynomial coefficients are initialized as $a_{jk}^i = 0,\ j \neq 0$ and $a_{0k}^i = y_k \forall k$ and can be subsequently updated using the standard RLMS algorithm and taking advantage of the instance that activates the rule R^i. Storage of a number of recent instances may be useful to guide alternative coefficient identification algorithms, for example, data chunks oriented algorithms. However, it comes with some additional cost concerning memory and processing time.

11.5.3 Choosing the Granularity

Values of ρ and σ set upper bounds of the level of abstraction of models. If ρ and σ are equal to 0, then the existing granules cannot be expanded. Conversely, when ρ and σ match 1, a single granule represents all the data. On trading off these extreme situations, we intermediate complexity and precision. Calculations involving imprecise data must consider the nature of the imprecision.

The size of a granule may be interpreted as its degree of detail. A simple procedure we use in IBeM to tune the maximum width ρ and σ of granules over time regards multiple views of the data and refining and coarsening of granules.

Let β be the number of rules created after a certain number of processing steps h_r. If the number of rules grows faster than a threshold value η, that is, $\beta > \eta$, then ρ and σ are increased as follows:

$$\rho, \sigma(\text{new}) = \left(1 + \frac{\beta}{h_r}\right) \rho, \sigma(\text{old}).$$

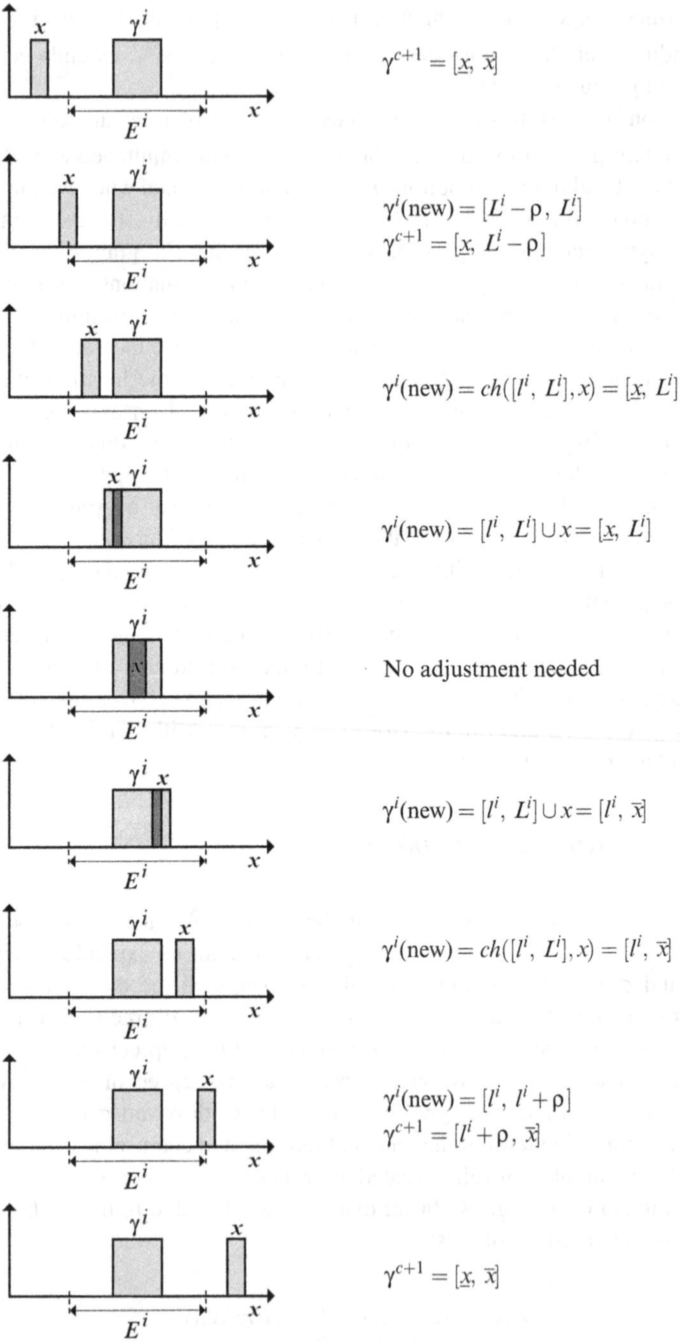

Fig. 11.6 Recursive adaptation of IBeM granules

Otherwise, if the number of rules grows at a rate smaller than η, that is, $\beta < \eta$, then ρ and σ are decreased as follows:

$$\rho,\sigma(\text{new}) = \left(1 - \frac{(\eta-\beta)}{h_r}\right)\rho,\sigma(\text{old}).$$

Decreasing the maximum width allowed may require adequate diet for some granules to get the new standard. The refinement of a granule is based on its midpoint and redefinition of its lower and upper bounds. The mechanism to deal with data stream granularity is useful to let ρ and σ learn values for themselves and to avoid guesses on how fast and how often streams change.

11.5.4 Refining and Coarsening the Quotient Structure

Once granules are identified, IBeM analyzes the relationship among them and proceeds accordingly. Top–down and bottom–up structural operations support refining and coarsening of granules over time. Structural knowledge is generated to help visualization of relationships between different parts of the problem.

Top–down processes produce $\cap$-closure granular models splitting large granules into smaller, lower level granules. Situations in which the maximum width allowed for a granule reduces, see Sect. 11.5.3, may cause top–down refinements. Whenever the granularity dictated by ρ becomes finer, checking $\text{wdt}(\gamma^i) < \rho$ may return false. In these cases, the granule γ^i is split into

$$\gamma_j^{i1} = [l_j^i, \text{mp}(\gamma_j^i)], \text{ and}$$
$$\gamma_j^{i2} = [\text{mp}(\gamma_j^i), L_j^i],\ j = 1,\ldots,n.$$

The refining procedure is repeated until $\text{wdt}(\gamma^i) < \rho$ holds for $i = 1,\ldots,c$. Analogous approach is used for output variables k and granularity σ.

A $\cup$-closure granular model results from a bottom–up process that involves forming a large, higher level granule using small, lower level granules. Let

$$D = \begin{bmatrix} D(\gamma^1,\gamma^1) & \cdots & D(\gamma^1,\gamma^i) & \cdots & D(\gamma^1,\gamma^c) \\ \vdots & \ddots & \vdots & & \vdots \\ D(\gamma^i,\gamma^1) & \cdots & D(\gamma^i,\gamma^i) & \cdots & D(\gamma^i,\gamma^c) \\ \vdots & & \vdots & \ddots & \vdots \\ D(\gamma^c,\gamma^1) & \cdots & D(\gamma^c,\gamma^i) & \cdots & D(\gamma^c,\gamma^c) \end{bmatrix},$$

be a distance matrix relating any pair of granules. The matrix D is symmetric with zeros in the main diagonal. Neighbor granules can be located close enough to justify

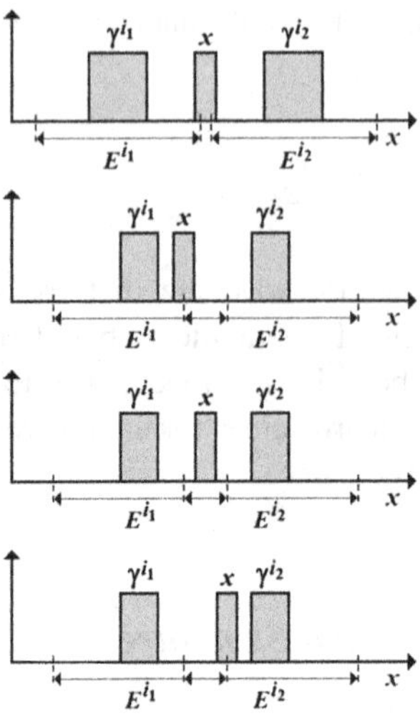

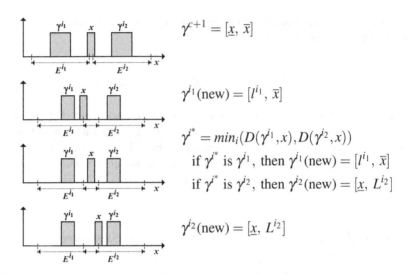

$$\gamma^{c+1} = [\underline{x},\ \overline{x}]$$

$$\gamma^{i_1}(\text{new}) = [l^{i_1},\ \overline{x}]$$

$$\gamma^{i^*} = min_i(D(\gamma^{i_1},x), D(\gamma^{i_2},x))$$
$$\text{if } \gamma^{i^*} \text{ is } \gamma^{i_1}, \text{ then } \gamma^{i_1}(\text{new}) = [l^{i_1},\ \overline{x}]$$
$$\text{if } \gamma^{i^*} \text{ is } \gamma^{i_2}, \text{ then } \gamma^{i_2}(\text{new}) = [\underline{x},\ L^{i_2}]$$

$$\gamma^{i_2}(\text{new}) = [\underline{x},\ L^{i_2}]$$

Fig. 11.7 Intergranular conflict and data accommodation

their combination into a unique, coarser granule. Coarsening evaluates whether the combination of the respective granules is possible or not based on the minimum entry of the matrix D. Combination is possible if the width of the convex hull of two granules, say γ^{i_1} and γ^{i_2}, returns a permissible granularity. Formally, if

$$\text{wdt}\left(\text{ch}\left(\gamma_j^{i_1}, \gamma_j^{i_2}\right)\right) \leq \rho_j,\ j = 1,\ldots,n,$$

then $\gamma^i = \text{ch}(\gamma^{i_1}, \gamma^{i_2})$ is coarsening of γ^{i_1} and γ^{i_2}. Coarsening produces more compact rule bases and contributes to eliminate spatial gaps between close enough granules. At the top level, we close the IBeM structure by the most general granule formed by the convex hull of all elementary granules.

11.5.5 Conflict of Interest

A requirement to be kept in mind when designing granular systems, such as IBeM, is the goal to include every information that assembles a solution. However, at the same time, it is desirable to keep the system as simple as possible. As learning occurs, conflicting situations may arise and adaptation procedures that result in narrower granules of data must be considered. Conflict of interest happens when two or more granules can be expanded to embrace the current input. Figure 11.7 shows four typical situations considering the current input x and two granules, say γ^{i_1} and γ^{i_2}; they are (*i*) $\underline{x} \in E^{i_1} = [L^{i_1} - \rho, l^{i_1} + \rho]$, but $\overline{x}$ does not. Conversely, $\overline{x} \in E^{i_2}$, but $\underline{x}$ does not; (*ii*) $E^{i_1} \cap E^{i_2} \cap x \neq \emptyset$, but $x \in E^{i_1}$; (*iii*) $x \subseteq (E^{i_1} \cap E^{i_2})$; and (*iv*) $E^{i_1} \cap E^{i_2} \cap x \neq \emptyset$, but $x \in E^{i_2}$. The respective adaptation procedures are shown in the figure.

In case (*i*), both granules cannot expand beyond ρ. Therefore, a new granule is created to include x. Cases (*ii*) and (*iv*) avoid redundancy and inconsistency neglecting the adaptation of the granule that cannot enclose x entirely. Case (*iii*) chooses the granule closer to x according to $D(.)$.

Intergranular conflict resolution helps to choose which IBeM rule to adapt and prevents overlapped intervals and contradiction. The tightest envelope for the data generates a more concise description about the information it carries.

11.5.6 Removing Granules

A granule should be removed from the IBeM structure if it is inconsistent with the current concept. Common removing strategies either (1) remove granules by age, (2) exclude the weakest granules based on error values, or (3) delete the most inactive granules. In IBeM, the strategy is to delete inactive granules by exclusion. Old granules may still be useful in the current environment, whereas weak granules are attempted to be strengthened by adjusting coefficients of local inclusion functions, see Sect. 11.5.7.

IBeM granules are deleted whenever they become inactive during a number of processing steps, h_r. If the application requires memorization of rare events, or if cyclical drifts are anticipated, then it may be the case to let the granules live forever. Removing inactive granules periodically helps to keep the rule set updated and concise.

11.5.7 Function Approximation

For each granule identified, its associated rule has the consequent function parameters adjusted using the RLMS algorithm as described next.

Let $(x,y)^{[h]}$ be the data pair available for training at instant h, and γ^i be the granule activated by the data pair. Local polynomials are estimated by linear equations

$$p_k^i = a_{0k}^i + \sum_{j=1}^{n} a_{jk}^i [l_j^i, L_j^i],\ k = 1, \ldots, m.$$

Using $(x,y)^{[h]}$ and assuming single output, without loss of generality, we get

$$y^{[h]} = a_0^i + \sum_{j=1}^{n} a_j^i x_j^{[h]}.$$

In the matrix form, we have

$$Y = XA^i,$$

where $Y = \left[y^{[h]}\right]$, $X = \left[1 \; x_1^{[h]} \; \ldots \; x_n^{[h]}\right]$, and $A^i = [a_0^i \; \ldots \; a_n^i]^T$ is a vector of unknown parameters. To estimate the coefficients a_j^i, we let

$$Y = XA^i + E,$$

where $E = [e^{[h]}]$ is the current modeling error. While in batch estimation, the rows in Y, X, and E increase agreeing with the number of available instances; in the recursive mode of the algorithm, only two rows are kept. We reformulate the state variables as

$$Y = \begin{bmatrix} y^{[h]} \\ y^{[h+1]} \end{bmatrix}, \; X = \begin{bmatrix} 1 \; x_1^{[h]} & \ldots & x_n^{[h]} \\ 1 \; x_1^{[h+1]} & \ldots & x_n^{[h+1]} \end{bmatrix}, \text{ and } E = \begin{bmatrix} e^{[h]} \\ e^{[h+1]} \end{bmatrix},$$

where the first and second rows refer to values before and just after adaptation, respectively. The RLMS algorithm sets A^i to minimize the functional

$$J(A^i) = E^T E.$$

Derived from [31], A^i can be estimated by

$$A^i = (X^T X)^{-1} X^T Y$$

to minimize the square error. Assuming $P = (X^T X)^{-1}$ and the matrix inversion lemma [31], similar to [32], we avoid inverting $X^T X$ at each processing step from the following recursion:

$$P(\text{new}) = P(\text{old}) \left[I - \frac{X X^T P(\text{old})}{1 + X^T P(\text{old}) X} \right],$$

where I is identity matrix. In practice, it is usual to choose large initial values for the main diagonal elements of P. In this chapter, we use $P^{[0]} = 10^3 I$ as default value.

After simple mathematical transformations, the vector of parameters is rearranged recursively as follows:

$$A^i(\text{new}) = A^i(\text{old}) + P(\text{new}) X \left(Y - X^T A^i(\text{old})\right).$$

Detailed derivations can be found in [33], and convergence proof in [34].

11.6 Application Examples

Experiments reported in this section consider singular and granular data streams to show the usefulness of the IBeM approach. In the first experiment, the IBeM system mines singular data concerning the level of rain precipitation in different European regions to form granules whose size and location reflect the essence of the data. IBeM is always requested to process flowing examples it has never seen before and that demand immediate response before being used for model training. The order of the streaming items is out of the control of the system, and data must not be stored nor retrieved. The second experiment delivers original interval data related to daily fluctuation of the price of an economic index. IBeM starts learning from scratch, and the data stream guides the creation and development of models freely. In both experiments, IBeM plays the role of an evolving predictor.

11.6.1 Rain Precipitation

Meteorological precipitations, for example, rainfall, occur when a portion of the atmosphere becomes saturated with water vapor; thus, the water condenses and falls under gravity. Prediction of rain precipitation concerns with estimating the amount of water to be accumulated over a time period in a definite region. Rainfall prediction is essential to prevent floods, droughts, food shortage; to assist decision making on agricultural crops, hydroelectric power plants, and dams; and to simulate the behavior of rivers, soil erosion, and ecosystems.

Data sets from the ECA&D project (available at http://eca.knmi.nl/download/mil-lennium/millennium.php) were considered for analysis. Measurements are recorded in millimeters of rain per month. We admit the meteorological stations 244 (Zurich/ Fluntern), 173 (Milan), and 378 (Athens) to evaluate the IBeM performance. Zurich is one of the wettest cities in Europe. Rainfall spreads throughout the year with the highest levels of precipitation recorded during summer months. Milan is a city known to have quite high humidity during the whole year. Its humid subtropical climate has four distinct seasons. Rainfall is relatively low in July but peaks by August. Rain generally falls in heavy outbursts during summer. During autumn and spring, it storms about half of the days, whereas in winter, rainfall lessens. Conversely, Athens is one of the driest cities in Mediterranean Europe and experiences a differentiated climatic pattern. Due to its location in relation to the Mount Parnitha, the Athenian climate is recognized quite dry with sparse precipitations during summer.

The Zurich, Milan, and Athens data sets consist of 1,314, 1,818, and 1,242 time indexed instances comprising millimeters of rainfall per month recorded from January 1901 to December 2010, January 1858 to December 2010, and January 1899 to December 2002, respectively. The task of IBeM is to predict the amount of rainfall in the subsequent month, $y^{[h+1]}$, using the last five observations, $x^{[h-4]}, \ldots, x^{[h]}$. IBeM scrutinizes the data only once to build its structure and tune local parameters. This is to reproduce a data stream. The rule base is initially empty, devoid of knowledge. However, the apprenticeship starts immediately after the first data pair is available.

Testing and training are performed concomitantly on a sample-per-sample basis. First, an estimation $p^{[h+1]}$ is provided for a given input $(x^{[h-4]}, \ldots, x^{[h]})$. One

Table 11.1 Performance of different algorithms in predicting levels of rainfall

	Zurich			Milan			Athens		
Model	Rules	*RMSE*	*NDEI*	Rules	*RMSE*	*NDEI*	Rules	*RMSE*	*NDEI*
MLP	–	0.3752	2.7032	–	0.2522	1.8170	–	0.1752	1.2622
eTS	9	0.1961	1.1193	9	0.1804	1.1168	6	0.1951	1.4056
xTS	10	0.1932	1.1029	9	0.1768	1.0946	8	0.1368	0.9861
$IBeM^1$	**6.37**	0.1996	1.1393	**5.53**	0.1820	1.1268	**5.22**	0.1390	1.0017
$IBeM^2$	9.22	0.1970	1.1248	14.24	0.1667	1.0318	8.39	**0.1037**	**0.7475**
$IBeM^3$	27.96	**0.1885**	**1.0760**	21.45	**0.1519**	**0.9401**	14.05	0.1057	0.7617

step ahead, the actual value $y^{[h+1]}$ becomes available and model adaptation is carried out if necessary. Training is necessary whenever an instance carries new information significantly mismatching the current knowledge. Sample-per-sample testing-before-training approach portrays the true online data stream context.

Performance evaluation is made based on the root mean square error:

$$\text{RMSE} = \sqrt{\frac{1}{H}\sum_{h=1}^{H}(y^{[h]} - p^{[h]})^2}$$

and the nondimensional error index

$$\text{NDEI} = \frac{\text{RMSE}}{\text{std}(y^{[h]}\forall h)},$$

which basically ponders the RMSE by the inverse of the standard deviation of the underlying data.

To evaluate the effect of different parameterizations, we conduct three experiments. Firstly, $IBeM^1$ prioritizes a more compact structure and adopts $\rho^{[0]} = \sigma^{[0]} = 0.5$, deletion threshold $h_r = 48$, and $\eta = 3$. $IBeM^3$ focuses on accuracy at the price of a larger structure and employs $\rho^{[0]} = \sigma^{[0]} = 0.4, h_r = 130$, and $\eta = 3$. $IBeM^2$ plays an intermediary role between the more compact $IBeM^1$ and the more precise $IBeM^3$. $IBeM^2$ uses $\rho^{[0]} = \sigma^{[0]} = 0.5, h_r = 60$, and $\eta = 3$. The multilayer perceptron (MLP), extended Takagi-Sugeno (xTS), and evolving Takagi-Sugeno (eTS) methods are used for performance comparison. Table 11.1 summarizes the results for the Zurich, Milan, and Athens monthly data.

Table 11.1 shows that rain precipitation in Zurich is more difficult to predict than in Milan and Athens according to the *RMSE* and *NDEI* indices provided by the algorithms. $IBeM^3$ evolves an average of 27.96 rules for the Zurich data to attain an *RMSE* index equal to 0.1885 and an *NDEI* value of 1.0760. However, using only 6.37 rules, $IBeM^1$ reached a performance of 0.1996 and 1.1393 for the *RMSE* and *NDEI*, respectively, which is slightly worse than the performance of $IBeM^3$. While a compact structure speeds up processing time and reduces memory usage, because the number of developed rules is smaller, input and output granules tend to be wider to pave the problem domain. Therefore, $IBeM^1$ granular predictions tend to be less significant. Analysis of the results for Milan and Athens is quite similar to the analysis for Zurich. However, we may notice that, even using fewer rules, the performance of IBeM^2 is better than the performance of IBeM^3 for the Athens data. Narrowing the bounds of the granules does not imply that the local-valued singular prediction will be necessarily better. Fewer instances are allocated to lower level smaller granules, and then the RLMS algorithm takes advantage of less information about the target function to adjust the corresponding parameters. The trend of a better singular approximation within tighter enclosures remains.

The importance of the incremental learning is clearly verified by comparing the accuracy of IBeM, eTS, and xTS (evolving algorithms) with the accuracy of an offline trained MLP neural network. Once the neural network has a fixed structure, it is limited in its ability to adapt to a new trend or concept.

By comparing evolving methods with each other, Table 11.1 shows that IBeM outperforms eTS and xTS in predicting Athens rainfall and that these algorithms are comparable in terms of accuracy and compactness for the Zurich and Milan data sets. However, we have noticed that the average per sample processing time of IBeM is the smallest. The IBeM algorithm consumed an average of 1.28 ms per item of the data streams on a dual-core 2.54-GHz processor with 4-GB RAM against 2.64 ms and 12.57 ms spent, respectively, by the xTS and eTS algorithms. This is explained by the easiness of acquiring and adapting upper and lower bounds of intervals from a data stream, and waiver of liability for adjusting parameters of fuzzy membership functions based on more refined clustering techniques.

Figures 11.8–11.10 detail IBeM one-step singular and granular predictions for the rainfall problem. The granular prediction $[u,U]$ allied to the more suggestive singular prediction p may assist decision making giving an idea about a range of values around p. Intervals here can be read as optimistic and pessimistic prediction values. Bounds of granules may enhance model acceptability. Figures also illustrate how the granules size, number of rules, and *RMSE* and *NDEI* indices vary over time. As evidenced in the figures, the IBeM algorithm does not profit from several granules and rules, but from a combination of ingredients concerning with structural premises, peculiar derivations of the learning algorithm, and interval granular framework and tools to achieve the performance. We notice that in all experiments, IBeM runs in linear time with respect to the length of the stream.

11.6.2 Bovespa BVSP Index

In this section, we address an economic time-series prediction problem using IBeM. Different from [8], where we deal with daily end-of-day forecast of the Brazil Bovespa BVSP Index, here, we investigate original interval data concerning with the range of values in which the price of the index fluctuates during a day. Data from January 2, 1998 to December 1, 2009, were obtained from the Yahoo! Finance website and used in the experiments. There are about 500 companies trading at BM&F BOVESPA, the Sao Paulo Stock Market, which is the fourth

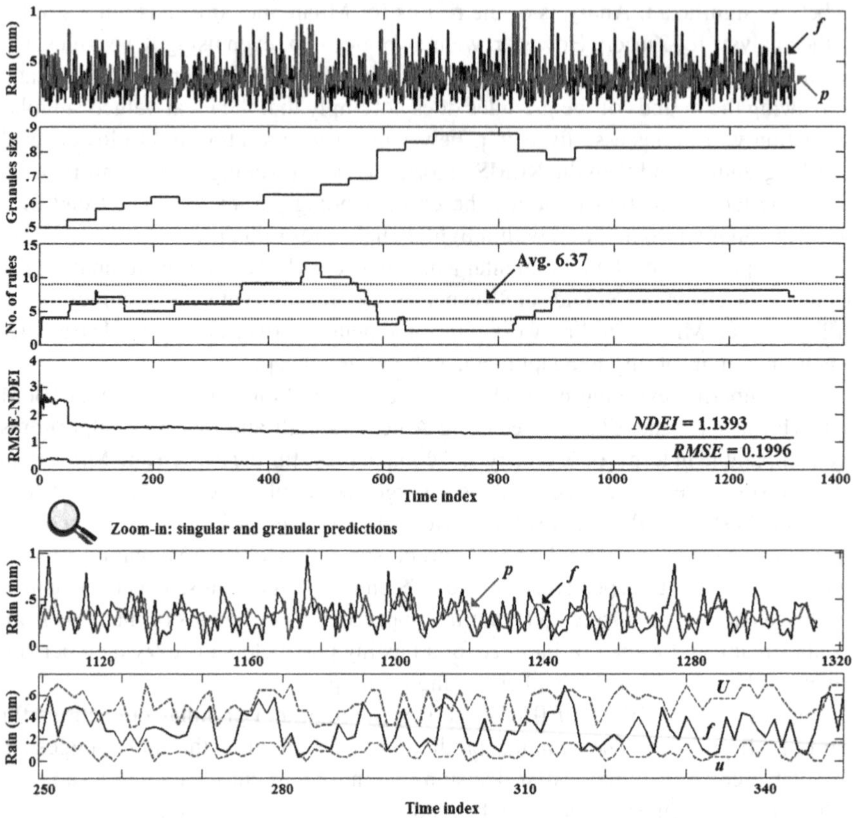

Fig. 11.8 IBeM prediction of Zurich Fluntern rainfall

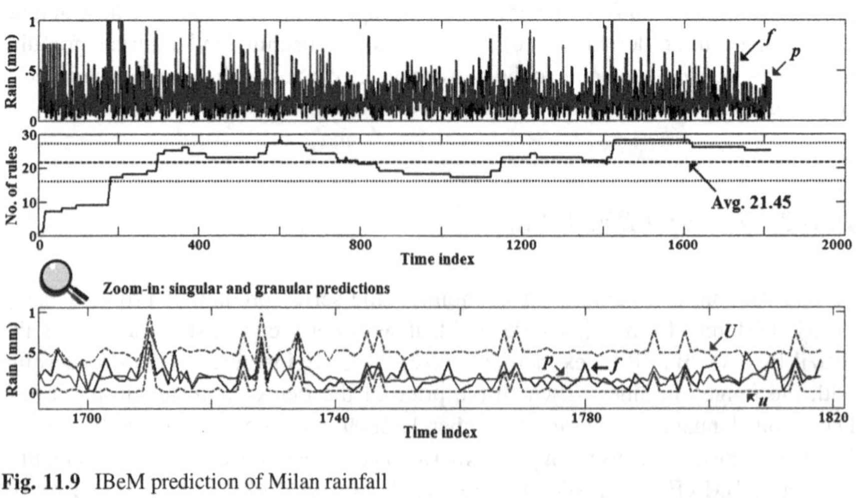

Fig. 11.9 IBeM prediction of Milan rainfall

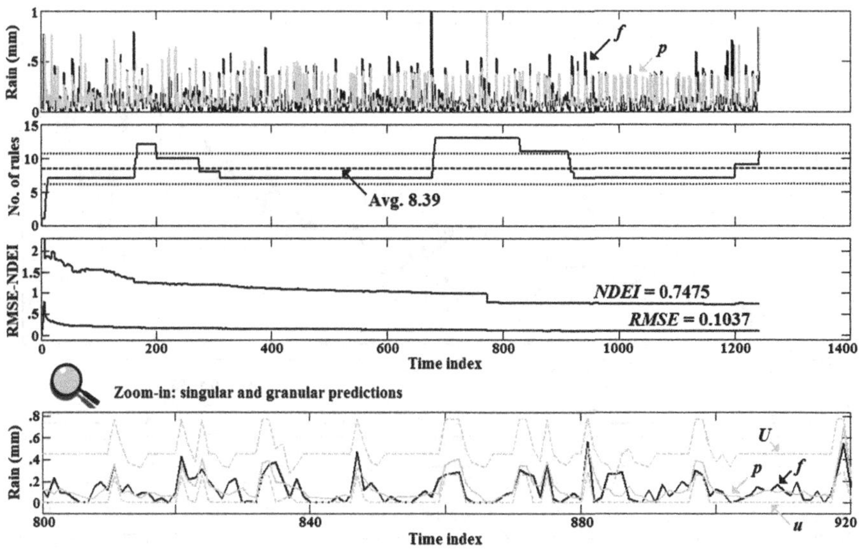

Fig. 11.10 IBeM prediction of Athens rainfall

largest stock exchange in the Americas in terms of market capitalization. The market capitalization used by IBovespa is the value of the publicly tradable part of the companies. The benchmark indicator of BM&F BOVESPA is the Bovespa BVSP index. BVSP price forecasts aim at giving information to support portfolio construction, risk management, and investment decisions.

The task of IBeM is to foresee the variation of the BVSP price in the next business day, $[\underline{y},\overline{y}]^{[h+1]}$, based on granular patterns observed in the last five business days $[\underline{x},\overline{x}]^{[h-4]},\ldots,[\underline{x},\overline{x}]^{[h]}$. Notice that the vast majority of machine learning algorithms cannot handle this type of data automatically and require off-line preprocessing steps and assumptions on how to curtail interval data into representative real numbers. IBeM inspects the data only once to mimic an online data stream. Its structure, initially null, grows on demand, steered by the information flow. The following parameter values were chosen to evaluate the IBeM behavior: $\rho^{[0]} = \sigma^{[0]} = 0.2$, deletion threshold $h_r = 300$, and $\eta = 3$. This parametrization stresses structural stability and a small number of rules. Figure 11.11 summarizes the results.

Figure 11.11 shows how the learning algorithm self-adapts the maximum width allowed for the granules during evolution. When the time series started bringing many new information and patterns due to the late-2000s economic recession (the great recession, which began in the United States, but affected the entire world economy), the IBeM learning algorithm automatically reduced the size of the granules to avoid losing information. The number of rules in the model structure increased accordingly to guarantee a complete coverage of the problem domain. Nonlinearities and novel behaviors were captured dynamically. Moreover, Fig. 11.11 illustrates the one-step granular interval forecast of the BVSP index,

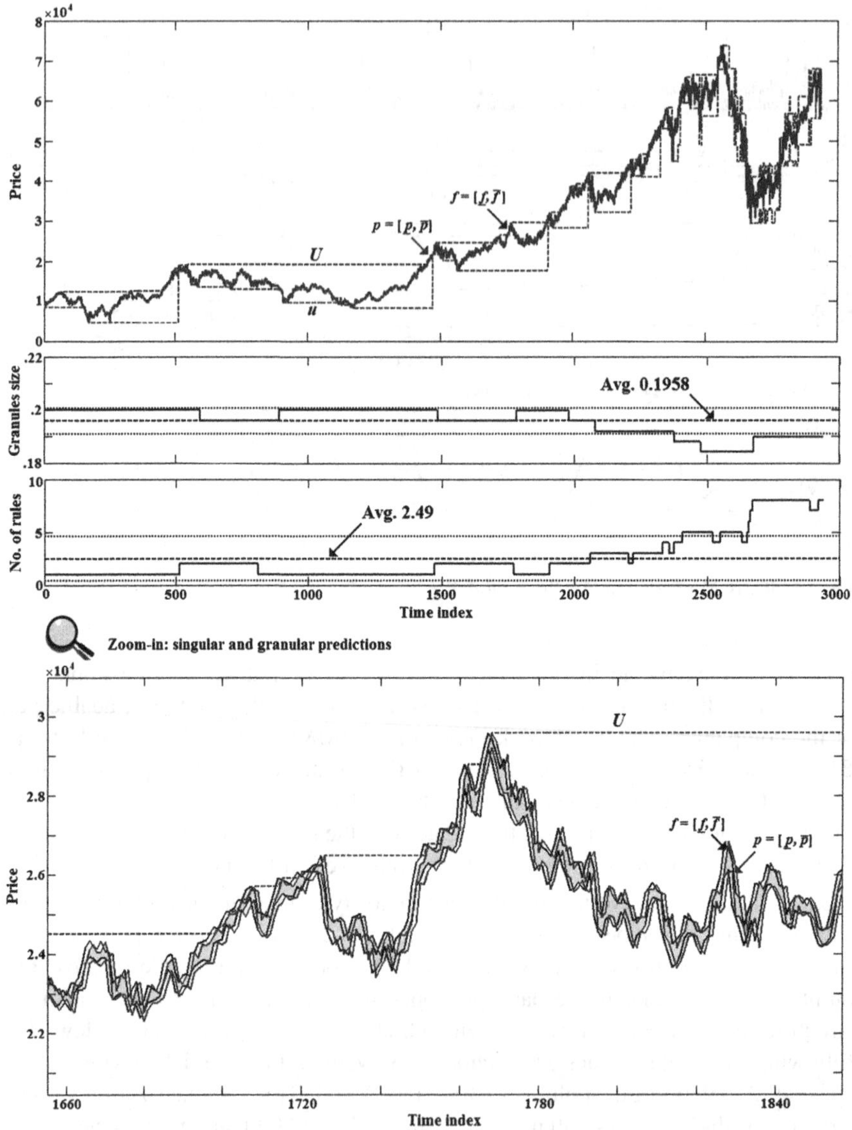

Fig. 11.11 IBeM one-step-ahead predictions for interval BVSP index data

$[\underline{p}, \overline{p}]$, and the outer approximation of the time series, $[u, U]$. We notice that the IBeM model provides accurate granular forecasts from the point of view of the *RMSE*, 0.0108, and *NDEI*, 0.0447, indices, and that it summarizes the content of the data stream into an average of 2.49 rules, with a maximum of eight rules. The interval enclosure in this experiment may be interpreted as optimistic and pessimistic bounds of the selling price, an important information which helps to reduce investment risks and stipulates portfolio return. We remark that IBeM runs in linear time with respect to the number of instances. The results illustrate the potential of evolving granular models to solve financial prediction problems that demand online incremental adaptability.

11.7 Conclusion

This work has introduced IBeM to assess the essence of heterogeneous data streams. The IBeM approach to granulation is based on changeable local models for evolving data structures. Focus was given on interval manifestation of data and on granular modeling framework. The IBeM approach for function approximation makes no specific assumption about the properties of the data sources but rather let the data stream guide the structural development and model learning freely. Application examples considering rainfall prediction and finance system have shown the usefulness of the approach. Further work shall address interval and fuzzy granular frameworks for interval and fuzzy data streams.

Acknowledgements The first author acknowledges CAPES, Brazilian Ministry of Education, for his fellowship. The second author thanks the Energy Company of Minas Gerais - CEMIG, Brazil, for grant P&D178. The last author is grateful to CNPq, the Brazilian National Research Council, for grant 304596/2009-4. The authors also thank the climate data sets from the ECA&D and EU-FP6 project Millennium and the economical data set from the Yahoo! Finance. Comments and suggestions of anonymous referees helped to improve the manuscript and are kindly acknowledged.

References

1. Kreinovich, V.: Interval computations as an important part of granular computing: an introduction. In: Pedrycz, W.; Skowron, A.; Kreinovich, V. (Eds.). Handbook of Granular Computing, pp. 1–31 (2008)
2. Pedrycz, W.: Evolvable fuzzy systems: some insights and challenges. Evolving Systems, **1** (2), pp. 73–82 (2010)
3. Bargiela A., Pedrycz, W.: Granulation of temporal data: a global view on time series. Int. Conf. of the North American Fuzzy Information Processing Society, pp. 191–196 (2003)
4. Angelov, P., Filev, D.: An approach to online identification of Takagi-Sugeno fuzzy models. IEEE Transactions on Systems, Man and Cybernetics — Part B: Cybernetics, **34** (1), pp. 484–498 (2004)
5. Leite, D., Gomide, F.: Evolving linguistic fuzzy models from data streams. To appear in: Trillas, E.; Bonissone, P.; Magdalena, L.; Kacprycz, J. (Eds.). Studies in Fuzziness and Soft Computing: A Homage to Abe Mamdani, Springer (2011)
6. Leite, D., Costa Jr., P., Gomide, F.: Evolving granular neural network for semi-supervised data stream classification. International Joint Conference on Neural Networks, pp. 1–8 (2010)
7. Leite, D.; Costa Jr., P.; Gomide, F.: Interval-based evolving modeling. IEEE Symposium Series on Computational Intelligence — Workshop ESDIS, pp. 1–8 (2009)

8. Leite, D.; Costa Jr., P.; Gomide, F: Granular approach for evolving system modeling. In: E. Hullermeier, R. Kruse, and F. Hoffmann (Eds.). Lecture Notes in Artificial Intelligence, **6178**, pp. 340–349, Springer (2010)
9. Yao, Y. Y.: Interpreting concept learning in cognitive informatics and granular computing. IEEE Transactions on Systems, Man, and Cybernetics — Part B: Cybernetics, **39** (4), pp. 855–866 (2009)
10. Yao, Y. Y.: Human-inspired granular computing. In: J. T. Yao (Ed.), Novel Developments in Granular Computing: Applications for Advanced Human Reasoning and Soft Computing (2010)
11. Yao, Y. Y.: The art of granular computing. International Conference on Rough Sets and Emerging Intelligent Systems Paradigms, LNAI **4585**, pp. 101–112 (2007)
12. Zadeh, L. A.: Toward a theory of fuzzy information granulation and its centrality in human reasoning and fuzzy logic. Fuzzy Sets and Systems, **90** (2), pp. 111–127 (1997)
13. Pedrycz, W., Skowron, A., Kreinovich, V. (Eds.): Handbook of Granular Computing. Wiley-Interscience, Hoboken, New Jersey (2008)
14. Bargiela A., Pedrycz, W.: Granular Computing: An Introduction. Kluwer Academic Publishers - Boston, Dordrecht, London (2003)
15. Jaulin, L., Keiffer, M., Didrit, O., Walter, E.: Applied Interval Analysis. Springer-Verlag, London (2001)
16. Shafer, J. L.: Analysis of Incomplete Multivariate Data. Chapman and Hall, London (1997)
17. Little, R. J. A., Rubin, D. B.: Statistical Analysis with Missing Data. $2^n d$ edition, Wiley-Interscience, Hoboken, New Jersey (2002)
18. Ozawa, S.; Pang, S.; Kasabov, N.: Incremental learning of chunk data for online pattern classification systems. IEEE Trans. on Neural Networks, **19** (6), pp. 1061–1074 (2008)
19. Last, M.: Online classification of nonstationary data streams. Intelligent data analysis, vol. **6** (2), pp. 129–147 (2002)
20. Yager, R.: Measures of specificity over continuous spaces under similarity relations. Fuzzy Sets and Systems, **159** (17), pp. 2193–2210 (2008)
21. Hansen, E. R.; Walster, G. W.: Global Optimization using Interval Analysis. Marcel Dekker, 2nd edition, New York - Basel (2004)
22. Moore, R. E.: Interval Analysis. Prentice Hall, Englewood Cliffs, N.J. (1966)
23. Moore, R. E.: Methods and Applications of Interval Analysis. SIAM, Philadelphia (1979)
24. Kearfott, R. B.; Kreinovich, V.: Applications of Interval Computations. Kluwer Academic Publishers (1996)
25. Neumaier, A.: Interval Methods for Systems of Equations. Cambridge University Press, Cambridge (1990)
26. Hickey, T., Ju, Q., van Emden, M. H.: Interval arithmetic: from principles to implementation. Journal of the ACM, textbf48 (5), pp. 1038–1068 (2001)
27. Moore, R. E., Kearfott, R. B., Cloud, M. J.: Introduction to Interval Analysis. SIAM, Philadelphia (2009)
28. Petkovic, M. S., Petkovic, L, D.: Complex Interval Arithmetic and Its Applications. Wiley - VCH, Germany (1998)
29. Hahn, G. J., Meeker, W. Q.: Statistical Intervals: A Guide for Practitioners. Wiley, USA (1991)
30. Moore, R. E.; Lodwick, W.: Interval analysis and fuzzy set theory. Fuzzy Sets and Systems, **135** (1), pp. 5–9 (2003)
31. Young, P. C.: Recursive Estimation and Time-Series Analysis: An Introduction. Springer-Verlag, Berlin, (1984)
32. Ballini, R., Mendonca, A., Gomide, F.: Evolving fuzzy modeling of sovereign bonds. The Journal of Financial Decision Making, special issue: The Fuzzy Logic in the Financial Uncertainty, **5** (2) (2009)
33. Astrom, K. J.; Wittenmark, B.: Adaptive Control. Addison-Wesley Longman Publishing Co., Inc., Boston, USA, 2 edition, (1994)
34. Johnson, C. R.: Lectures on Adaptive Parameter Estimation. Prentice-Hall, Inc., Upper Saddle River, USA (1988)

Part IV
Applications of Learning in Non-Stationary Environments

Chapter 12
Dynamic Learning of Multiple Time Series in a Nonstationary Environment

Harya Widiputra, Russel Pears, and Nikola Kasabov

Abstract This chapter introduces two distinct solutions to the problem of capturing the dynamics of multiple time series and the extraction of useful knowledge over time. As these dynamics would change in a nonstationary environment, the key characteristic of the methods is the ability to evolve their structure continuously over time. In addition, reviews of existing methods of dynamic single time series analysis and modeling such as the dynamic neuro-fuzzy inference system and the neuro-fuzzy inference method for transductive reasoning, which inspired the proposed methods, are presented. This chapter also presents a comprehensive evaluation of the performance of the proposed methods on a real-world problem, which consists of predicting movement of global stock market indexes over time.

12.1 Introduction

Time series data is a train of numerical data points in sequential order, usually recorded in uniform intervals. Thus, a time series consists of a sequence of numbers collected at regular intervals over a period of time. In statistics, signal processing, econometrics, and mathematical finance, a time series is described as a sequence of data points, measured typically at successive times spaced at uniform time intervals. Some common examples of time series are the daily closing value of an equity market, that is, the Dow Jones index or the annual flow volume of the Nile River in Egypt.

An obvious characteristic of time series data that distinguishes it from cross-sectional data is temporal ordering [38]. For example, given a time series data set on employment, the minimum wage, and other economic variables for a certain

H. Widiputra (✉) • R. Pears • N. Kasabov
The Knowledge Engineering and Discovery Research Institute, Level 7 350 Queen St, Auckland, New Zealand
e-mail: harya.widiputra@aut.ac.nz; rpears@aut.ac.nz; nkasabov@aut.ac.nz

M. Sayed-Mouchaweh and E. Lughofer (eds.), *Learning in Non-Stationary Environments: Methods and Applications*, DOI 10.1007/978-1-4419-8020-5_12,
© Springer Science+Business Media New York 2012

country, it is possible to learn that the data for year 1970 immediately precedes the data for 1971. A basic concept in analyzing time series data for a real-world phenomenon is then to recognize that the past can affect the future, but not vice versa.

Another difference between cross-sectional and time series data is more subtle. The cross-sectional data is viewed as random outcomes, which is fairly straightforward as different sample drawn from the population will generally yield different values of the independent and dependent variables. Yet, how should randomness in time series data be considered? Certainly, observations taken from the field of economics satisfy the intuitive requirement for being random variables. For instance, today's next closing value of an equity market is not known until the end of the trading day. Time series data of weather conditions also satisfy this intuitive requirement as the level of air pressure, wind speed, air humidity, etc. in a certain place at 6.00 a.m. tomorrow is not yet known today. Since the outcomes of these variables are not foreknown, they should clearly be viewed as random variables.

Formally, a sequence of random variables indexed by time is called a *stochastic process* or a *time series process* [38]. When time series data set is collected, one possible outcome or *realization* of the stochastic process is obtained. Only a single realization can be observed, since it is not possible to go back in time and start the process over again. However, if certain conditions in history had been different, generally, a different realization for the stochastic process will be obtained, and this is why a time series data is considered as the collection of the outcomes of a set of random variables.

These facts about randomness of a time series simply give a clear picture of how dynamic and nonstationary real-world phenomena are. Therefore, even though currently we are capable of estimating what might happen in the near future by constructing a model based on historical data, the real challenge is actually to be able to develop models that can dynamically learn and adapt to the new condition of these nonstationary environments as new information becomes available. Furthermore, as previous studies have revealed that dynamic relationships between series exist in multiple time series data relating to real-world phenomenona, that is, the biological and economic domains [2, 5, 6, 13, 23], it becomes imperative to be able to capture the dynamics of not just the individual variables but also how they relate to each other over time. In relation to this, this chapter outlines new methods of dynamic learning that are capable of extracting dynamic interactions between multiple time series data from real-world phenomenon.

We review the general concept of learning as well as existing methods of time series modeling before the new methods of dynamic learning of multiple time series are explained.

12.2 Time Series Analysis and Modeling

Time series analysis comprises methods for analyzing time series data in order to extract meaningful statistics and other characteristics of the data. There are two main goals of time series analysis: (a) identifying the nature of the phenomenon represented by the sequence of observations and (b) forecasting (predicting future values of the time series variable). Both of these goals require that the pattern of observed time series data is identified and more or less formally described. Once the pattern is established, it can be interpreted and integrated with other data.

Regardless of the depth of the understanding and validity of the interpretation of the phenomenon, the identified pattern can be extrapolated to predict future events. Time series forecasting is the use of a model to foresee future events based on known past events, that is, to predict data points before they are measured. An example of time series forecasting in econometrics is predicting the opening price of a stock based on its past performance.

Time series data have a natural temporal ordering. This makes time series analysis distinct from other common data analysis problems, in which there is no natural ordering of the observations (e.g., explaining people's wages by reference to their educational level, where the individuals' data could be entered in any order). Time series analysis is also distinct from spatial data analysis where the observations typically relate to geographical locations (e.g., accounting for house prices by suburb). A time series model will generally reflect the fact that observations closer together in time will be more closely related than observations further apart. Therefore, it is essential in time series analysis to build a model that can dynamically evolve its structure in relation to current behavior of the system.

12.2.1 *Methods of Reasoning*

As it has been explained in the previous section, one of the objectives of time series analysis is to identify the nature of the phenomenon represented by the sequence of observations. This process can be seen as a learning or *reasoning* process. In general, there are two reasoning methods that can be used in time series analysis, and these are the *inductive* reasoning and *transductive* reasoning.

12.2.1.1 Inductive Reasoning

Induction or inductive reasoning, sometimes called inductive logic or inductive learning, is the process of reasoning in which the premises of an argument is believed to support the conclusion but does not entail it [8]. Induction is a form of reasoning that makes generalizations based on individual instances. It is used to describe properties or relations to types based on an observed instance (i.e., on a number of observations or experiences), or to formulate laws based on limited observations of recurring phenomenal patterns. This method is concerned with the creation of a model (a function) from all available data representing the entire problem space, for example, a regression formula, a neural network of multi-layer perceptron (MLP), support vector machine (SVM), etc., and then is applied on new data (deductive) [29]. Another name given to this type of reasoning is global modeling [12].

12.2.1.2 Transductive Reasoning

Transductive reasoning or transductive inference, introduced by Vapnik in 1998 [32], is defined in contrast as a method that is used to estimate the value of a potential model (function) only for a single point of space (i.e., a new data vector) by utilizing additional information related to that vector. While the inductive approach is useful when a global model of the problem is needed in an approximate form, the transductive approach is more appropriate for applications where the focus is not on the model, but rather on every individual case. This relates to the common sense principle which states that, to solve a given problem, one should avoid solving a more general problem as an intermediate step [4].

12.3 Local Modeling for Knowledge Discovery

The common realization of the inductive reasoning is the construction of global models, that is, a regression formula, MLP [16], SVM [41], etc. Global models are built using all historical data and thus can be used to predict future trends. However, the trajectories that global models produce often fail to track localized changes that take place at discrete points in time. This is due to the fact that trajectories tend to smooth localized deviations by averaging the effects of such deviations over a long period of time [36] (Fig. 12.1).

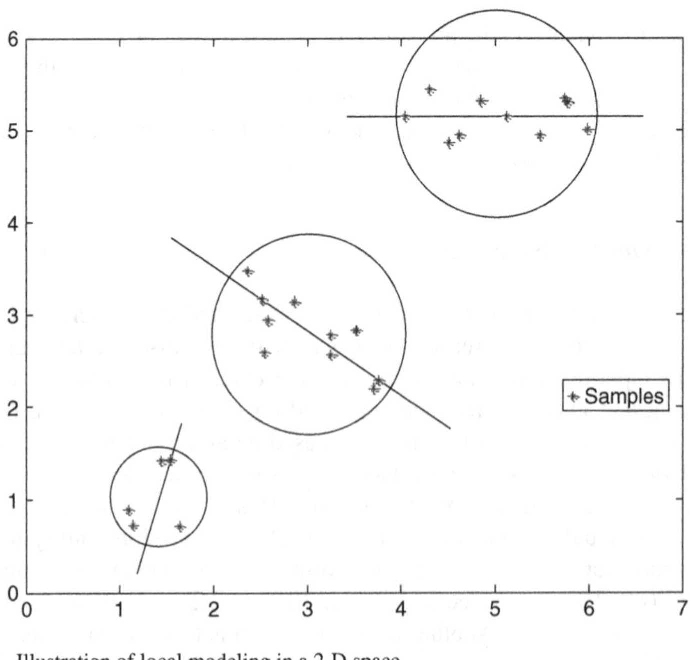

Fig. 12.1 Illustration of local modeling in a 2-D space

In reality, localized disturbances may be of great significance as they capture the conditions under which a time series deviates from the norm. For example, financial markets react very favorably when interest rates are cut or when better than expected economic fundamentals are announced by a government under which they operate. To accurately capture such phenomena requires a discontinuity in the global trajectory function, and this goes against the fundamental design philosophy behind the construction of global models. Furthermore, it is of interest to capture similar deviations from a global trajectory that take place repeatedly over time, in other words, to capture recurring deviations from the norm that are similar in shape and magnitude. Such localized phenomena can only be captured accurately by localized models that are built only on data that define the phenomenon under consideration and are not contaminated by data outside the underlying phenomenon.

Local models [11, 15, 18, 22, 24, 29, 40] is a type of model ensemble that breaks down the problem into many smaller subproblems, based on its position in the problem space. Local models can be built by grouping together data that has similar behavior. For example, when the value of a variable suddenly increases significantly and then maintains the increased value over a period of time, a natural cluster containing the time points that define this heightened activity can be defined. Different types of phenomena will define their own clusters. Models can then be developed for each cluster (i.e., local regressions) that will yield better accuracy over the local problem space covered by the model in contrast to a global model.

In local modeling, individual models are created to evaluate the output function for only a subset of the problem space, for example, a set of rules over a cluster or a set of local regressions, etc. Having a set of local models offers greater flexibility as predictions can be either on the basis of a single model or, if needed, at a global level by combining the predictions made by the individual local models [12]. Additionally, it is expected that local models would enable us to capture recent trends in the data and relate them to similar behavior from the past. This is in contrast to a global model that takes into account *all* past activity, thus resulting in diluting the effects of recent trends in the data [36].

12.3.1 Dynamic Evolving Neuro-Fuzzy Inference System

The dynamic evolving neural-fuzzy inference system, denoted as DENFIS, introduced and proposed by Kasabov and Song in 2002 [15], is a fuzzy inference systems for adaptive online learning and dynamic single time series analysis and prediction. DENFIS evolves through incremental, hybrid (supervised/unsupervised), learning and accommodates new input data, including new features, new classes, etc., through local element tuning. New fuzzy rules are created and updated during the operation of the system. At each time moment, the output of DENFIS is calculated through a fuzzy inference system based on the m-most activated fuzzy rules which are dynamically chosen from a fuzzy rule set.

12.3.1.1 Learning Processes in DENFIS

In DENFIS, the rules are created and updated at the same time with the input space partitioning using the specially designed evolving, online, maximum distance-based clustering method called the evolving clustering method and denoted as ECM [28].

The ECM is an evolving, online, maximum distance-based clustering method proposed by Song and Kasabov in 2001 [28] to implement a scatter partitioning of the input space for the purpose of creating fuzzy inference rules. ECM is a fast, one-pass algorithm for a dynamic estimation of the number of clusters in a set of data samples and for finding their current centers in the input data space. It is a distance-based clustering method where the cluster centers are represented by evolved nodes in an online mode. In any cluster, the maximum distance, *MaxDist*, between a data sample and the cluster center, is less than a threshold value, *Dthr*, that has been set as a clustering parameter. This parameter would affect the number of clusters to be created.

In the clustering process, the data samples come from a data stream, and this process starts with an empty set of clusters. When a new cluster is created, its cluster center, *Cc*, is located and its cluster radius, *Ru*, is initially set with a value 0. As new samples are presented one after another, new clusters may be created or some already created clusters will be updated through changing their centers' positions and increasing their cluster radii. Which cluster should be updated and how it should be changed depends on the position of the current data sample. A cluster will not be updated any more when its cluster radius, *Ru*, has reached the special value that is, usually, equal to the threshold value *Dthr* (Fig. 12.2).

In DENFIS, the first-order Takagi-Sugeno-type fuzzy rules [31] are employed, and the linear functions in the consequences are created using weighted linear least-square estimator (WLSE) and updated by recursive weighted linear least-square estimator (RWLSE) [7] with learning data. Each of the linear functions can be expressed as follows:

$$y = \beta_0 + \beta_1 x_1 + \beta_2 x_2 + \ldots + \beta_q x_q. \qquad (12.1)$$

The creation of the first m fuzzy rules in DENFIS is described as follow:

- Step 1: Take the first n_0 learning data samples from the learning data set.
- Step 2: Implement clustering using ECM to these n_0 data to obtain m cluster centers.
- Step 3: For every cluster center C_i, find p_i data samples from the learning data set whose positions in the input space are closest to the center, $i = 1, 2, \ldots, m$.
- Step 4: To obtain a fuzzy rule corresponding to a cluster center, create the antecedents of the fuzzy rule using the position of the cluster center and use either a triangular or Gaussian membership function. Using the weighted linear least-square estimator on p_i data samples, calculate the coefficients of the consequent function. The distances between p_i data samples and the cluster center are taken as the weights.

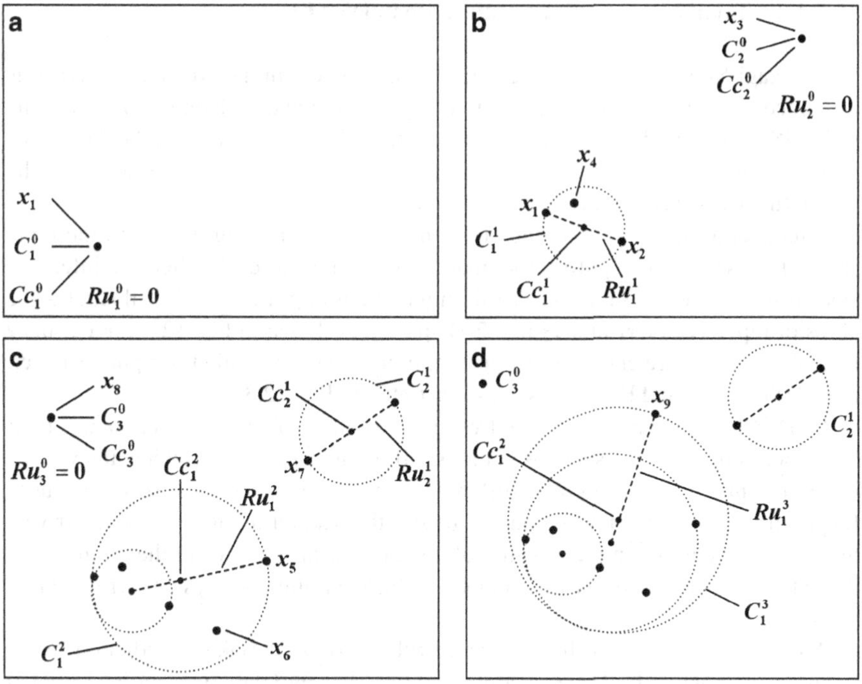

Fig. 12.2 Illustration of ECM clustering process in a 2-D space [28]

In the above steps, m, n_0, and p are the parameters of the DENFIS online learning model, and the value of p_i should be greater than the number of input variables, q.

As new data samples are presented to the system, new fuzzy rules may be created and existing rules updated. A new fuzzy rule is created if a new cluster center is found by the ECM. The antecedent of the new fuzzy rule is formed using either a triangular or Gaussian membership function with the position of the cluster center as a rule node. An existing fuzzy rule whose rule node is the closest to the new rule node is then found; the consequence function of this rule is then taken as the consequence function for the new fuzzy rule.

For every data sample, several existing fuzzy rules are updated using RWLSE if their rule nodes have distances to the data point in the input space that are not greater than $2 \times Dthr$ (the threshold value, a clustering parameter in ECM). The distances between these rule nodes and the data sample in the input space are taken as the weights. In addition to this, one of these rules may also be updated through changing its antecedent so that, if its rule node position (cluster center) is changed by the ECM, consequently, the fuzzy rule will then have a new antecedent.

For each input vector, a Takagi-Sugeno inference system with m activated rules is dynamically created. The rules are chosen based on the position of the input vector. Since in DENFIS the rules are updated continuously, two input vectors with the same values at different time points may have different inferences as the fuzzy rules may have been updated before the second input vector entered the system.

12.3.1.2 Takagi-Sugeno Fuzzy Inference in DENFIS

The Takagi-Sugeno fuzzy inference system utilized in DENFIS is a dynamic inference system. In addition to dynamically creating and updating fuzzy rules, the DENFIS online model has some other major differences with the other inference systems. Firstly, for each input vector, DENFIS chooses m fuzzy rules from the whole fuzzy rule set for forming a current inference system.

This operation depends on the position of the current input vector in the input space. In case of two input vectors that are very close to each other, the inference system may have the same fuzzy rule inference group. Figure 12.3 illustrates the cases of input vector x_1 and x_2 in a 2-D space. As shown in Fig. 12.3 for x_1, fuzzy rules A, B, and C are chosen to form an inference system, while for input vector x_2, fuzzy rules C, D, and E are chosen as illustrated in Fig. 12.3.

In DENFIS, however, even if two input vectors are exactly the same, their corresponding inference systems could be different. This happens when the vectors were presented to the system at different time moments and the fuzzy rules used for the first input vector was updated before the second input vector had arrived. Secondly, depending on the position of the current input vector in the input space, the antecedents of the fuzzy rules chosen to form an inference system for this input vector may vary.

An example of a set of three activated rules chosen to make a prediction for an input vector **x** when DENFIS is applied to the Mackey-Glass data set is presented in Fig. 12.4.

12.4 Instance-Based Learning for Knowledge Discovery

In contrast to learning methods that construct a general and explicit description of the target function when training examples are provided, transductive reasoning methods simply store the training examples. Generalizing beyond these examples is postponed until a new instance must be classified. A key advantage of this type of learning method is that instead of estimating the target function once for the entire instance space, this method is capable of constructing local and specific estimation models for each new instance that needs to be classified or predicted.

The k-NN [30] and WKNN algorithms, which fall under the category of *instance-based* learning, are well-known realizations of transductive reasoning. This type of learning offers the following benefits over the local model:

1. In a real-world problem where the amount of data increases on an ongoing basis, instance-based learning will only utilize that part of the data that is relevant to the new input vector.
2. Since only a relevant subset of the input vectors in the sample data set is used to derive the solution, it may reduce the effect of outliers, or incorrect identification of subproblems.

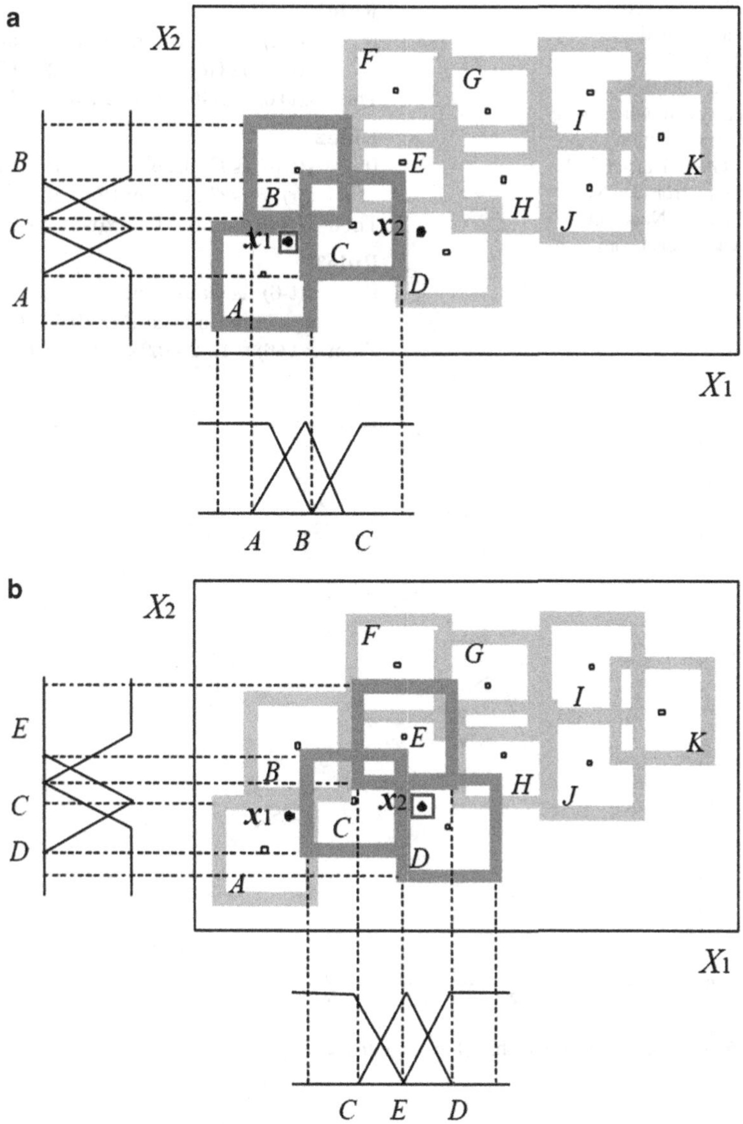

Fig. 12.3 Illustration of the construction of two fuzzy rule groups by DENFIS in a 2-D space for input vector x_1 and x_2 that is entered at a later time moment. Figure is extracted from [15]

The limitation of instance-based learning is in its reliance on good definition of problem space utilized to build the solution (Fig. 12.5). A good definition of problem space is important to every type of reasoning; however, it may be more so with instance-based learning through transductive reasoning. This is because the

Fig. 12.4 Three activated fuzzy rules created and chosen by DENFIS to construct a Takagi-Sugeno inference system when being applied for prediction of the Mackey-Glass data set. Rules are extracted from DENFIS available in the NeuCom (http://www.theneucom.com)

Rule1
If x(t-6) is GaussianMF (0.50 0.59)
x(t) is GaussianMF (0.52 0.73)
Then x(t+6) = 1.40 - 0.06*x(t-6) + 0.56*x(t)

Rule2
If x(t-6) is GaussianMF (0.49 0.73)
x(t) is GaussianMF (0.49 0.77)
Then x(t+6) = 1.36 - 0.96*x(t-6) + 1.39*x(t)

Rule3
If x(t-6) is GaussianMF (0.49 0.78)
x(t) is GaussianMF (0.50 0.59)
Then x(t+6) = 1.22 - 0.92*x(t-6) + 1.53*x(t)

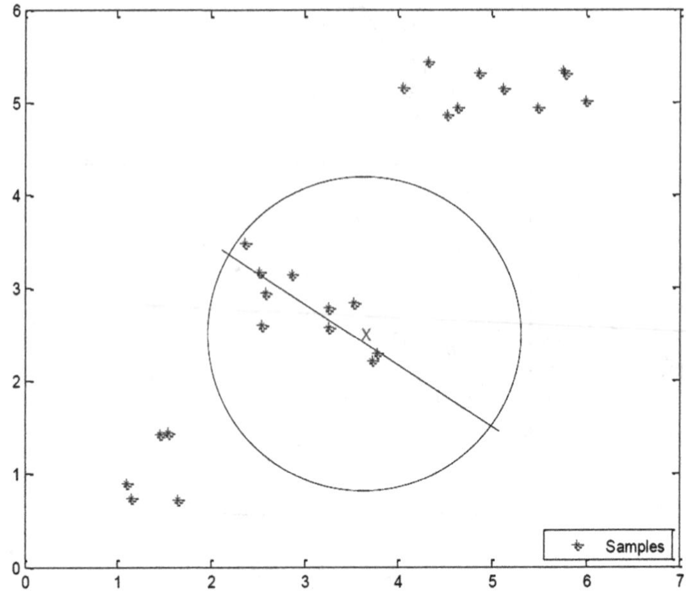

Fig. 12.5 Illustration of instance-based learning in a 2-D space

definition of problem space affects the performance of the similarity function used to identify the neighborhood, that is, a subset of input vectors in the training data which are relevant to the new test input vector.

Despite its limitations, instance-based learning has been widely used to solve classification problems such as text classification [9], heart disease diagnostics [39], synthetic data classification using graph-based approach [19], digit and speech recognition [10], promoter recognition in bioinformatics [14], image recognition [21] and image classification [25], microarray gene expression classification [33], and biometric tasks such as face surveillance [20].

Furthermore, this reasoning method is also used in prediction tasks such as finding if a given drug binds to a target site [33], evaluating prediction reliability in regression [4], and providing additional measures to determine reliability of predictions made in medical diagnosis [17]. However, the use of this learning method for time series analysis and prediction, in particular multiple time series, has not been widely studied except for the preliminary study by Widiputra et al. in 2009 [34] which investigated the possibility of using the WKNN in predicting movement of multiple stock market indexes [34].

12.4.1 Neuro-Fuzzy Inference Method

The neuro-fuzzy inference method for transductive reasoning denoted as NFI is a dynamic fuzzy inference system with local generalization proposed by Song and Kasabov in 2005 [29], in which, either the Zadeh-Mamdani [42]- or the Takagi-Sugeno [31]-type fuzzy inference is used. The local generalization means that in a subspace of the whole problem space (local area), a model is created from N_i training samples that are closest to the input vector $\mathbf{x}_i$ which is later used to generalize to the subspace.

In the Zadeh-Mamdani type of NFI model, Gaussian fuzzy membership functions are applied in each fuzzy rule for both antecedent and consequent parts, while for the Takagi-Sugeno type of NFI model, the consequent part is presented by a linear or nonlinear function. A back propagation/steepest descent [1] learning algorithm is used for optimizing the parameters of the fuzzy membership functions (in both Zadeh-Mamdani and Takagi-Sugeno types). The distance between two vectors x and y is measured in the NFI model as the *normalized Euclidean distance* defined as follows (the values range between 0 and 1):

$$\| \mathbf{x} - \mathbf{y} \| = \left(\frac{1}{q} \sum_{j=1}^{q} (x_j - y_j)^2 \right)^{1/2}, \tag{12.2}$$

where $\mathbf{x}, \mathbf{y} \in \Re^q$, and q is number of input variables.

To partition the input space N_i for creating and obtaining initial values of fuzzy rules, the ECM [28] is again applied as in DENFIS and the cluster centers and radii are taken as initial values for the centers and widths, respectively, for the Gaussian membership functions (for both Zadeh-Mamdani and Takagi-Sugeno types). For the Takagi-Sugeno type of NFI model, the training samples belonging to a cluster are used for creating a linear function as a local model for output function evaluation.

12.4.1.1 NFI Learning Algorithm

For each new input vector $\mathbf{x}_i$, the NFI model performs the following learning algorithm [29]:

- Step 1, search in the training data set based on the input space to find N_i training samples that are closest to $\mathbf{x}_i$. The value for N_i can be predefined based on experience, or optimized through the application of an optimization procedure. In the NFI model, the former approach is used. Here, N_i can be considered as number of k as in the k-NN algorithm.
- Step 2, calculate the distances $d_j; j = 1,2,...,N_i$ between each of these samples and $\mathbf{x}_i$ using the normalized Euclidean distance (as in (12.2)). Calculate the weights $w_j = 1 - (d_j - \min(\mathbf{d})); j = 1,2,...,N_i$ where $\min(\mathbf{d})$ is the minimum value in the distance vector $\mathbf{d} = (d_1, d_2, ..., d_{N_i})$.
- Step 3, use the ECM to cluster and partition the input subspace that consists of N_i selected training samples.
- Step 4, create fuzzy rules and set their initial parameter values according to the clustering results of the ECM; for each cluster, the cluster center is taken as the center of a fuzzy membership function (Gaussian function) and the cluster radius is taken as the width.
- Step 5, apply the steepest descent method (backpropagation) to optimize the parameters of the fuzzy rules in the local model LM_i.
- Step 6, calculate the output value y_i for the input vector $\mathbf{x}_i$, applying fuzzy inference over the set of fuzzy rules that constitute the local model LM_i.
- Step 7, end of the procedure.

The procedure of optimizing the parameters in the NFI model (step 5 in the above algorithm) is described as follows:

- Consider the system having q inputs, one output, and M fuzzy rules defined initially through the ECM clustering procedure, the lth rule would have the form of:

  $$\text{R}_l : \text{if } x_1 \text{ is } F_{l,1} \text{ and } ... \text{ and } x_q \text{ is } F_{l,q} \text{ then } y \text{ is } G_l \ (Zadeh - Mamdani)$$

  or,

  $$\text{R}_l : \text{if } x_1 \text{ is } F_{l,1} \text{ and } ... \text{ and } x_q \text{ is } F_{l,q} \text{ then } y \text{ is } n_l \ (Takagi - Sugeno).$$

  Here, $F_{l,q}$ are fuzzy sets defined by the following Gaussian-type membership function:

  $$\text{GaussianMF} = \alpha \exp\left(-\frac{(x-\mu)^2}{2\sigma^2}\right), \tag{12.3}$$

  where μ is the center of the fuzzy membership function and σ is the width. In the NFI model, the center of the fuzzy membership function is initially defined by the cluster center, while the width is defined by the cluster radius. For the

Zadeh-Mamdani type, G_l is of a similar type as $F_{l,q}$, while for the Takagi-Sugeno type, n_l is defined by a linear function as follows:

$$n_l = \beta_{l,0} + \beta_{l,1}x_1 + \beta_{l,2}x_2 + \ldots + \beta_{l,q}x_q.$$

- Using the modified center average defuzzification procedure, the output value of the system can be calculated for an input vector $\mathbf{x}_i = (x_{i,1}, x_{i,2}, \ldots, x_{i,q})$ as follows for the Zadeh-Mamdani type:

$$f(\mathbf{x}_i) = \frac{\sum_{l=1}^{M} \frac{G_l}{\delta_l^2} \prod_{j=1}^{q} \alpha_{lj} \exp\left(-\frac{(x_{ij} - \mu_{lj})^2}{2\sigma_{lj}^2}\right)}{\sum_{l=1}^{M} \frac{1}{\delta_l^2} \prod_{j=1}^{q} \alpha_{lj} \exp\left(-\frac{(x_{ij} - \mu_{lj})^2}{2\sigma_{lj}^2}\right)}, \tag{12.4}$$

or as follows for the Takagi-Sugeno type:

$$f(\mathbf{x}_i) = \frac{\sum_{l=1}^{M} n_l \prod_{j=1}^{q} \alpha_{lj} \exp\left(-\frac{(x_{ij} - \mu_{lj})^2}{2\sigma_{lj}^2}\right)}{\sum_{l=1}^{M} \prod_{j=1}^{q} \alpha_{lj} \exp\left(-\frac{(x_{ij} - \mu_{lj})^2}{2\sigma_{lj}^2}\right)}. \tag{12.5}$$

- Suppose the NFI model is given a training input–output data pair $(\mathbf{x}_i, o_i)$, the system minimizes the following objective function (a weighted error function):

$$E = \frac{1}{2} w_i \left(f(\mathbf{x}_i) - o_i\right)^2, \tag{12.6}$$

where w_i is defined in step 2 of the NFI learning algorithm. The steepest descent algorithm/backpropagation [1] is used then to obtain the formulas for the optimization of the parameters $G_l, \delta_l, \alpha_{lj}, \mu_{lj}$, and σ_{lj} of Zadeh-Mamdani-type NFI model such that the value of E from (12.6) is minimized.

When being applied for time series prediction, the algorithm above is executed for each new time series point. Therefore, as the NFI creates a unique submodel for each input vector, it usually needs more processing time than inductive models, especially in the case of large data sets. Furthermore, with the existence of new input vectors with exactly the same or very similar condition, the NFI model will create the same or similar models repeatedly. Consequently, time complexity of the method depends mainly on the *search algorithm*, employed for similar data to the new input vector from the complete set of data samples.

12.5 Local Modeling of Multiple Time Series

It is interesting to note that most of the research carried out in the field of time series modeling and prediction have based their approach on the concept of inductive reasoning [8], in which a number of historical data samples are used to construct a single global model covering the entire training data set space. Nevertheless, as argued previously, local modeling is needed to cover subsets of the problem space that the global model cannot cover with sufficient accuracy. Local model is another type of realization of inductive reasoning. A system can be represented by a collection of local models trained on a given data set. However, when applied to new data, only one or a subset of the relevant models will actually contribute to the solution.

In this section, we outline a methodology to construct local models for multiple time series containing profiles of relationships between series from different time localities [37]. The construction of local models in the proposed methodology consists of two main steps, which are (1) the continuous extraction of profiles of relationships between time series over time and (2) the detection and clustering of recurring trends of movement in time series when a particular profile emerges.

The principal objective of the methodology is to construct a repository of profiles and recurring trends whose structure will dynamically evolve as changes take place in the observed nonstationary environment. This repository will then be utilized as *knowledge-based* as a key data resource to learn and understand the underlying behavior of the system and to estimate future states of the system's variables, that is, to perform a multiple time series prediction. To realize such an objective, a 2-level local modeling process is utilized within the proposed methodology.

The first level of local modeling deals with the extraction of profiles of relationships between series in a subspace of the given multiple time series data in which the methodology utilizes a cross-correlation analysis to elucidate the existence of relationships between pairs of time series that influence each other. The second level of local modeling is used to capture and cluster recurring trends of movement that take place in time series when a particular profile is emerging. Here, the methodology employs a nonparametric regression analysis, in combination with the ECM [28]. Detailed explanation of this local modeling method, which termed as the *localized trends model* and denoted as LTM, is outlined in the upcoming sections of the chapter.

12.5.1 Extracting Profiles of Relationship of Multiple Time Series

Most of the work in clustering time series data has concentrated on *sample clustering* rather than *variable clustering* [27]. However, one of the key tasks in this methodology is to group together series or variables and *not* samples that are highly correlated and have similar shapes of movement, as it is considered that multiple local models representing clusters of similar profiles will provide a better basis than a single global model for predicting future movements of the multiple time series.

For instance, in predicting movement of five global stock market indexes (i.e., New Zealand, Australia, Hong Kong, Japan, and United States), if one is able to learn that at the current time-point New Zealand and Australia are moving together collectively, Hong Kong and Japan are progressing mutually, while the United States travels by itself, then it would be relevant to use only data of stock market indexes from the past which possesses the same profiles of relationships to predict future values of these stock market indexes, rather than to use the entire data set.

Algorithm 1 outlines the scheme for clustering together similar time series. The first step in extracting profiles of relationships between multiple time series is the computation of cross-correlation coefficient between the observed time series using Pearson's correlation analysis. Yet, only statistically significant correlations, which are determined through the use of the *t-test* with a confidence level of 95%, are used. After the most significant correlations between time series have been identified, the RNOMC, *rooted normalized one-minus correlation* coefficients [27], known henceforth as *normalized correlation* in this manuscript, is calculated to assess the degree of dissimilarity between a pair of time series (a,b). The normalized correlation is given by:

$$\mathrm{RNOMC}(a,b) = \sqrt{\frac{1 - corr(a,b)}{2}}. \tag{12.7}$$

The normalized correlation coefficient ranges from 0 to 1, in which 0 denotes high similarity and 1 signifies the opposite condition.

Thereafter, the last stage of the algorithm is to extract profiles of relationships from the normalized correlation matrix. The methodology used in this step is outlined in line 3 to 24 of Algorithm 1. The whole process of extracting profiles of relationships is illustrated in Fig. 12.6. In any case, the fundamental concept of this algorithm is to group multiple time series with comparable fashion of movement while validating that every time series belong to the same cluster are correlated and hold significant level of similarity.

The underlying concept of Algorithm 1 is closely comparable to the CAST, *clustering affinity search technique*, clustering algorithm [3]. However, Algorithm 1 works by dynamically creating new clusters, deleting and merging existing clusters as it evaluates the coefficient of similarity between time series or observed variables. Therefore, Algorithm 1 is considerably different to CAST which creates a single cluster at one time and performs updates by adding new elements to the cluster from a pool of elements, or by removing elements from the cluster and returning it to the pool as it evaluates the affinity factor of the cluster in which the elements belong.

After the profiles have been extracted, then the next step of the methodology is to mine and cluster trends of movement from each profile. This process is outlined and explained in the next section. Additionally, as the time complexity of Algorithm

Pearson Correlation

	TS-1	TS-2	TS-3	TS-4
TS-1	1	0.4741	0.8637*	0.7208*
TS-2	0.4741	1	0.5483	0.8718*
TS-3	0.8673*	0.5483	1	0.7534*
TS-4	0.7208*	0.8718*	0.7534*	1

Normalized Correlation

	TS-1	TS-2	TS-3	TS-4
TS-1	-	-	0.26	0.37
TS-2	-	-	-	0.25
TS-3	0.26	-	-	0.35
TS-4	0.37	0.25	0.35	-

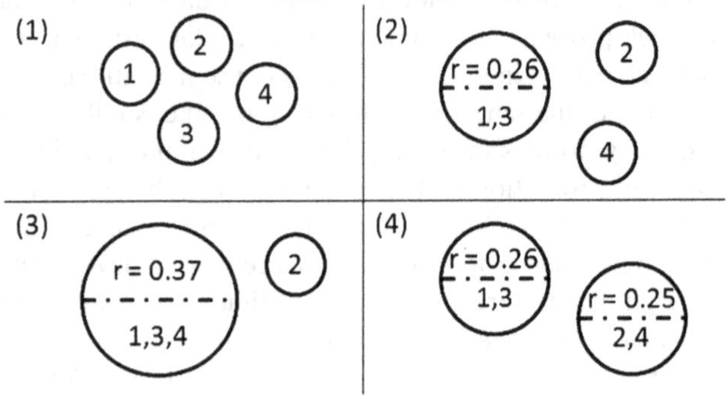

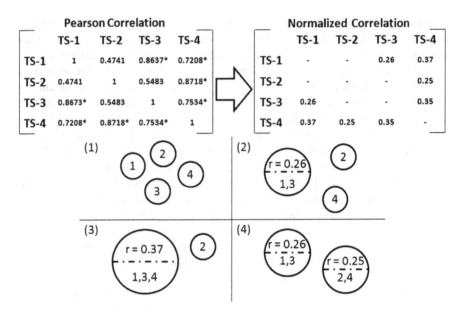

Fig. 12.6 The Pearson's correlation coefficient matrix is calculated from a given multiple time series data (TS-1,TS-2,TS-3,TS-4), and then converted to *normalized correlation*, (12.7), before the profiles are finally extracted. Statistically significant correlation coefficients are marked with *, and only these values are used to form the normalized correlation matrix whereas the insignificant coefficients are ignored. Equation (12.7) is used to calculate the normalized correlation coefficient. Figure is extracted from [37]

1 is $O(\frac{1}{2}(n^2 - n))$, to avoid expensive recomputation and extraction of profiles; extracted profiles of relationships are stored and updated dynamically instead of being computed on the fly.

12.5.2 *Clustering Recurring Trends of a Time Series*

Maintaining profiles of relationships between multiple time series allows the ability to identify which time series most influence movement of other time series in a particular time locality. However, this type of knowledge does not offer any predictive power to estimate future values of multiple time series.

To predict future values of multiple time series simultaneously, information about different shapes of movement across a group of correlated multiple time series needs to be acquired and maintained. Therefore, the methodology groups similar trends of movement into clusters which are then used to construct local models to predict future trends of movement in the time series involved.

Algorithm 5 Extracting profiles of relationship of multiple time series

Require: X, where $X_1, X_2, \ldots, X_n$ are observed time series
Ensure: profiles of relationships between multiple time series
calculate the *normalised correlation* coefficient [Equation (12.7)] of X
for each time series $X_1, X_2, \ldots, X_n$ **do**
  //pre-condition: X_i,X_j do not belong to any cluster
  if (X_i,X_j *are* correlated) AND (X_i,X_j do not belong to any cluster) **then**
    allocate X_i,X_j together in a new cluster
  end if
  //pre-condition: X_i belongs to a cluster; X_j does not belong to any cluster
  if (X_i,X_j *are* correlated) AND (X_i belongs to a cluster) **then**
    if (X_j *is* correlated with all X_i cluster member) **then**
      allocate X_j to cluster of X_i
    else if (X_i,X_j correlation > max(correlation) of X_i with its cluster member) AND (X_j *is not* correlated with any of X_i cluster member) **then**
      remove X_i from its cluster; allocate X_i,X_j together in a new cluster
    end if
  end if
  //pre-condition: X_i and X_j belong to different cluster
  if (X_i,X_j *are* correlated) AND (X_i,X_j belong to different cluster) **then**
    if (X_i *is* correlated with all X_j cluster member) AND (X_j *is* correlated with all X_i cluster member) **then**
      merge cluster of X_i,X_j together
    else if (X_i,X_j correlation > max(correlation) of X_j with its cluster member) AND (X_j *is* correlated with all X_i cluster member) **then**
      remove X_j from its cluster; allocate X_j to cluster of X_i
    else if (X_i,X_j correlation > max(correlation) of both X_i,X_j with their cluster member) AND (X_i *is not* correlated with one of X_j cluster member) AND (X_j *is not* correlated to any of X_i cluster member) **then**
      remove X_i,X_j from their cluster; allocate X_i,X_j together in a new cluster
    end if
  end if
end for
return clusters of multiple time series

12.5.2.1 General Principles

Widiputra et al. [35] proposed an algorithm to detect and cluster recurring trends of movement from localized sets of time series based on a polynomial regression function. In order to eliminate the limiting assumption of normality of data required by polynomial regression, we use a nonparametric version of regression in this research.

12.5.2.2 Learning Algorithm of Clustering Trends with Kernel Regression

The first step of the learning algorithm is to define the size of data chunk or snapshot window from which the trend of movement will be extracted using the autocorrelation analysis. This is done by applying autocorrelation analysis to the time series under examination. The next step is to extract trends of movements by performing a bootstrap sampling process through all available data chunks. This process of extracting trends of movement is achieved by utilizing the kernel regression method as explained in the previous section of this chapter. Consequently, as an outcome of the kernel regression analysis, the computed kernel weight vectors are then used as the features vectors to represent trends of movements in this methodology.

Thereafter, the algorithm implements a clustering process to group similar and recurring trends of movement. Recurring trends are grouped based on a modified version of the ECM [28], where the *correlation coefficient* is used in place of the Euclidean distance to measure similarity between a kernel weight vector and a cluster center. Additionally, in this methodology, a cluster center represents the mean of trends of movement calculated as an average value of all kernel weight vectors which belong to the same cluster. As new observations become available, new data chunks or snapshots are presented to the system. Accordingly, new clusters containing new trends of movement may be created while some existing clusters are updated. A new cluster is created when the algorithm recognizes that a new noncomparable trend of movement has emerged. Conversely, existing clusters are updated when a data chunk or snapshot with recurring trends of movement is identified.

Clusters of trends of movement are then stored in each extracted relationship profile. This information about relationships between series and trends of movements will then be exploited through *knowledge repository* to perform simultaneous multiple time series prediction. A detailed algorithm for clustering recurring trends of a time series based on the use of kernel regression is outlined as follows:

- Step 1, perform the autocorrelation analysis to the time series data set from which trends of movement will be extracted and clustered. Number of lag, as outcome of the autocorrelation analysis where $lag > 0$, with highest correlation coefficient is then taken as the size of data chunk or snapshot window n. The process will then progress by performing a bootstrap sampling process through all data chunks or snapshots.
- Step 2, create the first cluster C_1 by simply taking $\mathbf{w}_1$, which is the trend of movement of the first data chunk or snapshot $\mathbf{X}^{(1)} = (X_1^{(1)}, X_2^{(1)}, ..., X_n^{(1)})$, from the input stream as the first cluster center Cc_1 and set the cluster radius Ru_1 to 0. In this methodology, the ith trend of movement represented by the kernel weight vector $\mathbf{w}_i = (w_{i1}, w_{i2}, ..., w_{in})$ as outcome of the nonparametric regression analysis, is calculated using the Nadaraya-Watson kernel weighted average formula defined as follows:

$$\hat{X}_j^{(i)} = f_j(\mathbf{x}_j^{(i)}, \mathbf{w}_i) = \frac{\sum_{k=1}^{n} w_{ik} x_{jk}}{\sum_{k=1}^{n} x_{jk}}. \tag{12.8}$$

Here, $\mathbf{x}_j^{(i)} = (x_{j1}^i, \ldots, x_{jk}^i)$ is the extended smaller value of the original data $\mathbf{X}^{(i)}$ at domain j and certain small step dx where $j = 1, 2, \ldots, (\frac{n}{dx} + 1)$. $\mathbf{x}_j = (x_{j1}, \ldots, x_{jk})$ is calculated using the Gaussian MF equation as follows:

$$x_{jk} = K(x_j, k) = \exp\left(-\frac{(x_j - k)^2}{2\alpha^2}\right), \tag{12.9}$$

where $x_j = dx \times (j-1), k = 1, 2, \ldots, n$, and α is a predefined kernel bandwidth. The kernel weight, $\mathbf{w}_i$, is estimated using the common OLS (ordinary least square) such that the following objective functions is minimized:

$$\text{SSR} = \sum_{k=1}^{n} (X_k^i - \hat{X}_j^{(i)}), \forall \hat{X}_j^{(i)} \text{ where } x_j = X_k. \tag{12.10}$$

To gain knowledge about upcoming trend of movement when a particular trend emerge in a locality of time, the algorithm also model next trajectories of a data chunk or snapshot defined by,

$$\hat{X}_j^{(i)(u)} = f_j\left(\mathbf{x}_j^{(u)}, \mathbf{w}_i^{(u)}\right) = \frac{\sum_{k=1}^{n+1} w_{ik}^{(u)} K\left(x_j^{(u)}, k\right)}{\sum_{k=1}^{n+1} K\left(x_j^{(u)}, k\right)}, \tag{12.11}$$

where $x_j^{(u)} = dx \times (j^{(u)} - 1); j^{(u)} = 1, 2, \ldots, (\frac{n+1}{dx} + 1); k = 1, 2, \ldots, n+1$ and the kernel weights $\mathbf{w}_i^{(u)} = (w_{i1}^{(u)}, w_{i2}^{(u)}, \ldots, w_{i(n+1)}^{(u)})$.

- Step 3, if there is no more data chunk or snapshot, then the process stops (go to Step 7); else next data chunk or snapshot, $\mathbf{X}^{(i)}$, is taken. Trend of movement from $\mathbf{X}^{(i)}$ is then extracted as in Step 2, and distances between current trend and all m already created cluster centers are calculated by

$$D_{i,l} = 1 - \text{CorrelationCoefficient}(\mathbf{w}_i, Cc_l), \tag{12.12}$$

where $l = 1, 2, \ldots, m$. If found cluster center Cc_l where $D_{i,l} \leq Ru_l$, then current trend joins cluster C_l and the step is repeated; else, continue to next step.

- Step 4, find a cluster C_a (with center Cc_a and cluster radius Ru_a) from all m existing cluster centers by calculating the values of $S_{i,a}$ given by

$$S_{i,a} = D_{i,a} + Ru_a = \min(S_{i,l}), \tag{12.13}$$

where $S_{i,l} = D_{i,l} + Ru_l$ and $l = 1, 2, \ldots, m$.

- Step 5, if $S_{i,a} > 2 \times Dthr$, where $Dthr$ is a clustering parameter to limit the maximum size of a cluster radius, then current trend of $\mathbf{X}^{(i)}$, $\mathbf{w}_i$, does not belong to any existing clusters. A new cluster is then created in the same way as described in Step 2, and the algorithm returns to Step 3.

- Step 6, if $S_{i,a} \leq 2 \times Dthr$, current trend of $\mathbf{X}^{(i)}$, $\mathbf{w}_i$, joins cluster C_a. Cluster C_a is updated by moving its center, Cc_a, and increasing the value of its radius, Ru_a. The updated radius Ru_a^{new} is set to $S_{i,a}/2$, and the new center Cc_a^{new} is now the mean value of all trends of movement belong to cluster C_a. Distance from the new centre Cc_a^{new} to current trend $\mathbf{w}_i$, is equal to Ru_a^{new}. The algorithm then returns to Step 3.
- Step 7, end of the procedure.

In the procedure of clustering trends with the kernel regression, the following indexes are used:

- Number of data chunks or snapshots: $i = 1,2,...$
- Number of clusters: $l = 1,2,...,m$
- Number of input and output variables: $k = 1,2,...,n$

12.5.3 LTM for Multiple Time Series Modeling and Prediction

Figure 12.7 illustrates how a repository containing profiles of relationships and recurring trends (the knowledge repository) is built and maintained. Using data from the first data chunk or snapshot, the algorithm extracts two profiles of relationship in the multiple time series by creating two clusters. The first cluster represents a profile whereby time series #1 and time series #3 are correlated and moving together, while the second cluster is a profile of relationship whereby time series #2 and time series #4 are progressing in a similar fashion.

Trends of movement of each time series that belongs to a particular profile are then extracted and kept within the profile. As illustrated in Fig. 12.7, after extracting trend of movement from each time series in the first profile denoted by Cluster-1[TS-1,TS-3], the algorithm creates and stores two other clusters in Cluster-1[TS-1,TS-3], denoted by TS-1 and TS-3. Here TS-1 and TS-3 represent trends of movement of time series #1 and #3 when they are correlated. The same process is then applied to the second profile of time series #2 and time series #4 denoted by Cluster-2[TS-2,TS-4].

As the second data chunk or snapshot becomes available, the algorithm applies the same procedure to extract profiles of relationship in the multiple time series. As it retains same profiles from the second data chunk or snapshot which are [TS-1,TS-3], and [TS-2,TS-4], the algorithm does not create any new cluster in the knowledge repository. However, as it extracts trends of movement from each time series, the algorithm finds that the second data chunk or snapshot holds different type of behavior compared to the first data chunk or snapshot.

Consequently, the algorithm updates the information about trends of movement of each time series in all existing profiles. New clusters of trends are then created and stored in Cluster-1[TS-1,TS-3] as well as in Cluster-2[TS-2,TS-4] to represent the new behavior exhibited by the second data chunk or snapshot.

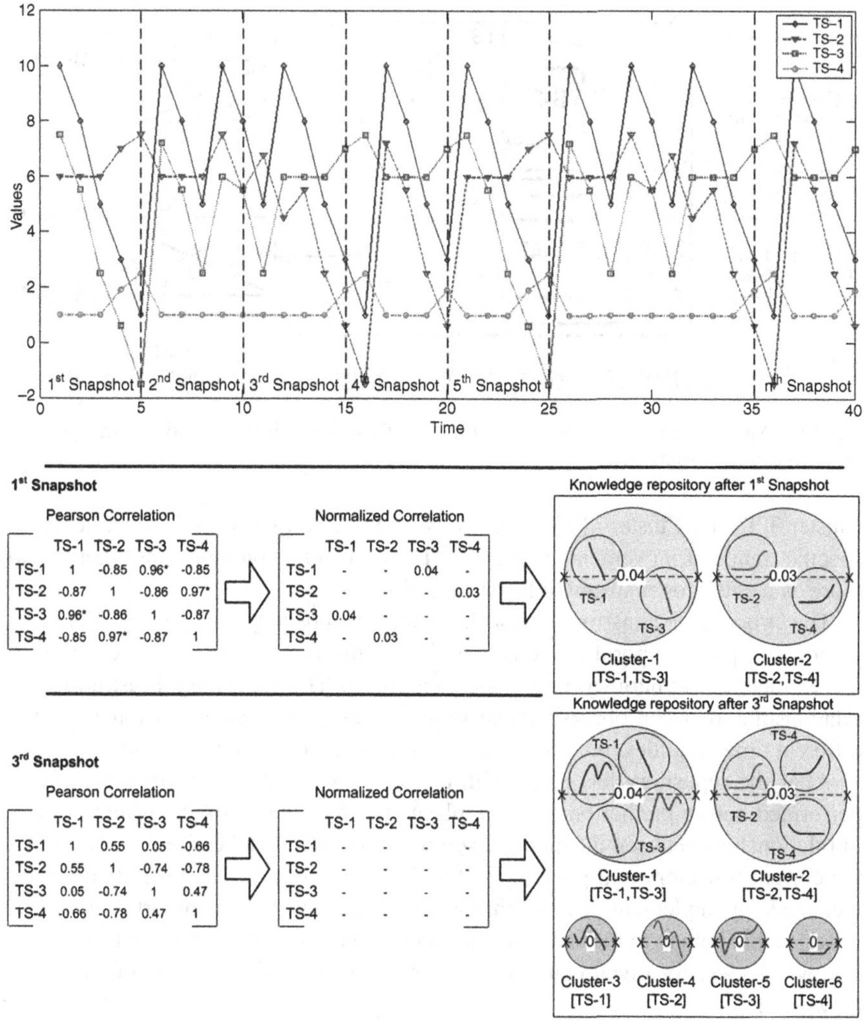

Fig. 12.7 Creation of knowledge repository (profiles of relationships and recurring trends). Figure is extracted from [37]

For Cluster-1[TS-1,TS-3], two instances are created to represent a new form of relationship between the pair of time series #1 and #3 that differs from the one which exists in the first data chunk or snapshot, whereas for Cluster-2[TS-2,TS-4] only one a new instance is created. This is because the trend of movement of time series #2 in the second data chunk or snapshot is comparable to the existing instance and therefore it joins the cluster.

Additionally, as the algorithm processes the third data chunk or snapshot, it realizes that, within this locality of time, the four series are uncorrelated and moving individually. As a result, new profiles represented by four new clusters:

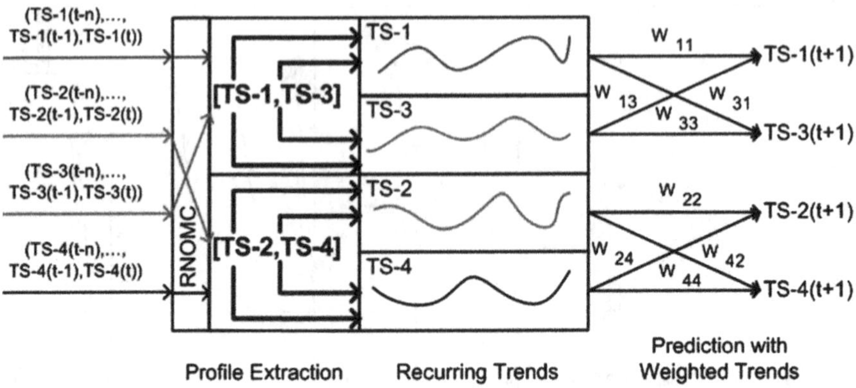

Fig. 12.8 Multiple time series prediction using profiles of relationships and recurring trends. Figure is extracted from [37]

Cluster-3[TS-1], Cluster-4[TS-2], Cluster-5[TS-3], and Cluster-6[TS-4], denoting specific trends of movement are created. The procedure continues until there is no more data chunk or snapshot to be processed.

The process of constructing the knowledge repository can be considered as a form of spatiotemporal modeling, whereby different shapes of trends (spatio) are extracted continuously over time (temporal). The repository illustrates how relationships between observed time series or variables change dynamically over different time localities, retaining different shapes of movement (trends).

After the repository has been built, there are two further steps that need to be performed before prediction can take place. The first is to extract current profiles of relationships between the multiple series. Thereafter, matches are found between the current trajectory and previously stored profiles from the past. Predictions are then made by implementing a weighting scheme that gives more importance to pairs of series that belong to the same profile and retain comparable trends of movement. The weight $w_{i,j}$ for given pair i,j of series, is given by the distance of similarity between them.

The prediction process is illustrated in Fig. 12.8, while the procedure of predicting movements of multiple time series simultaneously using the knowledge repository is outlined as follows:

- Step 1, after the knowledge repository **KR** has been initialized using the training data set, as new data $\mathbf{x}_t = (x_t^{(1)}, x_t^{(2)}, \ldots, x_t^{(i)})$ becomes available, where t is current time point and i is the number of series, new data set $\mathbf{X}_t^{'}$ is constructed as follows:

$$\mathbf{X}_t^{'} = \begin{bmatrix} x_{t-(n-1)}^{(1)} & x_{t-(n-2)}^{(1)} & \cdots & x_t^{(1)} \\ x_{t-(n-1)}^{(2)} & x_{t-(n-2)}^{(2)} & \cdots & x_t^{(2)} \\ \vdots & \vdots & \vdots & \vdots \\ x_{t-(n-1)}^{(i)} & x_{t-(n-2)}^{(i)} & \cdots & x_t^{(i)} \end{bmatrix},$$

where n is the size of data chunk or snapshot window used in the process to extract profiles of relationship and cluster the recurring trends.

- Step 2, extract profiles of relationship $\mathbf{p}_t = (p_t^1, p_t^2, ..., p_t^k)$ where $1 \leq k \leq i$, from current data set using $\mathbf{X}_t^{'}$ as described in previous section.
- Step 3, find profiles $\mathbf{p}_{kr} = (p_{kr}^1, p_{kr}^2, ..., p_{kr}^k)$ from previously constructed knowledge repository **KR** where,

$$\mathbf{p}_{kr} = \mathbf{p}_t. \tag{12.14}$$

- Step 4, for each series $x^{(i)} \in p_t^k$, extract its current trend of movement $\mathbf{w}_t^{(i)}$ as described in previous section.

  For each series $x^{(i)} \in p_t^k$, find j cluster centers of recurring trends in p_{kr}^k, where $j = 1, 2, ..., m$ is the number of series belongs to profile p_{kr}^k by calculating minimum distances between $\mathbf{w}_t^{(i)}$ to all existing cluster centers of recurring trends in p_{kr}^k as follows:

$$D_{i,j} = 1 - \max(\text{CorrelationCoefficient}(\mathbf{w}_t^{(i)}, Cc_j^l)), \tag{12.15}$$

  where $l = 1, 2, ...$ is the number of clusters of recurring trends of series j in p_{kr}^k.
- Step 5, calculate next value of $x^{(i)}$ using j found cluster centers of recurring trends, by giving more weight w to cluster centers which are closer to $\mathbf{w}_t^{(i)}$. *Note: In this methodology, cluster centers Cc_j of recurring trends represent trends of movement of time series in a particular profile.*

  The weight $w_{i,j}$ that gives more importance to cluster center j when predicting next value of $x^{(i)}$ is calculated as follows;

$$w_{i,j} = \frac{\max(\mathbf{D}) - (D_{i,j} - \min(\mathbf{D}))}{\max(\mathbf{D})}, \tag{12.16}$$

  where $\max(\mathbf{D})$ and $\min(\mathbf{D})$ are the maximum and minimum values of distance vector $\mathbf{D} = (D_{i,1}, D_{i,2}, ..., D_{i,j})$.

  In addition, next value of $x^{(i)}$ is given by

$$x_{t+1}^{(i)} = \frac{\sum_{m=1}^{j} w_{i,m} Cc_m}{\sum_{m=1}^{j} w_{i,m}}. \tag{12.17}$$

- Step 6, update the knowledge repository **KR** using current data set that has just been processed for prediction.
- Step 7, end of the procedure.

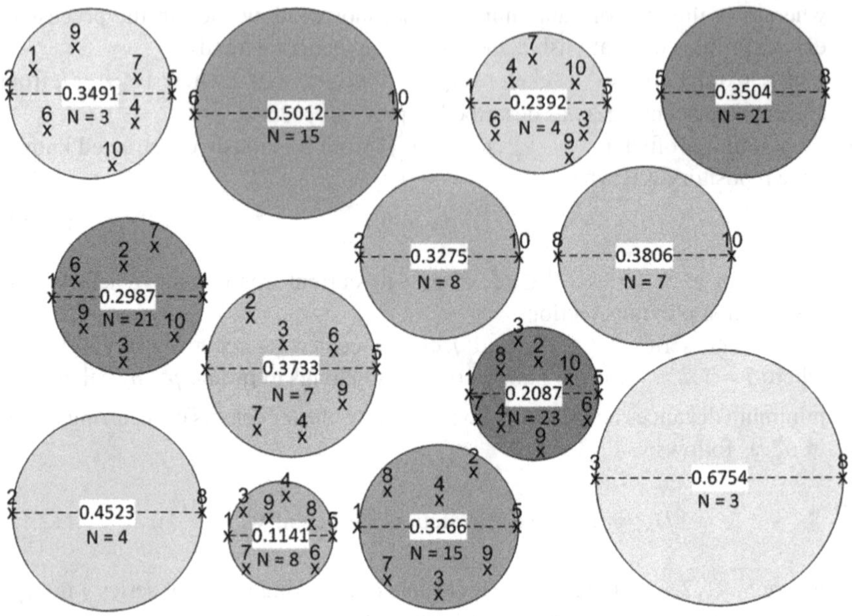

Fig. 12.9 Profiles of relationships in the knowledge repository constructed by LTM using 100 weeks index of ten stock markets in the Asia Pacific region. Index #1, #2, #3, #4, #5, #6, #7, #8, #9, #10 represent the NZ50, AORD, HSI, JSX, KLX, KOSPI, Nikkei 225, SSEC, STI, and TSEC, respectively

12.5.4 Dynamic Learning of Stock Market Indexes with LTM

The globalized security markets of today form the basis of a case study to demonstrate the ability of LTM in performing dynamic learning of multiple time series from a nonstationary environment. Additionally, as previous study had suggested that the globalized security markets are characterized by interdependencies among stock markets and often demonstrate dynamic contagious behavior in different periods [26], analysis of how LTM responds to such behavior is also outlined in this section of the chapter.

The financial data set comprising the globalized security markets used in this experiment includes time series indexes of ten stock markets in the Asia Pacific region, available from http://finance.yahoo.com/intlindices?e=asia spanning 161 weeks from 1st June 2007 to 30th June 2010. The weekly aggregated values of the stock market indexes are considered here. The ten selected market indexes are NZ50 (New Zealand), AORD (Australia), HSI (Hong Kong), JSX (Indonesia), KLX (Malaysia), KOSPI (South Korea), Nikkei 225 (Japan), SSEC (China), STI (Singapore), and TSEC (Taiwan).

Figure 12.9 illustrates the constructed knowledge repository after 100 points (i.e., 100 weeks) of stock market indexes of the ten selected stock market in the

Asia Pacific region that are conferred to LTM. Figure 12.9 shows the existence of strong relationship between Australia (AORD), Hong Kong (HSIX), South Korea (KOSPI), Singapore (STI), and Taiwan (TSEC). Extracted profiles of relationships in the knowledge repository reveal that these five stock markets are grouped together in the same cluster frequently. This outcome is in agreement with previous findings by Masih and Masih in 2001 [23] in their research on the dynamics of stock market interdependency which found that the five stock markets are interdependent with each other.

Additionally, in Fig. 12.9, cluster radius represents the farthest correlation between time series in the same cluster while relative positioning of the labels indicates the degree of similarity in behavior. For instance, in the cluster of NZ50, HSI, JSX, KLX, KOSPI, N225, STI, TSEC, and KOSPI is positioned closer to NZ50. This indicates that similarity between KOSPI and NZ50 is higher compared to similarity between KOSPI and the other markets.

This initial result confirms the ability of LTM to capture the existence of diverse profiles of relationships that exist in the globalized security markets. However, the other imperative was to perform dynamic learning of multiple time series. Therefore, to evaluate the effectiveness of LTM in meeting this requirement, another 50 points of weekly stock market indexes are used in addition.

Figure 12.10 illustrates the states of extracted profiles of relationships in the knowledge repository after the total of 150 weeks of the ten stock market indexes entered the system. It is clearly seen that new profiles of relationships have emerged in the repository, and existing profiles, in terms of the cluster radius, have been updated. This result confirms that LTM is capable to perform dynamic learning of multiple time series by capturing the dynamics of relationships between the series. Finally, the rest 11 weeks stock market indexes are presented to construct the final knowledge repository for 161 weeks of ten stock markets from the Asia Pacific region as illustrated in Fig. 12.11.

As mentioned before, the main objective of constructing a knowledge repository with the capability to learn dynamically facilitates the simultaneous prediction of multiple time series in a nonstationary environment. Figure 12.12 shows the prediction results of the ten selected stock market indexes in the Asia Pacific region. The prediction is made for 46 weeks and done simultaneously for the ten stock markets using the process that was explained in the previous section.

The initial LTM's knowledge repository was constructed by utilizing the training data set, and throughout the experiments conducted in this work predictions for the ten stock market, indexes are made for only one-step ahead. Nevertheless, as new data/time series points become available, the method updates its knowledge repository incrementally.

It is again clearly seen that the prediction results match closely with the actual trajectory of the ten stock markets. This results confirm that, by being able to learn dynamically from multiple time series of a nonstationary environment, complete understanding of the underlying behavior of the observed environment can be constructed and a simultaneous multiple time series prediction can be performed with a high degree of accuracy.

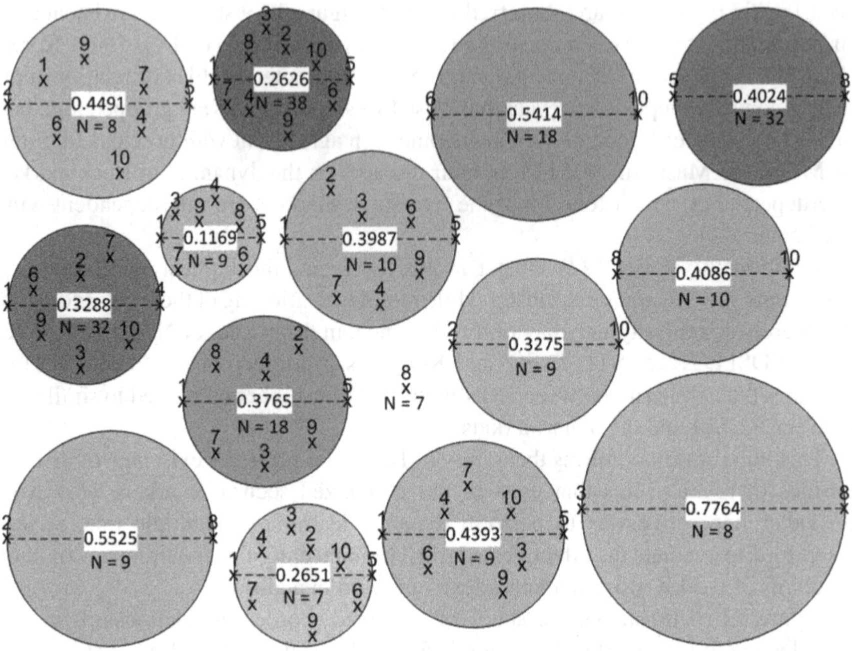

Fig. 12.10 Profiles of relationships in the knowledge repository constructed by LTM using 150 weeks index of ten stock markets in the Asia Pacific region. Indexes #1, #2, #3, #4, #5, #6, #7, #8, #9, #10 represent the NZ50, AORD, HSI, JSX, KLX, KOSPI, Nikkei 225, SSEC, STI, and TSEC, respectively

12.6 Instance-Based Learning of Multiple Time Series

As multiple data streams consist of various variables producing examples continuously over time, the basic idea behind the methodology outlined in this section is simply to find and model relationships between these streams of data at a particular time point and then to search for similar patterns of relationships from the past. The relationships found will then be utilized to constitute a specific model (i.e., weighted localized linear regression, localized fuzzy rules, etc.) to predict future values of multiple time series simultaneously. Instead of constructing a single model or a number of local models using a fixed size training data set, this methodology creates and updates local models dynamically whenever new data arrives.

The use of transductive reasoning for multiple time series analysis is inspired by the NFI proposed by Song and Kasabov in 2005 [29], see Sect. 12.4.1, which develops further some ideas from DENFIS [15], see Sect. 12.3.1. However, the NFI was designed to work only as a single time series prediction algorithm. Therefore, to cope with multiple time series modeling and prediction, some adjustments need

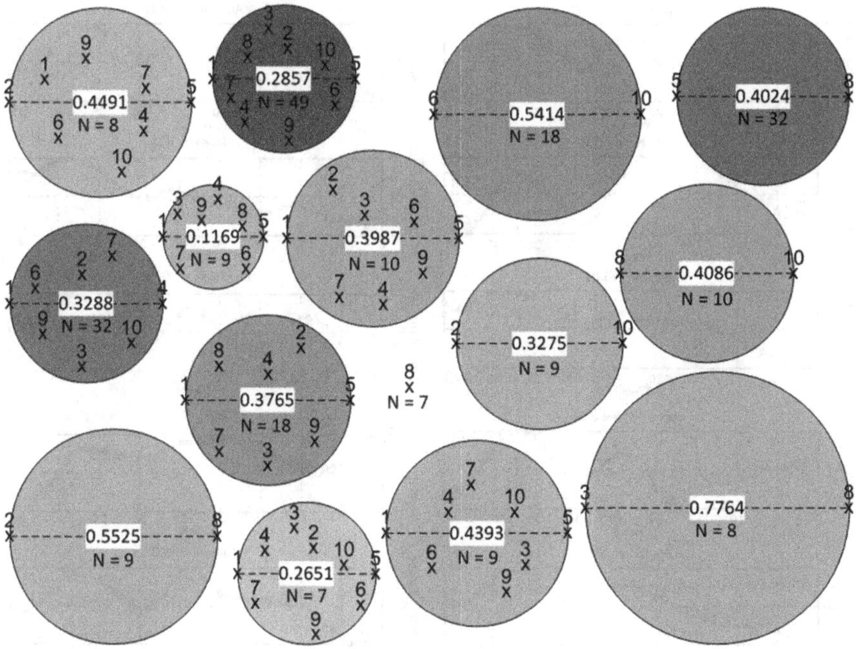

Fig. 12.11 Profiles of relationships in the knowledge repository constructed by LTM using 161 weeks index of ten stock markets in the Asia Pacific region. Indexes #1, #2, #3, #4, #5, #6, #7, #8, #9, #10 represent the NZ50, AORD, HSI, JSX, KLX, KOSPI, Nikkei 225, SSEC, STI, and TSEC, respectively

to be made. Furthermore, there are some issues that also need to be addressed in implementing the methodology. First, what are the features to be used to describe relationships between multiple time series? Second, how are these relationships going to be modeled? Third, how to find similar conditions from the past and to calculate the final prediction results of multiple time series?

12.6.1 Multivariate Transductive NFI

This section of the chapter outlines a transductive approach for multiple time series analysis and modeling. The *multivariate transductive neuro-fuzzy inference system*, denoted as mTNFI, introduced and explained in this section, is an extension of the NFI model (see Sect. 12.4.1), in which modifications were made so that the new methodology is capable of performing multiple time series data analysis and modeling.

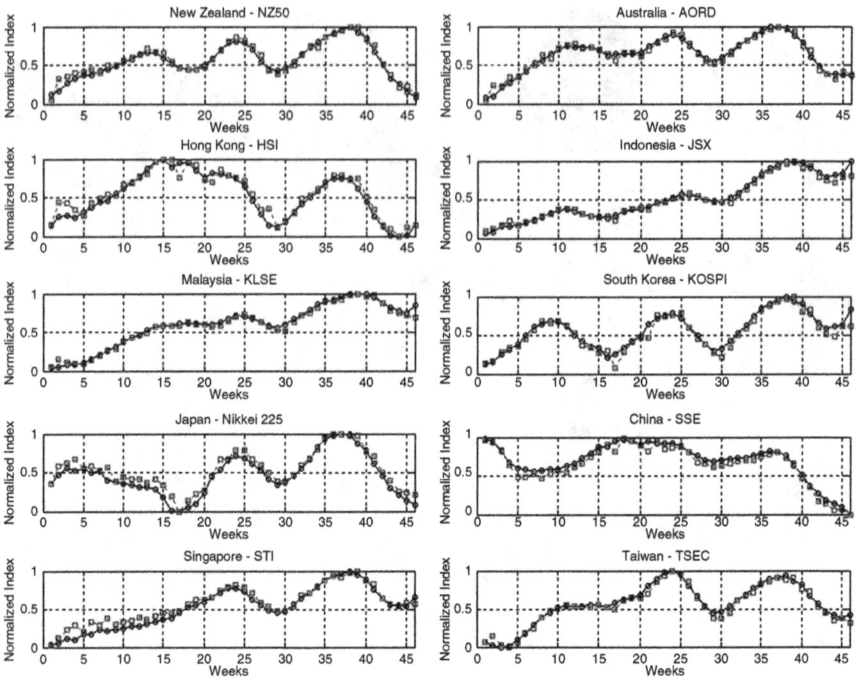

Fig. 12.12 Prediction results [□] of the LTM against the actual trajectory [○] of ten stock market in the Asia Pacific region

12.6.1.1 General Principles

In general, mTNFI employs the same principles as the NFI model, where a dynamic neural-fuzzy inference system with Gaussian membership function is constructed from a set of nearest neighbors of a new input vector. However, as the mTNFI model is intended to perform multiple time series analysis and modeling, some alterations had to be made to the NFI model.

The first modification made to the NFI model is the use of a different distance metric. The NFI model considers input vector $\mathbf{x}_i$ as feature vectors and uses the normalized Euclidean distance (as in (12.2)) to find the closest N_i training samples. Yet, as previous studies have found that dynamic relationships exist in multiple time series from a specific setting, the basic idea behind the mTNFI model is to use the state of relationships in $\mathbf{x}_t$, where $\mathbf{x}_t = (x_{1t}, x_{2t}, \ldots, x_{qt})$ and x_{qt} to measure the expression level of a time series q at time point t as a feature vector instead of its actual values. Please note that in mTNFI an input vector is denoted as $\mathbf{x}_t$ instead of $\mathbf{x}_i$ to represent the temporal aspect of the data set (Fig. 12.13).

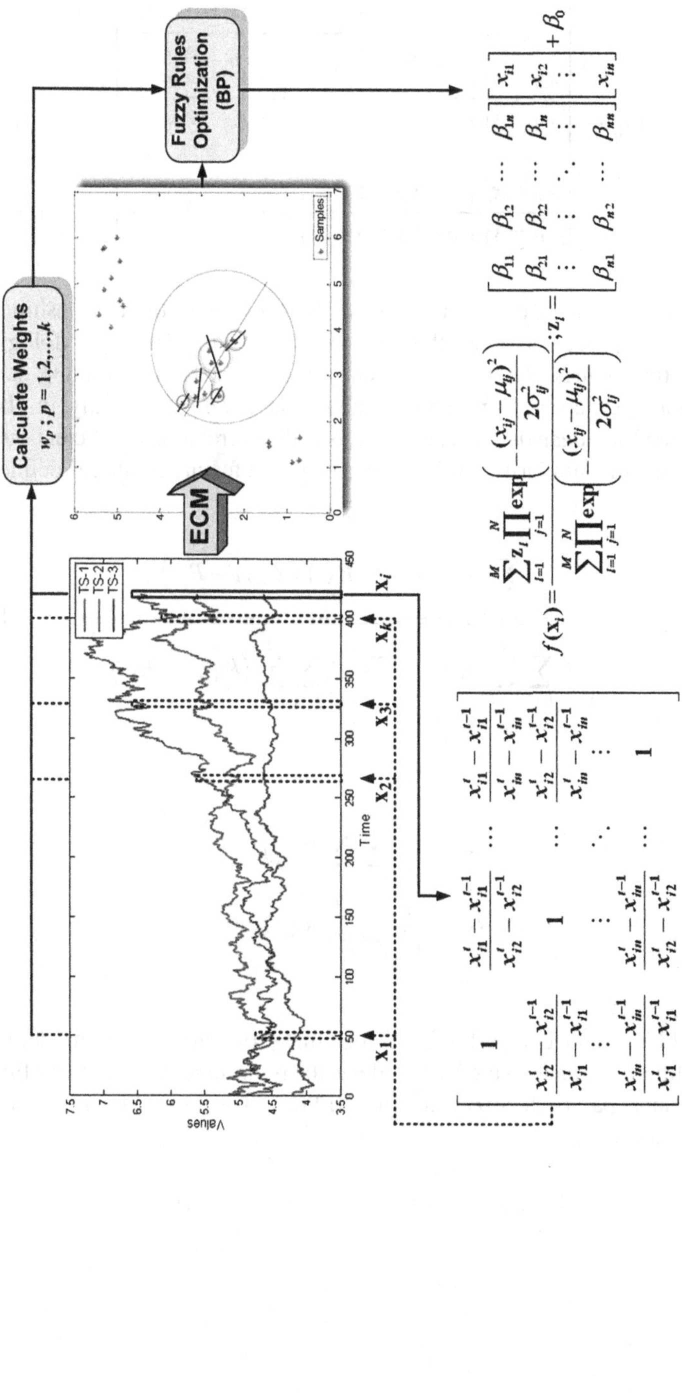

Fig. 12.13 Illustration of finding the nearest neighbors, constructing the fuzzy inference system, and output calculation in mTNFI

In the mTNFI model, the state of relationship between multiple time series at a particular time point t, is defined by calculating the *ratio of first-order rate of changes* denoted as $R_{\mathbf{X}_t}$, from multiple time series under examination described as follows:

$$R_{\mathbf{X}_t} = \begin{bmatrix} 1 & \frac{x_{1t} - x_{1(t-1)}}{x_{2t} - x_{2(t-1)}} & \cdots & \frac{x_{1t} - x_{1(t-1)}}{x_{qt} - x_{q(t-1)}} \\ \frac{x_{2t} - x_{2(t-1)}}{x_{1t} - x_{1(t-1)}} & 1 & \cdots & \frac{x_{2t} - x_{2(t-1)}}{x_{qt} - x_{q(t-1)}} \\ \vdots & \vdots & \ddots & \vdots \\ \frac{x_{qt} - x_{q(t-1)}}{x_{1t} - x_{1(t-1)}} & \frac{x_{qt} - x_{q(t-1)}}{x_{2t} - x_{2(t-1)}} & \cdots & 1 \end{bmatrix}, \qquad (12.18)$$

where $R_{\mathbf{X}_t}$ now is the features matrix describing the state of relationship in $\mathbf{x}_t$. Additionally, the process assumes that $x_{i0} = 0; i = 1, 2, ..., q$. The methodology then employs feature matrix $R_{\mathbf{X}_t}$ to find N_t closest or most related training samples which form a subdata set D_t from an existing data set D. Additionally, in place of the normalized Euclidean distance, mTNFI uses the correlation coefficient distance measure to quantify similarity level between different features matrices defined by

$$S_{R_{\mathbf{X}_t} R_{\mathbf{X}_i}} = 1 - \frac{\sum_{j=1}^{q}\sum_{k=1}^{q} \left(R_{\mathbf{X}_t j,k} - \overline{R}_{\mathbf{X}_t}\right)\left(R_{\mathbf{X}_i j,k} - \overline{R}_{\mathbf{X}_i}\right)}{\sqrt{\left(\sum_{j=1}^{q}\sum_{k=1}^{q} \left(R_{\mathbf{X}_t j,k} - \overline{R}_{\mathbf{X}_t}\right)^2 \sum_{j=1}^{q}\sum_{k=1}^{q} \left(R_{\mathbf{X}_i j,k} - \overline{R}_{\mathbf{X}_i}\right)^2\right)}}, \qquad (12.19)$$

where

$$\overline{R}_{\mathbf{X}_t} = \frac{1}{q^2}\sum_{j=1}^{q}\sum_{k=1}^{q} R_{\mathbf{X}_t j,k},$$
$$\overline{R}_{\mathbf{X}_i} = \frac{1}{q^2}\sum_{j=1}^{q}\sum_{k=1}^{q} R_{\mathbf{X}_i j,k},$$

and $i = 1, 2, \ldots, t-1$.

As mTNFI is intended to perform multiple time series analysis and modeling, the second modification made to the NFI model is the replacement of the linear function in the consequent part of the fuzzy rule (mTNFI model uses the Takagi-Sugeno-type fuzzy rule) defined by

$$f(\mathbf{x}_t) = \hat{x}_{t+1} = \beta_0 + \beta_1 x_{1t} + \beta_2 x_{2t} + \ldots + \beta_q x_{qt},$$

with a different form of linear function as follows:

$$f(\mathbf{x}_t)\begin{cases}\hat{x}_{1(t+1)} = \beta_0 + \beta_{11}x_{1t} + \beta_{12}x_{2t} + \ldots + \beta_{1q}x_{qt} \\ \hat{x}_{2(t+1)} = \beta_0 + \beta_{21}x_{1t} + \beta_{22}x_{2t} + \ldots + \beta_{2q}x_{qt} \\ \vdots \\ \hat{x}_{p(t+1)} = \beta_0 + \beta_{p1}x_{1t} + \beta_{p2}x_{2t} + \ldots + \beta_{pq}x_{qt}\end{cases}, \tag{12.20}$$

where p is the number of dependent variables and q is the number of explanatory variables. However, in mTNFI, $p = q$ as the number of time series being explained is the same as the number of the explanatory time series. Equation (12.20) can be represented in a more general and simplified form as follows:

$$\mathbf{y}_t = \beta_0 + \beta \mathbf{x}_t, \tag{12.21}$$

where $\hat{\mathbf{x}}_{t+1}$ is a vector of multidependent variables, $\mathbf{x}_t$ is a vector of multiindependent variables, and β is the coefficients matrix that maps $\mathbf{x}_t$ to $\hat{\mathbf{x}}_{t+1}$. Representing the consequent part of the fuzzy rule by a linear function with multidependent and independent variables gives rise to the ability of modeling interactions between observed variables and performing multiple time series prediction at a particular time point.

Other than the two modifications outlined above, the mTNFI model utilizes the same process as the NFI model. As such, for every new input vector, the algorithm dynamically constructs a neural-fuzzy inference system with local generalization. As in the NFI model, the mTNFI model also employs the ECM proposed by Kasabov and Song in 2002 [15], to partition the input subspace that consists of N_t selected training samples. A local model LM_t for input vector $\mathbf{x}_t$ will then be constituted in the form of a fuzzy inference system using a set of created fuzzy rules, derived from the clustering process.

12.6.1.2 mTNFI Learning Algorithm

For each new input vector $\mathbf{x}_t$ of multiple time series, the mTNFI model performs the following learning algorithm:

- Step 1, construct the ratio of first-order rate of changes from input vector $\mathbf{x}_t$ to form features matrix $R_{\mathbf{X}_t}$, using (12.18).
- Step 2, search in the training data set, based on the input space, N_t training samples that are closest to $\mathbf{x}_t$ which form a subdata set $D_t = (\mathbf{x}_1, \mathbf{x}_2, \ldots, \mathbf{x}_j); j = 1, 2, \ldots, N_t$, by utilizing features matrix $R_{\mathbf{X}_t}$ and calculating features matrices $R_{\mathbf{X}_i}; i = 1, 2, \ldots, t-1$ from all training samples. Closest training samples in mTNFI are defined using the Correlation Coefficient distance measure, as described in (12.19). Additionally, in mTNFI, the value for N_t is predefined based on experience, where N_t can be considered as number of k when being related to the k-NN algorithm.

- Step 3, calculate the distances $d_j; j = 1,2,...,N_t$ between each of the training samples in D_t and input vector $\mathbf{x}_t$ and calculate the weights $w_j = 1 - (d_j - \min(\mathbf{d})); j = 1,2,...,N_t$ where $\min(\mathbf{d})$ is the minimum value in the distance vector $\mathbf{d} = (d_1, d_2, ..., d_{N_t})$.
- Step 4, use the ECM clustering algorithm to cluster and partition the input subspace D_t that consists of N_t selected training samples.
- Step 5, create Takagi-Sugeno-type fuzzy rules by representing the consequent part of the rules as a linear function with multidependent and independent variables (as in (12.21)) and set their initial parameter values according to the clustering results of the ECM.

  For each cluster, the cluster center is taken as the center of a fuzzy membership function (Gaussian function) μ and the cluster radius is taken as the width σ.

  Consider at time point t the system under examination has q inputs and outputs, where q is the amount of time series being observed. As the outcome of this step, M fuzzy rules are defined initially through the ECM clustering procedure, and the lth rule has the form of

$$R_l : \text{if } x_{1t} \text{ is } F_{l1} \text{ and } x_{2t} \text{ is } F_{l2} \text{ and ... and } x_{qt} \text{ is } F_{lq}, \text{ then } \hat{\mathbf{x}}_{t+1} \text{ is } f_l(\mathbf{x}_t).$$

  Here, F_{lk} are fuzzy sets of x_k in cluster l, where $k = 1,2,...,q$, defined by the following Gaussian-type membership function:

$$\text{GaussianMF}_l(x_k) = \alpha_{lk} \exp\left(-\frac{(x_k - \mu_{lk})^2}{2\sigma_{lk}^2}\right). \tag{12.22}$$

  Additionally, $f_l(\mathbf{x}_t)$ in R_l is represented as linear function with multidependent and independent variables as follows:

$$f_l(\mathbf{x}_t) = \begin{cases} \hat{x}_{1(t+1)} = \beta_0 + \beta_{11}x_{1t} + \beta_{12}x_{2t} + ... + \beta_{1q}x_{qt} \\ \hat{x}_{2(t+1)} = \beta_0 + \beta_{21}x_{1t} + \beta_{22}x_{2t} + ... + \beta_{2q}x_{qt} \\ \vdots \\ \hat{x}_{p(t+1)} = \beta_0 + \beta_{p1}x_{1t} + \beta_{p2}x_{2t} + ... + \beta_{pq}x_{qt} \end{cases}.$$

  The M created fuzzy rules are then utilized to constitute the local model LM_t in the form of Takagi-Sugeno inference system.
- Step 6, apply the steepest descent method (backpropagation) to optimize the parameters of the fuzzy rules in the local model LM_t.

  Suppose the mTNFI model is given a training input–output data pair $(\mathbf{x}_t, \mathbf{y}_t)$, the parameters are being optimized by minimizing the objective function (a weighted error function) as follows:

$$E = \frac{1}{2} w_t \sum_{k=1}^{q} (\hat{y}_{kt} - y_{kt})^2, \tag{12.23}$$

where w_t is the weight of the input–output training data pair, $\hat{y}_t = f(\mathbf{x}_t)$, and q is again the amount of time series being observed.

In mTNFI, the training input–output data pairs used to optimize the fuzzy rules' parameters are the N_t selected training samples from step 2, $(\mathbf{x},\mathbf{y}) = ((\mathbf{x}_1,\mathbf{y}_1),(\mathbf{x}_2,\mathbf{y}_2),...,(\mathbf{x}_j,\mathbf{y}_j)), \text{and} j = 1,2,...,N_t$. The weight of each input–output training data pairs w_j, is defined in step 3.

- Step 7, calculate the output value $f(\mathbf{x}_t)$, for the input vector $\mathbf{x}_t$, applying fuzzy inference over the set of fuzzy rules that constitute the local model LM_t using the modified center average defuzzification procedure as follows:

$$f(\mathbf{x}_t) = \frac{\sum_{l=1}^{M} f_l(\mathbf{x}_t) \prod_{j=1}^{q} \alpha_{lj} \exp\left(-\frac{(x_{jt}-\mu_{lj})^2}{2\sigma_{lj}^2}\right)}{\sum_{l=1}^{M} \prod_{j=1}^{q} \alpha_{lj} \exp\left(-\frac{(x_{jt}-\mu_{lj})^2}{2\sigma_{lj}^2}\right)}.$$

- Step 8, end of the procedure.

The procedure of optimizing the parameters $\beta_l, \alpha_{lj}, \mu_{lj}$, and σ_{lj} of the Takagi-Sugeno type in the mTNFI model (step 6 in the mTNFI algorithm) is carried out using the steepest descent method such that the value of E from (12.23) is minimized. The optimization equations for each parameter are then defined as follows:

$$\beta_{l0}(k+1) = \beta_{l0}(k) - \eta_\beta w_t \Phi(\mathbf{x}_t) \left(\frac{1}{q}\sum_{i=1}^{q}(\hat{y}_{it}(k) - y_{it})\right), \tag{12.24}$$

$$\beta_{li,j}(k+1) = \beta_{li,j}(k) - \eta_\beta x_{jt} w_t \Phi(\mathbf{x}_t)(\hat{y}_{it}(k) - y_{it}), \tag{12.25}$$

$$\begin{aligned} \alpha_{li}(k+1) = \alpha_{li}(k) - \frac{\eta_\alpha}{\alpha_{li}(k)} w_t \Phi(\mathbf{x}_t) \left(\frac{1}{q}\sum_{i=1}^{q}(\hat{y}_{it}(k) - y_{it})\right) \\ \times \left(\frac{1}{q}\sum_{i=1}^{q}(\hat{y}_{lit}(k) - \hat{y}_{it}(k))\right), \end{aligned} \tag{12.26}$$

$$\begin{aligned} \mu_{li}(k+1) = \mu_{li}(k) - \frac{\eta_\mu}{\sigma_{li}^2(k)} w_t \Phi(\mathbf{x}_t) \left(\frac{1}{q}\sum_{i=1}^{q}(\hat{y}_{it}(k) - y_{it})\right) \\ \times \left(\frac{1}{q}\sum_{i=1}^{q}(\hat{y}_{lit}(k) - \hat{y}_{it}(k))\right)(x_{it} - \mu_{li}(k)), \end{aligned} \tag{12.27}$$

$$\sigma_{li}(k+1) = \sigma_{lj}(k) - \frac{\eta_\sigma}{\sigma_{li}^3(k)} w_t \Phi(\mathbf{x}_t) \left(\frac{1}{q} \sum_{i=1}^{q} (\hat{y}_{it}(k) - y_{it}) \right) \times \left(\frac{1}{q} \sum_{i=1}^{q} (\hat{y}_{lit}(k) - \hat{y}_{it}(k)) \right) (x_{it} - \mu_{li}(k))^2, \tag{12.28}$$

where

$$\hat{y}_t(k) = f^{(k)}(\mathbf{x}_t); \hat{y}_{lt}(k) = f_l^{(k)}(\mathbf{x}_t)$$

and

$$\Phi(\mathbf{x}_t) = \frac{\prod_{i=1}^{q} \alpha_{li} \exp\left(-\frac{(x_{it}(k) - \mu_{li}(k))^2}{2\sigma_{li}^2(k)} \right)}{\sum_{l=1}^{M} \prod_{i=1}^{q} \alpha_{li} \exp\left(-\frac{(x_{it}(k) - \mu_{li}(k))^2}{2\sigma_{li}^2(k)} \right)},$$

$\eta_\beta, \eta_\alpha, \eta_\mu$, and η_σ are learning rates for updating the parameters $\beta_l, \alpha_{li}, \mu_{li}$, and σ_{li} respectively. Additionally, in the mTNFI learning algorithm, the following indexes are used:

- Training data samples: $t = 1, 2, ..., N$
- Input variables: $i, j = 1, 2, ..., q$
- Fuzzy rules: $l = 1, 2, ..., M$
- Learning epoch: $k = 1, 2, ...$

12.6.2 *Dynamic Learning of Stock Market Indexes with mTNFI*

As in Sect. 12.5.4, the weekly indexes of ten stock markets in the Asia Pacific region spanning 161 weeks from 1st June 2007 to 30th June 2010 is again utilized to evaluate mTNFI's capability to perform dynamic learning of multiple time series in a nonstationary environment.

However, as mTNFI falls under the category of transductive reasoning or instance-based learning, no training phase is required to construct global or local models as in LTM. Consequently, the training data set is put in place as the initial search space. A number of nearest neighbors that form a subdata set to construct specific estimation models will then be located from this search space. Additionally, the experiment employs an incremental testing process, which means that, whenever a new instance arrives, the accuracy of predictions is first tested before it is added to the training set or search space as a training example.

Table 12.1 Comparison of the proposed integrated framework prediction error rates against single time series prediction methods: MLR, MLP, and random walk model, in RMSE

No	Stock market	IMMF	MLR	MLP	Random walk
1	AORD	25.1345	171.7993	88.5213	85.7106
2	HSI	156.3394	460.3831	428.6957	320.0959
3	JSX	19.8513	77.3591	41.7204	47.5735
4	KLSE	5.2856	36.4490	26.6499	22.5799
5	KOSPI	13.5639	47.6204	28.5997	27.3852
6	Nikkei 225	61.8075	311.8844	293.4835	220.9357
7	NZ50	8.8974	126.1689	48.9904	46.8238
8	SSX	30.7801	175.5635	105.7773	113.7639
9	STI	16.1047	75.2957	60.7475	57.2612
10	TSEC	57.3583	198.6822	141.3183	108.3788

In our experimentation, 100 weeks of stock market indexes out of the 161 weeks of data is selected as the training data set. Furthermore, as mTNFI constructs specific solutions for every input vector, different fuzzy inference systems are extracted for input vectors corresponding to different time points.

Table 12.1 outlines the fuzzy rules created by mTNFI using the 30 nearest samples from the 100 weeks training data set when calculating output for input vector at time-point 101. Please note that throughout the experimentation, the number of selected nearest samples in mTNFI is set to a fixed value of 30 samples. Table 12.1 indicates that three clusters are created in the clustering process of the 30 nearest samples. These three clusters are then utilized to form the fuzzy rules (as described in the mTNFI learning algorithm) which in the end constructs the final fuzzy inference system as in Fig. 12.14.

To demonstrate mTNFI's capability to construct an individual local model that best fits a new input vector or problem, a number of fuzzy rules is again extracted when estimating output for input vector at time-point 110. Consequently, new fuzzy inference systems as presented in Fig. 12.15, consisting of different number of fuzzy rules is constructed by mTNFI for this particular input vector. Additionally, instead of having three fuzzy rules, the fuzzy inference system has only two fuzzy rules, indicating that in the learning process the 30 nearest samples are now being clustered to only two clusters. These results confirm that, by being able to construct specific individual local models for every new input vector or problem, mTNFI retains the capability to perform dynamic learning of multiple time series in a nonstationary environment.

By constructing different inference systems for every new input vector, the mTNFI is expected to be able to predict simultaneously movement of multiple time series of a nonstationary environment. To evaluate this, an experiment for predicting future index values for the ten stock markets is performed. The experiment utilizes the last 46 weeks of observed indexes as the test data set and the first 115 weeks of observed indexes as the search space. However, the experiment employs an

```
Rule1
If    RatioOfDiff NZ50(t)-AORD(t)  is GaussianMF (1.74  0.23)
      RatioOfDiff NZ50(t)-HSI(t)   is GaussianMF (2.33  0.21)
      ...
      RatioOfDiff TSEC(t)-STI(t)   is GaussianMF (0.59  0.19)
Then NZ50(t+1)  = 1.14*NZ50(t)   - 0.20*AORD(t) + 0.03*HSI(t) - 0.04*JSX(t) - 0.01*KLSE(t)
                 -0.28*KOSPI(t) + 0.04*N225(t)  + 0.01*SSX(t) - 0.16*STI(t) - 0.05*TSEC(t)
     AORD(t+1) = 0.32*NZ50(t)   + 0.63*AORD(t) + 0.07*HSI(t) - 0.10*JSX(t) - 0.42*KLSE(t)
                 -0.40*KOSPI(t) + 0.06*N225(t)  + 0.06*SSX(t) - 0.40*STI(t) + 0.12*TSEC(t)
     ...
     TSEC(t+1)  = 0.40*NZ50(t)   - 0.39*AORD(t) + 0.09*HSI(t) + 0.04*JSX(t) - 0.49*KLSE(t)
                 -0.36*KOSPI(t) + 0.08*N225(t)  + 0.17*SSX(t) - 1.00*STI(t) + 1.14*TSEC(t)
Rule 2
If    RatioOfDiff NZ50(t)-AORD(t)  is GaussianMF (2.46  0.19)
      RatioOfDiff NZ50(t)-HSI(t)   is GaussianMF (2.47  0.24)
      ...
      RatioOfDiff TSEC(t)-STI(t)   is GaussianMF (0.70  0.13)
Then NZ50(t+1)  = 1.12*NZ50(t)   - 0.10*AORD(t) - 0.01*HSI(t) - 0.01*JSX(t) - 0.00*KLSE(t)
                 -0.46*KOSPI(t) + 0.05*N225(t)  - 0.01*SSX(t) + 0.52*STI(t) + 0.02*TSEC(t)
     AORD(t+1) = 0.25*NZ50(t)   + 0.80*AORD(t) + 0.02*HSI(t) - 0.15*JSX(t) + 0.03*KLSE(t)
                 -0.90*KOSPI(t) - 0.07*N225(t)  - 0.01*SSX(t) + 0.61*STI(t) + 0.10*TSEC(t)
     ...
     TSEC(t+1)  = 0.15*NZ50(t)   - 0.04*AORD(t) - 0.05*HSI(t) - 0.25*JSX(t) + 1.16*KLSE(t)
                 -1.87*KOSPI(t) - 0.19*N225(t)  - 0.00*SSX(t) + 1.48*STI(t) + 1.14*TSEC(t)
Rule 3
If    RatioOfDiff NZ50(t)-AORD(t)  is GaussianMF (1.58  0.14)
      RatioOfDiff NZ50(t)-HSI(t)   is GaussianMF (2.74  0.27)
      ...
      RatioOfDiff TSEC-STI         is GaussianMF (0.70  0.19)
Then NZ50(t+1)  = 1.14*NZ50(t)   - 0.20*AORD(t) + 0.03*HSI(t) - 0.08*JSX(t) - 0.03*KLSE(t)
                 -0.24*KOSPI(t) + 0.02*N225(t)  + 0.01*SSX(t) - 0.03*STI(t) + 0.04*TSEC(t)
     AORD(t+1) = 0.28*NZ50(t)   + 0.61*AORD(t) + 0.07*HSI(t) - 0.23*JSX(t) - 0.08*KLSE(t)
                 -0.64*KOSPI(t) + 0.05*N225(t)  + 0.03*SSX(t) - 0.20*STI(t) + 0.13*TSEC(t)
     ...
     TSEC(t+1)  = 0.14*NZ50(t)   - 0.35*AORD(t) + 0.05*HSI(t) - 0.38*JSX(t) + 1.11*KLSE(t)
                 -1.07*KOSPI(t) + 0.06*N225(t)  + 0.08*SSX(t) - 0.37*STI(t) + 1.19*TSEC(t)
```

Fig. 12.14 Extracted fuzzy rules (first-order Takagi-Sugeno type) from mTNFI when predicting upcoming indexes of ten stock markets in the Asia Pacific on week 101. *RatioOfDiff* is the representation of variable $R_{\mathbf{X}_t}$ in (12.18)

incremental testing process, which means that whenever a new instance arrives the accuracy of predictions is first tested before it is added to the training set as a training example.

Figure 12.16 shows the performance of mTNFI when predicting movement of ten stock market indexes in the Asia Pacific region simultaneously. The plots indicate that the predicted trajectories closely track the actual ones. This result confirms mTNFI's capability to estimate upcoming movement of multiple time series by dynamically constructing a specific fuzzy inference system for every state or condition that emerges in a nonstationary environment.

Rule1

If RatioOfDiff NZ50(t)-AORD(t) is GaussianMF (1.24 0.19)
RatioOfDiff NZ50(t)-HSI(t) is GaussianMF (0.51 0.14)
...
RatioOfDiff TSEC(t)-STI(t) is GaussianMF (4.87 0.23)
Then NZ50(t+1) = 1.15*NZ50(t) – 0.16*AORD(t) + 0.04*HSI(t) + 0.05*JSX(t) – 0.24*KLSE(t)
–0.12*KOSPI(t) + 0.01*N225(t) + 0.00*SSX(t) – 0.21*STI(t) + 0.03*TSEC(t)
AORD(t+1) = 0.36*NZ50(t) + 0.74*AORD(t) + 0.09*HSI(t) + 0.15*JSX(t) – 0.02*KLSE(t)
–0.01*KOSPI(t) – 0.01*N225(t) + 0.03*SSX(t) – 0.52*STI(t) + 0.06*TSEC(t)
...
TSEC(t+1) = 0.44*NZ50(t) – 0.26*AORD(t) + 0.07*HSI(t) + 0.31*JSX(t) – 1.55*KLSE(t)
+0.40*KOSPI(t) + 0.01*N225(t) + 0.20*SSX(t) – 0.87*STI(t) + 1.05*TSEC(t)

Rule 2

If RatioOfDiff NZ50(t)-AORD(t) is GaussianMF (1.45 0.12)
RatioOfDiff NZ50(t)-HSI(t) is GaussianMF (0.55 0.27)
...
RatioOfDiff TSEC(t)-STI(t) is GaussianMF (4.48 0.19)
Then NZ50(t+1) = 1.12*NZ50(t) – 0.10*AORD(t) – 0.00*HSI(t) – 0.01*JSX(t) + 0.03*KLSE(t)
–0.60*KOSPI(t) – 0.06*N225(t) – 0.02*SSX(t) + 0.59*STI(t) + 0.03*TSEC(t)
AORD(t+1) = 0.26*NZ50(t) + 0.85*AORD(t) – 0.00*HSI(t) – 0.13*JSX(t) – 0.03*KLSE(t)
–0.81*KOSPI(t) – 0.12*N225(t) – 0.02*SSX(t) + 0.91*STI(t) + 0.07*TSEC(t)
...
TSEC(t+1) = 0.10*NZ50(t) + 0.16*AORD(t) – 0.09*HSI(t) – 0.14*JSX(t) + 0.95*KLSE(t)
–2.24*KOSPI(t) – 0.30*N225(t) – 0.03*SSX(t) + 2.33*STI(t) + 1.11*TSEC(t)

Fig. 12.15 Extracted fuzzy rules (first-order Takagi-Sugeno type) from mTNFI when predicting upcoming indexes of ten stock markets in the Asia Pacific on week 110. *RatioOfDiff* is the representation of variable R_{X_t} in (12.18)

12.7 Integrated Framework of the LTM and mTNFI

Local models are capable of capturing local patterns valid for subsets of the problem space, and the transductive models are capable of constructing local and specific estimation models for each new instance that needs to be classified or predicted. These two approaches are useful for complex modeling tasks, and both of them provide complementary information and knowledge learned from the data. Integrating the two approaches into a single multimethodological approach would be a useful and challenging task.

Integrating different types and levels of knowledge about the dynamics of the relationships in a multiple time series under examination is a key objective in this work. It is expected that by integrating different types of models, one should be able to constitute a comprehensive understanding about the underlying behavior of the dynamics of the system being investigated.

An integrated scheme to assimilate different types of knowledge has been introduced by Kasabov in 2007 [12] in the bioinformatics domain. In his study, Kasabov stated that every model has their own power in prediction, and by being able to combine models, a more powerful model for time series prediction can be realized. In his study, Kasabov proposed an integrated multimodel system that

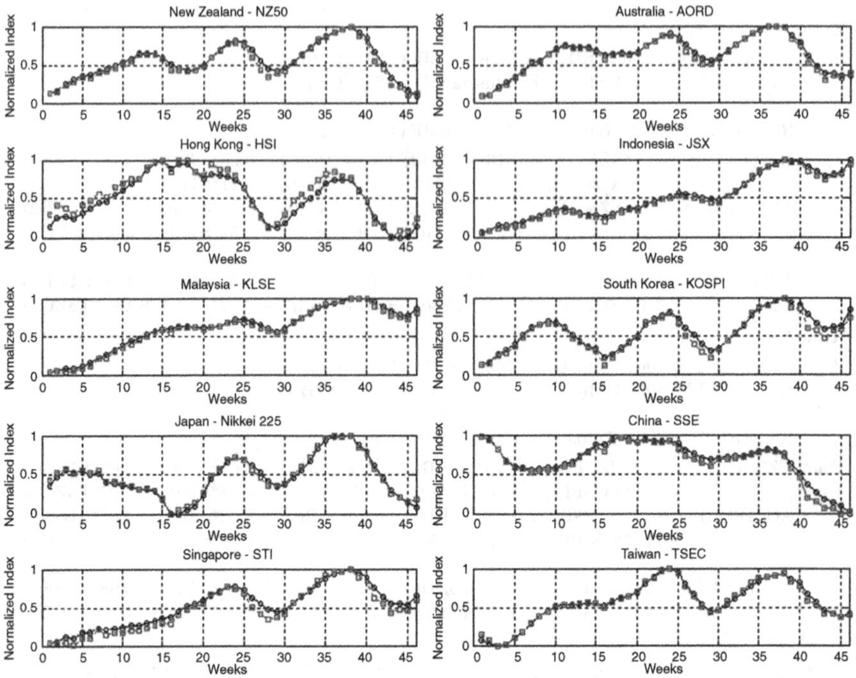

Fig. 12.16 Prediction results [□] of the mTNFI against the actual trajectory [○] of ten stock market in the Asia Pacific region

includes a global model, a local model, and a transductive model to increase the accuracy and power of prediction in gene expression data. However, in this work, we limit the realization of the integrated framework to only consist of the local and transductive models, as explained in previous sections.

The key idea of constructing the integrated framework is to estimate which model out of the local and transductive models should be trusted more in any given time point based on the characteristics of the series under observation. Completely, the whole structure of the integrated framework is illustrated in Fig. 12.17. The main component of the integrated framework is the *accumulator* module. The accumulator will calculate, based on performance of each model, weight values that will be associated with each model. The output of the accumulator represented by a is the final prediction formed by the weighted output of the local and transductive model defined by (12.29):

$$a = w_{\text{local}} * output_{\text{LTM}} + w_{\text{transductive}} * output_{\text{mTNFI}}, \tag{12.29}$$

where $output_{\text{LTM}}$ and $output_{\text{mTNFI}}$ represent predictions calculated by the LTM as the local model and the mTNFI as the transductive model.

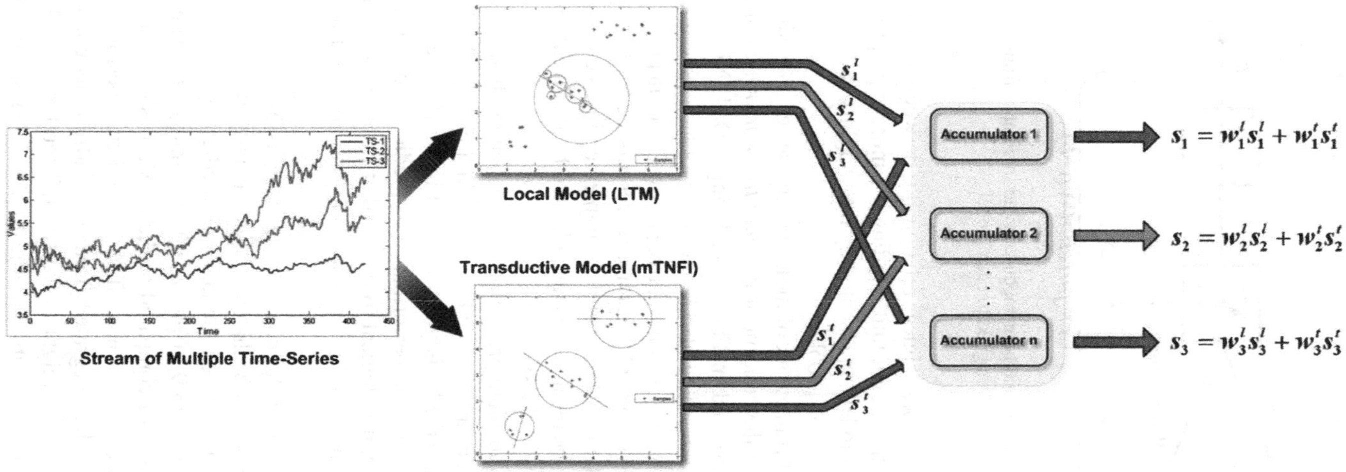

Fig. 12.17 Complete scheme of the integrated framework of local and transductive model for multiple time series analysis and prediction

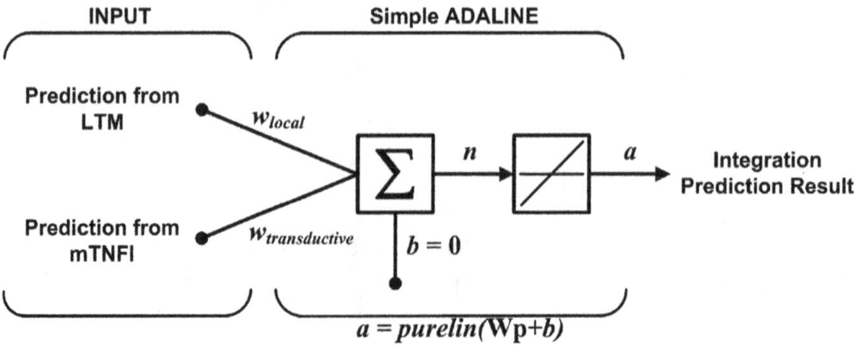

Fig. 12.18 ADALINE as the accumulator module in the integrated framework to find the most optimum weight distribution to be assigned to each model

Equation (12.29) describes a linear relationship between the values of the input units and the value of the output unit. Finding the most appropriate weight values to be assigned to each model (represented by the weight vector w) now amounts to solving a linear optimization problem.

Here, the integrated framework of multiple time series analysis and modeling is constructed by utilising the concept of the *adaptive linear neuron* network known as ADALINE and illustrated in Fig. 12.18. Using ADALINE learning rules, the accumulator module of the integrated framework would be able to calculate weight values for each model based on their relative performances in predicting movements of multiple time series. The weight vector w represents the trust values given to each model by the accumulator. Additionally, by implementing recursive learning rules, these weights can then be recalculated and adjusted based on current characteristics of the new data arriving in the system.

Comparison of the absolute prediction error, when predicting movement of the New Zealand NZ50 index as presented previously in Figs. 12.12 and 12.16, between the LTM and the mtNFI model is illustrated in Fig. 12.19. The plot reveals that in different localities of time, the performance of prediction of each model changes dynamically. Therefore, it is logical to expect that in different localities of time a particular model of multiple time series analysis should be trusted more than the other one.

Figure 12.20 shows how the contributing weights assigned by the integrated framework to each model are changing dynamically over time based on their performance. For instance, when the absolute prediction error of LTM is higher than mTNFI (time-point 1 to 9 in Fig. 12.19), the integrated framework assigns a larger weight to the mTNFI compared to the LTM, as showed in time-point 1 to 9 in Fig. 12.20. Consequently, when LTM shows better performance (time-point 10 to 12 in Fig. 12.19) the integrated framework adjusts its contributing weights structure by assigning more weight to the LTM. Therefore, the plot confirms that the integrated framework is capable of adjusting its level of trust by changing the contributing weight structure when it learns that the performance of that particular model is decreasing.

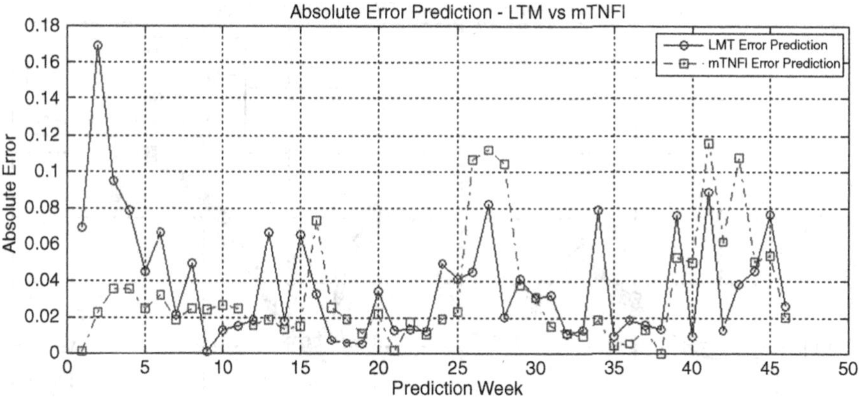

Fig. 12.19 Absolute prediction error of the New Zealand NZ50 index

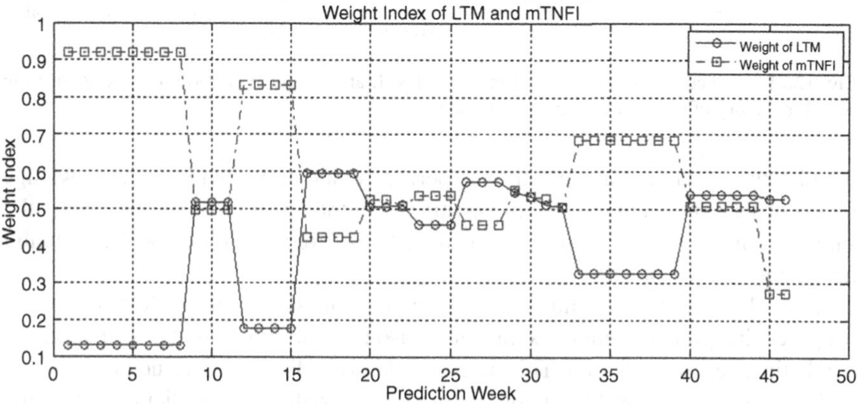

Fig. 12.20 Contributing weights assigned to the LTM and mTNFI by the integrated framework when calculating final prediction of the New Zealand NZ50 index

Additionally, Fig. 12.21, which compares the prediction trajectory produced by the integrated framework to the prediction trajectories produced by the LTM and the mTNFI, indicates that the integrated approach is superior.

To validate if forecasting movements of multiple time series simultaneously offers better prediction accuracy, a comparative analysis with multiple linear regressions (MLR), MLP, and random walk methods applied on single time series is conducted in this work. The random walk model is a time series analysis that assumes that next value of a time series is equal to current value. Here, the random walk without drift model defined simply by (12.30) is used:

$$x_{t+1} = x_t. \tag{12.30}$$

The random walk model in some cases might offer better prediction accuracy (in terms of sum squared residual). However, this model produces a shadow plot of

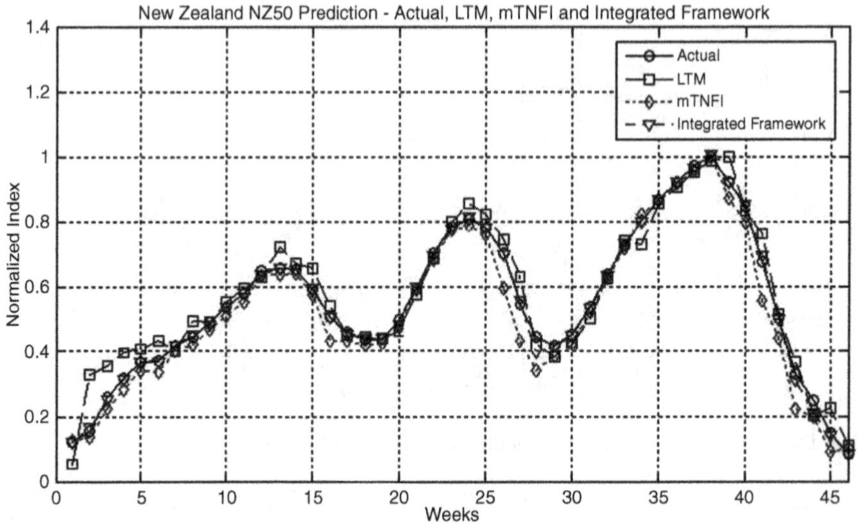

Fig. 12.21 Prediction results of the LTM, the mTNFI, and the integrated framework against the actual trajectory of the New Zealand NZ50 index

the observed data, lagging exactly one period behind and providing no knowledge on the observed system as it simply assumes that the upcoming value is exactly the same as current value. Therefore, it can be considered that this model is actually have no predictive power.

Table 12.1 shows the much smaller root mean square error (RMSE) of the proposed integrated framework in comparison to the other methods applied for single time series prediction. This outcome clearly indicates the value of extracting and exploiting relationships between multiple variables in prediction when the variables concerned are influencing each other in a dynamic fashion.

12.8 Conclusion

The chapter presented two novel approaches for multiple time series of a nonstationary environment analysis and modeling. The first approach, named the localized trends model and denoted by LTM, constructs local models which captures recurring specific behavior of the data set under observation. The recurring specific behavior in this model is described as recurring relationships between pairs of time series that influence each other, and recurring trends of movement within the series. Outcomes of conducted experiments using index values of ten selected stock markets in the Asia Pacific region prove that LTM demonstrates the ability to:

- Extract profiles of relationships and recurring trends from a multiple time series data.

- Perform simultaneous prediction of multiple time series with excellent precision.
- Evolve by continuing to extract profiles of relationships and recurring trends over time when new data samples become available.

The second approach, named the multivariate transductive neuro-fuzzy inference system denoted as mTNFI, develops an individual model over the new input vector (by considering the state of relationships between multiple time series in the new vector to the training samples or search space) and therefore provides a specific and better local generalization. Furthermore, outcomes of conducted experiments suggests that mTNFI is capable of performing multiple time series prediction accurately.

Additionally, an integrated framework that assimilates the capabilities of both LTM and mTNFI is also proposed and outlined in this chapter. Conducted experiments using the New Zealand NZ50 index as test data set reveals that the integrated framework is capable of changing its structure dynamically by in response to changes in the underlying movement of the multiple time series. The integrated framework dynamically assigns different weights to each contributing model based on their current predicting power. By being able to do so, the integrated framework produces a prediction trajectory that matches closer to the actual trajectory compared to the prediction trajectories calculated by the LTM and mTNFI individually.

Nevertheless, application of proposed methods is not limited to only the modeling of relationships between variables from financial domain, but also for other real-world problems, that is, to model interactions of climatology variables in our global weather system or to learn about profiles of relationships between genes in a living being from the gene expression level data sets. Conclusively, the proposed local and transductive model and the integrated framework indicate that by being able to dynamically learn and model changes that emerge in a nonstationary environment, one could have complete understanding of the underlying behavior of the observed environment and hence would help to estimate its future states more accurately.

Acknowledgements The authors would like to thank the Knowledge Engineering and Discovery Research Institute (KEDRI) and all the members for their supports, constructive discussions, and inspirational ideas. The authors would also like to thank the School of Computing and Mathematical Sciences of Auckland University of Technology, New Zealand for the scholarship granted to Harya Widiputra.

References

1. Amari, S.: Mathematical foundations of neuro-computing. Proceedings of the IEEE **78** (9), pp. 1443–1463 (1990)
2. Antoniou, A., Pescetto, G., Violaris, A.: Modelling international price relationships and interdependencies between the stock index and stock index futures markets of three EU countries: A multivariate analysis. Journal of Business Finance & Accounting, **30**, pp. 645–667 (2003)

3. Ben-Dor, A., Shamir, R., Yakhini, Z.: Clustering gene expression patterns. Journal of Computational Biology, **6** (3/4), pp. 281–297 (1999)
4. Bosnic, Z., Kononenko, I., Robnik-Sikonja, M., Kukar, M.: Evaluation of prediction reliability in regression using the transduction principle. In The IEEE region 8 EUROCON 2003, computer as a tool, **2**, pp. 99–103 (2003)
5. Collins, D., Biekpe, N.: Contagion and interdependence in African stock markets. South African Journal of Economics, **71** (1), pp. 181–194 (2003)
6. Friedman, L., Nachman, P.: Using Bayesian networks to analyze expression data. Journal of Computational Biology, **7**, pp. 601–620 (2000)
7. Hastie, T., Tibshirani, R., Friedman, J.: The elements of statistical learning: Data mining, inference, and prediction — second edition. Springer, New York Berlin Heidelberg, Germany (2009)
8. Holland, J., Holyoak, K., Nisbett, R., Thagard, P.: Induction processes of inference, learning, and discovery. Cambridge University Press, Cambridge, U.S.A. (1989)
9. Joachims, T.: Transductive inference for text classification using support vector machines. In Proceedings of the sixteenth international conference on machine learning, ICML 1999, pp. 200–209. San Francisco, CA, USA (1999)
10. Joachims, T.: Transductive learning via spectral graph partitioning. In International conference on machine learning (ICML), pp. 290–297. Washington, DC U.S.A. (2003)
11. Kasabov, N. (2001).: Evolving fuzzy neural networks for supervised/unsupervised online knowledge-based learning. IEEE Transactions on Systems, Man and Cybernetics, **31**, pp. 902–918 (2001)
12. Kasabov, N. (2007).: Global, local and personalised modelling and pattern discovery in Bioinformatics: An integrated approach. Pattern Recognition Letters, **28** (6), pp. 673–685 (2007)
13. Kasabov, N., Chan, Z., Jain, V., Sidorov, I., Dimitrov, D.: Gene regulatory network discovery from time-series gene expression data: a computational intelligence approach. In Lecture Notes in Computer Science **3316**, pp. 1333–1353. Springer Berlin / Heidelberg (2004)
14. Kasabov, N., Pang, S.: Transductive support vector machines and applications in Bioinformatics for promoter recognition. In Proceedings of the 2003 International Conference on Neural Networks and Signal Processing, **1**, pp. 1–6. IEEE Press (2003)
15. Kasabov, N., Song, Q.: DENFIS: dynamic evolving neural fuzzy inference system and its application for time-series prediction. IEEE Transactions on Fuzzy Systems, **10**, pp. 144–154 (2002)
16. Kim, T., Adali, T.: Approximation by fully complex multilayer perceptrons. Neural Computing, **15**, pp. 1641–1666 (2003)
17. Kukar, M.: Transductive reliability estimation for medical diagnosis. Artificial Intelligence in Medicine, **29** (1-2), pp. 81–106 (2003)
18. Lei, Z., Yang, Y., Wu, Z.: Ensemble of support vector machine for text-independent speaker recognition. International Journal of Computer Science and Network Security, **6** (5), pp. 163–167 (2006)
19. Li, C., Yuen, P.: Transductive learning: Learning iris data with two labelled data. In G. Dorffner, H. Bischof, K. Hornik (Eds.), Lecture Notes in Computer Science, Artificial neural networks, ICANN 2001, **2130**, pp. 231–236. Springer Berlin / Heidelberg (2001)
20. Li, F., Wechsler, H.: Watch list face surveillance using transductive inference. In D. Zhang A. Jain (Eds.), Lecture Notes in Computer Science, Biometric authentication, **3072**, pp. 1–15. Springer Berlin/Heidelberg (2004)
21. Li, J., Chua, C.: Transductive inference for color-based particle filter tracking. In Proceedings of international conference on image processing, ICIP 2003, **3**, pp. 949–952 (2003)
22. Lucks, M., Oki, N.: A radial basis function network (RBFN) for function approximation. In Proceedings of the 42nd Midwest symposium on circuits and systems **2**, pp. 1099–1101 (1999)
23. Masih, A., Masih, R.: Dynamic modeling of stock market interdependencies: An empirical investigation of Australia and the Asian NICs. Review of Pacific Basin Financial Markets and Policies, **4** (2), pp. 235–264 (2001)

24. Poggio, F.: Regularization theory, radial basis functions and networks. In From statistics to neural networks: Theory and pattern recognition applications, pp. 83–104. NATO ASI Series.
25. Proedrou, K., Nouretdinov, I., Vovk, V., Gammerman, A.: Transductive confidence machines for pattern recognition. In T. Elomaa, H. Mannila, H. Toivonen (Eds.), Lecture Notes in Computer Science, machine learning: ECML 2002, **2430**, pp. 221–231. Springer Berlin/Heidelberg (2002)
26. Psillaki, M., Margaritis, D.: Long-run interdependence and dynamic linkages in international stock markets: Evidence from France, Germany and the U.S. Journal of Money, Investment and Banking, **4**, pp. 59–73. EuroJournals Publishing (2008)
27. Rodrigues, P., Gama, J., Pedroso, J.: Hierarchical clustering of time-series data streams. IEEE Trans. on Knowl. and Data Eng., **20**, pp. 615–627 (2008)
28. Song, Q., Kasabov, N.: ECM - a novel on-line, evolving clustering method and its applications. In M. Posner (Ed.), Foundations of cognitive science, pp. 631–682. The MIT Press, Massachusetts, USA (2001)
29. Song, Q., Kasabov, N.: NFI: a neuro-fuzzy inference method for transductive reasoning. IEEE Transactions on Fuzzy Systems, **13** (6), pp. 799–808 (2005)
30. Soucy, P., Mineau, G.: A simple *k*NN algorithm for text categorization. In Proceedings IEEE international conference on data mining, ICDM 2001, pp. 647–649 (2001)
31. Takagi, T., Sugeno, M.: Fuzzy identification of systems and its applications to modelling and control. IEEE Transactions on Systems, Man, and Cybernetics, **15** (1), pp. 116–132 (1985)
32. Vapnik, V.: Statistical learning theory. Wiley-Interscience, Chichester (2008)
33. Weston, J., Perez-Cruz, F., Bousquet, O., Chapelle, O., Elisseeff, A., Scholkopf, B.: Feature selection and transduction for prediction of molecular bioactivity for drug design Bioinformatics, **19** (6), pp. 764–771 (2003)
34. Widiputra, H., Kho, H., Lukas, Pears, R., Kasabov, N.: A novel evolving clustering algorithm with polynomial regression for chaotic time-series prediction. In C. Leung, M. Lee, J. Chan (Eds.), Lecture Notes in Computer Science, neural information processing **5864**, pp. 114–121. Springer Berlin/Heidelberg (2009)
35. Widiputra, H., Pears, R., Kasabov, N.: Personalised modelling for multiple time-series data prediction: a preliminary investigation in Asia Pacific stock market indexes movement. In Proceedings of the 15th international conference on advances in neuro-information processing part I, ICONIP 2008, **5506**, pp. 1237–1244. Springer, Berlin Heidelberg (2008)
36. Widiputra, H., Pears, R., Kasabov, N.: Dynamic interaction networks versus local trend models for multiple time-series prediction. Cybernetics and Systems, **42**, pp. 1–24 (2011)
37. Widiputra, H., Pears, R., Kasabov, N.: Multiple time-series prediction through multiple time-series relationships profiling and clustered recurring trends. In Proceedings of the pacific asia conference on knowledge discovery and data mining, PAKDD (2011)
38. Wooldridge, J.: Introductory econometrics: a modern approach, 3rd edition. Cengage Learning Services, South Western College, Florence KY, USA (2005)
39. Wu, D., Bennett, K., Cristianini, N., Shawe-Taylor, J.: Large margin trees for induction and transduction. In Proceedings of the sixteenth international conference on machine learning, ICML 1999, pp. 474–483. Morgan Kaufmann Publishers Inc., San Francisco, CA (1999)
40. Yamada, T., Yamashita, K., Ishii, N., Iwata, K.: Text classification by combining different distance functions with weights. In Seventh ACIS international conference on software engineering, artificial intelligence, networking, and parallel/distributed computing, SNPD 2006, pp. 85–90 (2006)
41. Yang, H., Chan, L., King, I.: Support vector machine regression for volatile stock market prediction. In Third international conference on intelligent data engineering and automated learning, IDEAL 2002, pp.391–396. Springer (2002)
42. Zadeh, L.: Outline of a new approach to the analysis of complex systems and decision processes. IEEE Transactions on Systems, Man and Cybernetics, **3** (1), pp. 28–44 (1973)

[illegible]

27. [illegible]

28. [illegible]

[illegible]

32. [illegible]

[illegible]

38. [illegible]

[illegible]

40. [illegible]

[illegible]

Chapter 13
Optimizing Feature Calculation in Adaptive Machine Vision Systems

Christian Eitzinger and Stefan Thumfart

Abstract A classifier's accuracy substantially depends on the features that are utilized to characterize an input sample. The selection of a representative and—ideally—small set of features that yields high discriminative power is an important step in setting up a classification system. The features are a set of functions that transform the raw input data (an image in the case of machine vision systems) into a vector of real numbers. This transformation may be a quite complex algorithm, with lots of parameters to tune and consequently with much room for optimization. In order to efficiently use this additional room for optimizing the features, we propose an integrated optimization step that adapts the feature parameters in such a way that the separation of the classes in feature space is improved, thus reducing the number of misclassifications. Furthermore, these optimization techniques may be used to "shape" the decision boundary in such a way that it can be easily modeled by a classifier. After covering the relevant elements of the theory behind this automatic feature optimization process, we will demonstrate and assess the performance on two typical machine vision applications. The first one is a quality control task, where different types of defects need to be distinguished, and the second example is a texture classification problem as it appears in image segmentation tasks. We will show how the optimization process can be successfully applied in morphological and textural features that both offer a number of parameters to tune and select.

13.1 Introduction

In many classification tasks, the investigations start with a set of features that are the input to various machine learning structures, such as classifiers. In nonstationary learning environments,the classifier is adjusted to adapt to changes in the concepts

C. Eitzinger (✉) • S. Thumfart
Profactor GmbH Im Stadtgut A2, 4407 Steyr-Gleink, Austria
e-mail: christian.eitzinger@profactor.at; stefan.thumfart@profactor.at

M. Sayed-Mouchaweh and E. Lughofer (eds.), *Learning in Non-Stationary Environments: Methods and Applications*, DOI 10.1007/978-1-4419-8020-5_13,

© Springer Science+Business Media New York 2012

or rules that apply. However, quite often we find situations where the rules should remain the same, but the underlying processes that generate the input data (the features) have changed. This may quickly turn an initially well-posed classification task into an overly complex problem with weird decision boundaries. This may happen, for example, in a machine vision system, when the surface properties of the objects change. The statistical properties of the features will change, which may increase the complexity of the following classification tasks and may make it hard for the classifier to find reasonable decision boundaries. On-line adaptation and evolution of the classifiers (as applied in Chap. 7 for on-line surface inspection problems) may help to overcome these problems by permanent adjustments of the decision boundaries; however, the real class labels are not always available or may require high costs due to significant operators' efforts.

Another promising approach to counter such a change is to adapt the procedures that generate the features in such a way that they compensate for the change in surface properties. This chapter thus focuses on what can be achieved by adapting feature calculation. Features are always application specific and are often assumed to be carefully chosen by an expert, who makes sure that these features are relevant for the task. However, it has to be understood that the resulting feature vector is just a very low-dimensional representation of the object and that much—possibly relevant—information is already lost by converting the raw data of the object into a set of features. In fact, the feature calculation is the step that performs the largest reduction in the dimension of the problem. A typical image used in machine vision applications is several thousand pixels wide and thus may be considered a data vector coming from a 10^6- to 10^7-dimensional space. Clearly, the gray values of neighboring pixels are highly correlated, and images used in typical machine vision applications are usually very far from filling up this huge space. Instead, they are restricted to a comparably small subspace, which allows a compression of the information by means of features. Feature calculation thus reduces this high-dimensional space to a representation with dimensions in a typical range of 20–200. Depending on the number of samples that are available as training data and also on the properties of the features, the dimension needs to be reduced further before applying machine learning methods and classifiers. Therefore, a feature selection step is used that typically reduces the number of features down to 5–15 features.

Essentially, feature selection tries to select a subset of features that is optimal for the task at hand. Optimal here means that the subset contains low redundancy and that correlated features are removed while preserving most of the relevant information. This approach depends on various hypotheses about the distribution of the features and is called "filter approach". Alternatively, one may directly select a subset in such a way that the classification accuracy is optimized. This "wrapper approach" does not require any additional hypotheses, but depends on the classifier that is used. Some classifiers have a built-in feature selection process, for example, decision trees that select a single feature for the decision that is to be made at each node.

Table 13.1 Reduction of dimension when processing an image

Processing step	Data representation	Dimension
Raw (image) data	Image	10^6–10^7
Initial feature space	Feature vector	20–200
Feature space after selection	Feature vector	5–15
Classification result	Discrete value	1

The classifier then performs the final data reduction step and reduces the feature space to a one-dimensional space of a small, discrete set of classes. Table 13.1 illustrates this data reduction.

Within this chapter, we want to focus on the first processing step that converts the raw input data to an initial set of features. This processing step builds the basis for the downstream processing, and it also performs the most significant reduction and compression of data. The algorithms used for this reduction may be highly complex and are domain specific. For example, in the analysis of time signals, one may apply spectral methods to characterize the signals, whereas in image processing, texture analysis might be appropriate to calculate a set of relevant features. In any case, these algorithms include a large number of parameters that have to be chosen and that can be tuned to a particular application. This tuning process is often left to the expert in the field and is sometimes done on an intuitive basis coming from past experience and from the particular requirements of the task. We claim that these parameters can be used with great effect to improve the accuracy of downstream classification by directly optimizing feature calculation during the off-line and on-line adaptation of classification systems.

At this point, we would also like to make a clear distinction between feature selection and feature optimization. Feature selection converts a high-dimensional set of features into a smaller set [8] while maintaining most of the relevant information. In the case of a filter approach, various hypotheses are used that lead to optimization criteria based on distance, information, consistency, or dependency [7, 15, 27]. In the case of wrapper approaches [18], the goal is to directly improve classification accuracy. At the heart of the problem is a subset selection task that has a runtime of $O(2^N)$, but good approximations can be obtained using heuristic methods with a runtime of $O(N^2)$. A wide range of algorithms have been developed for this task, such as RELIEF [19, 28], the decision tree method [4], or branch and bound [25]. A recent survey lists 42 different algorithms [14, 24]. Feature selection may thus be considered a projection of the features to a low-dimensional space. This transformation is continuous, and its main property is that objects that were close together in the initial feature space will also be close together in the reduced feature space. If these two objects belong to different classes, then the margin between the two classes will be narrow no matter how the features are selected.

Feature optimization, on the other hand, has access to the original raw data for adapting the features. Even if two objects are identical in the initial feature space, they will not necessarily be so in the space of raw data. If these objects belong to different classes, then feature optimization may be used to create or tune features to put particular emphasis on this difference in the raw data and thus optimize feature calculation. It thus has the potential of increasing the margin between the two classes beyond what is possible with projections or other continuous transformations of feature space.

13.2 Parameterized Image Features

In the following, we will focus on two different types of features, both of which are widely used in image processing. The first set of features includes shape descriptors that are used to characterize the properties of image regions (so-called "blobs"). "Blobs" usually refer to small-image regions that are darker or brighter than the background and that can be more or less easily detected using (locally adaptive) thresholding methods or edge detectors. In the area of surface inspection, these blobs often correspond to the defects that have to be found and analyzed. For this purpose, the defects are characterized by a set of descriptors such as the total area of the blob, the position of the blob in the image, the ratio of area and circumference of the blob, or the inner structure of the blob.

The second set contains texture features that are used for image segmentation. Texture is a small-scale, visible surface structure that is characterized by local similarity. Small patches of a single texture, so-called texels, thus share a set of properties or features that are similar for all texels coming from this texture. These features may thus be used to characterize the texture and to enable texture segmentation. By using a classifier, one may determine which texels belong to the texture and which do not to establish a boundary between regions of different texture. There is a huge variety of texture features among which Gabor features are a popular choice for segmentation, classification, and image retrieval.

13.2.1 Blob Features

The notion "blob" is a commonly used abbreviation for "binary linked objects". "Linked objects" mean that the objects are usually connected in the sense that there is a path from one pixel of the object to any other pixel of the object that is fully inside the object. Such linked objects can be easily extracted from the image using various (sometimes recursive) algorithms, for example, by following the edge of the object. "Binary" refers to the fact that the algorithms are often applied to bi-valued images, where pixel values of 0 corresponds to background and 1 corresponds to pixels inside the blob. Such binarization is obtained using thresholding algorithms that search for dark or bright areas in the image. This threshold value is also a very important parameter for optimizing feature calculations. This is demonstrated in Fig. 13.1.

If we consider a simple "area" feature that describes the number of pixels covered by the blob, then this feature will substantially depend on the threshold value that

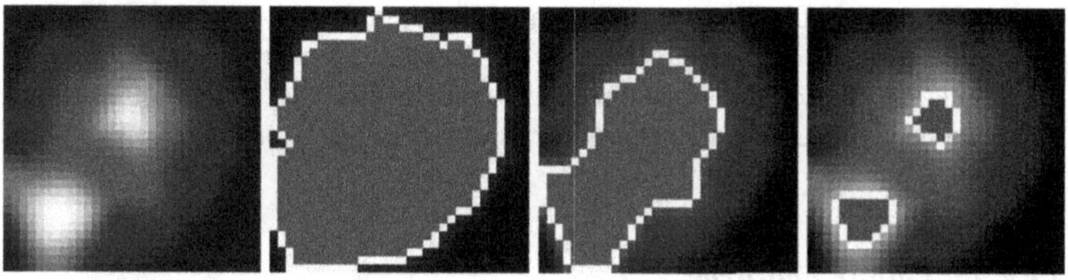

Fig. 13.1 "Area" feature for different threshold values in an 8-bit gray-level image. From *left* to *right*: original image; segmented area using a threshold of 55, 115, and 200. With increasing threshold value, the area is becoming smaller, and may even split into two disconnected regions

Table 13.2 Typical shape features used in blob analysis

Feature	Description	Parameter(s)
Area	Number of pixels inside the blob	Threshold value
Bounding rectangle	Width and height of the bounding rectangle	Threshold value, parameters dealing with outliers, for example, percentage of outlying pixels not considered
Roundness	Ratio of the principal axis to the secondary axis of the circumscribing ellipse	Threshold value, parameters dealing with outliers
Second-order moments	Second-order moments calculated along an axis (column-wise, row-wise, or arbitrary angle)	Threshold value, angle of the axis relative to the principal axis
Circumference	Number of pixels along the edge of the blob	Threshold value
Perimeter	Perimeter of a circle covering the same area as the blob	Threshold value, parameters dealing with outliers
Compactness	Ratio of the total area of the blob and the area of the circular disc that fully covers the blob	Threshold value, parameters dealing with outliers
p-Percentile 1	Gray value for which p percent of the total number of pixels inside the blob are above that value	p
p-Percentile 2	Percentage of pixels inside the blob below/above a certain threshold value	Gray value threshold

is chosen for binarization of the image during feature calculation. The region may even split into two separate blobs for higher threshold values. The correct choice of the threshold value is not immediately clear, and obviously, there is room for optimization.

There is a huge variety of features used in blob analysis, most of which describe the shape of the blob, but there are also features that provide information about the contrast or the inner structure (texture) of the blob. Table 13.2 below lists 9 examples of such features, including a description of the feature and parameters that can possibly be used for optimization. We have not included formulas or algorithms for the calculation of each feature and refer the reader to the vast amount of literature on basic image processing algorithms, for example, Chap. 10 of [9].

13.2.2 Gabor Features

A wide variety of features has been developed for the characterization of textures. The goal of these features is to capture the local properties of textures and to enable higher-level processing of textures for applications such as image segmentation, content-based image retrieval, or texture defect detection. The main approach in these tasks is to use the features as input to a machine learning structure such as classifiers or regression models. In the following, we describe Gabor features as an example of parameterized texture features and later show how these parameters can be tuned to a specific task.

Gabor features are often used in the above mentioned applications, and their importance also comes from the fact that in some way they resemble processing steps going on in the human visual cortex. However, Gabor features are just one example of a huge set of texture features, and we use them purely for demonstrating the general concept of feature adaptation. For the purpose of texture segmentation, which will be the example that we investigate here, the Gabor features are calculated for every image pixel to obtain a feature map. Thus, each image location $\omega_i \in \Omega$ is characterized by a d-dimensional vector of Gabor features. The actual value of d depends on the number of Gabor filters that form the applied Gabor filter bank and the type of Gabor features.

The calculation of a Gabor feature map for the image requires an intermediate step, in which a set of Gabor filters (a Gabor filter bank) is applied to the image. The feature map is calculated from the individual filter responses. A two-dimensional Gabor filter $g(x,y),(x,y) \in \Omega$ is a sinusoidal plane wave, modulated by a Gaussian envelope given by

$$g_{\lambda,\theta,\gamma,\varphi}(x,y) = \mathrm{e}^{-\frac{x'^2+\gamma^2 y'^2}{\sigma^2}} \cos\left(2\pi j \tfrac{x'}{\lambda} + \varphi\right),$$

where

$$x' = x\cos\theta + y\sin\theta,\ y' = -x\sin\theta + y\cos\theta \quad \text{and}$$
$$\sigma = c_\sigma \lambda.$$

This filter acts as an oriented local-band-pass filter that is optimal in terms of joint localization and resolution in image and frequency domain [16]. The parameters that can be tuned to obtain a selective Gabor filter are:

λ : Wavelength of the sinusoidal plane wave
θ : Orientation of the Gabor filter

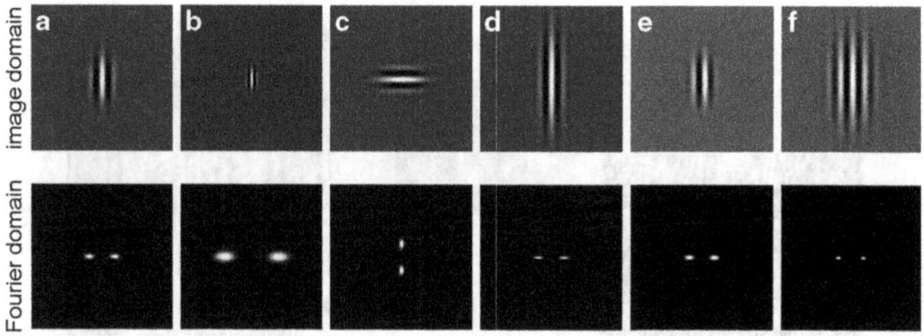

Fig. 13.2 Gabor filters in image and Fourier domain for various filter parameters. Compared to the mother wavelet filter ($\lambda = 1/23, \theta = 0, \gamma = 0.5, \varphi = 0, c_\sigma = 0.56$) in (**a**), the filters (**b**) to (**f**) are different in one parameter setting: (**b**) $\lambda = 1/49$, (**c**) $\theta = \pi/2$, (**d**) $\gamma = 0.25$, (**e**) $\varphi = -\pi/2$, (**f**) $c_\sigma = 1.0$. All filters in image domain are cropped and zoomed by factor 2 to increase visibility

γ : Elongation of the Gaussian envelope
φ : Shift of the sinusoidal plane wave that determines whether the filter is symmetric ($\varphi = 0$) or antisymmetric ($\varphi = \frac{\pi}{2}$)
c_σ : Factor that determines the size of the Gaussian envelope

A plot of typical Gabor filters for different parameters is shown in Fig. 13.2. In Sect. 13.5.3.2, we will demonstrate the importance of tuning these parameters to obtain highly accurate segmentation results.

By calculating the convolution of a Gabor filter g with an image I as in

$$r(x,y) = \iint_{\Omega} I(x,y) g(x-\xi, y-\eta) \mathrm{d}\xi \mathrm{d}\eta,$$

a new image (the filter response r) is obtained as shown in Fig. 13.3.

Usually, one does not only apply a single Gabor filter, but a family of filters—often called a filter bank—generated by varying the wavelength and orientation of a mother wavelet filter. A typical value is to use six orientations and four different scales, resulting in 24 different Gabor filters and thus in 24 filtered images. Spatial frequencies (scale) are chosen to cover the relevant frequencies in the image.

In many applications, these images are then converted into so-called Gabor energy maps. The energy map $e(x,y)$ of a Gabor filter [13] is calculated by considering filter responses to symmetric and antisymmetric Gabor filters:

$$e_{\lambda,\theta,\gamma}(x,y) = \sqrt{r^2_{\lambda,\theta,\gamma,0}(x,y) + r^2_{\lambda,\theta,\gamma,-\pi/2}(x,y)}.$$

The Gabor energy e is related to the local power spectrum p as follows:

$$p_{\lambda,\theta,\gamma}(x,y) = \mathrm{e}^2_{\lambda,\theta,\gamma}(x,y),$$

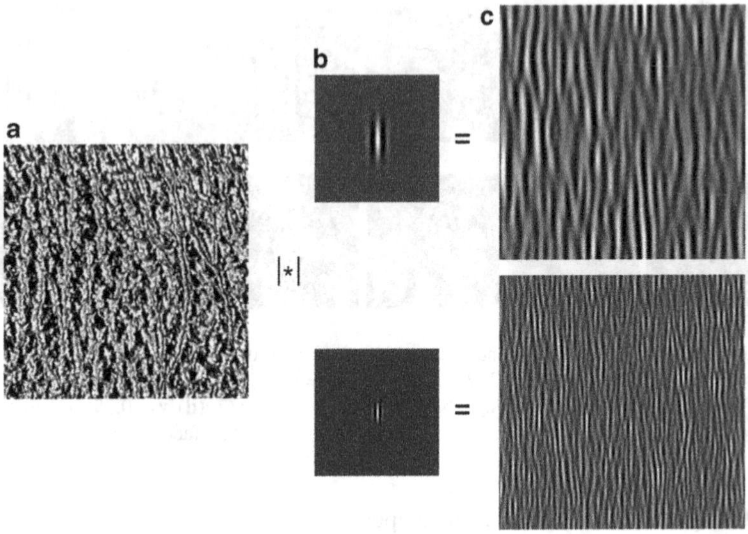

Fig. 13.3 The Gabor filter responses r for an image I from the Brodatz texture album [3]. The operator $|*|$ represents a convolution, which is typically done in Fourier domain to reduce the computational load for larger images

as the Gabor filters are essentially a Fourier transformation with a Gaussian windowing function. The resulting image contains an energy value per pixel and describes the local energy contained in the different spatial frequencies and directions. In order to finally obtain a single feature value, some kind of aggregated information is calculated for the image. This can be done in several different ways, some of which are described in the following:

The *sum of Gabor orientation energy difference* (SGOED) [17] is obtained by summing over all pixels of the energy map:

$$E_{\lambda,\theta,\gamma} = \iint_{\Omega} p_{\lambda,\theta,\gamma}(x,y)\mathrm{d}x\mathrm{d}y.$$

Assuming that we operate with a filter bank with Θ different orientations and Λ different scales, we can further sum up the entries of all energy maps $E_{*,*,\gamma}$ along scales or orientations such that we obtain two vectors

$$E_\Theta = \sum_{i=1}^{\Lambda} E_{i,\theta,\gamma}$$

and

$$E_\Lambda = \sum_{i=1}^{\Theta} E_{\lambda,i,\gamma},$$

with Θ and Λ elements, respectively. The SGOED [17] is then found by

$$f_{\mathrm{SGOED}} = \sum_{i=1}^{\Theta} |E_\Lambda(\theta_i) - E_\Lambda(\theta_{i+1})|,$$

where the index θ_i is to be understood as being calculated $\theta_i = \mod(i, \Theta)$. SGOED generates large values for abrupt orientation changes between energy values. The features thus measure in some sense whether the texture is isotropic (low value for SGOED) or whether it has a strong directional structure (high values for SGOED) in a certain spatial frequency range. In a similar fashion, one can derive the sum of Gabor scale energy differences (SGSED) by

$$f_{\text{SGSED}} = \sum_{i=1}^{\Lambda} |E_\Theta(\lambda_i) - E_\Theta(\lambda_{i+1})|,$$

where the index λ_i is to be understood as being calculated $\lambda_i = \mod(i, \Lambda)$. The SGSED is high if the image under investigation shows a dominant texture scale, while it is low if multiple texture scales are present.

While the aggregation of Gabor energy map entries as described so far has been applied successfully, for example, to distinguish natural from man-made objects in [17] or to texture classification and content-based retrieval [31], their basic idea can hardly be transferred to texture segmentation as these features aim to capture properties of larger image areas.

For pixel-wise segmentation, the energy map entries $e_{\lambda,\theta,\gamma}(x,y)$ for a fixed image location ω_i but different filter parameters (i.e., the energy maps obtained from a Gabor filter bank) are directly used to build a Gabor energy feature vector f_{ω_i}. Grigorescu et al. [13] evaluated the segmentation accuracy for more complex Gabor filter-based features, inspired by the early processing in human vision system and could show that Gabor energy features outperform complex moment features [13]. Superior performance was reported for grating cell operator features that are computed in a two-stage process. The first stage, based on a Gabor filter bank as described above, detects the presence of three parallel bars at any image location. The second stage integrates the output of the first within a certain surrounding and therefore detects the presence of multiple combinations of parallel bars.

13.3 Feature Adaptation Concepts (Off-line)

In the following, we first outline existing approaches to feature adaptation for blob and Gabor features. Next, we describe the general concept of feature optimization for off-line processing which will also be the basis for the on-line adaptation. To motivate the necessity for feature adaptation, let us consider a simplified inspection task, where we have to detect elongated scratches of predefined orientation and size. These scratches are present on a milled surface that shows an oriented surface texture itself. Given an image of a part with a scratch orientation of 10° and a surface texture orientation of 35°, the energy response for two Gabor filters with $\theta = 10°$ and $\theta = 35°$ with an appropriate scale λ would be highest. For an image without scratch, only the Gabor filter with $\theta = 35°$ would yield a high energy response. Obviously, the energy for the Gabor filter with $\theta = 10°$ can be used to distinguish defective and defect-free surface patches. However, if we would rely on a fixed

Gabor filter set with $\theta = \{0°, 45°, 90°, 135°\}$, the distinction between defective and defect-free areas cannot be based on a single filters' energy response any more. In real-world inspection tasks, we can never be sure that the properties of a (defect free) surface or the scratches remain constant over time (e.g., wear of the machine tools, material differences, temperature difference, etc.). Thus, it is crucial to adapt the feature parameters to discriminate between defective and defect-free surface areas.

13.3.1 Blob Feature Adaptation

For blob features, the main approach to feature adaptation is a manual selection of the relevant parameters by an expert. Automatic adaptation is rarely implemented, because of the perceived simplicity of the features. In some applications, however, such as tracking of objects, the features will change, for example, because the object that needs to be tracked is seen from a different viewpoint. Consequently, the relevance of features may change depending on the angle of view, because certain parts of the object will not be visible any more. Some recent results that cover this topic can be found in [5] and [6]. An adaptive feature transformation is described in [1], where the feature vector is postprocessed by an adaptive transformation matrix that is used to make the features invariant to environmental changes. The joint optimization of classifiers and features is investigated in [20], but also in this case, the optimization of features is done by a selection process rather than by adapting parameters inside the feature calculation.

13.3.2 Gabor Feature Adaptation

The filter bank which is required to derive Gabor energy map features obviously offers a set of parameters that can be used for optimizing the subsequent processing steps. These parameters include the scale λ, the orientation θ, as well as the number of scales Λ and directions Θ used. Gabor filter banks are usually designed to optimally represent the texture signal of the image. Optimality is measured by the mean squared error between the reconstructed and the original signal [21]. The goal of texture segmentation, however, is not to optimally represent the single textures but to divide the image into regions that contain the different textures. The focus of the optimization should thus be on identifying those parameters for the Gabor filters that allow an optimal discrimination between the different textures.

The second aspect that needs to be considered is computing time. Calculating the convolution of a Gabor filter with a whole image is computationally costly, and one is thus interested in minimizing the number of filters used and on choosing those with the highest discriminative power.

For the optimization, we may either focus on a single filter or the whole filter bank. Both approaches are slightly different.

In the filter bank design, one generally uses a set of directions that cover 180° (360° is not required, because the filter is symmetric), and the only question that needs to be answered is how many directions we want to consider. This decision is made purely on computational cost, and typical values are in the range of four (0°, 45°, 90°, 135°) to eight directions. The scale parameter of the Gaussian function is generally selected intuitively and assumed to be a constant. In [11], guidelines for selecting values for the scale parameter are proposed, but quite often, human intervention is still required to assist in selecting the appropriate filter parameters for texture segmentation. The selection of filters is then done by visually inspecting the filtered images, to identify those filters that best characterize different textures in the image.

For filter design, the focus is on selecting one or very few filters with parameters that are optimal for the task. This avoids some of the problems of filter banks, especially the high computational effort that is required for calculating the convolution with a larger number of filters. In many cases, the choice of parameters for the single filters is made based on a priori knowledge about the texture. Such knowledge may include typical dominating directions in the textures or the visual coarseness of the textures. This may be put on a more objective basis by performing a Fourier analysis on the whole image to identify the most significant (or discriminative) spectral components to deduce the scale and orientation for the Gabor filter. An explicit method for the selection of scale and direction parameters is proposed in [32]. The main idea is to solve a max-min optimization problem that maximizes the minimal ratio of Gabor energies contained in the different textures. The idea has a strong similarity to Fisher discriminant analysis [10] (Sect. 3.8.2) and to the optimization criteria used in support vector machines. It basically tries to maximize the margin between the two Gabor energy representations of the textures. A more recent approach presented in [29] aims at maximizing the fraction of separable harmonic signal pairs in a given frequency range. Separable here means that the filter responses are disjoint in at least one component of the response vector. This method proved to be more efficient than previous methods while achieving the same accuracy on texture segmentation tasks. For solving the resulting optimization problem, a range of methods has been applied such as genetic algorithms [22], simulated annealing [10] and any other kind of optimization method that can deal with the discrete nature of the optimization problem.

13.3.3 General Feature Adaptation Concept

Optimization of features as proposed in this context is an optimization step that is done once the classification system is already in place. The assumption is that a set of features has been selected and that a classifier has been trained off-line to achieve reasonable accuracy. At this stage, it makes sense to reconsider and optimize the feature calculation in order to make the classification system more robust and to further increase its accuracy. Optimizing features "from scratch" is not likely

Table 13.3 Different target functions J for commonly used classifiers

Classifier	Within-class scatter s_W	Between-class scatter s_B	J
Nearest neighbor	$\frac{1}{n_0}\sum_{j=1...n_0}\left\|\mathbf{F}(x_{0,j},\mathbf{p})-\mu_0\right\|^2 + \frac{1}{n_1}\sum_{j=1...n_1}\left\|\mathbf{F}(x_{1,j},\mathbf{p})-\mu_1\right\|^2$	$\left\|\mu_0-\mu_1\right\|^2$	$\frac{s_W(\mathbf{p})}{s_B(\mathbf{p})}$
Linear classifier	$\frac{1}{n_0}\sum_{j=1...n_0}\left(\mathbf{c}^T\left(\mathbf{F}(x_{0,j},\mathbf{p})-\mu_0\right)\right)^2 + \frac{1}{n_1}\sum_{j=1...n_1}\left(\mathbf{c}^T\left(\mathbf{F}(x_{1,j},\mathbf{p})-\mu_1\right)\right)^2$	$\left(\mathbf{c}^T\left(\mu_0-\mu_1\right)\right)^2$	$\frac{s_W(\mathbf{c},\mathbf{p})}{s_B(\mathbf{c},\mathbf{p})}$
Decision tree	$\frac{1}{n_0}\sum_{j=1...n_0}\left(\mathbf{e}_i^T\left(\mathbf{F}(x_{0,j},\mathbf{p})-\mu_0\right)\right)^2 + \frac{1}{n_1}\sum_{j=1...n_1}\left(\mathbf{e}_i^T\left(\mathbf{F}(x_{1,j},\mathbf{p})-\mu_1\right)\right)^2$	$\left(\mathbf{e}_i^T\left(\mu_0-\mu_1\right)\right)^2$	$\sum_{i=1...m}\frac{s_W(\mathbf{e}_i,\mathbf{p})}{s_B(\mathbf{e}_i,\mathbf{p})}$

to succeed, because the search space is huge and features are in the danger of degenerating into measurements for properties that are totally different from the originally intended meaning of the feature. This would impair the interpretability of the classification system and should thus be avoided.

For the adaptation of a set of features $F_i, i = 1, \ldots, m$, which we collect in a feature vector $\mathbf{F} \in \mathbb{R}^m$ and which depend on a parameter vector $\mathbf{p}$, we try to minimize the within-class scatter s_W in relation to the between-class scatter s_B. Feature adaptation should lead to a decision boundary that can be easily reproduced by the classifier at hand and that maximizes the distance between the classes perpendicular to the decision boundary. The way how we quantify the scatter thus depends on the classifier that is used. We will discuss this for three commonly used types of classifiers: nearest neighbor classifiers, linear classifiers, and decision trees.

Nearest neighbor classifiers use a distance measure $\|.\|$ to determine the distance of a feature vector to the class center. Classification performance will improve if the feature parameters are adapted to maximize the distance between the classes with respect to this distance measure. Linear classifiers create a separating hyperplane in feature space. This hyperplane is usually characterized by its normal vector $\mathbf{c}^T$. Feature parameters should thus be chosen to optimize the distance perpendicular to this hyperplane. Adaptation requires a joint optimization of feature parameters and classifier. Finally, for decision trees, the decision boundaries are in general parallel to the feature space axes $\mathbf{e}_i^T$. Thus, the scatter perpendicular to the feature space axes needs to be considered. The different methods of calculating the within- and between-class scatter and the target function J are shown in Table 13.3, where μ_0 and μ_1 define the average feature vectors over the class 0 (with n_0 representatives) and class 1 (with n_1 representatives), respectively. The target function aims at maximizing the distance between the classes, while minimizing the scatter within each class. The distance measure is selected in such a way that it favors the separation perpendicular to boundaries that can be easily modeled by the different classifiers.

Clearly, it will be beneficial for the simplicity of the optimization if the features can be optimized independent of the current parametrization of the classifier, but this will not always be the case. We can thus think of three different strategies:

- For decision-tree classifiers and also for the nearest neighbor classifiers using the L2-norm ($\|\mathbf{x}\|_2 = \sqrt{\mathbf{x}^T\mathbf{x}}$) as a distance measure, one can optimize the features independent of the current classifier. The optimization process thus consists of calculating the target function and applying a numerical optimization method.
- For linear classifiers, the target function depends on the particular parametrization of the classifiers, as the direction of the decision boundary may change. A similar situation is found if the Mahalanobis distance is used instead of the L2 norm for the nearest neighbor classifiers. The Mahalanobis distance accounts for the variance differences of the individual features and scales the features accordingly:

$$\|\mathbf{x}\|_{\text{Mahalanobis}} = \sqrt{(\mathbf{x}-\mu)^T\Sigma^{-1}(\mathbf{x}-\mu)},$$

  where Σ is the covariance matrix of $\mathbf{x}$.

  In this case, the distance measure (and thus the target function) will change depending on the distribution of the features. Clearly, this distribution will change whenever the feature calculation is adapted. These changes have to be included in the target function during the optimization. Alternatively, one may think of optimizing the features for a fixed classifier (i.e., fixed direction of the decision boundary or fixed distance measure) coming from the initial off-line training step. This clearly simplifies the optimization process, but probably gives away some potential for improvement.
- For some types of classifiers, such as neural networks, the decision boundary is so complex that the basic concept of optimizing the scatter perpendicular to the decision boundary cannot be directly applied. In this case, we propose to implement a strategy similar to the "wrapper" approach that directly optimizes classification accuracy. This joint feature and classifier optimization will lead to a quite complex optimization problem, which is limiting its applicability.

Independent of the particular choice of strategy the resulting optimization problems are usually nonsmooth and—considering the discrete nature of the threshold—even discontinuous. If there are only few parameters to tune, an exhaustive search is possible. For more complicated data sets, we found that gradient descent methods with a simple numerical estimation of the gradient work well. As usual, only a local minimum can be guaranteed in this case. More details of the algorithms are presented in [12].

13.4 Feature Adaptation Concept (On-line)

The basic assumption for the on-line adaptation of features is that a reasonable initial set of parameters has been found and that also an initial set of training data is available from which we can estimate certain feature statistics. The general approach will then be a gradient-based optimization that uses a numeric estimate of the gradient. This is due to the discrete nature of many parameters, especially of the threshold values that are usually chosen from a discrete set of $0, 1, \ldots, 255$ for 8-bit gray-level images.

Referring to Table 13.3, we find for the linear classifier that the gradient of the target function J with respect to the classifier's parameter $\mathbf{c}$ and the feature parameters $\mathbf{p}$ can be calculated analytically. For the derivation, we reformulate s_W and s_B of $J(\mathbf{c},\mathbf{p}) = \frac{s_W(\mathbf{c},\mathbf{p})}{s_B(\mathbf{c},\mathbf{p})}$, compute the partial derivatives $\frac{\partial s_W(\mathbf{c},\mathbf{p})}{\partial \mathbf{c}}$, $\frac{\partial s_W(\mathbf{c},\mathbf{p})}{\partial \mathbf{p}}$, $\frac{\partial s_B(\mathbf{c},\mathbf{p})}{\partial \mathbf{c}}$, $\frac{\partial s_B(\mathbf{c},\mathbf{p})}{\partial \mathbf{p}}$, and combine them using the quotient rule

$$\frac{\partial J(\mathbf{c},\mathbf{p})}{\partial \cdot} = \frac{\frac{\partial s_W(\mathbf{c},\mathbf{p})}{\partial \cdot} s_B(\mathbf{c},\mathbf{p}) - s_W(\mathbf{c},\mathbf{p}) \frac{\partial s_B(\mathbf{c},\mathbf{p})}{\partial \cdot}}{s_B(\mathbf{c},\mathbf{p})^2}. \tag{13.1}$$

By introducing the per class feature covariance matrices

$$\Sigma_k(\mathbf{p}) = \frac{1}{n_k} \left(\mathbf{F}(x_k,\mathbf{p}) - \mu_k\right) \left(\mathbf{F}(x_k,\mathbf{p}) - \mu_k\right)^T, \tag{13.2}$$

we can rewrite the scatter terms into the quadratic forms

$$s_W(\mathbf{c},\mathbf{p}) = \mathbf{c}^T \left(\Sigma_0(\mathbf{p}) + \Sigma_1(\mathbf{p})\right) \mathbf{c} \tag{13.3}$$

and

$$s_B(\mathbf{c},\mathbf{p}) = \mathbf{c}^T (\mu_0 - \mu_1)(\mu_0 - \mu_1)^T \mathbf{c}. \tag{13.4}$$

The vector-valued derivatives are then

$$\frac{\partial s_W(\mathbf{c},\mathbf{p})}{\partial \mathbf{c}} = 2\left(\Sigma_0(\mathbf{p}) + \Sigma_1(\mathbf{p})\right)\mathbf{c} \tag{13.5}$$

$$\frac{\partial s_B(\mathbf{c},\mathbf{p})}{\partial \mathbf{c}} = 2(\mu_0 - \mu_1)(\mu_0 - \mu_1)^T \mathbf{c} \tag{13.6}$$

$$\frac{\partial s_W(\mathbf{c},\mathbf{p})}{\partial \mathbf{p}} = \mathbf{c}\left(\frac{\partial \Sigma_0(\mathbf{p})}{\partial \mathbf{p}} - \frac{\partial \Sigma_0(\mathbf{p})}{\partial \mathbf{p}}\right)\mathbf{c}^T \tag{13.7}$$

$$\frac{\partial s_B(\mathbf{c},\mathbf{p})}{\partial \mathbf{p}} = 2\mathbf{c}(\mu_0 - \mu_1)\left(\frac{\partial \mu_0}{\partial \mathbf{p}} - \frac{\partial \mu_1}{\partial \mathbf{p}}\right)\mathbf{c}^T. \tag{13.8}$$

Regarding the on-line adaptation, the key issue is to avoid the repeated processing of a large number of test images. We thus have to make some compromises with respect to the quality of the estimates of the second-order statistics and the gradients.

For the second-order statistics, we create two buffers in which we store samples. These buffers are of fixed length with a typical size of 25–100 samples. Each buffer corresponds to one of the two classes (accept/reject). Depending on the label

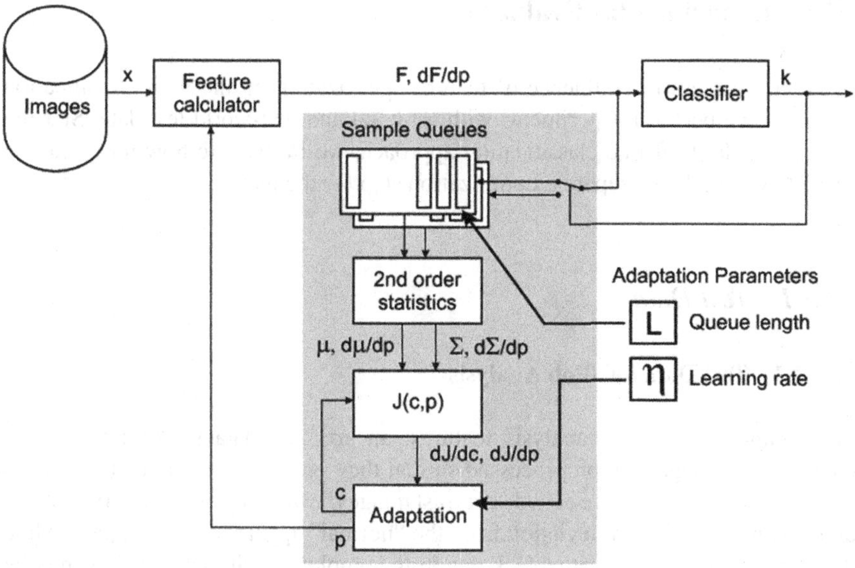

Fig. 13.4 Overview of the concept for on-line feature optimization

obtained from the classifier or from an expert, the feature vectors **F** from the images as well as their gradients with respect to the feature parameters **p** are routed to one of the two buffers. The buffers can also be thought of as a sliding window over the sequence of sample images. Based on the buffered samples, second-order statistics (the average μ and the variances Σ) of the features and the feature gradients are computed along with their derivatives with respect to their parameters. From the statistics and the classifier's parameters **c**, the target function J and its gradients with respect to **c** and **p** can be computed. Incremental optimization of the quality criterion is then performed by updating the **p** and **c** in the direction of the negative gradient of J. The step length is controlled by a fixed learning rate η:

$$\mathbf{p}_{n+1} = \mathbf{p}_n - \eta \nabla_p J_n$$
$$\mathbf{c}_{n+1} = \mathbf{c}_n - \eta \nabla_c J_n.$$

An overview of the whole concept is shown in Fig. 13.4.

A basic evaluation of this method on a simple test case showed that learning rates of 0.001–0.01 lead to good results and that a buffer size of 25–100 is required. After about 500 to 1,000 iterations (samples), the parameters were reasonably close to their optimal values. It should be noted, however, that depending on the length of the queue, there will also be fluctuations in the parameters.

13.5 Experimental Evaluation

In order to assess the influence of feature optimization on the final classification accuracy, we perform experiments with artificial and real-world test data. Starting from a standard off-line classification approach, we investigate how much can be gained by including a separate optimization step for the features.

13.5.1 Test Data

13.5.1.1 Test Data for Blob Analysis

For optimizing the blob-analysis features, an artificial data set of images was created. The images are preprocessed so that they only show the potential defects (objects). These images are called "contrast images," whose (gray scale) pixel values depend on the degree of deviation from the "normal" appearance of the part (white denoting complete similarity, black denoting complete dissimilarity). This may be considered an abstraction of a surface inspection task in machine vision. By using contrast images, we remove the application-specific low-level image processing from our consideration and focus purely on the classification problem. Figure 13.5 shows an example that represents a deviation image from a printing process: blobs that correspond to potential defects are highlighted in different gray levels.

We used five sets of artificial test data, each with 20,000 images, which were labeled automatically either as good (accept) or as bad (reject) with about 10,000 images in each class. In order to generate the labels, a set of rules was used for

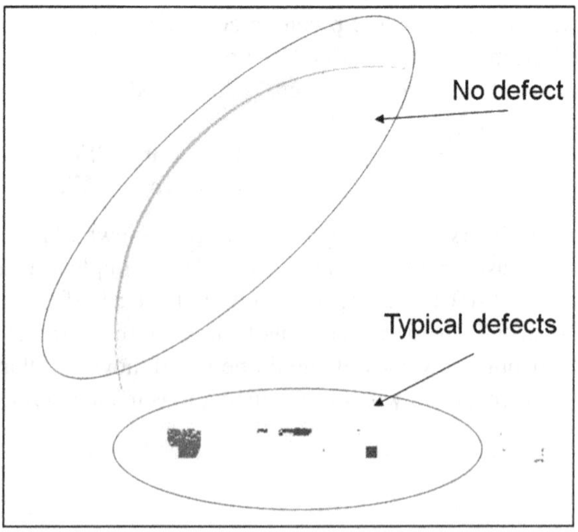

Fig. 13.5 For testing, we use contrast images. Only deviations from the normal appearance are shown in the image, the background is removed

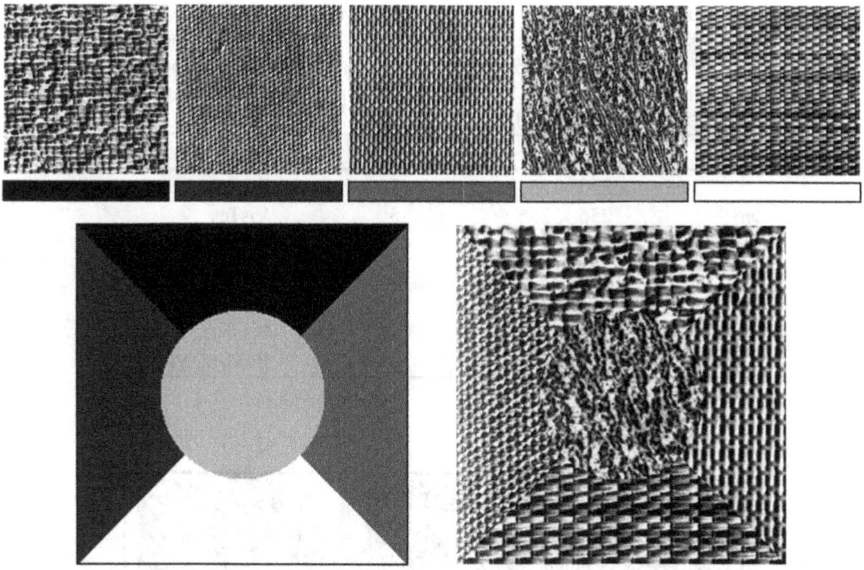

Fig. 13.6 Texture mosaic m_1 (*right*) built from five source textures (*top row*) according to the ground truth map (*left*)

each set of test images. The rules were based on descriptions that are regularly found in quality control instructions, such as "part is bad, if there is a fault with size >1.5 mm." The rules also included more complicated combinations, such as "part is bad, if there is a cluster of 4 faults, each with a size >0.75 mm." Three to five such rules were logically combined for each set of images. The images and the rules were chosen to have some resemblance to inspection of machined parts.

13.5.1.2 Test Data for Texture Segmentation

In order to assess the discriminative power of texture features, we perform a texture segmentation experiment on so-called texture mosaics. A texture mosaic $m(x,y)$ is built from n source textures $s_i(x,y), i = 1,\ldots,n$. For each pixel of $m(x,y)$, the ground truth map $t(x,y)$ defines its source texture index s_i. Figure 13.6 shows the source textures, ground truth map, and the resulting mosaic for $n = 5$.

In order to investigate the adaptation properties of Gabor-feature-based segmentation, we utilize mosaics combined from Brodatz [3], VisTex [23], and MeasTex [30] source textures to ensure diverse texture properties. Diversity of the source textures is crucial for our experiments as it calls for Gabor filter (banks) that adapt to the specific properties of each source texture. Table 13.4 contains detailed information about the selected mosaics. Figure 13.7 shows all texture mosaics and their ground truth information, respectively.

Table 13.4 All evaluated texture mosaics. The mosaics m_1 to m_5 are composed from five source textures each and are available on-line [26]. Mosaics m_6 and m_7 are composed from two source textures in order to obtain reasonable test data for single Gabor filter adaptation. All texture mosaics are shown in Fig. 13.7

ID	Size	n	Source of textures
m_1	256×256	5	Brodatz
m_2	256×256	5	VisTex
m_3	256×256	5	VisTex
m_4	256×256	5	VisTex
m_5	256×256	5	MeasTex
m_6	256×256	2	Brodatz, Vistex
m_7	256×256	2	Brodatz, Vistex

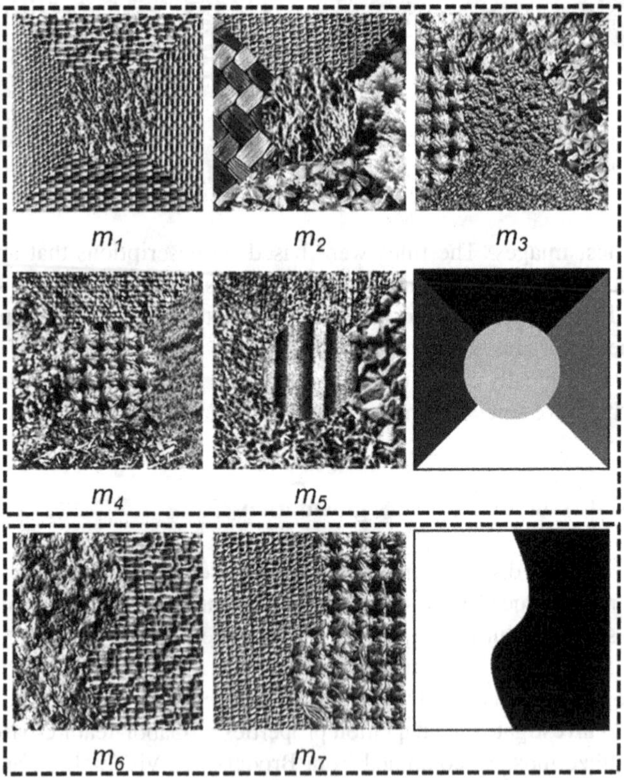

Fig. 13.7 Texture mosaics m_1 to m_7 and the corresponding ground truth maps

13.5.2 *Classification with Adaptive Features*

This section describes how the classification experiment based on adaptive features was conducted. The quantitative results of these experiments are discussed in Sect. 13.5.3.

13.5.2.1 Blob Analysis Features

For those features that allow an optimization, one parameter was chosen, which in most cases was a gray-level threshold value. If the pixel was above the threshold, it was considered as belonging to the fault and otherwise as belonging to the background. A reasonable range for the threshold was between 0 and 255 (as we deal with 8-bit gray-level images, these are the minimum and maximum values of the pixels). For the sake of optimization, this range was scaled to an interval of $[0,1]$. The goal was to adapt these thresholds in such a way that the classification accuracy is improved and that the decision boundary can be more easily reproduced by the classifier.

Regarding the computational complexity, it should be noted that the computational effort of optimizing the feature parameters can be significant. This is caused by the fact that for each iteration, the features of all objects in all images need to be calculated with the current parameter settings. Depending on the size and number of the images, this optimization may take several hours. On the other hand, once the optimal parameters are found, the feature calculation and classification takes the same amount of time as for any other method. With respect to the application of surface inspection, this is important because the "on-line" processing of the images, which usually has tight constraints on computing time, is not affected. It is just the "off-line" training process that becomes quite time-consuming.

13.5.2.2 Gabor Features

We choose a fixed feature extraction method, classification, and postprocessing setup in order to investigate the influence of feature adaptation on the segmentation accuracy. These subsequent processing steps are briefly outlined below as it is not the aim of our evaluation to optimize classification or postprocessing. A detailed discussion of Gabor filter (bank) optimization can be found in Sect. 13.5.3.2.

For *feature extraction*, we choose Gabor energy features to compute a feature map as described in Sect. 13.2.2 as their computational costs are reasonably low to perform multiple segmentation runs. Thus, we obtain a feature vector $\mathbf{F}$ of dimensionality d for each texture mosaic location $\omega_i = (x,y)$, where d is the number of Gabor filters in the filter bank. The training feature matrix is created by randomly selecting 200 feature vectors per source texture.

Table 13.5 Change of classification accuracy achieved by optimizing the feature calculation for different classifiers

Classifier	Initial (%)	Optimized (%)
C4.5	74.2	88.08
CART	73.8	87.77
Cluster-based	74.0	77.9
kNN	68.7	87.8

For the *pixel classification*, we choose two methods. We compute the mean Fisher criterion as described above to obtain a score that is independent from the classifier and describes the separation between the different texture classes for the current feature parameters. In order to obtain mosaic class-label results, we train a Random forest (RF) [2] with 300 trees on the source textures in order to classify each pixel of the texture mosaic. We rely on the R-package of the Random forrest available at http://cran.r-project.org/web/packages/randomForest/.

It is well known [13] that *filtering* of the class label improves the segmentation accuracy which is measured as ratio of misclassified pixels divided by the total number of mosaic pixels.

To remove isolated pixels and to smooth the boundaries between the segmented regions, we apply a filtering approach [13] of the classification result. Therefore, we assign that class label to a pixel that occurs most frequently in its 24×24 neighborhood.

For all experiments, we postprocess our class labels by applying a filter that replaces each label with the mode of its surrounding (size 24×24).

13.5.3 Results

The results are discussed separately for blob and Gabor features as these tasks differ in terms of classification and postprocessing of the results. Despite these differences, we show that both machine vision tasks benefit from feature adaptation.

13.5.3.1 Blob Analysis Features

We have chosen four different classifiers that performed well on surface inspection tasks and used target functions of Table 13.3 for optimization. Even though this required the repeated processing of 10,000 images an exhaustive search was performed in order to be sure that the global optimum is found. The improvement in classification accuracy that could be achieved is shown in Table 13.5.

For some classifiers, the change is quite substantial, and we find improvements of more than 10%. It should be noted however that the gain that can be achieved highly depends on the initial parameter settings. If these parameters are preset by a machine vision expert, then the improvement may be minimal, if any. Feature

Table 13.6 Parameter ranges that constitute the parameter grid for single Gabor filter optimization. The reciprocal of the filter wavelength $(1/\lambda)$ denotes the number of cycles of the plane wave for the whole image

Parameter	Range
Wavelength: λ	$[1/11, 1/47]$
Orientation: θ	$[0, \pi]$
Elongation: γ	$[0.3, 0.7]$
Size of Gaussian: c_σ	$[0.3, 0.7]$

Table 13.7 This table shows the filter parameters that resulted in the highest classification accuracy per mosaic. It lists the average Fisher criterion, the accuracy for classification with density estimation, and the corresponding filter parameters for single Gabor filter optimization

Mosaic ID	Fisher criterion	Accuracy (%)	Parameters
m_1	0.946	61.5	$\lambda = 0.026, \theta = 0.000, \gamma = 0.5, c_\sigma = 0.7$
m_2	0.858	53.7	$\lambda = 0.029, \theta = 1.396, \gamma = 0.3, c_\sigma = 0.7$
m_3	0.959	49.9	$\lambda = 0.091, \theta = 0.000, \gamma = 0.7, c_\sigma = 0.5$
m_4	0.788	46.2	$\lambda = 0.091, \theta = 0.000, \gamma = 0.7, c_\sigma = 0.5$
m_5	0.691	40.2	$\lambda = 0.091, \theta = 1.047, \gamma = 0.7, c_\sigma = 0.5$
m_6	0.950	92.5	$\lambda = 0.032, \theta = 1.396, \gamma = 0.7, c_\sigma = 0.7$
m_7	1.968	98.2	$\lambda = 0.091, \theta = 0.000, \gamma = 0.7, c_\sigma = 0.7$

adaptation, however, may help for those features that have multiple parameters. The results given as "initial" in Table 13.5 were obtained for parameter values chosen as a first guess by an expert without further manual optimization.

13.5.3.2 Gabor Features

Gabor filters offer many tuning knobs that are likely to influence the discriminative power of any Gabor feature. These parameters (λ, θ, γ, c_σ) were already discussed in Sect. 13.2.2. The parameter φ, which determines the shift of the sinusoidal plane wave will not be tuned as we require both the symmetric and the antisymmetric filter response to compute Gabor energy features. Table 13.6 contains the parameters' ranges that form our search grid for single Gabor filter optimization.

As we are dealing with a one-dimensional feature space in single filter optimization, we do not apply a complex RF but a classifier that fits a Gaussian density to each texture class and assigns a mosaic pixel to the class with highest probability for the pixels' feature value. The results for single filter optimization are listed in Table 13.7 and visualized in Fig. 13.8. The classification accuracy is substantially above the baseline accuracy (i.e., 20% for m_1 to m_5 and 50% for m_6, m_7) for all mosaics. We can also observe that the Fisher criterion scores do not necessarily correlate with the classification accuracy because of the postprocessing

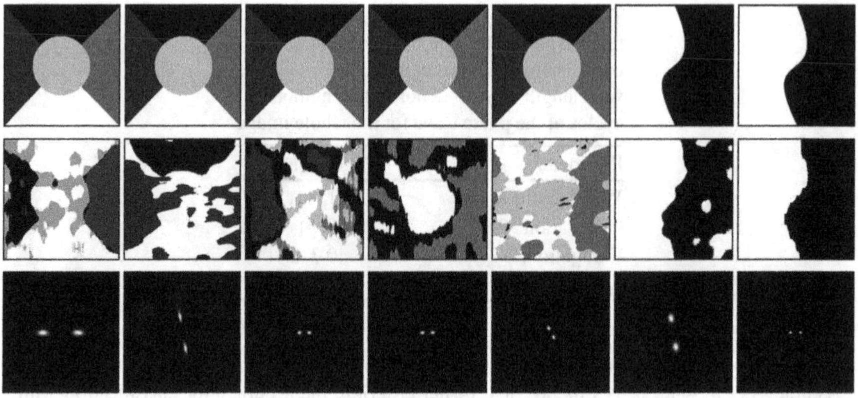

Fig. 13.8 Results for single Gabor filter optimization. The *top row* shows the ground truth map, and the *middle row* contains the best segmentation result obtained for the Gabor filter show in the *bottom row* (Fourier domain)

Table 13.8 Parameter ranges that constitute the parameter grid for filter bank optimization

Parameter	Range	Comments
Wavelength: λ_l	$[1/11, 1/47]$	Wavelength of the filter with lowest frequency
Orientation offset: o_θ	$[0, \pi/\Theta]$	o_θ is added to the equidistant orientation values
Elongation: γ	0.5	Not tuned to reduce the computational costs
Size of Gaussian: c_σ	$[0.3, 0.7]$	
No. scales: Λ	$[1, 4]$	For example $\Lambda = 3$, $\lambda_l = 11$ then $\lambda \in \{1/11, 1/19, 1/35\}$
No. orientations: Θ	$[2, 8]$	Orientation of filter j: $\theta_j = (j-1) * (\pi/\Theta)$

of the segmentation result. In general, we can observe that a single filter is not sufficient to distinguish between five texture classes while a two-class segmentation problem can be solved with reasonable accuracy.

However, a single Gabor filter is likely to be insufficient for real-world segmentation tasks as the classification decision is obtained on the basis of a one-dimensional feature space. Thus, Gabor filter banks that contain a family of self-similar Gabor filters are typically applied for classification, segmentation, and retrieval systems. The optimization of a whole Gabor filter bank introduces several additional parameters such as the number of filter orientations Θ and scales Λ. In order to reduce the computation cost for optimization, it is worthwhile to consider interdependencies between different parameters. For instance, the filter orientation which can be chosen in the range $[0, \pi]$ for a single feature can be treated as orientation offset o_θ in case of a filter bank that alters the filters orientation that is typically chosen equidistant to cover the range of $[0, \pi]$. The tuned parameters are listed in Table 13.8.

The results for the Gabor filter bank optimization are listed in Table 13.9 and visualized in Fig. 13.9. We compare our results to the baseline, obtained for the Gabor filter bank proposed in [13] that is based on experimental findings on early

Table 13.9 This table shows the filter bank parameters that resulted in the highest RF classification accuracy per mosaic. It lists the average Fisher criterion, the baseline accuracy obtained for the filter bank proposed in [13], the RF accuracy, and the corresponding Gabor filter bank parameters

Mosaic ID	Fisher criterion	Baseline accuracy (%)	RF accuracy (%)	Parameters
m_1	2.520	95.3	97.0	$\lambda_l = 0.029, o_\theta = 0.196, c_\sigma = 0.5, \Lambda = 3, \Theta = 8$
m_2	0.524	70.8	77.0	$\lambda_l = 0.037, o_\theta = 0.000, c_\sigma = 0.3, \Lambda = 3, \Theta = 2$
m_3	1.178	69.4	75.6	$\lambda_l = 0.053, o_\theta = 0.000, c_\sigma = 0.5, \Lambda = 3, \Theta = 6$
m_4	1.129	65.8	72.8	$\lambda_l = 0.053, o_\theta = 0.000, c_\sigma = 0.3, \Lambda = 3, \Theta = 6$
m_5	1.979	77.1	82.0	$\lambda_l = 0.029, o_\theta = 0.000, c_\sigma = 0.7, \Lambda = 3, \Theta = 8$
m_6	1.390	98.9	99.1	$\lambda_l = 0.029, o_\theta = 0.000, c_\sigma = 0.5, \Lambda = 3, \Theta = 4$
m_7	2.251	98.9	99.5	$\lambda_l = 0.037, o_\theta = 0.000, c_\sigma = 0.5, \Lambda = 3, \Theta = 8$

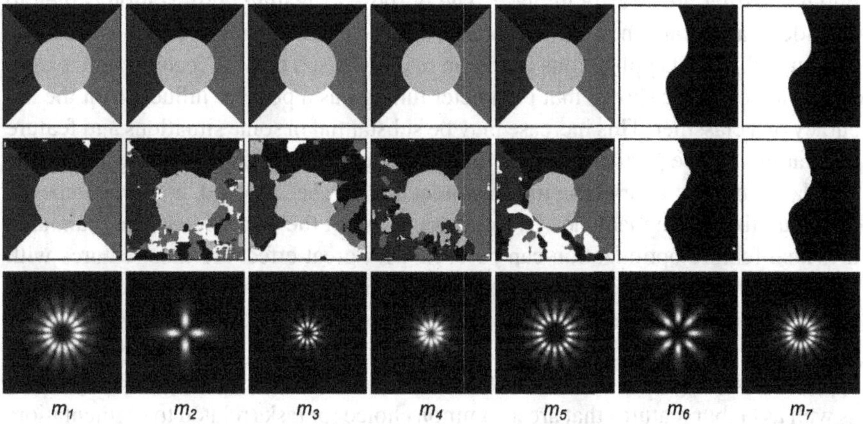

Fig. 13.9 Segmentation results for an optimized Gabor filter bank. The *top row* shows the ground truth map, and the *middle row* contains the best segmentation result obtained for the Gabor filter bank shown in the *bottom row* (Fourier domain)

human vision processes. Therefore, this baseline is no arbitrary choice of filter bank parameters but a setting that is hard to improve without excessive knowledge about Gabor filter design and the application domain.

Compared to the baseline, a substantial gain in segmentation accuracy could be achieved with Gabor filter bank optimization. Interestingly, the best filter banks differ with respect to many parameters. For instance, the best segmentation for m_2 was obtained for a filter bank with $\Theta = 2$, whereas for m_5, eight different orientations gave the best result. In general, a substantial reduction of the segmentation error rate (between 54.6% and 18.2%, average 27.5%) could be achieved. This clearly demonstrates that Gabor filter optimization should be conducted on a common basis, as feature selection and classifier hyperparameter selection, whenever a segmentation system is designed.

13.6 Conclusion

Features are a way of compressing raw data into a set of real numbers that represent the relevant information contained in the original data.

The properties of the raw data that are relevant cannot be determined a priori, but are application dependent. To tune features to the particular application, one may use the parameters available in feature calculation to optimize the features in such a way that the margin between the classes in a classification problem is increased. Furthermore, it may also shape the decision boundary so that it can be more easily reproduced by the classifier. This optimization may take place in an off-line mode or in an on-line mode.

In most cases, the assumption is that an initial set of features and possibly also an initial classifier are already in place. The purpose of feature optimization is then to provide an additional increase in robustness and classification performance. Tests on two quite different applications (decision making based on blob features and texture segmentation) have shown that parameter tuning has a positive influence on the accuracy of a classifier. This increase may be substantial in some situations and feature adaptation may be particularly helpful if the underlying processes that generate the raw data are instationary. The improvement that can be achieved, however, depends on the ability of the machine vision expert to preset the parameters. The main area in which feature optimization can be used with great effect are thus features with a larger number of tunable parameters and complex machine learning problems, where the effects of parameter changes cannot be easily assessed.

In this chapter, we demonstrated the effect of feature adaptation for two kinds of applications using blob features that are common in real-world inspection systems, as well as Gabor features that are a common choice for tasks related to segmentation, classification and retrieval of (textured) images. For both scenarios, we demonstrated a substantial reduction of the classification error rate that underpins the need for feature adaptation whenever a feature-based machine vision system is designed.

References

1. Azimi-Sadjadi, M.R.D.Y., Dobeck, G.J.: Adaptive feature mapping for underwater target classification. In: IJCNN '99. International Joint Conference on Neural Networks, vol. 5, pp. 3221–3224 (1999)
2. Breiman, L.: Random forests. Machine Learning **45**(1), 5–32 (2001)
3. Brodatz, P.: A Photographic Album for Artists and Designers. Dover Publications, New York (1966)
4. Cardie, C.: Using decision trees to improve case-based learning. In: Proceedings of 10th International Conference on Machine Learning, pp. 25–32 (1993)
5. Chen, H.T., Liu, T.L., Fuh, C.S.: Probabilistic tracking with adaptive feature selection. In: 17th International Conference on Pattern Recognition (ICPR'04), volume 2, pp. 736–739 (2004)
6. Collins, R., Liu, Y.: On-line selection of discriminative tracking features. In: Proc. of the 2003 International Conference of Computer Vision (ICCV 03), pp. 346–352 (2003)

7. Costanza, C.M., Afifi, A.A.: Comparison of stopping rules in forward stepwise discriminant analysis. Journal Amer. Statist. Assoc. **74**, 777–785 (1979)
8. Dash, M., Liu, H.: Feature selection for classification. International Journal of Intelligent Data Analysis **1**, 131–156 (1997)
9. Demant, C., Streicher-Abel, B., Waszkewitz, P.: Industrielle Bildverarbeitung. Springer-Verlag, Berlin Heidelberg New York (1998)
10. Duda, R.O., Hart, P.E., Stork, D.G.: Pattern Classification, 2nd edition. John Wiley & Sons, New York (2001)
11. Dunn, D., Higgins, W., Wakeley, J.: Texture segmentation using 2-d Gabor elementary functions. Pattern Analysis and Machine Intelligence, IEEE Transactions on **16**(2), 130 –149 (1994). DOI 10.1109/34.273736
12. Eitzinger, C., Gmainer, M., Heidl, W., Lughofer, E.: Increasing classification performance with adaptive features. In: A. Gasteratos, M. Vincze, J. Tsotsos (eds.) Proceedings of ICVS 2008, *LNCS*, vol. 5008, pp. 445–453. Springer, Santorini Island, Greece (2008)
13. Grigorescu, S.E., Petkov, N., Kruizinga, P.: Comparison of texture features based on gabor filters. In: IEEE Trans. on Image Process., vol. 11, pp. 1160–1167 (2002)
14. Guyon, I., Elisseeff, A.: An introduction to variable and feature selection. Journal of Machine Learning Research **3**, 1157–1182 (2003)
15. Hand, D.J.: Discrimination and classification. Wiley Series in Probability and Mathematical Statistics, Wiley, Chichester, UK (1981)
16. Jain, A.K., Farrokhnia, F.: Unsupervised texture segmentation using gabor filters. Pattern Recogn. **24**(12), 1167–1186 (1991). DOI http://dx.doi.org/10.1016/0031-3203(91)90143-S
17. Kim, M., Park, C., Koo, K.: Natural / man-made object classification based on gabor characteristics. In: W.K. Leow, M. Lew, T.S. Chua, W.Y. Ma, L. Chaisorn, E. Bakker (eds.) Image and Video Retrieval, *Lecture Notes in Computer Science*, vol. 3568, pp. 550–559. Springer Berlin / Heidelberg (2005)
18. Kohavi, R., John, G.: Wrappers for feature subset selection. Artificial Intelligence **97**(1–2), 273–324 (1997)
19. Kononenko, I.: Estimating attributes: Analysis and extensions of relief. In: Proceedings of ECML-94, pp. 171–182. Springer Verlag, Catania, Sicily (1994)
20. Krishnapuram, B., Hartemink, A.J., Carin, L., Figueiredo, M.A.T.: A Bayesian approach to joint feature selection and classifier design. IEEE Transactions on Pattern Analysis and Machine Intelligence **26**(9), 1105–1111 (2004)
21. Lee, T.S.: Image representation using 2d gabor wavelets. Pattern Analysis and Machine Intelligence, IEEE Transactions on **18**(10), 959–971 (1996). DOI 10.1109/34.541406
22. Li, M., Staunton, R.: Optimum gabor filter design and local binary patterns for texture segmentation. Pattern Recognition Letters **29**(5), 664–672 (2008). DOI 10.1016/j.patrec.2007.12.001
23. MIT Media Laboratory Cambridge: Vistex - Vision Texture Database. http://vismod.media.mit.edu/pub/VisTex/ (1995)
24. Molina, L.C., Belanche, L., Nebot, A.: Feature selection algorithms: A survey and experimental evaluation. In: ICDM '02: Proceedings of the 2002 IEEE International Conference on Data Mining, pp. 306–311. Maebashi City, Japan (2002)
25. Narendra, P., Fukunaga, K.: A branch and bound algorithm for feature subset selection. IEEE Transactions on Computer **26**(9), 917–922 (1977)
26. University of Oulu, C.S., Laboratory, E.: Supervised texture segmentation mosaics. http://www.cse.oulu.fi/MVG/SupervisedTextureSegmentation/ (2011)
27. Rao, C.R.: Linear statistical inference and its applications. John Wiley & Sons, Inc., NY, U.S.A. (1965)
28. Reisert, M., Burkhardt, H.: Feature selection for retrieval purposes. In: Proceedings of the ICIAR'06, Vol. 1, pp. 661–672. Pavoa do Varzim, Portugal (2006)
29. Sandler, R., Lindenbaum, M.: Optimizing gabor filter design for texture edge detection andclassification. International Journal of Computer Vision **84**, 308–324 (2009). DOI 10.1007/s11263-009-0237-x

30. Smith, G., Burns, I.: Measuring texture classification algorithms. Pattern Recognition Letters **18**(14), 1495–1501 (1997). Http://www.cssip.uq.edu.au/staff/meastex/meastex.html
31. Thumfart, S., Heidl, W., Scharinger, J., Eitzinger, C.: A quantitative evaluation of texture feature robustness and interpolation behaviour. In: X. Jiang, N. Petkov (eds.) Computer Analysis of Images and Patterns, *Lecture Notes in Computer Science*, vol. 5702, pp. 1154–1161. Springer Berlin / Heidelberg (2009)
32. Tsai, D.M., Wu, S.K., Chen, M.C.: Optimal gabor filter design for texture segmentation using stochastic optimization. Image and Vision Computing **19**(5), 299–316 (2001). DOI 10.1016/S0262-8856(00)00078-0

Chapter 14
Online Quality Control with Flexible Evolving Fuzzy Systems

Edwin Lughofer, Christian Eitzinger, and Carlos Guardiola

Abstract This chapter is dealing with the application of flexible evolving fuzzy systems (described in Chap. 9) in online quality-control systems and therefore also provides a complete evaluation of these on (noisy) real-world data sets. Hereby, we are tackling with two different types of quality-control applications:

- The first one is based on visual inspection of production items and therefore can be seen as a postsupervision step whether items or parts of items are ok or not, laying the basis for sorting out of bad products and decreasing customers' claims.
- The second one is conducted directly during the production process as dealing with a plausibility analysis of process measurements (such as temperatures, pressures, etc.) and therefore opens the possibility of an early intervention for product improvement (internal correction or external reaction).

In both scenarios, permanent update of nonlinear fuzzy models/classifiers during online operation based on data streams is an essential issue in order to cope with changing system dynamics, range extensions of measurements and features, and the inclusion of new operating conditions (e.g., fault classes) on demand without requiring time-intensive retraining phases. In the result section of this chapter, we will explicitly highlight the performance gains achieved when using flexible evolving fuzzy systems (EFS) in both quality-control paths.

E. Lughofer (✉)
Department of Knowledge-Based Mathematical Systems, Johannes Kepler University Linz, Altenbergerstrasse 69, A-4040 Linz, Austria
e-mail: edwin.lughofer@jku.at

C. Eitzinger
Profactor Research GmbH, A-4407 Steyr-Gleink, Steyr, Austria
e-mail: christian.eitzinger@profactor.at

C. Guardiola
CMT-Motores Térmicos/Universidad Politécnica de Valencia, Camino de Vera s/n, Valencia 46022, Spain
e-mail: carguaga@mot.upv.es

M. Sayed-Mouchaweh and E. Lughofer (eds.), *Learning in Non-Stationary Environments: Methods and Applications*, DOI 10.1007/978-1-4419-8020-5_14,

© Springer Science+Business Media New York 2012

14.1 Introduction

During the last two decades, a significant growth of the size and complexity of the technological installations in the automotive, power, chemical, and food industries has been observed [12]. A side effect of this growth is the increase in the concentration of measuring, processing, and control devices. The likelihood of appearance of a fault that may lead to a breakdown of a component or the whole system increases with the complexity of the system [10]. Faults can affect the system itself, the measuring and monitoring devices, or the control system (which also modifies the system behavior). Faults may not only spoil production items (or parts thereof), but in severe cases also may lead to dangerous situations for operators working at the systems (e.g., consider emission gas gushing out of a leakage in a pipe). Quality control is also a nonproductive part of the production process aiming to prevent the production of defective products. Quality control is thus a necessary part of production lines for increasing both product quality and process safety. There is a trend of integrating quality control earlier in the production process which leads to two main lines of quality control:

- Postprocessing quality control by product inspection (conducted visually with machine vision techniques [14, 20])
- Process quality control by plausibility analysis of process measurement data [37]

Postprocessing style quality control is centered in the evaluation of the product characteristics. Depending on the considered product, different tests can be used and usually combined. Ordinary quality tests include visual inspection, meteorological verification, physical characteristics evaluation, or performance tests, where the individual item is tested according to its use. In some cases, tests are of destructive nature, as in the case of aging tests. Depending on the intensity, cost, and destructiveness of the tests to be run and also of the quality standards to be satisfied, postprocessing style quality control can be performed to all the produced items, or to a selection of a representative sample of the production [35]. In some cases, postprocessing quality control can be fully automated, as is the case of the visual inspection of the product surface. For that, images showing the surfaces of production items are automatically analyzed towards the appearance of any untypical occurrences and reflections (e.g., scratches, pinholes, weak colors, dents, renouncements, etc.) once they have been produced. An automatic decision is then made whether the product is accepted or rejected. Occasionally, there is a third class of product items, called "rework," that includes those products that may be repaired in some way. The latter is particularly relevant for high-value products at the end of the production process. In any case, the ultimate goal of quality control is to prevent the production of rejects. This necessitates feedback from quality control to the production process. Currently, such feedback is often provided manually by experts who analyze the data and deduce changes in the production process. The feedback is thus usually long term and focused on gradual improvements of the processes rather than dealing with short-term fluctuations.

Process quality control operates on one stage earlier by examining the correctness of process parameters. The input for this quality-control system approach are process data rather than data about the product itself, and this has the advantage that weak indications of defects can be detected. Furthermore, it has the potential of recognizing upcoming faults at an early stage. The quality control is based in the assumption that product faults result from measurable deviations in the production process. However, this cannot always be ensured in all production processes, then postprocessing quality control is necessary. Beyond the monitoring of the production process, process quality control can provide valuable data for the correction and control of the production process and opens the possibility for preventing bad production items by the inclusion of feedback control for automatically correcting system parameters. Hence, this quality-control approach has a substantial overlap with feedback process control. For many production systems, the feedback can also be fully automated, and in case explicit process model exists, model predictive control techniques can be used [2, 8].

14.1.1 Motivation for Evolving Components

In many quality-control applications, supervision and failure analysis approaches act in so-called dynamic environments, where the data is collected over time within an online (production) process. This means that data are not stored as persistent tables or feature matrices, but are rather processed in a transient stream-like manner. This is especially the case of large sensor networks or multiparallel production lines where the data (stream) is massive [5] and hence have to be handled on demand appropriately. For the failure detection algorithms, this means that they have to be able to cope with online data which also may permanently include changing system states, varying operating conditions, or different types of product items which trigger different characteristics of the collected process or feature data (extracted from signals or images).

In principle, fault models may be developed off-line and simply applied to the online data to produce fault indicators, symptoms or warnings [24, 27] without any model adaptation cycles; however, this may lead to several shortcomings:

- A significant amount of process or feature data has to be collected in advance in order to cover as many different system states and behaviors as possible (for guaranteeing models with high-process safety). However, this triggers a high-effort in designing excitation signals and annotating samples (e.g. assigning fault classes).
- Also, if doing so, still new operating conditions, changing system behaviors, or new fault classes may arise during online operation (e.g., some environmental influences change). Often, a system is affected by a so-called concept drift [26], where the underlying data distribution may change over time and makes older learned relations/dependencies obsolete.

In order to overcome these deficiencies, incremental and adaptive learning techniques for the fault models are required which (1) adapt some important parameters and (2) evolve structural components (e.g., rules, neurons, leafs, and new mathematical expressions) on demand based on the characteristics of the current streaming data blocks. This guarantees more accuracy of the fault indicators and furthermore higher product quality and even process safety in case of faults being dangerous for human beings. In this chapter, we will clearly demonstrate how evolving components in terms of flexible evolving fuzzy systems (EFS) and classifiers can actually increase the detection rate of faults and failures in various online quality-control applications (for both, process-based and postproduction-based quality checks).

14.1.2 Our Contribution

In the subsequent two sections, we will describe two machine learning oriented approaches for tackling two variants of quality control:

- The first one acts on images showing the surfaces of production items. Images are usually recorded using matrix or line scan cameras at resolutions of 0.1 mm per pixel or lower. In more complex applications, different light sources are used and combined to acquire complementary information about the surface, such as an approximation of the 3D shape of the surface and standard grey-value texture information. Such combination requires complex preprocessing steps including precise image registration, illumination models, and texture analysis for segmentation. After such preprocessing, a set of features is extracted, used as main input to the classifiers. In this sense, a reliable combination of machine vision and machine learning is the key driver of our framework: (1) the machine vision parts are settled on the image preprocessing phase including a substraction of newly recorded images to the master (called *deviation image* or also *contrast image*), recognizing the regions of interest in the deviation images, and extracting a wide variety of reasonable features describing the shape, outlook, appearance, distributions, etc. of the single regions of interests (objects); (2) the machine learning parts are basically dealing with appropriate feature preprocessing steps, including the concept of adaptive features (see also Chap. 7), handling different levels of details and operators' experiences/skills, and finally, with classifier training and evaluation stages.
- The second one directly acts as plausibility analysis engine on the (multichannel) measurement data recorded at the production process, reflecting specific operation modes and system characteristics. Thus, it is able to detect system failures at an early stage, preventing faults and defective production items. Usually, dependencies and relations within (parts of) the system can be modeled by means of system identification methods using data-driven modeling approaches based on the measurement data [36]. For instance, there may exist certain correlations

between pressures and temperatures in the system. Failure detection is then conducted by means of measuring the degree of deviation (called *residuals*) between newly recorded measurements and the identified models at hand (see also [3]) according to some model reliability criteria (extracted from the data in order to measure regions where the models are more or less certain) and examine the development of these residuals over time with statistical approaches [32].

Our main contribution handled in this chapter basically consists of extending conventional data-driven modeling tools in quality-control systems with online learning and adaptation capabilities in order to appropriately react on changing system dynamics and new operating conditions upcoming at these systems (see Sect. 14.1.1). For postprocessing quality control, this is achieved by introducing evolving image classifiers including the fuzzy methodology (fuzzy model architectures) and some enhanced concepts as described in Chap. 9. The image classifiers are able to classify production items into "good" and "bad" parts on image level and also to assign different fault classes to different parts on the items' surfaces on object level. For the quality control directly conducted at the production process, we propose evolving fuzzy-regression-based models using the Takagi–Sugeno fuzzy model architecture (refer to Chap. 9, Sect. 9.2.1) as fault-free reference oracle. Any deviation from these resp. from the adaptive local error bars surrounding these models as kind of confidence bands (see Sect. 9.3.5) and serving as model reliability criterion can be treated as a potential fault candidate.

14.2 Postprocessing Online Quality Control

After introducing our online image preprocessing and classification framework, we provide a summary of the machine vision and machine learning key components used in these (Sect. 14.2.1)—for further details refer to [18,34]. The applicability of the framework will be validated based on four different concrete surface inspection problems (CD imprint, eggs, rotor parts, and bearings) defined in Sect. 14.2.2; the results are presented in Sect. 14.2.3, where a specific focus will be placed on the impact of the online evolution of image classifiers in order to include dynamic changes in the actual process and therefore to significantly boost their predictive performance.

14.2.1 Methodology

In order to circumvent manual tuning, long-time developments, and application-dependent components in visual inspection systems, we designed an image classification which is applicable to a wider range of applications. This is achieved by removing application-dependent elements and applying machine vision and

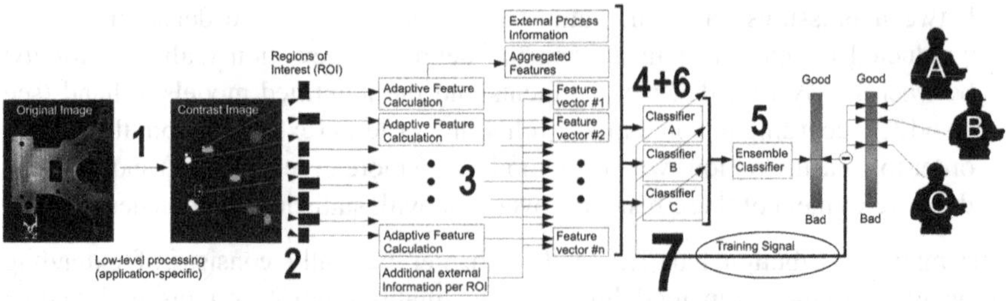

Fig. 14.1 Online image classification framework at surface inspection systems

learning approaches based on image descriptors (also called features) extracted fully automatically from the images. In this way, the problem of a stepwise deduction of classification rules (based on analytical and expert knowledge) is shifted to the problem of gaining sufficient training samples. The latter can be done fully automatically during production phases, the annotation of the training samples is made as easy as possible by a user-friendly GUI front-end integrated in an annotation wizard (see also [34]). Furthermore, the framework contains a lot of other aspects for extending the applicability, improving the user-friendliness and the accuracy of visual inspection systems and providing a coherent technology for joining various components and responding a unique accept/reject decision. Figure 14.1 visualizes the core components of the framework, whose functionality can be summarized as follows (the numbers in Fig. 14.1 correspond to the enumeration points and the feedback loop (issue 7) is specifically underlined by a bigger font as this part is the major focus under study in detail in this chapter):

1. Low-level processing on the images for removing the application dependent elements (contrast image): hereby, the basic assumption is that a fault-free master image is available; for newly recorded images during the production process, the deviation to these master images (deviation image) is calculated by subtraction ($\pm$ a threshold for an upper and lower allowed bound). The pixels in a deviation image represent potential fault candidates, but need not indicate necessarily a failure in the production item. This depends on the structure, density, and shape of the distinct pixel clouds (also called regions of interest). The degree of deviation is reflected in the brightness of the pixels. For color images, the color-wise (RGB) deviations from the master are calculated and averaged over the absolute values of the deviations to produce one grey-level image.
2. Recognition of regions of interest (objects) in the contrast image: the deviation pixels belonging to the same regions are grouped together; therefore, various clustering techniques were exploited which can deal with arbitrary shape of objects and arbitrary number of objects, ranging from hierarchical type to density- and spectra-based approaches [30, 47]. A specific examination was dedicated to a sensitivity analysis among these methods with respect to the classification accuracy of the final classifiers trained on the features extracted from the found clusters (wrongly extracted regions of interest = clusters may cause a bias in the classifiers), see [38].

3. Extraction of features with a feature calculation component: object features characterizing single objects (potential fault candidates) and aggregated features characterizing images as a whole are extracted. Aggregated features are important to come to a unique final accept/reject decision and are, for example, the number of objects, the maximal local density of objects in an image or the average brightness of objects. A key aspect was the extraction of as many features as possible in order to cover a wider range of different application scenarios and problem settings. Hence, a feature selection step during classifier training is included in order to reduce the curse of dimensionality. Adaptive feature concepts as described in the previous chapter (Chap. 13) were applied prior to classifier training.
4. Building of high-dimensional classifiers based on the extracted features and label information on the images (or even single objects) provided by one or more operators. In some cases, the label information contained not only a good/bad label for an image but also additional information like, for instance, the uncertainty in his/her decision (during annotation). The training process consisted of three steps:

   - Dimension reduction (deletion of redundancies and filter feature selection approaches [22]) in order to reduce the high-initial set of features.
   - Best parameter grid search coupled with 10-fold cross-validation (in order to elicit the optimal parameter setting for final classifier training).
   - Training of the final classifier with all training samples and the optimal parameter setting achieved in the previous step.

5. Classifier fusion methods for resolving contradictory input among different operators: various operators may label the same training examples or give feedback during online production mode on new samples. Hence, some contradicting feedback may arise, especially when the skills of the operators vary. Fusion methods are able to resolve these contradictions by performing a voting, democratic decision, or more complex decision templates—how this can be calculated in online mode or dynamic changing environment is handled in more detail in Chap. 7.
6. Early prediction of success or failure of a classifier: it was examined how classification accuracies behave with an increasing number of samples and at which point of time the classifier cannot be improved further by feeding more samples into the training algorithm—this issue is handled in more detail and in a more generic context for arbitrary ML classifiers in Chap. 6.
7. Finally, within the scope of this chapter, we are interested in and concentrate on the impact of image classifier updates during the online classification phase with the help of incremental evolving (fuzzy) classification techniques. This part of the component is highlighted by the feedback loop to the image classifiers based on the operator's feedback upon the classifiers decisions (circumvented by an ellipsoid in Fig. 14.1). This feedback is necessary, especially when the classifier is uncertain in its response or the operator even disagrees with the classifier decision: in this case, the likelihood is high that the classifier trains

its own errors into its structure and parameters, deteriorating its performance over time. In Sect. 14.2.4, we will show that updating the classifiers (instead of keeping them static once they are built-in batch off-line phase) is a central aspect for integrating new data and for improving the performance of the classifiers set up during the off-line phase. Incremental learning steps during the online operation mode are necessary as a regular rebuilding of the classifiers from scratch with all samples seen so far did not terminate in real-time—consider massive data streams from production process to be analyzed as (parts of) items are produced with a high-frequency and may change their characteristics, etc. (see also Sect. 14.1.1). The evolving fuzzy classifier (EFC) approach *FLEXFIS-Class MM* and *AP* (short for FLEXible Fuzzy Inference Systems for Classification using Multi-Model resp. All-Pairs architecture) and the evolving clustering-based classifier *eVQ-Class* (as spin-off from *FLEXFIS-Class SM* and short for *evolving Vector Quantization for Classification*) were used to achieve this goal (see Chap. 9 for a detailed description of these). We will also demonstrate how a dynamic soft dimensionality reduction by including incremental feature weights and a rule merging/pruning step to eliminate local redundancies increase the performance and reduce the complexity of the evolved classifiers, evaluating the advanced concepts for EFS proposed in Sect. 9.3.

For further details on the components of the image classification framework, see [18, 34].

14.2.2 Experimental Setup

14.2.2.1 Surface Inspection Scenarios

The whole framework and its image classification components were applied to four real-world surface inspection problems:

- CD imprint inspection: the task was to identify faults on the compact discs caused by the imprint system, for example, a color drift during offset print, a pinhole caused by a dirty sieve (→ color cannot go through), occurrence of colors on dirt, palettes running out of ink, and distinguish them between so-called pseudo-faults, like, for instance, shifts of CDs in the tray (disc not centered correctly) or masking problems at the edges of the image or illumination problems (causing reflections). An example of a typical deviation image is shown in Fig. 14.2.
- Egg inspection: the main task was to identify broken eggs on a conveyor belt by recognizing some spoors of yolk and scratches on them. Here, the main problem was to distinguish these spoors from dirt on the eggs. Whereas yolk is always a defect, dirt may or may not be considered a defect, for example, depending on the countries in which the eggs are sold.
- Inspection of bearings: The main challenge in this application is whether the trainable system is able to learn how to distinguish between different types of

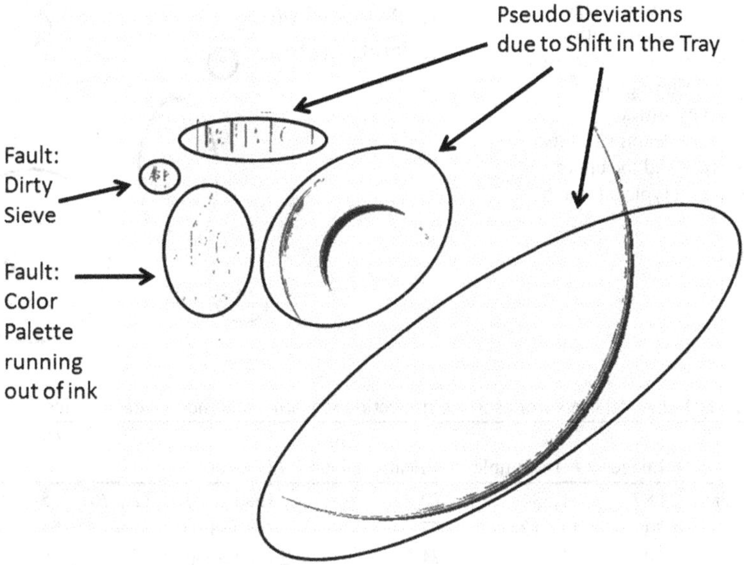

Fig. 14.2 Deviation image from the CD imprint production process, different grey levels and pixel groups represent different regions of interest circumscribed by ellipses (five in sum, two faulty, three nonfaulty); note that most of the image is white which means that the deviation to the fault-free master is concentrated in small parts of the whole pixel space; the faulty and nonfaulty regions of interest are exclusively marked as such: clearly, it requires some enhanced object descriptors in order to be able to distinguish between faulty and nonfaulty regions of interest

faults with only a good/bad label for the whole image. This is of particular importance as quite often the quality expert can only spend very short time on each part and is not able to provide more detailed training input. The main types of defects found on the surface are small dents and scratches. Even if they do not lead to an immediate failure of the product, they will lead to noise and possibly reduce the lifetime of the bearing. Figure 14.3 visualizes a typical deviation image from this process.

- Inspection of metal rotor parts: main problems are gas bubbles in the die cast part (so-called link holes or pores). As the excess material is cut away during milling or grinding, the gas bubbles become visible on the surface. They usually appear as dark spots on the surface with a mostly circular shape. They are 3D deformations of the surfaces and cause problems particularly on sealing or bearing areas, since casings will not be tight any more.

14.2.2.2 Data Collection

The data collection for the four application scenarios was conducted in a way that the recorded data was stored onto hard disk in the same order as it appeared during

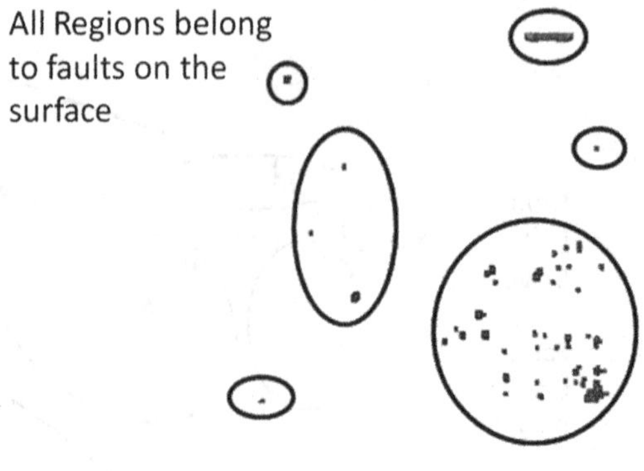

Fig. 14.3 Deviation image from the bearing production process, all regions of interest (found by clustering and highlighted by surrounding ellipsoids) are denoting faults on the surface and the upper right denotes a typical scratch

Table 14.1 Image data sets from surface inspection scenarios and their characteristics

	# Images	# Tr. samples	# Feat.	Obj. labels	# Classes	Class distr. in %
CD imprint	1,687	1,534	29/74	Yes	2 (bad/good) + 12 (6/6)	15–20/80–85
Eggs	4,341	11,312	57	Yes	2 (yolk/dirt)	51.4/48.6
Rotor	225	225	74	No	2 (bad/good)	66.6/33.3
Bearing	997	997	74	No	2 (bad/good)	69.8/30.2

the online production process. In this sense, after preprocessing the raw images, recognizing the regions of interest in the deviation images and the feature extraction process, the order of the features in the feature matrix belongs to the order of the image recordings. This finally means that we were able to simulate the online modeling process in MATLAB by taking feature vector per feature vector together with the class labels from the stored matrices and send these into the image classifier evolution algorithm (using *FLEXFIS-Class MM* and *eVQ-Class*). The class labels are coming from the annotation process conducted by an operator, supported by an own developed software (called annotation wizard) in order to make the annotation process as comfortable and as fast as possible [34].

The number of collected image samples varied from application to application and are summarized in Table 14.1: these images can be all seen as critical images along the borderline between being "good" or "bad". Images for which the deviation to the master was 0 (or lied within an acceptable range) are the most usual case and could be neglected for training purposes as can be trivially classified as good ones. Object labels (resulting in various fault and fault-free classes) could only be collected for CD imprint and egg data sets.

The features extracted from the images (denoted as aggregated features) were selected based on several discussions with experts. These features are summarized in Table 14.2.

The list of selected objects features (57 in sum) includes statistical values such as skew, kurtosis, grey-level histograms, as well as shape and density descriptors (due to space reasons we neglect a complete listing at this stage).

Table 14.2 List of aggregated features used. These features have been chosen because of their relevance for a very wide range of surface inspection applications

No.	*Description*	*No.*	*Description*
1	Number of objects	10	Max. grey-value in the image
2	Average minimal distance between two objects	11	Average grey-value of the image
3	Minimal distance between any two objects	12	Total area of all objects
4	Size of largest objects	13	Sum of grey-values in all objects
5	Center position of largest objects, x-coord	14	Maximal local density of objects
6	Center position of largest objects, y-coord	15	Average local density of objects
7	Maximal intensity of objects	16	Average area of the objects
8	Center position of object with max. intensity, x-coord	17	Variance of the area of the objects
9	Center position of object with max. intensity, y-coord		

14.2.3 Some Results

Summarizing the off-line results (for further details, see [18]), (averaged) accuracies from a 10-fold cross-validation process (with 10 different shuffles of the training data) ranged from 93% to 98% for CD imprint data, from 70% to 95% for bearing data, from 94% to 96% for egg production, and from 84% to 92% for the metal rotor parts, depending on the classifier used. An essential issue was to apply a specific feature preprocessing step including operators' levels of detail and confidence in their rankings (see also [34]): this could improve the accuracies up to 5% in most cases. The sensitivity with respect to the folds in the 10-fold CV as well as w.r.t. shuffles of data is nearly negligible, as laid around $\pm 1\%$ for most of the classifiers and data sets. *FLEXFIS-Class MM* and *eVQ-Class* applied in batch mode could compete with renowned machine learning methods such as k-nearest neighbor algorithm, classification and regression trees (CART) [7], support vector machines (SVMs) [40, 43], *AdaBoost* [13], *bagging (Bagg)* [15], or *possibilistic neural networks (NN)* [45] (most of these included in the hall of fame of top 10 data mining methods [46]). Taking into account that usually 96% of the deviation images are black and therefore trivially classified as good ones, a final CV accuracy rate of about 99.5–99.8% could be achieved, which was inline the expectations of the manufacturers.

14.2.4 Impact of Online Evolution of Image Classifiers

The results for batch off-line trained classifiers, however, significantly deteriorated on separate online test data sets, which could be verified by taking the last third of the online collected data as test data set (including new images in an online setup). This means the samples shown in Table 14.1 were divided into three parts: the first third was used as training and evaluation set for the off-line CV, the second set for simulating the online update of the classifiers, and the third for verifying the

Table 14.3 Comparison of the accuracies (in %) between static image classifiers built on the first half of the training data and sample-wise evolved image classifiers with the second half of the training data for the three surface inspection problems of CD imprints (operator #2 labels), eggs, and rotor parts

	CD imprint (#2)	*Eggs*	*Rotor*	*Bearing*
Static image classifiers (trained in off-line mode)				
eVQ-Class variant *A*	75.69	91.55	66.67	63.75
eVQ-Class variant *B*	88.82	90.11	66.67	64.65
EFC SM	78.82	95.20	66.67	60.73
EFC MM	73.53	95.89	54.67	55.59
k-NN	79.61	91.51	53.33	58.30
CART	78.82	91.78	52.00	65.26
Evolved image classifiers (updated in online mode)				
eVQ-Class variant *A*	89.22 (+13.5)	91.12(−0.4)	86.67 (+20)	67.67 (+3.9)
eVQ-Class variant *B*	90.39 (+1.6)	93.33 (+3.2)	86.67 (+20)	67.98 (+3.3)
EFC SM	78.82 (+0.0)	96.21 (+1.0)	64.00 (−2.6)	63.14 (+2.4)
EFC MM	87.65 (+14.1)	97.19 (+1.3)	78.67 (+24)	65.56 (+10.0)
k-NN (retrained)	90.98 (+11.4)	96.06 (+4.6)	74.67 (+21.3)	59.52 (+1.2)
CART (retrained)	90.59 (+11.8)	97.02 (+5.2)	52.00 (+0.0)	69.18 (+3.9)
+max %	14.1	5.2	21.3	10.0

accuracy on separate test data for off-line trained, retrained, and online updated classifiers—we also used *k*-NN and *CART* as batch off-line methods retrained on each data block containing 100 new data samples for comparison purposes with the evolving fuzzy classifiers: this gives us a feeling how close our incremental methods can achieve the accuracies of batch classifiers.

Table 14.3 shows the results when using *FLEXFIS-Class* (as described in Chap. 9) as *evolving fuzzy classifier (EFC)* method. By comparing the accuracies in the first part of the table with those in the second part, it can be realized that an evolution of the image classifiers is necessary in order to boost the predictive accuracy and to guide the predictive accuracies to a reasonable range, especially for CD imprint and rotor data sets; also, for eggs and bearings, the increase is significant as up to 5% (for eggs) and 10% (for bearings). Furthermore, the table also shows us that a retraining of batch classifiers does not pay off in terms of significantly higher accuracies, only in one case (bearings) *CART* could outperform all evolving fuzzy classifiers when retrained on all samples; in case of rotors, both batch classifiers are significantly behind *EFC MM*. The maximal expected performance increase is stated in the last row of Table 14.3. However, the computation times are significantly higher for batch approaches, as *CART* takes about 1,000 times more computation time than *eVQ-Class* and *EFC SM* and about 150 times more computation time than *EFC MM* for CD imprint and egg data sets when updated = retrained for each new incoming sample. This finally means that an evolving classifier installed

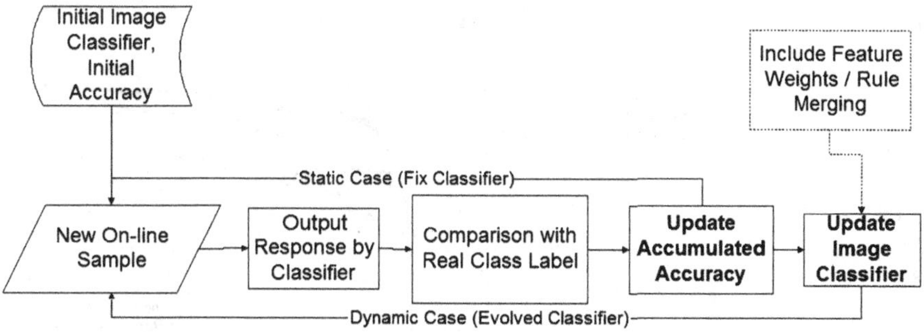

Fig. 14.4 Work-flow of updating accumulated (sample-wise) accuracy and classifier update during online processing—note the two back-path options, the upper indicates that no update of the classifiers is performed (update process is excluded) and the lower includes the update and evolution phase

at a machine vision system can be seen as an essential component in order to achieve high-qualitative predictions and products with high-quality during online (data stream) operation mode.

Another test concerned simulating the impact of updating classifiers when directly predicting the next samples in the production process. This is achieved by measuring their performance on online accumulated one-step-ahead prediction scenarios in an interleaved test-and-then-train manner (as also recommended in the MOA framework for massive online analysis, see [5]), that is, a new sample comes in, the classifier classifies this sample to a certain class, which is compared to the real class (e.g., obtained from operator's feedback) and based on this comparison the classifier accuracy is updated. Afterwards, the classifier is updated based on this new sample with the real class label. In this sense, we compare the one-step-ahead accuracies of permanently updated classifiers with the accuracies on the online samples achieved by static (not updated) classifiers. This procedure is shown in Fig. 14.4.

We used the *EFC (FLEXFIS-Class) MM* approach for this purpose and compared the evolution of the accumulated one-step ahead prediction accuracy of the evolved (indicated by the lower back-path in Fig. 14.4) with those of static classifiers (indicated by the upper back-path in Fig. 14.4). Furthermore, we examined the effect of rule merging and feature weighting during classifier update (as described in Sects. 9.3.2 and 9.3.2.1, Chap. 9). The accuracy improvement lines are shown in Fig. 14.5a–d for the four data sets.

For all data sets, the impact of classifier update and evolution is significant as clearly outperforming the accumulated accuracy of static classifiers (dotted line); in fact, when not using any update approach, the basic accuracy trend of the classifier is even deteriorating when more and more new online samples are loaded, especially for CD imprint data and rotor data sets (for egg data set the accuracy stays constant after 1,500 online samples). This means that an update of classifier is even strictly necessary in these cases in order to keep the accuracy on a reasonable level.

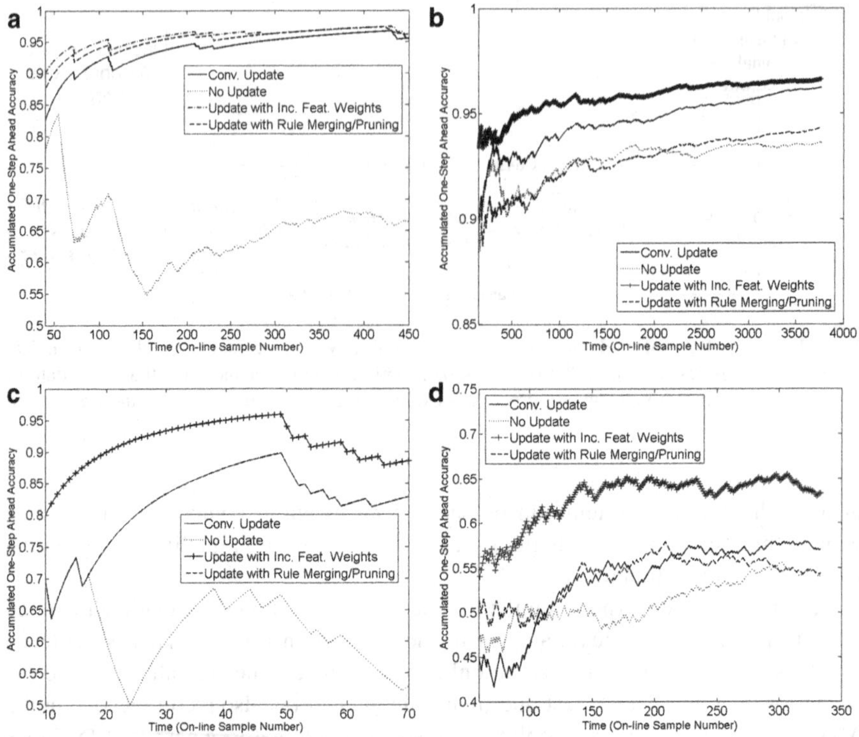

Fig. 14.5 Accumulated accuracy lines over the number of online samples for CD imprint (**a**), egg (**b**), rotor (**c**), and bearing (**d**) data sets; in each plot, four lines are shown according to static image classifiers (no update) (*dotted line*), evolved image classifiers with conventional update (*solid line*), evolved image classifiers with incremental feature weighting included (*dashed line with crosses*) and evolved image classifiers with rule merging/pruning steps included

The inclusion of incremental feature weights for dynamic soft dimension reduction brings an improvement in three of four cases and stays at a similar level as the conventional update for one case (CD imprint data set). Rule merging/pruning steps after each incremental learning cycle provides similar accuracy trends in case of bearing and egg data and exactly the same in case of rotor data (no pruning is required as no upcoming redundancies exist) and is able to increase classifier performance for CD imprint data. The reduction of the number of rules in final classifiers over conventional update is significant: from 29 to 15 in case of bearing, from 19 to 2 in case of CD imprint, and from 6 to 3 in case of egg data (for bearing always 7 rules were present), while achieving same accuracy levels.

Finally, Table 14.4 shows the accuracy results obtained on object labels for CD imprint data set which is a 12-class classification problem. For this purpose, the data set were shuffled 10 times and the whole online modeling process executed for all 10 shuffles, the accuracies averaged over all shuffles and σ denotes the variance of the shuffles, that is, the sensitivity of the methods for different shuffles of data.

Table 14.4 Performance comparison of evolved and batch-trained image classifiers on multiclass classification scenario

Data set	*EFC SM*	*EFC MM*	*EFC AP*	*eVQ-Class A*	*eVQ-Class B*	*CART*	*k*-NN
CD imprint	62.0±2.5	73.1±1.1	82.6±1.5	64.1±2.4	74.9±1.6	75.81±2.1	73.85±2.9

We compare the results with those obtained by *CART* and *k*-NN using all samples at once in batch training mode. Obviously, the recently introduced all-pairs (AP) technique in connection with EFC [31] (see Chap. 9) is able to outperform all other evolving methods and even the batch-trained classifiers (with optimal selected parameters) significantly. *eVQ-Class B* and *EFC-MM* show similar performance as *CART* and *k*-NN.

14.3 Online Plausibility Analysis of Process Data (Process Quality Control)

Opposed to the previous section which was dealing with online quality control in a postprocessing manner (the products are first produced and inspected, afterwards—upon recognizing failures on their surfaces they are sorted out and not sent to the customer(s)), this section deals with process quality control, where measurement data recorded at the production process (therefore, also called process data) is supervised. The basic assumption is that any anomalies in the measured data point to failures during the production system, which further on leads to faulty production items. In this sense, process quality control can be seen as a kind of generalization of postprocessing quality control, as not only faults on the surface of the items, but also in their inner structure can be detected. In connection with feedback control, that is, reacting to any detected systems failure by appropriately controlling the process, process quality control has even the ability to prevent faulty items and system failures. Such failures may even be dangerous for system components and operators.

14.3.1 Methodology

The section is divided into subsections, the first one dealing with a univariate quality-control approach, where times series in form of data streams are supervised with respect to showing any untypical behaviors or uncommon trends. Update of simple models (weak learners) with forgetting is necessary in order to track local trends. The second one describes a multivariate fault detection framework, where high-dimensional data-driven models are used as fault-free reference situation. These models are either generated based on past recorded historic data or with first dozens/hundreds of online measurements. A permanent update and evolution of the models during the online production/recording phase is necessary in order to include new system states and operating conditions in the models. This improves the fault detection performance.

The advantage of the first approach is that it is very fast and does not require any search which measurement channels/variables are affected by a fault. This is often important for a subsequent fault reasoning process in order to perform a control feedback/change to the production system adequately. The disadvantage of the first approach is that it is a simple one-dimensional view and does not take any interrelation between measurement variables into account. Therefore, faults which are manifested in a combination of variables cannot be detected. This can be achieved by the multivariate approach.

14.3.1.1 Univariate Approach using EFS

The univariate approach focuses preliminary on the detection of untypical patterns and behaviors in the data streams viewed as independent time series data. Opposed to many state-of-the-art (off-line, batch) approaches for analyzing time series signals toward some anomaly content (see [4, 11] or [25]), our approach sees the data in a sample-wise snapshot manner. This means that based on the past behavior of the signal, the anomaly content of one single new incoming sample can be provided. No inspection of future recordings and trends is necessary.

Untypical signal occurrences in data streams can be roughly categorized in two stages:

- Sudden untypical occurrences in form of peaks, jumps, intense drifts, or other form of significant outliers. These types of anomalies usually represent typical fault cases at the system.
- More complex upcoming untypical patterns in the time series curves, often also called discords. These types of anomalies usually represent more intrinsic, not so obvious fault cases.

An example of the former is type is shown in Fig. 14.6a, and an example of the latter in Fig. 14.6b. In both images, the anomalies are marked by surrounding ellipses.

For detecting both types of anomalies, we exploit the evolving fuzzy modeling component for regression problems *FLEXFIS* as described in Chap. 9. In the first case, it is sufficient to describe the local trend of the time series by a sliding fuzzy regression model (using Takagi–Sugeno model architecture), in most simple form including only one rule (capturing the basic linear trend) and in a more complex form with 2, 3, up to 5 rules (also capturing some nonlinear trend). Locality can be achieved by including a forgetting factor, which outdates older trends of the time series signal. We use the methodology described in Sect. 9.3.1 for achieving a gradual smoothed outdating of older learned relationship. Opposed to Sect. 9.3.1, where this was performed for reacting on drifts in data streams, here, the purpose is to model the latest local trend of the time series. Hence, the value of the forgetting factor λ depends on the window size specifying the degree of locality of the model.

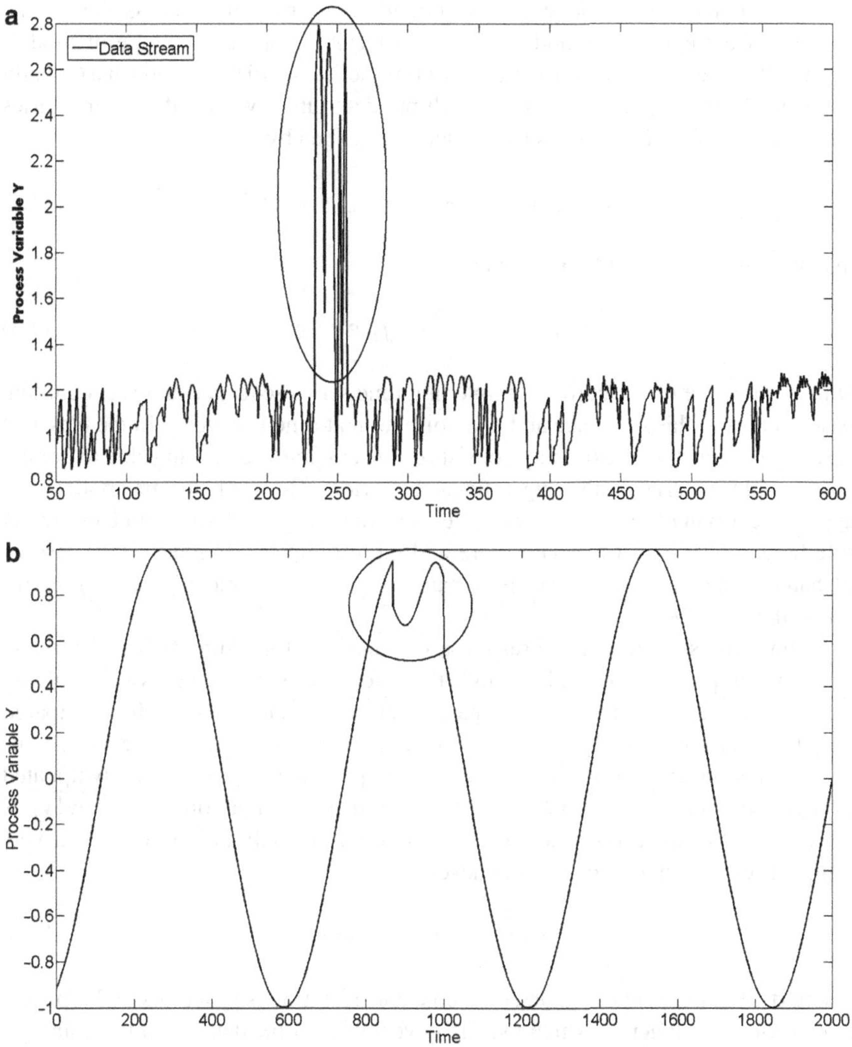

Fig. 14.6 (**a**) anomaly as sudden jump, (**b**) anomaly as untypical pattern (discord) in the basic frequency

Assuming to have a window size M given and assuming that the sample at the border of this window, that is, the sample, which lies M time steps back, should have a very low-weight $\varepsilon > 0$ when included into the fuzzy model training, then the wished forgetting factor λ can be automatically elicited by:

$$\lambda = e^{\frac{\log \varepsilon}{M}}. \tag{14.1}$$

For the detection of more complex untypical patterns in the process signals, the concept of autoregressive models (AR) is a feasible option. This type of models extracts the relevant patterns from the past time series in order to establish a (usually one-step ahead) prediction model, which predicts future values of the time series based on past ones. Formally, such models are defined by:

$$y_t = f(y_{t-1}, y_{t-2}, \ldots, y_{t-m}) + \varepsilon_0 \text{ and} \tag{14.2}$$

in extended form (ARMA) defined by:

$$y_t = f_1(y_{t-1}, y_{t-2}, \ldots, y_{t-m}) + f_2(\varepsilon_{t-1}, \varepsilon_{t-2}, \ldots, \varepsilon_{t-m}), \tag{14.3}$$

with y_t the value of the current sample in a time series sequence and y_{t-l} denoting past samples in this sequence and ε_t a noise term at time instance t. In the classical case [6], the function f explicitly consists of linear parameters, being able to predict only simple patterns (following a linear behavior). However, in our applications applying evolving (TS-type) fuzzy systems as function of f resp. f_1 and letting the rule base evolve over time with more and more samples recorded was necessary to handle nonlinearities in the past patterns properly (achieving *autoregressive evolving fuzzy models*).

In both cases, first, a model response is obtained—for sliding fuzzy regression, the new sample is processed through the inference, for autoregressive evolving fuzzy models, a prediction $\hat{y}(t)$ from past signal values is made—which is compared with the measured signal value $y(t)$ by taking into account the confidence regions surrounding the evolved models and which are permanently incrementally updated as well (see Sect. 9.3.5.2): whenever the confidence region $\text{conf}_f = \sqrt{\text{cov}\{\hat{y}\}}$ is narrow, the tolerance for a new sample to be an anomaly is low. In fact, a new sample is classified as anomaly whenever

$$|y(t) - \hat{y}(t)| > \sigma * \text{conf}_f, \tag{14.4}$$

where σ is a threshold operator. A feasible consideration is to set this to 2:2 means a $2-\sigma$ area is triggered, which usually covers 96% of the data (used for training). Thus, with a confidence level of 1–0.04, a sample can be confirmed as a fault when exceeding (14.4). A value of 3 would lead to a more optimistic threshold (less samples are recognized as faults), but to a more firm confidence level of around 1–0.005. We had to use larger values in our applications cases due to high-noise levels, see Sect. 14.3.2 (experimental setup).

14.3.1.2 Multivariate Approach using EFS

The multivariate approach is based on the identification of high-dimensional models from multichannel measurement data characterizing internal dependencies and relations between certain system variables. Hence, these models are able to express the fault-free characteristics of a system and any significant deviation from them are potential fault candidates. Therefore, faults affected and becoming apparent within a combination of variables can be also detected, which is not the case for the univariate approach (as performing a single dimension-wise view on the data streams). Several data-driven fault detection methods have been proposed in the past (see [12, 27]), also in connection with data-driven nonlinear estimators such as fuzzy models [29] and neural networks [39], usually based on fixed thresholding principles on the *residuals* (= deviation of new samples to the data-driven models) and/or off-line batch training phases.

Key Aspects of Our Approach

Specific characteristics of the approach presented in this section are:

1. It is based on a dynamic data-driven modeling approach with the help of flexible TS fuzzy systems (see Chap. 9), which are able to dynamically change its structure with newly recorded measurements and hence able to include flexibly and quickly dynamic changes or extensions in the system behavior. In this sense, it is more flexible than static analytical or knowledge-based models and also than recursive parameter estimation methods (as, e.g., used in [44]).
2. It uses the concept of adaptive local error bars which are updated synchronously to the flexible fuzzy systems and denoting a confidence region for fuzzy model predictions according to uncertainties in the process, noise in the data, and extrapolation effects.
3. It uses an adaptive threshold concept (for fault warnings) through statistical analysis of residual signals.
4. It does not require any time-intensive annotation phase for classifying samples into faulty and fault-free (as is necessary in case of pattern recognition and classification approaches [12] and also for the postprocessing approach as described in Sect. 14.2). This is because faults are recognized by measuring the deviation of new samples to the regression models, where the targets are also measurement channels and therefore by-measured. This also means that an operators' feedback is only required in case of detected faults. On the other hand, it can be only discriminated between fault and nonfault situations, so no categorization into different fault classes/modes can be made.
5. It does not focus on one global system model, but is cascadable in the sense that a larger collection of systems models can be integrated (describing various dependencies in the system). The intention of this aspect of the framework is to cover the detection of as many faults as possible which may arise in any of the P variables. Only models with sufficient quality (>0.8) will be used in the fault detection component, omitting responses from uncertain or even nonreasonable models—in the fully automatic all-coverage approach, each system variable is selected as target and a model is built by mapping a subset of the other variables onto the target.

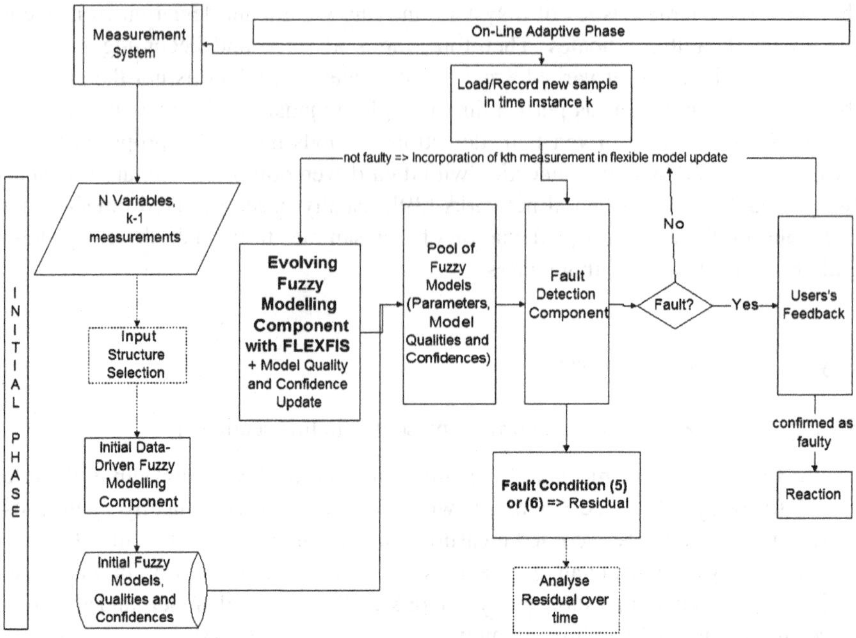

Fig. 14.7 Online fault detection framework in an online measurement system using evolving fuzzy systems

The self-explanatory framework joining these components to one process work-flow is shown in Fig. 14.7. It applies an initial off-line modeling step and further evolution of models using fuzzy system architecture. The optional components are shown in dotted boxes: input structure selection is only necessary when P is large and analysis of the residual signals over time can be seen as add-on for bringing in more flexibility regarding setting appropriate thresholds for fault warnings.

A feedback of operator(s) is required in case when a fault warning of the system is produced, otherwise it is likely that real faults (warnings confirmed as faulty) are trained into the models which may spoil their correctness and deteriorate their performance further on. If a measurement is not confirmed as faulty, at least one model's decision was wrong (usually due to a new operating condition or due to a badly generalizing model) and the measurement is incorporated into the evolution process in order to improve and extend the fuzzy models to new system states. In general, each data-driven model component can be used in the framework, which is able to permanently update the models with new incoming measurements and to provide some confidence regions for the fault detection logic. In our case, we are focusing on flexible (evolving) fuzzy systems as described in Chap. 9 as data-driven design component. Therefore, the initial modeling phase is performed by flexible fuzzy inference systems using batch mode training (clustering first and learning of consequents in least squares mode afterward) and Takagi–Sugeno fuzzy systems architecture which are claimed to be universal approximators [48] (i.e., able to approximate any nonlinear behavior up to a certain accuracy degree). In case of a high-input dimensionality, we apply an own modification of forward selection [21] for eliciting a set of most important variables in order to explain each target: the variables are ranked according to their importance (most important first, etc.) and the additional contribution of each further variable measured in terms of quality increase; those variables are selected which provide a significant contribution to this increase. The online evolving phase is performed by an incremental training procedure included in *FLEXFIS*, see Chap. 9. Model qualities are obtained by applying the r-squared-adjusted measure [23], confidence regions are achieved by adaptive local error bars according to Sect. 9.3.5.2.

Basic and Advanced Fault Condition

The native fault condition which decides whether a new measurement is affected by a fault or not includes the sensor inaccuracy of the target channel ε and incorporates the expected prediction error of the model "model_error" measured by a combination of bias error (error due to low-model flexibility) and variance error (error due to high-noise in the data) (for details of the derivation see [33]):

$$\exists m: \frac{\hat{f}_{\mathrm{fuz},k,m} - x_{k,m} - \varepsilon_{x_m}}{\mathrm{model\_error}_m} > \mathrm{thr}$$

$$\vee \quad \frac{\hat{f}_{\mathrm{fuz},k,m} - x_{k,m} + \varepsilon_{x_m}}{\mathrm{model\_error}_m} < -\mathrm{thr}, \tag{14.5}$$

with $\hat{f}_{\mathrm{fuz},k,m}$ the estimated value for variable x_m from the mth evolved fuzzy model f at time instance k—note that there are $m = 1,\ldots,M$ reasonable models (models with quality higher than 0.8) which are checked through (14.5), and if one of these is violated, then it is sufficient to provide a fault warning. This is simply because an error in one single variable may already indicate a fault or system failure. The tunable parameter is "thr" which controls the sensitivity of the fault condition as it provides the width of the tolerance band (as a kind of confidence region) around the model. A too large value of "thr" yields a very optimistic fault detection system, where usually small errors are overseen. A too small value produces too much fault warnings, hence many overdetections, can be expected, which reduces the trustability of the operators in the system. In accordance to the consideration in the univariate approach (below (14.4)), we considered to apply a value of 2 or 3 for "thr", which however resulted in unsatisfactory performance and was very application dependent, that is, especially depending on the characteristics of noise levels in the data sets. Thus, a dynamical analysis of residual signals produced by (14.5) over time was developed which could compensate a weakly tuned "thr", that is, omitting more or less completely its necessity—see paragraph below.

The main drawback of condition (14.5) is its global constant nature not changing its width due to the actual data distributions in different regions. In fact, there may be holes or extrapolation regions in the feature space not covered with any data, where the model produces more uncertain predictions than in regions with a high-training data density. Hence, the above fault condition is extended by using the concept of adaptive local error bars, where the final fault condition becomes (for details, see [32]):

$$\begin{aligned} \exists m : & \frac{\hat{y}_{\text{fuz},k,m} - x_{k,m} - \varepsilon_{x_m}}{\sqrt{\text{cov}\{\hat{y}_{\text{fuz},k,m}\}}} > \text{thr} \\ \vee & \frac{\hat{y}_{\text{fuz},k,m} - x_{k,m} + \varepsilon_{x_m}}{\sqrt{\text{cov}\{\hat{y}_{\text{fuz},k,m}\}}} < -\text{thr}, \end{aligned} \tag{14.6}$$

that is, the constant band width model_error$_m$ is substituted by the locally changing error bar $\sqrt{\text{cov}\{\hat{y}_{\text{fuz},k,m}\}}$ in dependency of the actual (the kth) measurement (see Chap. 9, Sect. 9.3.5.2 for the concrete formula), where $\hat{y}_{\text{fuz},k,m}$ is the estimated output value for the kth measurement of the mth fuzzy model (approximating the mth (measurement) variable x_m). Similar considerations as above for setting "thr" can be made.

Towards Automatic Thresholding Concept

Although the adaptive local error bars are allowing for more flexibility to model appropriate confidence bands with different width in different regions of the feature space (following the distribution and the noise level of the samples therein), still a fixed threshold value for deciding on a fault alarm has to be set in (14.6). However, to our best experience, for different applications the optimal threshold for producing a high-detection rate while keeping the overdetection rate at a low-level varies. In this sense, we propose to use a kind of automatic thresholding concept by analyzing the behavior of the residual signals over time: each (normalized) residual (forming the fault condition (14.6)),

$$\text{res}_{i=1,\ldots,k-1;m} = \frac{\min\left(\left|\hat{y}_{\text{fuz},i,m} - x_{i,m} - \varepsilon_{x_m}\right|, \left|\hat{y}_{\text{fuz},i,m} - x_{i,m} + \varepsilon_{x_m}\right|\right)}{\sqrt{\text{cov}\{\hat{y}_{\text{fuz},i,m}\}}}, \tag{14.7}$$

denotes one sample in the residual signal for the mth model. Now, for the kth data sample (measurement), the residuals $\text{res}_{k;m}$ are computed for all m models with (14.7) and are not checked versus a fixed defined threshold, but whether they denote an anomaly in the current time instance k or not. Therefore, the same univariate approaches as discussed in Sect. 14.3.1.1 are used, relying on the past history of all the residual signals extracted from the m models and using local error bars as confidence regions "conf_region$_k$" for the sliding regression or ARMA models as defined in (14.4). A fault alarm is triggered whenever the following condition is fulfilled in one of the m residual signals:

$$\text{res}_k > \hat{\text{res}}_k + \text{conf\_region}_k, \tag{14.8}$$

with $\hat{\text{res}}_k$ the estimated and res_k the actual residual at time instance k.

If the fault alarm is not confirmed as correct, the measurement can be assumed as fault-free and is taken into account for the evolution of the fuzzy models as outlined by the back-path in Fig. 14.7. In this case, a new operating condition is more likely and the sliding regression models updated with a decreased λ in order to adapt faster to the new residual signal behavior and therefore to reduce false alarms. On the other hand, if the fault alarm is confirmed as correct, the corresponding residual res_k is not sent into the parameter update and the measurement is not taken into account for the evolution of the fuzzy models.

14.3.1.3 Fault Isolation, Reasoning, and Feedback to the Production Process

This section deals with the problem to allow more automatization in fault detection systems. Fault detection is a very useful and often necessary option to call attention to failures in the system. In severe cases, an operator has to be informed about these failures, which in turn may switch off some system components or perform some manual control operations in order to prevent damages on production items and system tools. In other cases, the system should automatically correct itself, reducing significant supervision workload for the operators. Therefore, it is a challenge to analyze the origin of a fault in a first step (fault reasoning) and then to react on this by performing an adequate feedback to the production process.

Fault Isolation

In order to find the origin of a fault, it is important to know which variables are affected by the fault. In case of univariate approach as discussed in Sect. 14.3.1.1, this is automatically achieved, as each time series (belonging to one process variables) is analyzed as data stream separately and completely independently. In multivariate approaches, this issue gets more complicated, as (most) "responsible" variables for triggering a fault alarm need to be isolated (hence, the process is called *fault isolation*). If for instance in $m_1 < m$ (out of m) residual signals, an untypical behavior could be observed, indicating a fault in the system, and all measurement channels occurring in these m_1 models are potential candidates to be affected by the fault. This means in our problem setting, fault isolation reduces to rank all the involved measurement channels in the m_1 models according to the likelihood that they were affected by the fault. A fault isolation value in $[0,1]$ indicates this likelihood ($1 =$ very likely, $0 =$ not likely). A two-dimensional example of the fault isolation problematic is shown in Fig. 14.8, where it is not obvious to decide whether

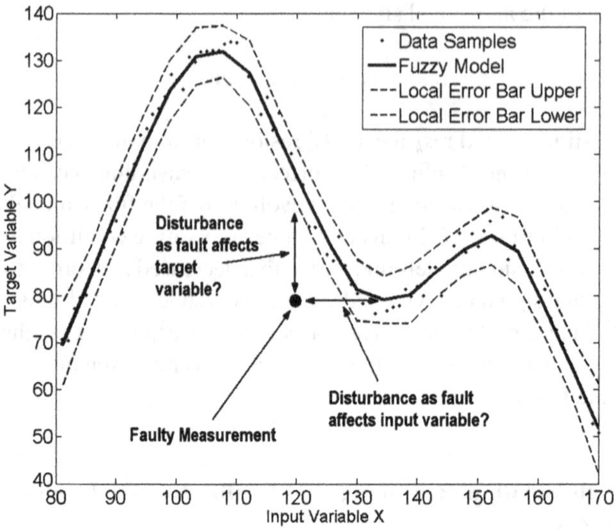

Fig. 14.8 A two-dimensional visualization of the fault isolation problematic: a new measurement affected by a fault (dark dot) appears significantly outside the error bar of a two-dimensional relationship (modeled by a fuzzy model, shown as solid line); it is not obvious whether this "disturbance" occurs because the input channel or because the target channel was affected by the fault (or even both)

the sample (lying significantly outside the confidence regions) is affected by a fault in the target (vertical deviation) or by a fault in the input channel (horizontal deviation).

A fault isolation approach which solves this task within our problem setting is demonstrated in [17], which exploits the idea of calculating a sensitivity vector as proposed in [19]. The sensitivity vector includes the normalized partial derivative at the current measurement (for which a fault alarm was triggered) with respect to each variable in m_1 models, denoted as $\frac{\partial f_l}{\partial x_j}$ for the partial derivative of the jth variable in the lth violated model. The normalization is achieved by the proportion of the ranges between input variable x_j and target variable x_l. Hence, the sensitivity vector for model l is defined by [17]:

$$s_j = \begin{cases} \max(1-\beta_j, 0) & l \neq j \in \{1,\ldots,p\} \\ 1 & l = j \in \{1,\ldots,p\}, \end{cases} \tag{14.9}$$

with

$$\beta_j = \gamma \frac{\text{range}(x_l)}{\text{range}(x_j)} \frac{1}{\left|\frac{\partial f_l}{\partial x_j}\right|} \tag{14.10}$$

and γ usually set to 1. The partial derivative of a Takagi–Sugeno fuzzy model using Gaussian membership functions (as defined in Chap. 9) with respect to a single variable x_j can be calculated by:

$$\frac{\partial f_l}{\partial x_j} = \sum_{i=1}^{C} w_{ij}\Psi_i + l_i \frac{\partial \Psi_i}{\partial x_j}. \tag{14.11}$$

A faster (local) calculation of the derivative for online operation mode can be numerically achieved through Newton's difference quotient. Furthermore, the model qualities measured by means of r-squared-adjusted values are included in the fault isolation likelihood, as multiplied with the sensitivity information of each variable, see also [16]: the lower the model quality the lower the fault isolation likelihood. Finally, the likelihoods $L_{ij}, i = 1,\ldots,$ and m_1 from all violated m_1 models are summed up for each variable j (note that some variables may appear in more than one model as input channel) which yields an overall likelihood $L_j = \Sigma_{i=1}^{m_1} L_{ij}, j \in \{1,\ldots,p\}$ for each variable.

Fault Reasoning and Feedback to the Production Process (Outline)

Once the fault isolation is finished, a fault reasoning mechanism can be initiated which takes into account the fault likelihoods of the single variables and the intensity of the faults according to the maximal deviation from the confidence band of the m_1 violated models. Usually, the reasoning process is conducted based on expert knowledge, which can be automatized in an expert (fuzzy) system or in a symptom-fault map [42]. Important input features for a fuzzy-based expert system are the degree of likelihood of each variable (or an amalgamated value of this such as the maximal likelihood or the sum of likelihoods over all variables), the (amalgamated) intensity of the symptom (deviation from the model(s)' confidence bands), and the intensity of the anomaly in the residual signal.

Based on the elicited type of the fault in the fault reasoning process, a feedback to the production system can be initiated in form of control actions for preventing the system (1) from further failures and (2) from increasing severity of an upcoming failure (e.g., a drift of the system). Such control actions usually require a process model to make sure that the process remains stable. The actions that are taken fully depend on the particular production process and also on the actuators that are available for making automatic adjustments.

14.3.2 Experimental Setup and Results

14.3.2.1 For Univariate Approach

The methods described in Sect. 14.3.1.1 were developed for anomaly detection in a variety of injection molding machines when producing different die cast parts.

There, data is sampled from the available channels once for each part produced, and contains information about configuration settings, timing, forces and pressures, temperatures, speed, and dimensions, among others. In normal operation mode, the sampled data from most of the process variables show a quite constant behavior, varying around a specific value in a fluctuating manner. Therefore, sliding fuzzy regression models (showing the latest trend of the pseudoconstant behavior) with only two rules equally partitioning the one-dimensional input space represented a sufficient complexity. The parameters (forgetting factor, maximal time delay, and the σ-multiplicator of the confidence band) were tuned in an extensive tuning phase based on various sets of collected measurements from the process using grid search techniques and automatic evaluation procedures (based on annotated samples by experts). Finally, a quite common optimal setting could be achieved:

- A forgetting factor of $\lambda = 0.9$ enforcing a strong forgetting.
- For autoregressive evolving fuzzy models (in order to detect more complex abnormal patterns), a time delay of 10 turned out to be the best performing choice.
- The optimal σ-multiplicator for the confidence bandwidth varied in a small range of 12–15.

The models implemented in this way were applied to real online production process, where a detection rate of between 80% and 90% could be elicited while keeping the overdetection rate (falsely detected anomalies) at a low-level around 5%. Currently, the system is installed in half-automatic manner, that is, whenever an anomaly is detected, the operator is informed by a red light to check the system. This finally means that when the process is running properly for one day, it can be expected that an operator is superfluously informed within a time frame of around 20 min.

Other experiments with evolving autoregressive fuzzy models were conducted within the scope of an audio inspector software. There, the task was to find, among others, dropouts, clicks, crackles, and other high-frequent distortions in music hull curves (stereo, 44.1 KHz, 16 bit) recorded from compact discs and magnetic tapes as part of a larger audio inspection tool (see http://www.audioinspector.com/ under rubric *Features* for more information about the complete functionality). These types of faults also represent anomalies in the current music content and therefore in the corresponding one-dimensional curves (left and right stereo channels are usually treated as completely independent). Sliding fuzzy regression model technique was used to find dropouts, holds, and mutes in the music content, all characterized by unintended, abrupt pauses in the waveforms. Hence, the gradient of the latest model trend served as reliable indicator for a fault hint, for which the characteristics of its surrounding was further analyzed by an IF–THEN fuzzy rule base. In case of clicks and crackles, half autoregressive evolving fuzzy models helped to find the patterns of the typical audio content, not only based on time-based but also on frequency-based features. A typical dropout example is shown in Fig. 14.9a, a typical click/crackle example in Fig. 14.9b, both in 1-to-1 zoom.

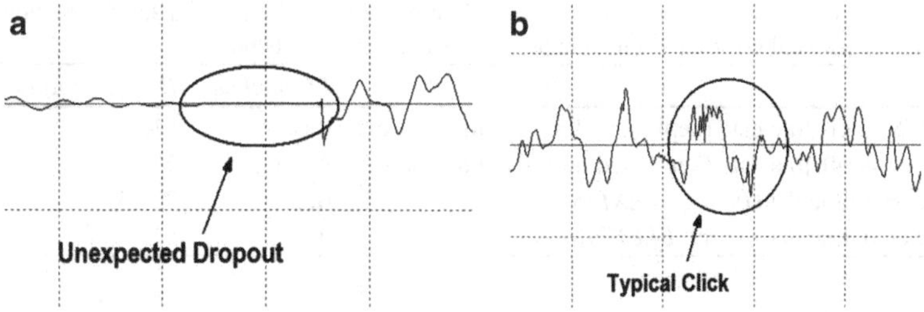

Fig. 14.9 (**a**) A typical dropout (small unexpected pause) in music; (**b**) typical click (unexpected high-frequency part)

After extensive parameter tuning phases the following results could be achieved using evolving fuzzy models as internal pattern/trend model:

- detection rate (80%) and 5% overdetection rate for (significant) dropouts on analogues tapes
- detection rate (98%) and 1% overdetection rate for (significant) dropouts on compact discs
- detection rate (85%) and 5% overdetection rate for (significant) clicks and crackles on analogues tapes
- detection rate (95%) and 2% overdetection rate for (significant) clicks and crackles on compact discs
- overdetection rate (1.5%) for all detectors on sum on fault-free customized music discs

The detection rates for analogue media are significantly lower than for digital media, as the occurrence of the faults is much less distinct.

14.3.2.2 For Multivariate Approach

The multivariate fault detection framework was tested on measurement data recorded online at engine test benches (courtesy of AVL List GmbH). The task was to perform an online plausibility analysis of the measurements arising during test and evaluation phases of new engine developments. Based on these measurements, it is possible to detect faults and failures in an early stage which may cause some incorrect conclusions about that state/behavior of the engine. Categories of faults are:

- Sensor overheatings resulting in sensor drifts
- Broken interfaces
- Pipe leakages
- Erratic behavior of the engine
- Undesired environmental conditions

Table 14.5 Comparison of basic (using constant error bars) and advanced fault detection logic (using adaptive local error bars) on simulated faults at engine test bench

Test case	*FD method*	*Detection rates* (in %)
FLEXFIS on off-line data (kept fixed during online phase)	Basic	61.36
FLEXFIS on off-line data (kept fixed during online phase)	Advanced	75
Update on online data (using *FLEXFIS*)	Basic	70.55
Update on online data (using *FLEXFIS*)	Advanced	81.90

Especially, leakages of pipes for emissions and erratic behaviors of the engine may lead to severe failures which are dangerous for operators working at the engine test bench. In this system, up to a few hundreds different channels may be recorded and sampled in parallel, both in dynamic mode (yielding dynamic measurements) where around each millisecond a new value is sampled and in steady-state mode (yielding static data) averaging the dynamic measurements over 30 s. For the latter, one stationary measurement is achieved per minute, and it is necessary to wait 30 s before averaging over another 30 s in order to allow the system to move from a transient phase to a steady-state phase.

According to the large variety of measurement channels, a dimensionality reduction algorithm is necessary. We used our own method of a modified version of forward selection [21] based on prerecorded/simulated data or based on the first few dozens of online measurements. The idea was not only to apply fuzzy models in off-line mode and letting them fixed during the whole online FD process but also to further evolve and adapt the models based on the response of the fault detection component, that is, to adapt the models only in fault-free cases, according to the FD framework as shown in Fig. 14.7, in order to extend the model on the fly to new operation conditions (here manifested by new regions in the engine map defined over the "control variables" rotation speed and torque).

Table 14.5 demonstrates the results achieved when performing fault detection with the help of evolving fuzzy models on measurement data collected during the evaluation and test phase of a BMW diesel engine (1,180 stationary points, 160 affected by real faults).

The overdetection rates (samples are falsely classified as faults) stayed at around 1–2%, that is, at a very low (same) level in all cases. The results in Table 14.5 were achieved by using a fixed threshold of 12 in the fault conditions and show that the more flexibility brought in by the adaptive local error bars (outlined as advanced FD method) pays off in terms of 10–15% more accuracy. Furthermore, when the initial models are built up on off-line data and further on kept fix during the whole online process (simulated by loading samples from a prestored online collected data matrix, see Sect. 14.2.2.2 for a detailed description), that is, not updated with new online information, the accuracy suffers to be lower by about 6–10%. Based on expert knowledge, we could detect all major faults when updating *FLEXFIS* on online data and using advanced fault condition; the miss of 18.1% are all corresponding to tiny faults, mostly lying beyond the noise level. In fact, the threshold in fault conditions (14.5) and (14.6) is important for the performance as it controls the trade off between

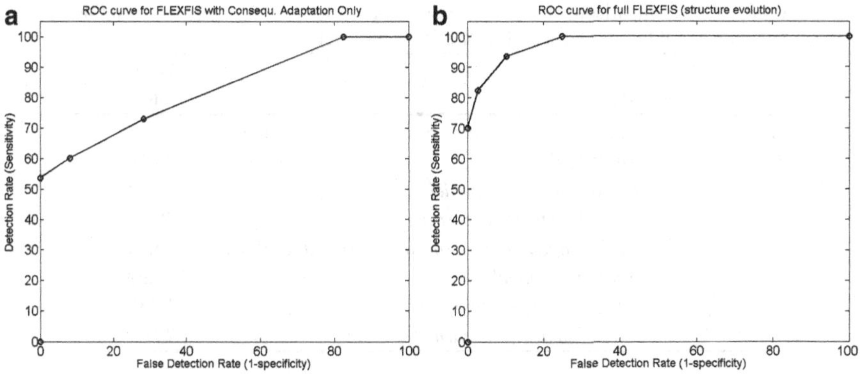

Fig. 14.10 (**a**) ROC curve for *FLEXFIS* using only consequent adaptation; (**b**) ROC curve for *FLEXFIS* using its full version including structure evolution

detection and overdetection rate: lower thresholds will generally provide a high-detection rate while also increasing the overdetection rate. Therefore, we varied the threshold to see the effect on the detection and overdetection rates and plot this as ROC curve in Fig. 14.10, showing the x-axes as the false detection rate (1-specificity) and the y-axes as the detection rate (sensitivity). The larger the area under this curve is, the better the method performs as achieving higher detection rates with lower accompanied overdetection rates.

For comparison purposes, Fig. 14.10a shows the ROC curve when using only adaptation of consequent parameters (with recursive fuzzily weighted least squares) without any structure evolution on demand, whereas Fig. 14.10b shows the full performance of *FLEXFIS* approach. A severe drop in the area under the ROC curve can be observed, which means that structural evolution during online phase significantly boosts fault detection performance.

In order to get an impression about the achievable bounds of our approach, we studied the performance on tiny faults with levels of about 5% and 10% deviation from the normal situation. Therefore, we disturbed some values in channels of simulated engine data (including a very low-noise level) by a small fraction of $\pm 5\%$ and $\pm 10\%$. In order to circumvent threshold tuning, we used the value of 12 leading to the results for the BMW diesel engine as shown in Table 14.5 and found out that no faults could be detected at all, neither for 5% nor for 10% disturbed data. Then, we applied the statistical residual analysis as described in Sect. 14.3.1.2 and obtained significant detection rates of 26% for 5% fault levels and 66% for 10% fault levels. For the 5% fault levels, the detection rate seems to be quite low; however, engine experts usually request to detect faults with a fault level $>= 10\%$, according to some expert standard. Therefore, using our FD approach, reasonable detection rates (>50%) can be expected in-line with the expert standard. In Table 14.6, EFS is compared with other data-driven modeling techniques such as multivariate linear regression [23] and lazy learning [1] and with analytical fault models (deduced from physical laws), best performing method for each fault level is highlighted in bold font.

Table 14.6 Comparison of fault detection approaches on simulated car engine data with small built in errors (10% and 5% fault levels)

Fault level (in %)	Method	Detection rates (in %)
10	*EFS, adaptive error bars with residual analysis*	**66.23**
10	*Analytical models*	35.13
10	*Linear regression models*	20.67
10	*Local linear (lazy learning)*	49.08
5	*EFS, adaptive error bars with residual analysis*	**26.23**
5	*Analytical models*	23.92
5	*Linear regression models*	12.88
5	*Local linear (lazy learning)*	20.04

14.4 Conclusion and Future Directions

This chapter deals with two major lines of online quality control approaches, postproduct inspection quality control and process quality control. While the former is based on visual inspection based on images showing the surfaces of production items, the latter directly uses the measurement data recorded from the processes and tries to identify the process behavior in conventional (nonfaulty) state. Compared to state-of-the-art quality control with fixed analytical [10, 41] or data-driven fault models [27] resp. classifiers [14], our approach is able to adapt dynamically to process changes, operating conditions and varying environmental influences by updating the reference models based on which fault indicators (residual signals resp. classification statements together with uncertainty levels) are extracted. The importance of evolving models for providing reliable and high-qualitative indicators is underlined by several application scenarios and online collected data streams from these. For that purpose, a specific variant of evolving models, the so-called flexible EFS are applied, whose integrated methodologies are described in Chap. 9. These are compared with fixed models which are set up during an off-line development phase and kept static during online phase.

In case of process quality control, a two-sided approach is presented where regression-based evolving fuzzy models are used for discriminating between faults and nonfaults. At this stage, this approach is not able to handle different types of faults or in general different operation modes. An approach which integrates this possibility is presented in [28], however requiring operator's feedback (effort) in case of new operation modes. Furthermore, an open point is the feedback control to the production process in an online manner. This may require some adaptive open-loop control approaches which are able to update their control structure. First investigations into this direction within the scope of adaptive fuzzy control are made in [9].

Acknowledgements This work was funded by the Austrian fund for promoting scientific research (FWF, contract number I328-N23, acronym IREFS). It reflects only the authors' views.

References

1. Aha, D.: Lazy Learning. Kluwer Academic Publishers, Norwell, Massachusetts (1997)
2. Allgöwer, F., Re, L.D., Glielmo, L., Guardiola, C., Kolmanovsky, I.: Automotive Model Predictive Control: Models, Methods and Applications. Springer, Berlin Heidelberg (2010)
3. Angelov, P., Giglio, V., Guardiola, C., Lughofer, E., Luján, J.: An approach to model-based fault detection in industrial measurement systems with application to engine test benches. Measurement Science and Technology **17**(7), 1809–1818 (2006)
4. Bay, S., Saito, K., Ueda, N., Langley, P.: A framework for discovering anomalous regimes in multivariate time-series data with local models. In: Symposium on Machine Learning for Anomaly Detection. Stanford, USA (2004)
5. Bifet, A., Holmes, G., Kirkby, R., Pfahringer, B.: Moa: Massive online analysis. Journal of Machine Learning Research **11**, 1601–1604 (2010)
6. Box, G., Jenkins, G., Reinsel, G.: Time Series Analysis, Forecasting and Control. Prentice Hall, Engelwood Cliffs, New Jersey (1994)
7. Breiman, L., Friedman, J., Stone, C., Olshen, R.: Classification and Regression Trees. Chapman and Hall, Boca Raton (1993)
8. Camacho, E., Bordons, C.: Model Predictive Control. Springer Verlag, London (2004)
9. Cara, A., Lendek, Z., Babsuka, R., Pomares, H., Rojas, I.: Online self-organizing adaptive fuzzy controller: Application to a nonlinear servo system. In: Proc. of the 2010 International Conference on Fuzzy Systems, pp. 1–8. Barcelona, Spain (2010)
10. Chen, J., Patton, R.: Robust Model-Based Fault Diagnosis for Dynamic Systems. Kluwer Academic Publishers, Norwell, Massachusetts (1999)
11. Cheng, H., Tang, P., Potter, C., Klooster, S.: Detection and characterization of anomalies in multivariate time series. In: Proc. of the 2009 SIAM International Conference on Data Mining, pp. 413–424. Sparks, Nevada (2009)
12. Chiang, L., Russell, E., Braatz, R.: Fault Detection and Diagnosis in Industrial Systems. Springer, London Berlin Heidelberg (2001)
13. Collins, M., Schapire, R., Singer, Y.: Logistic regression, adaboost and bregman distances. Machine Learning **48**(1–3), 253–285 (2002)
14. Demant, C., Streicher-Abel, B., Waszkewitz, P.: Industrial Image Processing: Visual Quality Control in Manufacturing. Springer Verlag, Berlin, Heidelberg (1999)
15. Duda, R., Hart, P., Stork, D.: Pattern Classification—Second Edition. Wiley-Interscience (John Wiley & Sons), Southern Gate, Chichester, West Sussex, England (2000)
16. Efendic, H., Re, L.D.: Automatic iterative fault diagnosis approach for complex systems. WSEAS Transactions on Systems **5**(2), 360–367 (2006)
17. Efendic, H., Schrempf, A., Re, L.D.: Data based fault isolation in complex measurement systems using models on demand. In: Proceedings of the IFAC-Safeprocess 2003, pp. 1149–1154. IFAC, Washington DC, USA (2003)
18. Eitzinger, C., Heidl, W., Lughofer, E., Raiser, S., Smith, J., Tahir, M., Sannen, D., van Brussel, H.: Assessment of the influence of adaptive components in trainable surface inspection systems. Machine Vision and Applications **21**(5), 613–626 (2010)
19. Fang, C., Ge, W., Xiao, D.: Fault detection and isolation for linear systems using detection observers. In: R. Patton, P. Frank, R. Clark (eds.) Issues of Fault Diagnosis for Dynamic Systems, pp. 87–113. Springer Verlag (2000)
20. Graves, M., Batchelor, B.: Machine Vision for the Inspection of Natural Products. Springer Verlag, Berlin, Heidelberg (2003)
21. Groißböck, W., Lughofer, E., Klement, E.: A comparison of variable selection methods with the main focus on orthogonalization. In: M. Lopéz-Díaz, M. Gil, P. Grzegorzewski, O. Hryniewicz, J. Lawry (eds.) Soft Methodology and Random Information Systems, Advances in Soft Computing, pp. 479–486. Springer, Berlin, Heidelberg, New York (2004)
22. Guyon, I., Elisseeff, A.: An introduction to variable and feature selection. Journal of Machine Learning Research **3**, 1157–1182 (2003)
23. Harrel, F.: Regression Modeling Strategies. Springer, New York, USA (2001)

24. Isermann, R.: Fault Diagnosis Systems: An Introduction from Fault Detection to Fault Tolerance. Springer, Berlin Heidelberg (2009)
25. Keogh, E., Lonardi, S., Chiu, W.: Finding surprising patterns in a time series database in linear time and space. In: Proc. of the Eighth ACM SIGKDD Int. Conf. on Knowledge Discovery and Data Mining, pp. 550–556. Edmonton, Alberta, Canada (2002)
26. Klinkenberg, R.: Learning drifting concepts: example selection vs. example weighting. Intelligent Data Analysis **8**(3), 281–300 (2004)
27. Korbicz, J., Koscielny, J., Kowalczuk, Z., Cholewa, W.: Fault Diagnosis—Models, Artificial Intelligence and Applications. Springer Verlag, Berlin Heidelberg (2004)
28. Lemos, A., Caminhas, W., Gomide, F.: Fuzzy multivariate gaussian evolving approach for fault detection and diagnosis. In: E. Hüllermeier, R. Kruse, F. Hoffmann (eds.) Proc. of the 13th International Conference on Information Processing and Management of Uncertainty, IPMU 2010, Part II (Applications), *CCIS*, vol. 81, pp. 360–369. Springer, Dortmund, Germany (2010)
29. Li, X., Li, H., Guan, X., Du, R.: Fuzzy estimation of feed-cutting force from current measurement—a case study on tool wear monitoring. IEEE Transactions Systems, Man, and Cybernetics Part C: Applications and Reviews **34**(4), 506–512 (2004)
30. Liao, T.: Clustering of time series data—a survey. Pattern Recognition **38**, 1857–1874 (2005)
31. Lughofer, E., Buchtala, O.: Reliable all-pairs evolving fuzzy classifiers. IEEE Transactions on Fuzzy Systems in revision (2012)
32. Lughofer, E., Guardiola, C.: On-line fault detection with data-driven evolving fuzzy models. Journal of Control and Intelligent Systems **36**(4), 307–317 (2008)
33. Lughofer, E., Klement, E., Lujan, J., Guardiola, C.: Model-based fault detection in multi-sensor measurement systems. In: Proceedings of IEEE IS 2004, pp. 184–189. Varna, Bulgaria (2004)
34. Lughofer, E., Smith, J.E., Caleb-Solly, P., Tahir, M., Eitzinger, C., Sannen, D., Nuttin, M.: On human-machine interaction during on-line image classifier training. IEEE Transactions on Systems, Man and Cybernetics, part A: Systems and Humans **39**(5), 960–971 (2009)
35. Montgomery, D.: Introduction to Statistical Quality Control (6th Edition). John Wiley & Sons, Hoboken, New Jersey (2008)
36. Nelles, O.: Nonlinear System Identification. Springer, Berlin (2001)
37. Ott, E., Schilling, E., Neubauer, D.: Process Quality Control: Troubleshooting And Interpretation of Data. ASQ Quality Press, Milwaukee (2005)
38. Raiser, S., Lughofer, E., Eitzinger, C., Smith, J.: Impact of object extraction methods on classification performance in surface inspection systems. Machine Vision and Applications **21**(5), 627–641 (2010)
39. Samanta, B.: Gear fault detection using artificial neural networks and support vector machines with genetic algorithms. Mechanical Systems and Signal Processing **18**(3), 625–644 (2004)
40. Schölkopf, B., Smola, A.: Learning with Kernels—Support Vector Machines, Regularization, Optimization and Beyond. MIT Press, London, England (2002)
41. Simani, S., Fantuzzi, C., Patton, R.: Model-based Fault Diagnosis in Dynamic Systems Using Identification Techniques. Springer Verlag, Berlin Heidelberg (2002)
42. Tang, Y., Al-shaer, E.S., Boutaba, R.: Active integrated fault localization in communication networks. In: Proceedings of the IEEE/IFIP International Symposium on Integrated Network Management (IM'2005), pp. 543–556. Nice, France (2005)
43. Vapnik, V.: Statistical Learning Theory. Wiley and Sons, New York (1998)
44. Wang, X., Kruger, U., Lennox, B.: Recursive partial least squares algorithms for monitoring complex industrial processes. Control-Engineering-Practice **11**(6), 613–632 (2003)
45. Wasserman, P.: Advanced Methods in Neural Computing. Van Nostrand Reinhold, New York (1993)
46. Wu, X., Kumar, V., Quinlan, J., Gosh, J., Yang, Q., Motoda, H., MacLachlan, G., Ng, A., Liu, B., Yu, P., Zhou, Z.H., Steinbach, M., Hand, D., Steinberg, D.: Top 10 algorithms in data mining. Knowledge and Information Systems **14**(1), 1–37 (2006)
47. Xu, R., Wunsch, D.: Survey of clustering algorithms. IEEE Transactions on Neural Networks **16**(3), 645–678 (2005)
48. Zhang, Y.Q.: Constructive granular systems with universal approximation and fast knowledge discovery. IEEE Transactions on Fuzzy Systems **13**(1), 48–57 (2005)

Chapter 15
Identification of a Class of Hybrid Dynamic Systems

Moamar Sayed-Mouchaweh, Nadhir Messai, Omar Ayad, and Sofiane Mazeghrane

Abstract The behavior of hybrid dynamic systems (HDS) switches between several modes with different dynamics involving both discrete and continuous variables in the course of time. Their identification aims at finding an accurate model of the system dynamics based on its past inputs and outputs. The identification can be achieved by two steps: the clustering and the regression. The clustering step aims at the estimation of the mode (discrete state) of each input–output data point as well as the switching sequence among these modes. The regression step determines the sub-models controlling the dynamic (continuous states) in each mode. In Pattern Recognition (PR) methods, each mode is represented by a set of similar patterns forming restricted regions in the feature space, called classes. A pattern is a vector built from past inputs and outputs. In this chapter, we propose to use an unsupervised PR method to realize the clustering step of the identification of switched linear HDS. The determination of the number of modes as well as the switching sequence does not require any information in advance about the modes, for example, their distribution, their shape, ..., or their number.

15.1 Introduction

Pattern recognition (PR) is the study of how machines can learn from experience to make sound decisions about the categories or classes of patterns of interest. In statistical PR methods [10], patterns are described as random variables, from which

M. Sayed-Mouchaweh (✉)
Ecole des Mines de Douai, Computer Science and Automatic Control Lab, EMDouai-IA, Douai, France
e-mail: moamar.sayed-mouchaweh@mines-douai.fr

N. Messai • O. Ayad • S. Mazeghrane
Centre de Recherche en STIC (CReSTIC), UFR Sciences Exactes et Naturelles, Universit de Reims Champagne-Ardenne (URCA), Reims, France
e-mail: nadhir.messai@univreims.fr; om_ayad@yahoo.fr; sofiane.life61@live.fr

M. Sayed-Mouchaweh and E. Lughofer (eds.), *Learning in Non-Stationary Environments: Methods and Applications*, DOI 10.1007/978-1-4419-8020-5_15,
© Springer Science+Business Media New York 2012

class densities can be inferred. These variables carry discriminating information about patterns. They are called features, which are usually quantitative observations, or measurements, about patterns. Therefore, a pattern is represented by a set of d features so it can be viewed as a d-dimensional feature vector in the feature space.

PR involves two stages: preprocessing and classification. The aim of the preprocessing is to find features in such a way that patterns belonging to different classes occupy different regions of the feature space. The classification stage is a mapping of a pattern from the feature space into the decision one. The latter is defined by a set of predefined classes. This mapping is achieved using a classifier. The latter is a method or algorithm which generates a class membership function in order to classify unlabeled incoming patterns into one of the predefined classes. Depending on the information available for classifier training, one can distinguish between supervised [21] and unsupervised [4, 11, 12] learning. In the first case, called also classification, there exists a set of patterns with their class assignment or label, called learning set. The goal of supervised learning is to learn a set of membership functions that allows the classification of new patterns into one of the existing classes. The problem of unsupervised learning, also called clustering, arises if clusters', that is, classes, memberships of available patterns, and perhaps even the number of clusters, are unknown. In such cases, a classifier is learned based on similar properties of patterns: patterns belonging to the same cluster should be as similar as possible and patterns belonging to different clusters should be clearly distinguishable. Hence, the goal of clustering is to partition a given set of patterns into clusters based on their similarity.

One of the applications of PR is the identification of hybrid dynamic systems (HDS). The latter are characterized by the interaction between continuous time dynamics and discrete events or logic rules [6,15]. The identification aims at obtaining an accurate model of the system dynamics based on its past inputs and outputs. The problem of obtaining a model of a hybrid system from a given set of input–output data has attracted, since few years, the attention of several researchers. Many models have been proposed to describe them as piecewise autoRegressive with exogenous inputs (PWARX), switched AR (SAR), switched ARX (SARX), switched nonlinear ARX (SNARX) and PW nonlinear ARX (PWNARX) ones [2,14,22].

Generally, the identification is divided into two steps: clustering and regression. In the clustering step, the discrete modes, that is, classes, that each input–output data point belongs to as well as the switching sequence among these modes are estimated. The regression step aims at finding the models governing the continuous dynamic in each mode.

In this chapter, we propose an approach to achieve the clustering step of the identification of switched HDS described by SAR or SARX models. The latter are a particular class of HDS [14]. In this approach, the number of discrete modes, classes, and the switching sequence among them are estimated using an unsupervised PR method. This estimation is achieved without the need to any prior information about these modes, for example, their shape or distribution, or their number.

This chapter is organized as follows. In Sect. 15.2, the principles of the proposed approach are detailed. Then, its performance is evaluated using two examples. The first one is an example of HDS, modeled as SARX, switching among 3 modes. The second example exploits the acoustic signals used to detect a leak in the steam generator unit of the nuclear power generator prototype fast reactors (PFR). These signals record the noises in the steam generator unit. The latter functioning, described as SAR model, switches between two modes (normal and faulty representing a leak) in several time instants. The advantages and the drawbacks of the proposed approach according to the ones of literature are discussed in Sect. 15.4. We finish the chapter with a conclusion and the future work.

15.2 Proposed Approach for the Identification of HDS

In this section, we present an approach to achieve the clustering step of the identification of switched HDS. This approach determines the number of modes or classes, $i = 1,\dots,c$ and the switching sequence $\lambda_j, j = 1,\dots,N$ using a historic of N observations of the system input u_j and output $y_j, j = 1,\dots,N$.

The proposed approach is based on two phases: the feature space construction and the modes estimation ones. The first phase aims at finding the features, based on the input–outputs data points, leading to well separate the modes in the feature space. The second phase uses the unsupervised fuzzy pattern matching (FPM) as a clustering method to determine the number of modes and to learn their membership functions. The performances of the mode estimation phase are evaluated by the closeness of the number of modes and of the time instants of switching among them to the real ones. These performances depend on the discrimination power of the feature space. Better the modes are separated in the feature space, better the modes estimation is. Figure 15.1 illustrates these phases of the proposed approach.

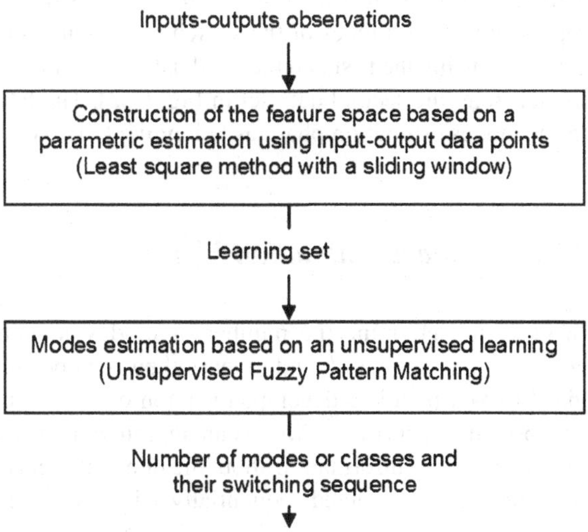

Fig. 15.1 Phases of the proposed approach

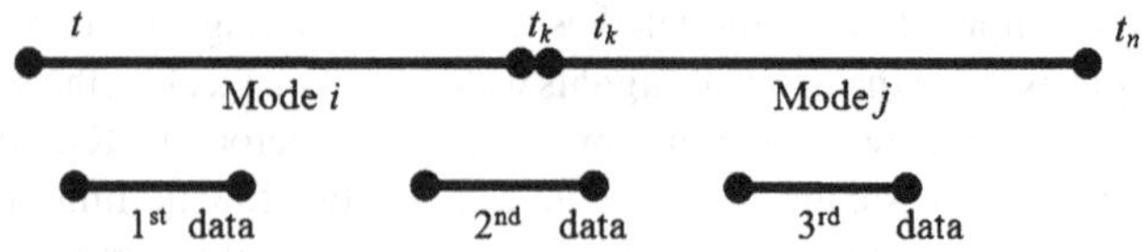

Fig. 15.2 Illustration of the feature space construction

15.2.1 Feature Space Construction Phase

Let $[u_0\ y_0, u_1\ y_1, \ldots, u_{k-1}\ y_{k-1}, u_k\ y_k]$ be the set of the system past and actual input–output observations, where $u_k \in \mathbb{R}$ and $y_k \in \mathbb{R}$ are, respectively, the input and output at time step k. The goal of this phase is to build the feature space from past inputs and outputs. In the obtained feature space, the pattern at the actual time step k is defined as follows: $x_k = \Phi(u_{k-1}, y_{k-1})$, where Φ is the mapping from the observation space, that is, input–output space, to feature one. In this chapter, we propose to use the Least Square Method (LSM) with a sliding window [17] to estimate the parameters of the continuous dynamic of each mode. In this method, the parameters about the system dynamic modes are estimated at instant t so that it minimizes the sum of the squares of the differences between the output of the system and the one of the prediction model over a sliding window of t measurements. These parameters are then used as features to represent the different modes or classes.

Hence, in order to construct the feature space, we propose to estimate the sub-models parameters using a sliding window as shown in Fig. 15.2. The latter shows clearly that two cases are possible: the first one corresponds to the use of the data points of the current mode (i.e., the first and the third data sets) and the second one corresponds to the case when the data points of the current and the successor modes are used to estimate the parameters (i.e. the second data set). Thus, the parameters estimation procedure will provide different sets of parameters. Some of these sets represent the real modes of the switching system (for example the parameters sets estimated using the first and the third data sets) and other sets represent some biased models (e.g. the second data set in Fig. 15.2). The latter case will be distinguished in the modes estimation phase as belonging to two different modes.

15.2.2 Modes Estimation Phase

In order to determine the number of modes contained in the learning set $X = [x_N, \ldots, x_k, x_{k-1}, \ldots, x_1]^T$ and to learn their membership functions, we use unsupervised FPM which is a developed version of the original supervised FPM [7]. The proposed unsupervised FPM has an agglomerative characteristic. Thus, it does not require any prior information about the number of classes. The classes' membership functions are constructed sequentially with the patterns arrival. According to the ratio $r = \frac{L}{U+L}$ of the number L of labeled points to the one U of unlabeled points, the proposed method can be totally supervised, $r = 1$, or totally unsupervised, $r = 0$. The functioning of unsupervised FPM is divided into detection, adaptation and fusion steps.

15.2.2.1 Classes Detection Step

Let $x = (x^1, x^2, \ldots, x^d) \in \mathbb{R}^d$ be a given pattern vector in a feature space constituted of d parameters or attributes. Statistically, the features are assumed to have a probability density function (pdf) conditioned on the pattern class. Thus, a pattern vector x belonging to the class C_i is viewed as an observation drawn randomly from the class-conditional probability function $p(x/C_i)$. Each attribute is divided into equal intervals defining the bins of the histogram according to this attribute. This histogram is used to estimate the conditional probability density for the class that x is driven from. Let $X^j_{\min}$ and $X^j_{\max}$ be, respectively, the lower and upper borders of the histogram according to the attribute j. These borders can be defined as the minimal and maximal values of all the patterns of the learning set X according to each attribute or parameter. Let h be the number of histogram bins, then each bin, according to the attribute j, has the larger:

$$\Delta^j = \frac{X^j_{\max} - X^j_{\min}}{h}, \quad j \in \{1, 2, \ldots, d\}. \tag{15.1}$$

Thus, the limits of these bins are defined as follows:

$$\begin{aligned} b^j_1 &= [X^j_{\min}, X^j_{\min} + \Delta^j], \quad b^j_2 = [X^j_{\min} + \Delta^j, X^j_{\min} + 2\Delta^j] \\ b^j_h &= [X^j_{\min} + (h-1)\Delta^j, X^j_{\max}], \quad j \in \{1, 2, \ldots, d\}. \end{aligned} \tag{15.2}$$

Generally, the histogram or the distribution of probability

$$\left\{ p^j_i(b^j_{ik}), i \in \{1, 2, \ldots, c\}, j \in \{1, 2, \ldots, d\}, k \in \{1, 2, \ldots, h\} \right\}$$

for a class C_i according to the attribute j is determined by calculating the probability $p^j_i(b^j_{ik})$ of each bin b^j_{ik}:

$$p^j_i(b^j_{ik}) = \frac{n^j_{ik}}{N_i}, \tag{15.3}$$

where n^j_{ik} is the number of points of the class C_i which are in the bin b^j_{ik} and N_i is the total number of points of the class C_i. The resulting distribution of probability is transformed into a distribution of possibility

$$\left\{ \pi^j_i(b^j_{ik}), i \in \{1, 2, \ldots, c\}, j \in \{1, 2, \ldots, d\}, k \in \{1, 2, \ldots, h\} \right\}$$

by using the transformation of Dubois and Prade [9]:

$$\pi_i^j(b_{ik}^j) = \sum_{z=1}^{h} \min(p_i^j(b_{iz}^j), p_i^j(b_{ik}^j)). \tag{15.4}$$

A membership function can be generated by considering the possibility distribution numerically equal to the fuzzy membership function [23]. The possibility distribution is more adapted than the probability one to estimate membership functions in the case of data infected by noises and uncertainties related to the features estimation [9]. Finally, the density of possibility Π_i^j of the class C_i according to the attribute j is obtained by a linear interpolation of the bins centers of the histogram of possibility.

The first incoming pattern x will be considered as the point prototype of the first class: $C_1 \leftarrow x, c \leftarrow 1$. If x is located in the bin $b_k^j, k \in \{1,2,\ldots,h\}$, then the probability histogram of C_1 according to the attribute j is: $\pi_1^j = \{p_{11}^j = 0, p_{12}^j = 0, \ldots, p_{1k}^j = 1, \ldots, p_{1h}^j = 0\}$. The possibility histogram will then be computed using (15.2). Since there is just one pattern, the possibility histogram is equal to the probability one. The possibility density of the class C_1 is obtained by a linear linking between the center of the bin b_k^j, which has the height 1, and the ones of its left b_{k-1}^j and right b_{k+1}^j neighbors, which have both at present the height 0. Generally, if $C = \{C_1, C_2, \ldots, C_c\}$ is the set of learned classes at present. Let x be a new incoming pattern which is not assigned to any of the learned classes (membership rejection). The detection strategy is defined as follows:

$$\pi_i(x) = 0,\ \forall i \in \{1,2,\ldots,c\} \Rightarrow c \leftarrow c+1, C_c = \{x\}, \pi_c = \left\{\pi_c^1, \ldots, \pi_c^j, \ldots, \pi_c^d\right\}. \tag{15.5}$$

15.2.2.2 Classes Adaptation Step

The local adaptation step aims at updating the classes' possibility densities after the classification of each new pattern in order to take into account the information carried by the new classified patterns in the class.

Let x' be a new pattern classified in the class $C_i, \forall i \in \{1,2,\ldots,c\}$. This classification is obtained by a projection of the pattern on the possibility density Π_i^j of the class C_i according to each attribute j and then merging the values according to all attributes using the aggregation operator "minimum." The point x will be assigned to the class for which it has the highest membership value. If the membership value $\pi_i(x')$ of x' to the class C_i is different of zero, then this pattern will be assigned to the class C_i and the possibility densities of this class according to each attribute will be updated. The goal is to take benefit of the information carried by the new classified pattern for the classification of the next incoming ones. To establish an incremental update of possibility densities, let $p_i^j = \left\{p_{i1}^j, p_{i2}^j, \ldots, p_{ik}^j, \ldots, p_{ih}^j\right\}$ and $\pi_i^j = \left\{\pi_{i1}^j, \pi_{i2}^j, \ldots, \pi_{ik}^j, \ldots, \pi_{ih}^j\right\}$ define, respectively, the probability and possibility histograms of the class C_i according to the attribute j. Let $p_i^{\prime j} = \left\{p_{i1}^{\prime j}, p_{i2}^{\prime j}, \ldots, p_{ik}^{\prime j}, \ldots, p_{ih}^{\prime j}\right\}$ and $\pi_i^{\prime j} = \left\{\pi_{i1}^{\prime j}, \pi_{i2}^{\prime j}, \ldots, \pi_{ik}^{\prime j}, \ldots, \pi_{ih}^{\prime j}\right\}$ define, respectively, the updated probability and possibility histograms of the class C_i according to the attribute j after the assignment of x' to the class C_i. Let us suppose for the simplicity that: $p_{ih}^{j} < p_{i(h-1)}^{j} < \ldots < p_{i1}^{j}$, then these new probabilities can be computed incrementally by [21]:

$$x^{\prime j} \in b_k^j, \forall k \in \{1, \ldots, h\} \Rightarrow p_{ik}^{\prime j} = p_{ik}^{j} * \frac{N_i}{N_i+1} + \frac{1}{N_i+1}$$

$$p_{iz}^{\prime j} = p_{iz}^{j} * \frac{N_i}{N_i+1}, \forall z \in \{1, \ldots, h\}, z \neq k. \tag{15.6}$$

Then the new possibilities can be computed using Dubois and Prade transformation defined by (15.2). Thus, the local adaptation step is defined as follows:

$$\pi_i(x') = \max_{z \in \{1, \ldots, c\}} (\pi_z(x')) \Rightarrow C_i \leftarrow \{C_i, x'\}, \pi_i' = \left\{\pi^{\prime 1}, \pi^{\prime 2}, \ldots, \pi^{\prime j}, \ldots, \pi^{\prime d}\right\}. \tag{15.7}$$

The flow chart of the detection and local adaptation steps of unsupervised FPM is presented in Fig. 15.3.

15.2.2.3 Classes Merging Step

The occurrence order of incoming patterns influences the final constructed clusters. This may lead to obtain several different partitions or number of clusters. Thus, several clusters can represent the same functioning mode. These clusters must be merged into one cluster to obtain one partition and one membership function. This fusion can be done using a similarity measure. The latter measures the overlap or closeness between constructed clusters. There are different similarity measures in the literature. Most of them are based on the computation of the degree of overlapping of clusters or the distance between clusters' centers. The clusters overlapping degree is based on the number of ambiguous patterns, belonging to several clusters, and their membership values to these clusters. If the number of these ambiguous patterns is large enough and their membership values to several clusters are high, then these clusters cannot be considered as heterogeneous anymore and must be merged. An interesting similarity criterion which takes into account at the same time the number of ambiguous patterns as well as their membership values is defined by [12]:

$$\delta_{iz} = 1 - \frac{\sum_{x \in C_i \vee x \in C_z} |\pi_i(x) - \pi_z(x)|}{\sum_{x \in C_i} \pi_i(x) + \sum_{x \in C_z} \pi_z(x)}. \tag{15.8}$$

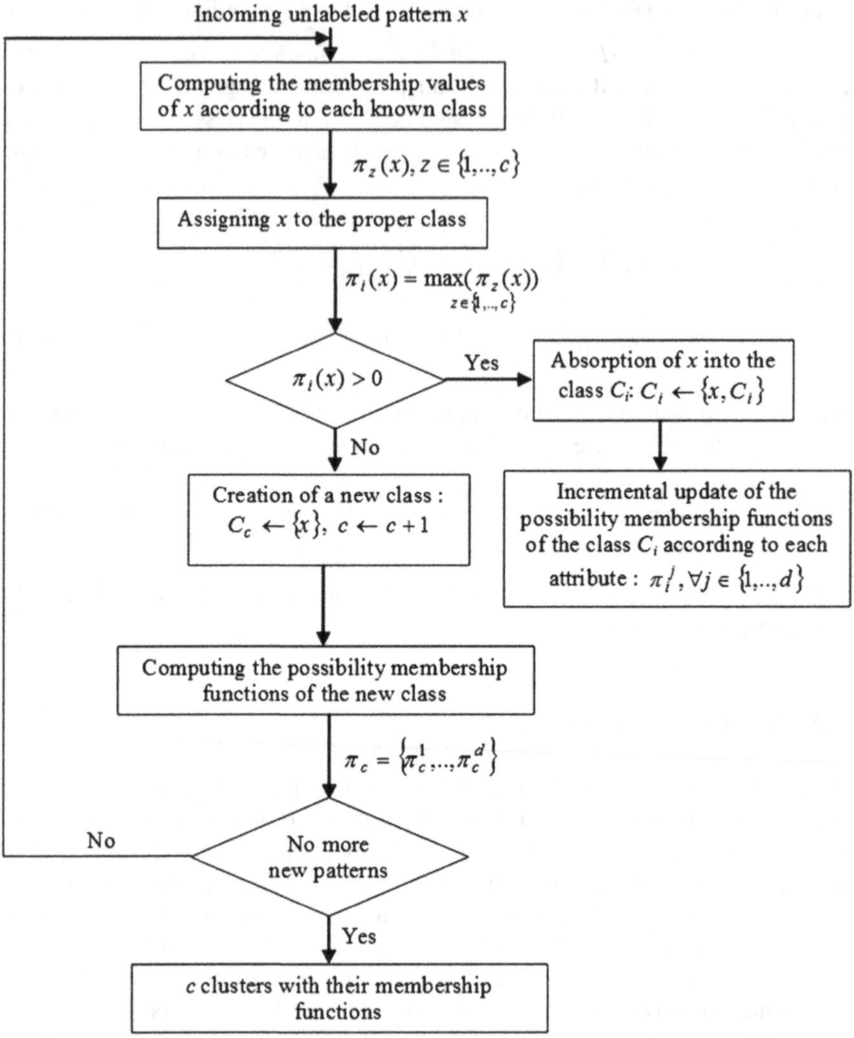

Fig. 15.3 Flow chart of unsupervised FPM for the detection and local adaptation steps

Where $\pi_i(x)$ and $\pi_z(x)$ are, respectively, the membership values of x to C_i and C_z. δ_{iz} is the similarity measure between the two classes. More the similarity value is close to 1, more the two classes are similar and must be merged. Figure 15.4 shows the values of this similarity measure according to the closeness of two Gaussian classes.

The clusters are merged when this measure reaches a predefined threshold. In general, a value of the similarity measure greater than 0.1 is enough to merge two clusters. Indeed, starting from this value, two clusters begin to be partially overlapped in the feature space. Figure 15.5 shows the flow chart of unsupervised FPM for the merging step.

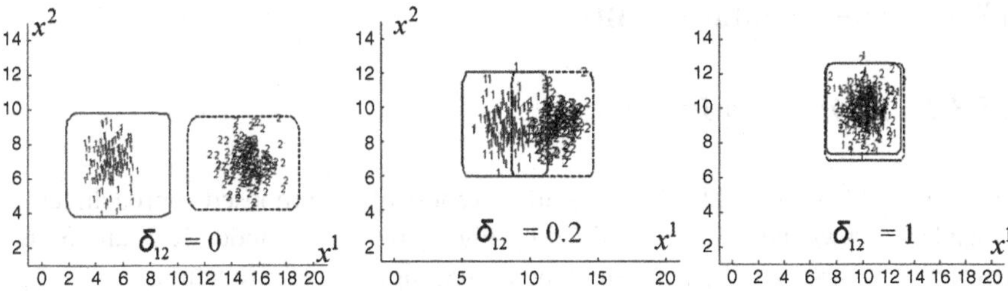

Fig. 15.4 Similarity measure between two classes

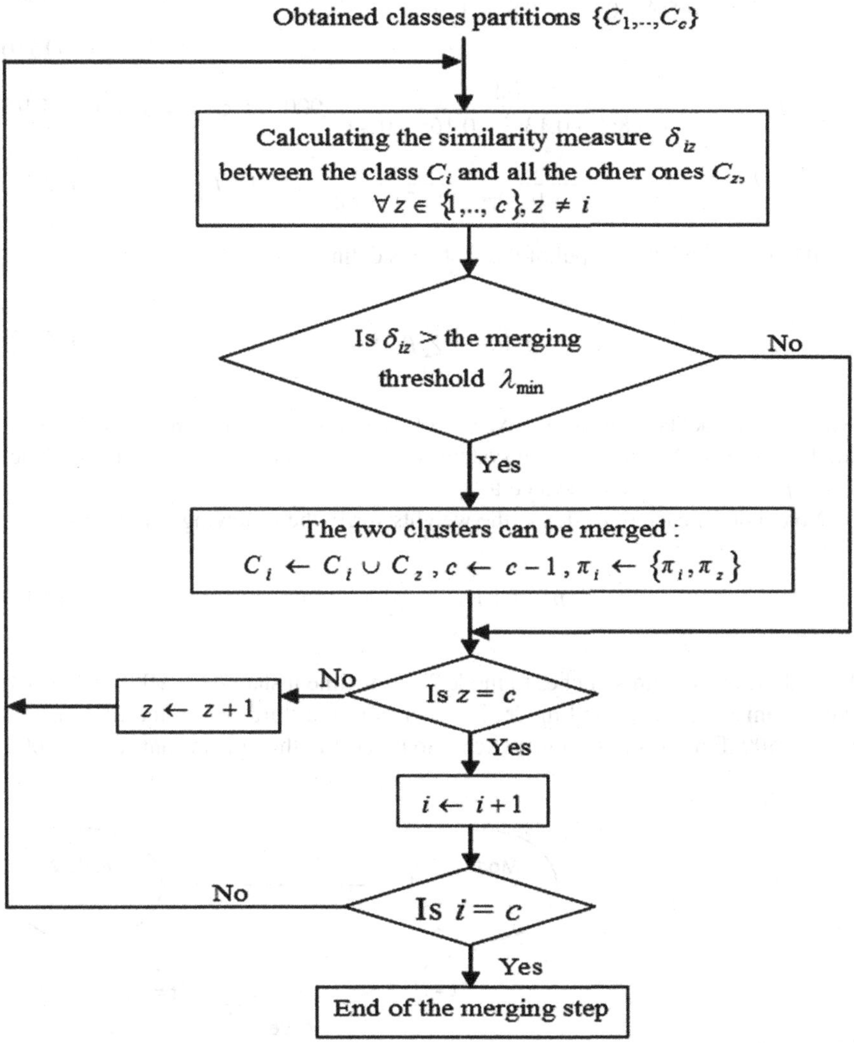

Fig. 15.5 Flow chart of unsupervised FPM for the merging step

15.3 Experimental Results

15.3.1 Simulation Example

In order to illustrate and to show the effectiveness of the proposed approach, let us consider the switched HDS described by Fig. 15.6. Where modes 1, 2, and 3 are, respectively, described by the following discrete-time transfer functions:

$$F_1(z) = \frac{z+0.5}{2z^4+0.5z^3+0.8z^2-0.3z+0.9} \quad 0 \leq t < t_{12} = 900 \wedge 3{,}800 \leq t \leq 5{,}000 \tag{15.9}$$

$$F_2(z) = \frac{z+0.9}{z^4-0.8z^3+0.13z^2-0.16z+0.45} \quad 900 \leq t < t_{23} = 2{,}500 \tag{15.10}$$

$$F_3(z) = \frac{z+0.2}{z^4-0.4z^3+0.29z^2-0.65z-0.2} \quad 2{,}500 \leq t_{31} < 3{,}800. \tag{15.11}$$

On the other hand, the output of the system is defined by:

$$y_c(k) = \sum_{i=1}^{3} p_{i,k} y_i(k), \tag{15.12}$$

where k represents the time index, $i = \{1,2,3\}$ represents the index of the local mode, $y_c \in \mathbb{R}$ is the output of the system, $y_i(k) \in \mathbb{R}$ is the output of the local model i, and $p_{i,k}$ is the weight associated to $y_{i,k}$.

Note that for each time step k, the weights verify the following conditions:

$$p_{i,k} \in \{0,1\}, \quad \sum_{i=1}^{n} p_{i,k} = 1. \tag{15.13}$$

In mode 1, the system switches to mode 2, at the time instant $t_{12} = 900$. Then when the system is in the second mode, it switches to the third one at the time instant $t_{23} = 2{,}500$. Finally, the system switches to mode 1 at the time instant $t_{31} = 3{,}800$.

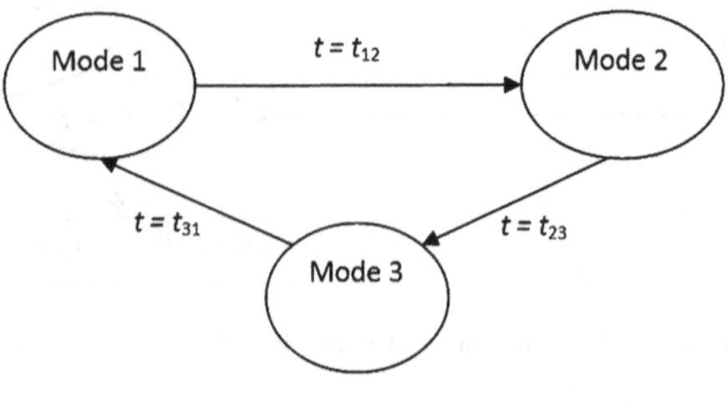

Fig. 15.6 Switching sequence for the simulation example

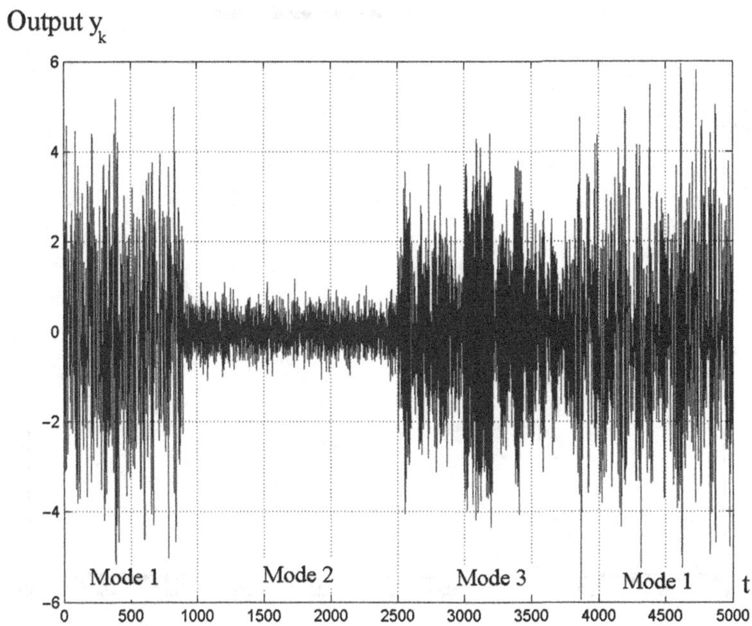

Fig. 15.7 Signal output of the simulation example of Fig. 15.6

Table 15.1 Similarity measure between the classes obtained by the application of the proposed approach on the learning set of the example of Fig. 15.6

Class	1	2	3	4	5	6
1	–	0	0	0	0	0
2	0	–	0	0	0	0
3	0	0	–	0.17	0	0.21
4	0	0	0.17	–	0.33	0
5	0	0	0	0.33	–	0
6	0	0	0.21	0	0	–

In order to simulate the system, an input–output identification data set has been generated. The output signal contains 5,000 data points and is generated using a pseudo random binary sequence (PRBS) as an input. Figure 15.7 shows the output signal of this example with the course of time. The feature estimation using the LSM, developed in Sect. 2.1, with a sliding window of 100 data points shifted by one time unit are used to determine the 4,900 patterns. This allows obtaining the learning set X containing 4,900 patterns in a feature space of 6 parameters. The number of parameters depends on the transfer function order. We have applied unsupervised FPM on the learning set X of the example of Fig. 15.6. Six classes are obtained. The similarity measures between these classes are shown in Table 15.1. If the fusion threshold δ is equal to 0.17, then we can obtain the following three classes: $\{C_1\}, \{C_2\}, \{C_3\} \leftarrow \{\{C_3\} \cup \{C_4\} \cup \{C_5\} \cup \{C_6\}\}$.

Figure 15.8 shows the distribution of the patterns of the learning set in each of the obtained three classes according to their occurrence time. Table 15.2 shows the switching time obtained for each mode or class. We can see that the error of time

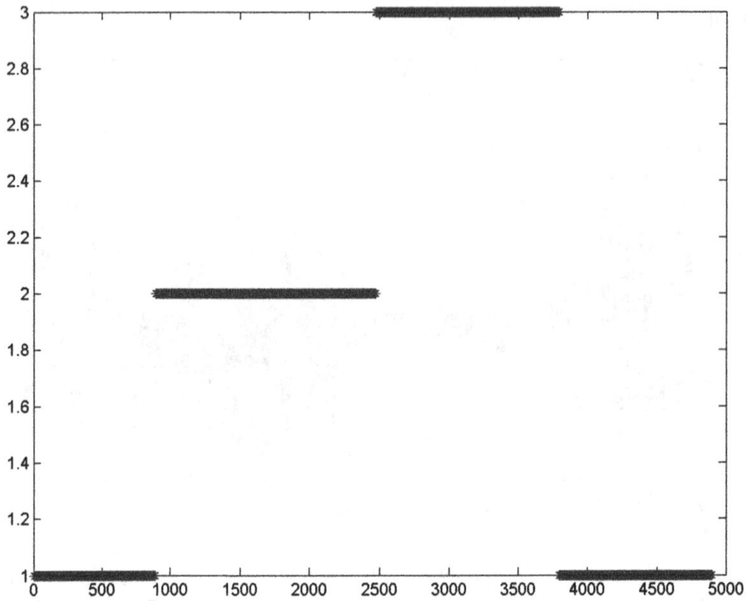

Fig. 15.8 Data points classes according to their occurrence time for the example of Fig. 15.6

Table 15.2 Switching time between the modes of the example of Fig. 15.6 obtained by the proposed approach

	Real switching time	Estimated switching time	Estimation error
t_{12}	900	901	1
t_{23}	2,500	2,495	5
t_{31}	3,800	3,794	6

switching is small (six time steps in the worst case). The proposed approach has a low computational complexity and low learning or classification time which depend both on the number of attributes and not on the number of patterns in the data set. This classification time for each pattern of the learning set is equal to 3.4×10^{-4} s using a computer with Pentium 4 2.8 GHz.

However, the performance of the proposed approach depends on the separability between the different modes, that is, classes, in the feature space. These different modes occupy separated regions in the feature space if their parameters are properly estimated. This needs a suitable size of the sliding time window in order to include enough of input–output data points. Thus, in order to find a suitable time window size, the proposed approach is applied using several time window sizes. We choose the smallest window size which minimizes the similarity measure (maximizes the separability) between the obtained classes. For the example of Fig. 15.6, we can notice that mode 3 has a very close behavior to the one of mode 1. While mode 2 has a clear different behavior of the one of other modes. Thus, the sliding window must have a sufficient size to well separate modes 1 and 3. Figure 15.9 shows the

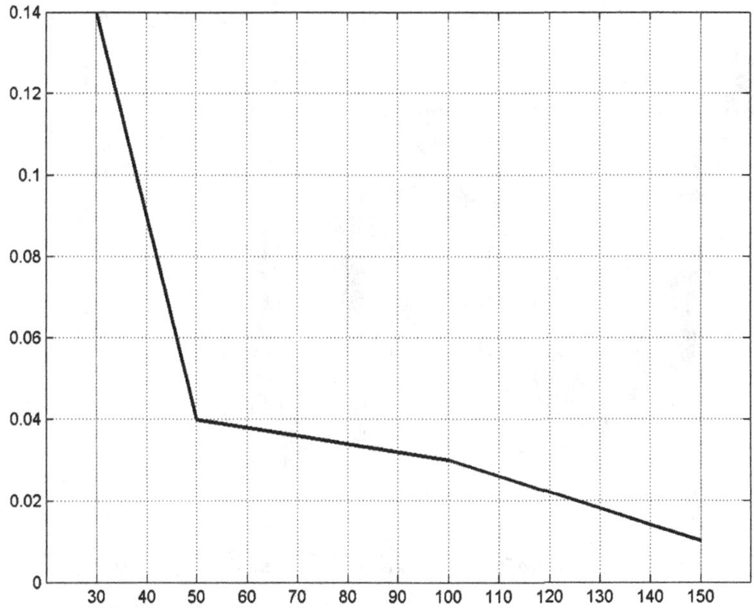

Fig. 15.9 Similarity measure between modes 1 and 3 for different sizes of the sliding window used to estimate the parameters of each mode

similarity measure between modes 1 and 3 for different sizes of the sliding window. We can notice that a sliding window of 30 input–output data points is not sufficient to enough separate modes 1 and 3. Indeed, the similarity measure indicates that these two modes belong to the same class or mode. Thus, in this case, modes 1 and 3 are merged into one mode and unsupervised FPM provides two modes which is an erroneous result. While a sliding window of 50 points or plus is sufficient to well separate the classes and to obtain the right number of modes. Figures 15.10 and 15.11 show the values of all the patterns of the learning set X according to each attribute for two sliding windows of sizes 30 and 100, respectively. We can observe that, for the case of a window size of 100 points, features 3 and 4 separate well modes 1 and 3, and the other features separate mode 2 from the other modes. In the case of a window of 30 points, mode 1 has a very close behavior to the one of mode 3 according to each one of the 6 features. This entails to consider these two modes as one mode by unsupervised FPM.

15.3.2 Application Example

PFR are used to produce nuclear power from nuclear fuel. They are cooled by metal liquid sodium. Indeed, water is difficult to use as a coolant for a fast reactor because collisions with the hydrogen nuclei in water quickly remove most of the kinetic

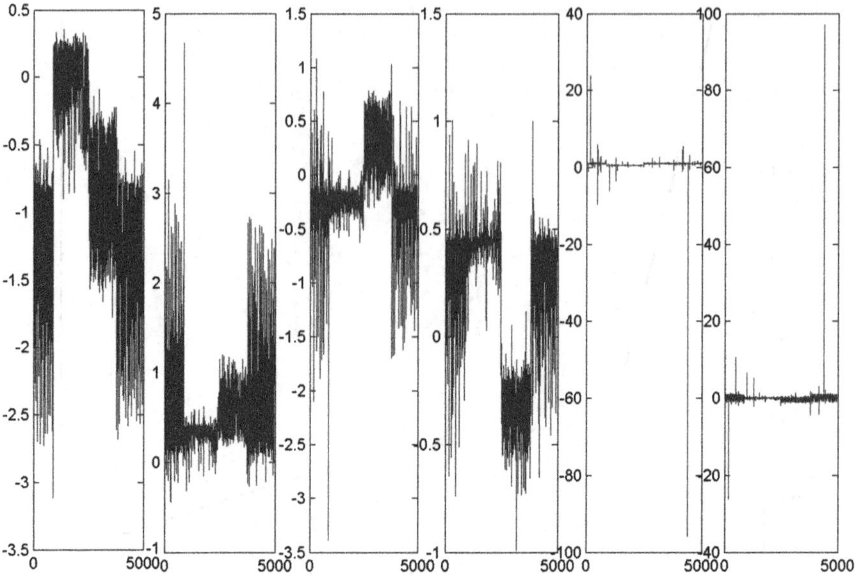

Fig. 15.10 Classes features estimated using a sliding window with 30 input–output data points

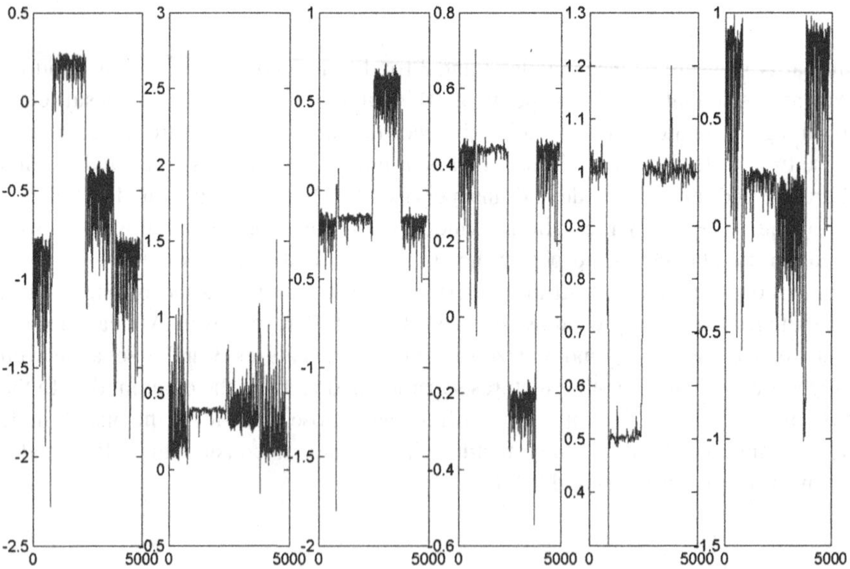

Fig. 15.11 Classes features estimated using a sliding window with 100 input–output data points

energy from the neutrons [8]. In contrast, sodium atoms are much heavier than both the oxygen and hydrogen atoms found in water, and therefore the neutrons lose less energy in collisions with sodium atoms. Sodium also does not need to be pressurized since its boiling point is higher than the reactor's operating temperature. However, a disadvantage of sodium is its violent chemical reactivity, which requires special precautions. If sodium comes into contact with water it explodes.

Actually, in-sodium hydrogen detectors are used in PFR to detect steam leaks in steam generator. However, they have long detection times of the order of two minutes. This is because hydrogen needs to transport from the leak site to the detector location [19]. Therefore, leaks will grow to a large leak which may cause serious damage through an explosion due to an increase of in-sodium gas pressure.

This limitation can be avoided using acoustic signals recording the background noises in the steam generator. Thus, the objective of this application is to design an acoustic leak detection to detect at early stage (faster than the hydrogen detectors) the reactions sodium/water. This acoustic leaks detection can be used as a supplementary tool besides the hydrogen detectors to detect steam leaks.

The available acoustic signals were recorded using data from background noise measurements on the steam generator from the end-of-life of PFR at United Kingdom. In these experiments, argon was injected into sodium, and acoustic noises were measured. Indeed, experimental results have shown that steam and argon injections give similar acoustic noise output at a given mass flow rate [20]. Figure 15.12 shows an acoustic signal recorded in response to an injection command. The signal records the noises resulting of the injection of argon in the steam generator unit of PFR. This injection simulates a fault occurred by a leak in the steam generator unit. Thus, the functioning of the steam generator unit switches between two modes (normal: non-injection and faulty: argon injection) in several time instants as it is shown in Fig. 15.12. The signal is sampled at the frequency 2,048 Hz.

We apply the proposed approach on the acoustic signal of Fig. 15.12. The signal is considered as the output of switched autoregressive (SAR) system. Therefore, the feature space is defined by the estimated parameters (coefficients) ai of the model AR. AIC criterion [1] has been used in order to select the order of the AR dynamic model. $d = 152$ is the AR model order which minimizes AIC criterion (Fig. 15.13). These features change with time. In order to capture this change, these features are calculated during a sliding time window. The latter size must include a sufficient number of data points in order to properly estimate the parameters of each mode. We have tested several sizes of time window. We have selected the one which maximizes the discrimination power between the different modes in the feature space, that is, obtaining compact and separated classes. This experimentation leads to select a sliding window with an initial length $\Delta t_{\mathrm{f}} = 8,192$ data points and a shift length Δt_{s} equal to 2,048 data points. Therefore, to define a pattern in the feature space, a time window containing 8,192 data points is required. Consequently, to determine the functioning mode (injection or non-injection), a delay time of 4 s is needed.

It is useful to reduce the feature space defined by the coefficients of the dynamic parametric model AR (d). The reduction operation aims at keeping the

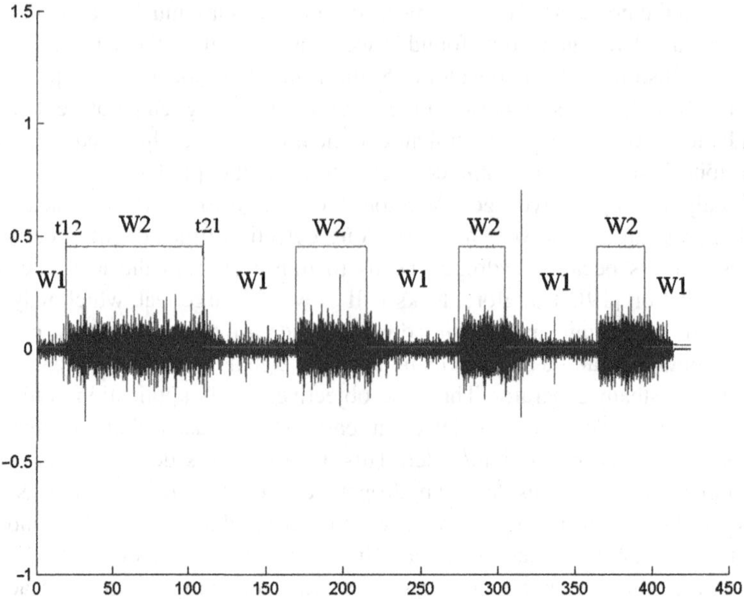

Fig. 15.12 Acoustic signal in response to argon command signal. W1: non-injection class, W2: injection class, t12: time of switching from W1 to W2, t21: time of switching from W2 to W1

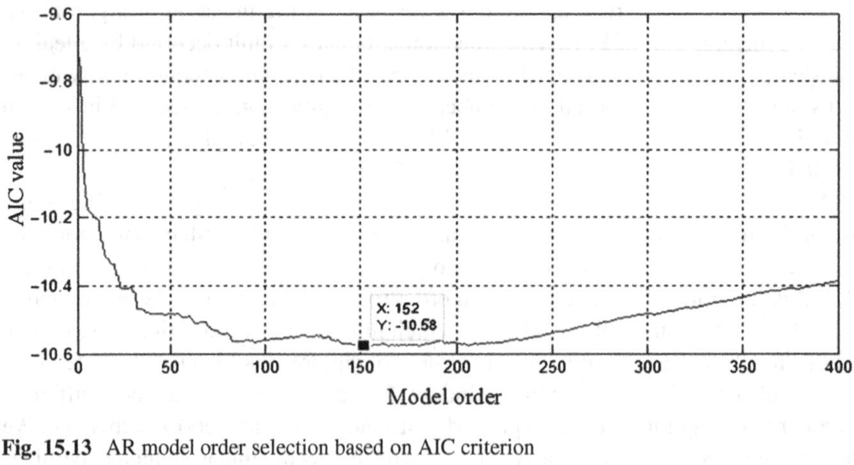

Fig. 15.13 AR model order selection based on AIC criterion

distinguishing features leading to separate as well as possible the different classes. We used the principal component analysis (PCA) to extract from the set of features the ones which are uncorrelated. 13 parameters have been selected because they are independent and conserve about 82% of the complete inertia carried by the 152 parameters. Then, we selected from this set of independent features the ones which have a combination leading to obtain the lowest error of classification. Two independent and discriminative AR model coefficients (coefficients 3 and 5) were conserved to define the feature space.

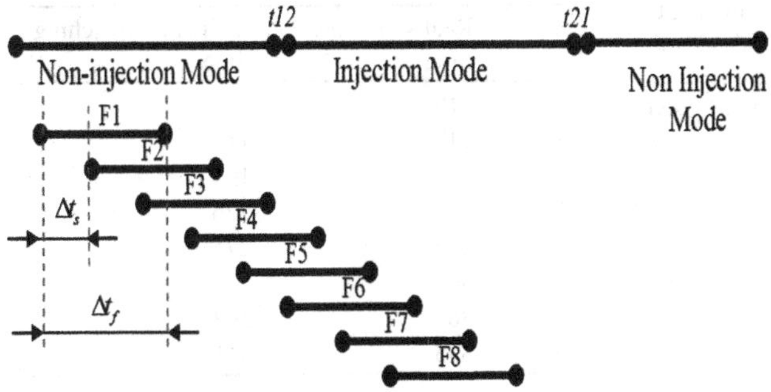

Fig. 15.14 Estimation of the feature space parameters using a sliding time window

Table 15.3 Similarity measure between the three obtained classes for the example of Fig. 15.12

Class	1	2	3
1	–	0.4	0
2	0.4	–	0.17
3	0	0.17	–

Feature space parameters represent the estimated coefficients of SAR model. These parameters, features, change with time. In order to capture this change, these features are calculated during a sliding time window (Fig. 15.14). However, the sliding window size must include a sufficient number of data points to properly estimate the parameters of each mode. We have tested several sizes of time window. We have selected the one which maximizes the discrimination power between the different modes in the feature space, that is, obtaining compact and separated classes. This experimentation leads to select a sliding window with a fixed length $\Delta t_f = 8{,}192$ data points and a shift length Δt_s equal to 2,048 data points. We apply the proposed approach on the acoustic signal of Fig. 15.12. The obtained results of clustering are shown in Table 15.3. Table 15.4 shows the switching times between the modes for the example of Fig. 15.12. The similarity, obtained by unsupervised FPM, between C_1 and C_2, is very important (equal to 0.40). Thus, the classes C_1 and C_2 must be merged. These two classes correspond to the non-injection class. The class C_3 corresponds to the injection class. The similarity value between C_2 and C_3 indicates that the class C_2 is a transitory one between C_1 and C_3.

Table 15.4 shows the switching time obtained for each mode, class. We can notice that the activation of the injection command several times leads to increase the delay required to detect the switching from injection to non-injection classes. This is due to the fact that the time required to allow the attenuation of the excitation resulted by the injection command increases with the number of activation of the injection command. However, there is no time delay to detect the switching from non-injection to injection classes. This advantage is very useful in a critical system

Table 15.4 Switching times between the modes for the example of Fig. 15.12 obtained by the proposed approach

	Real switching time	Estimated switching time	Estimation error
t_{12}	19	19	0
t_{21}	112	119	7
t_{12}	169	169	0
t_{21}	215	226	11
t_{12}	275	275	0
t_{21}	305	314	9
t_{12}	363	363	0
t_{21}	395	411	16

as the nuclear reactor since the transition from non-injection to injection modes simulates a leak in the steam generator. This leak must be detected as soon as possible to avoid an explosion.

15.4 Discussion and Related Work

Identification approaches of the literature are generally divided into clustering-based, Bayesian, bounded-error, algebraic-geometric, and optimization-based ones. Each of these approaches has its own advantages and drawbacks according to the assumptions needed on the number of modes, their order, the computational complexity, the dwell time in each mode, the achievable performance, the possibility to achieve on-line and/or off-line identification, etc.

In [13], the authors proposed a clustering-based approach that partitions the regressor space into regions on which a linear local model is valid. Then, it estimates each model parameters by standard least squares regression tools. However, this approach requires the knowledge of the ARX sub-models orders, and it stills suboptimal since the convergence depends strongly on the initialization step. In [5], a clustering-based method using the evidential theory is proposed. This approach supposes that each input–output data point is a cluster, that is, model. Then, the evidential theory is used for grouping data points that are more likely to have been generated by the same mode. The advantage of this approach is that it does not require the number of modes to be known a priori. However, the number of modes obtained by this approach depends strongly on a tuning parameter which is the number of neighbors. Moreover, the models order needs to be known a priori.

In [16], a Bayesian approach is proposed. This approach treats the parameters to be identified as random variables described with their pdfs. The data classification problem is posed as the problem of computing the a posteriori pdf of the model parameters, and the data are clustered in a suboptimal way using Bayesian inferences. However, this approach needs a priori knowledge about: the number of the modes, the ARX sub-models order and the pdfs of the parameters. In [3], a bounded-error approach is proposed. The main feature of this approach is to impose that the identification error is bounded by a given bound for all the samples in the data set. The approach consists of three main steps: initialization, refinement, and region estimation. At the initialization step, the estimation of the number of sub-models, data classification, and parameter estimation are performed simultaneously by partitioning a set of linear complementary inequalities derived from data into a minimum number of feasible subsystems. Then, a refinement procedure is proposed in order to reduce misclassifications and to improve parameter estimates. Region estimation is finally performed via two class or multi-class linear separation techniques. However, this approach requires the orders of the ARX sub-models to be known a priori. Moreover, the bound needs to be properly adjusted in the identification procedure in order to find the desired trade-off between model complexity and fit quality.

In [22], the authors proposed a solution for the identification of noiseless PWARX models with unknown and different orders. The presented algorithm is based on an algebraic approach in which homogeneous polynomials are used to realize a segmentation of the regression space into regions that correspond to the discrete states. [18] proposed an identification approach that considers the plant as a nonlinear black-box and uses feedforward neural networks to predict the continuous outputs of the given HDS. In the same context, [2] proposed an on-line identification that uses an adaptive growing and pruning radial basis function neural network.

In this chapter, we considered the identification of Switched linear autoregressive (SAR) and switched linear autoregressive with exogenous inputs (SARX) models. In this class of HDS, the system switches arbitrary from one mode to another one. The proposed approach is a clustering-based one. It does not require the knowledge about the number of modes, the model parameters, and the switching sequences. In addition, this approach can be used to achieve both off-line and on-line identification. This is possible thanks to the low classification time and to the agglomerative and recursive character of the proposed approach. Finally, this approach overcomes the problem of initialization thanks to the use of a similarity measure to merge the clusters belonging to the same mode.

15.5 Conclusions

In this chapter, a clustering-based approach is proposed for the identification of switched linear autoregressive (SAR) and Switched linear autoregressive with exogenous inputs (SARX) models of hybride dynamic systems (HDS). The goal is to determine the number of modes as well as the switching sequence among them. The estimation of the number of modes is achieved using unsupervised FPM. The latter is based on a competitive agglomerative technique which allows the detection of new clusters sequentially without the need to any prior information about these clusters, that is, modes, or their number. Then, the clusters membership functions are refined sequentially with the assignment of new unlabeled patterns. Since the order of patterns' occurrences can be different according to the switching sequence and there is no information about the clusters positions, several different portioning or clusters can be obtained. Thanks to the use of a similarity measure, the clusters, which are close to each other so that they cannot be considered anymore heterogeneous, are merged. The complexity and the computation time of the proposed approach are low and depend only on the dimension of the feature space. However, the proposed approach requires a discriminate feature space in order to separate the classes or the modes. Thus, a LSM with a sliding window is used to estimate the parameters of each mode or class. This estimation enhances the discrimination among classes. However, the sliding window must include enough of output data points in order to well separate the modes which have close dynamic behavior. We are developing this approach to be operant for the other classes of HDS as the piecewise autoregressive exogenous (PWARX), switched nonlinear ARX (SNARX), and PW nonlinear ARX (PWNARX) ones. In addition, we aim at relaxing the prior knowledge of the order of the transfer function for the construction of feature space. The goal is to be able to realize the identification of HDS containing subsystems or modes of different orders.

Acknowledgements This work is supported and integrated in the TEESC project. The authors would like to thank the scientific interest group surveillance, safety and security of the big systems.

References

1. Akaike, H.: A new look at the statistical model identification. IEEE Trans Autom Cont. **19**, 716–723 (1974)
2. Alizadeh, T., Salahshoor, K., Jafari, M.R., Alizadeh, A., Gholami, M.: On-line identification of hybrid systems using an adaptive growing and pruning rbf neural network. In: Proceedings of the IEEE Emerging Technologies and Factory Automatization, pp. 257–264 (2007)
3. Bemporad, A., Garulli, A., Paoletti, S., Vicino, A.: Data classification and parameter estimation for the identification of piecewise affine models. In: Proceedings of the 43rd IEEE Conference on Decision and Control, pp. 20–25. Paradise Island, Bahamas (2004)
4. Bezdek, J.: Pattern Recognition with Fuzzy Objective Function Algorithms. Kluwer Academic/ Plenum Publishers, U.S.A. (1981)
5. Boukharouba, K., Bako, L., Lecoeuche, S.: Identification of piecewise affine systems based on dempster-shafer theory. In: Proceeding of 15th IFAC Symposium on System identification, pp. 1662–1667. Saint-Malo, France (2009)
6. Branicky, M.: General hybrid dynamic system modelling: analysis and control. In: Hybrid System III, *Lecture Notes in Computer Science*, vol. 4066, pp. 187–200. Springer, Heidelberg, Germany (1996)
7. Cayrol, M., Farreny, H., Prade, H.: Fuzzy pattern matching. Kybernetes **11**, 103–116 (1982)
8. Cochran, T., Feiveson, H., Patterson, W., Pshakin, G., Ramana, M., Schneider, M., Hippel, T.S.F.: A research report of the International Panel on Fissile Materials (2010). ISBN 978-0-9819275-6-5
9. Dubois, D., Prade, H., Sandri, S.: On possibility/probability transformations. In: Proceedings of the IFSA conference 1993, pp. 103–112. Seoul (1993)
10. Duda, R., Hart, P., Stork, D.: Pattern Classification - Second Edition. Wiley-Interscience (John Wiley & Sons), Southern Gate, Chichester, West Sussex, England (2000)

11. Frigui, H., Krishnapuram, R.: A robust algorithm for automatic extraction of an unknown number of clusters from noisy data. Pattern Recognition Letters **17**, 1223–1232 (1996)
12. Frigui, H., Krishnapuram, R.: Clustering by competitive agglomeration, pattern recognition. Pattern Recognition **307**, 1109–1119 (1997)
13. G. Ferrari-Trecate M. Muselli, D.L., Morari, M.: A clustering technique for the identification of piecewise affine and hybrid systems. Automatica **39**, 205–217 (2003)
14. Heemels, W., Schutter, B.D., Bemporad, A.: Equivalence of hybrid dynamical models. Automatica **37**(7), 1085–1091 (2001)
15. Hoffmann, I., Engell, S.: Identification of hybrid systems. In: Proceedings of the IEEE American Control Conference 1998, pp. 711–712. Philadelphia, USA (1998)
16. Juloski, A., Weiland, S., Heemels, W.: A bayesian approach to identification of hybrid systems. In: Proceedings of the 43rd IEEE Conference on Decision and Control, pp. 13–19. Paradise Island, Bahamas (204)
17. Ljung, L.: System identification: Theory for the user. Prentice Hall, Englewood Cliffs, N.J. (1987)
18. Messai, N., Zaytoon, J., Riera, B.: Identification of a class of hybrid dynamic systems with feed-forward neural networks: about the validity of the global model. Nonlinear Analysis: Hybrid Systems **2–3**, 773–785 (2008)
19. Ramakrishna, R., Anandaraj, M., Kumar, P.A., Thirumalai, M., Prakash, V., An-andbabu, C., Kalyanasunda-ram, P., Vaidyanathan, G.: Signal processing techniques for the detection of in-sodium leaks in steam generator using argon injection. In: Proceedings of the National Seminar & Exhibition on Non-Destructive Evaluation, pp. 362–365 (2009)
20. Srinivasan, G., Singh, O.P., Prabhakar, R.: Leak noise detection and characterization using statistical features. Annals of Nuclear Energy **27**(4), 329–343 (2004)
21. Therrien, C.: Decision Estimation and Classification: An Introduction to Pattern Recog-nition and Related Topics. John Wiley & Sons, New York (1989)
22. Vidal, R., Soatto, S., Ma, Y., Sastry, S.: Identification of pwarx hybrid models with un-known and possibly different orders. In: Proceedings of the IEEE American Control Conference 2004, pp. 547–552. Boston, Massachusetts (2004)
23. Zadeh, L.: Fuzzy sets as a basis for a theory of possibility. Fuzzy Sets and Systems **1**, 3–28 (1978)

Epilogue: Achievements, Open Problems, and New Challenges

The aim of this book is to provide the reader a round picture of the latest important developments and investigations within the field of learning in non-stationary environments. As such, it contains methodologies, concepts, algorithms, and innovations for dynamic evolution and adaptation of system models, which emerged during the last years in various fields of intelligent systems research. Therefore, this books intends to attract people likewise from the machine learning, soft computing, fuzzy systems, neural networks, data mining, and pattern recognition communities. In this sense, this volume comprehends:

1. *Unsupervised dynamic and evolving learning concepts (Part I)*, including single-pass updates of statistical measures taking into account drifting data distributions, evolvable granules in spatiotemporal environments equipped with the ability to build relational and cluster-type models from time-dependent data recorded at different locations, and an incremental clustering variant based on the spectral information concept.
2. *Dynamic and evolving learning concepts for supervised classification problems (Part II)*, including a new approach for semi-supervised learning based on dynamic fuzzy k-nearest neighbors, a concept how to predict the success or failure of classifiers at an early stage, thus addressing the problem whether classifiers accuracies can be increased by further updates, an instance-based learning approach dealing with an appropriate update of the reference (case) base for capturing the most important samples used for further predictions, and several incremental learning algorithms for trainable classifier fusion methods, which may exploit the diversity of single incremental base classifiers for achieving significant performance boost and improved stability.
3. *Dynamic and evolving learning concepts for supervised regression/function approximation problems (Part III)*, including flexible fuzzy systems updated sequentially and on the fly from on-line data streams and equipped with several enhanced concepts for improved performance, self-awareness in own predictions and reduced complexity, sequential neuro-fuzzy type systems with an enhanced rule pruning and evolution concepts based on statistical information

M. Sayed-Mouchaweh and E. Lughofer (eds.), *Learning in Non-Stationary Environments: Methods and Applications*, DOI 10.1007/978-1-4419-8020-5,

© Springer Science+Business Media New York 2012

criteria as well as exploiting an extended Kalman filter for robust updates of singleton consequents, and an interval granular modeling approach based on multi-dimensional interval analysis and arithmetic which is used for function approximation and time series forecasting purposes.

Furthermore, the volume deals with attractive and successful *real-world applications of dynamic learning methodologies (Part IV and some evaluation sections in the other parts)*, including dynamic forecasting of stock market indices based on granular and multiple time-series data (the latter using local and transductive modeling concepts), on-line optimization of feature calculation in machine vision systems (for improved classification of textures and images), on-line quality control in dynamically changing production lines by surface inspection (CD imprints, eggs, rotor parts, and bearings) and early fault detection/diagnosis (engine test benches, audio tapes), identification of hybrid dynamic systems containing switching sequences of modes and classes, monitoring of welding quality in an industrial welding system, prediction of maintenance actions for copiers, and the prediction of rain precipitation in different European regions.

Although, due to various reasons (especially space and time restrictions, unavailability), some substantial approaches could have not been included, the volume should lie a broad information and inspiration basis for ongoing future developments and further publications in this still emerging field of research. In this context, we finally want to mention some future challenges in dynamic and evolving learning issues which we see as important and, from our point of view, were not sufficiently handled so far:

- *Dynamic changes in the input structure*: Current methodologies for learning in non-stationary environments basically include two essential concepts: (1) permanent adaptation and refinement of model parameters (e.g., movement of decision boundaries or adjustments of clusters, hyper-planes to fit the natural data stream distribution appropriately) and (2) changes in the structural components (rules, granules, intervals, neurons, leafs, etc.). Very little attention is paid to dynamic changes in the input structure of the models in order to account for changing importance/impact of system variables, features over time. An attempt to handle this issue is presented in Chap. 9 (second part) for evolving fuzzy classifiers, where input features may get out-weighted according to their discriminative power between two or several classes, and other features may receive a higher weight, thus having a higher influence on the final prediction (hence, a soft dimensionality reduction effect is achieved). Although a variety of on-line subset selection algorithms exists in literature, their smooth combination with incremental, evolving models is still in its infants, especially was hardly handled in connection with other model architectures than evolving fuzzy systems (EFS).
- *Enhanced incremental optimization procedures*: a lot of state-of-the art methods apply the least squares optimization problem as basis for learning linear and non-linear parameters in a recursive manner (see, e.g., Chaps. 9–12), some use incremental clustering, granule extraction techniques for approximation local data distributions as close as possible (see, e.g., Chaps. 3–5). However, little

effort is investigated so far in the inclusion of model complexities during the optimization/extraction process in order to avoid over-fitting by applying for example, enhanced optimization problems with constraints or regularized incremental learning procedures.

- *On-line model-based design of experiments*: currently, models are evolved and updated based on incoming streams in the same order as samples from the streams are processed; however, no concepts are worked out how to steer the data acquisition process in order to incorporate new valuable information quickly into the models and to omit unnecessary samples containing no new information.
- *More clear focus on model interpretation*: currently, most of the techniques are conducting precise modeling, such that the evolved models have very little interpretable meaning as major attention is paid to achieve models with high predictive accuracy. This restricts the communication and interaction with operators at an enhanced level and prevents the inclusion of experts/operators knowledge aside the information provided by the data.
- *Self-awareness in model responses*: most of the methods provide predictions, classification statements for query points, forecasts, and cluster information, but little care is taken in the certainty/uncertainty of these models responses (some concepts are mentioned in Chap. 9, second part)—this could serve as another viewpoint of model interpretation.

Linz, Austria Edwin Lughofer
Douai, France Moamar Sayed-Mouchaweh

About the Editors

Moamar Sayed-Mouchaweh received his Master degree from the University of Technology of Compiegne-France in 1999. Then, he received his Ph.D. degree from the University of Reims-France in December 2002. He was nominated as Associated Professor in Computer Science, Control, and Signal processing at the University of Reims-France in the Research center in Sciences and Technology of the Information and the Communication (CReSTIC). In December 2008, he obtained the Habilitation to Direct Researches (HDR) in Computer science, Control, and Signal processing. Since September 2011, he is working as a full professor in the School of Mines of Douai-France (Ecole des Mines de Douai EMD) at the Department of Automatic Control and Computer Science (Informatique & Automatique IA). He supervised several defended Master and Ph.D. thesis as well as research projects in the field of Modeling, Monitoring, and Diagnosis in non-stationary environments. He published more than 60 journal and conference papers. He served as International Program Committee member for several International Conferences as well as a member in IEEE and IFAC technical committees. He also (co-)organized several special sessions and presented several tutorials. He is an associate editor of the Springer international journal "Evolving Systems".

Edwin Lughofer received his Ph.D. degree from the Department of Knowledge-Based Mathematical Systems, University of Linz, where he is now employed as post-doctoral fellow. During the past 10 years, he has participated in several international research projects as key researcher of the University of Linz, such as the EU-projects AMPA, DynaVis (www.dynavis.org), and Syntex (www.syntex.or.at). In this period, he has published around 60 journal and conference papers in the fields of EFS, machine learning and vision, clustering, fault detection, image processing, and human–machine interaction, including a monograph on "Evolving Fuzzy Systems" (Springer, Heidelberg 2011). In these research fields, he acts as a reviewer in peer-reviewed international journals and as (co-)organizer of special sessions and issues at international conferences and journals. He served as programme committee member in several international conferences and is the

M. Sayed-Mouchaweh and E. Lughofer (eds.), *Learning in Non-Stationary Environments: Methods and Applications*, DOI 10.1007/978-1-4419-8020-5,
© Springer Science+Business Media New York 2012

associate editor of the international Springer journal "Evolving Systems" and a member of two task forces in the area of machine learning and evolving systems.

In 2006, he received the best paper award at the International Symposium on Evolving Fuzzy Systems, and in 2008 the award at the 3rd Genetic and Evolving Fuzzy Systems workshop. In 2007, he received a Royal Society Grant for know-how exchange with Lancaster University in the field of Evolving Fuzzy Systems. In 2010, he initiated the bilateral FWF/DFG Project "Interpretable and Reliable Evolving Fuzzy Systems" and is currently key researcher in the national K-Project "Process Analytical Chemistry (PAC)" (2010–2014, including 17 industrial companies and research institutes), as well as in strategic research projects of the Austrian Center of Competence in Mechatronics (ACCM).

Research visits, stays, and guest lectures include the following locations: Department of Mathematics and Computer Science (Philipps-Universitt Marburg), Center for Bioimage Informatics and Department of Biological Sciences, Carnegie Mellon University, Pittsburgh (U.S.A.), Faculty for Informatics (ITI) at Otto-von-Guericke-Universitt, Magdeburg (Germany), Department of Communications Systems InfoLab21 at Lancaster University (UK), Department for Mechanics, Control, and Mechatronics at the University of Siegen (Germany), and Fraunhofer Institute ITWM at the University of Kaiserslautern.

Index

M. Sayed-Mouchaweh and E. Lughofer (eds.), *Learning in Non-Stationary Environments: Methods and Applications*, DOI 10.1007/978-1-4419-8020-5,

© Springer Science+Business Media New York 2012